*Edited by*
*Carl-Fredrik Mandenius*

**Bioreactors**

*Edited by*
*Carl-Fredrik Mandenius*

# Bioreactors

## Design, Operation and Novel Applications

WILEY-VCH
Verlag GmbH & Co. KGaA

**Editor**

***Prof. Carl-Fredrik Mandenius***
Linköping University
Division of Biotechnology, IFM
58183 Linköping
Sweden



**Library of Congress Card No.:** applied for

**British Library Cataloguing-in-Publication Data**
A catalogue record for this book is available from the British Library.

**Bibliographic information published by the Deutsche Nationalbibliothek**
The Deutsche Nationalbibliothek lists this publication in the Deutsche Nationalbibliografie; detailed bibliographic data are available on the Internet at <http://dnb.d-nb.de>.



**Print ISBN:** 978-3-527-33768-2
**ePDF ISBN:** 978-3-527-68337-6
**ePub ISBN:** 978-3-527-68338-3
**Mobi ISBN:** 978-3-527-68339-0
**oBook ISBN:** 978-3-527-68336-9

**Cover Design** Formgeber, Mannheim, Germany
**Typesetting** SPi Global, Chennai, India

# Contents

# Preface

Designing and operating bioreactors have challenged bioengineers over centuries. Continuing advancement of our ability to tame the forces of the cell to create commodities for the benefit for our wellbeing has motivated constant development and refinement of the bioreactors. From the antiquity, across the first steps in unraveling of the cell theory during the 1850s, till the reprogramming of cells, the bioreactors have had, and still have, an essential role as artificial harbors for growing and maintaining cell cultures. This motivates persistent efforts for reaching deeper understanding of the design and operations of bioreactors. Due to the permeant progress in our technology-driven society, the prerequisites for this ambition are favorable: tools for systems biology elucidate intracellular course of events in bioreactor cultures are now at hand, computational capacity strongly supports modeling and prediction of the dynamics of bioreactor systems at various levels, and micro-machining techniques allow us to miniaturize bioreactors for wide high-throughput studies.

The book gives an up-to-date overview of a variety of emerging techniques and methods being used for the design of bioreactors, their operations, and applications to emerging fields in biotechnology. By that, the book intends to provide scientists in industrial biotechnology with inspiration to apply novel approaches and methodologies in their own work. If the work is performed in academic environment, the chapters of the book can hopefully motivate further research for generating new ideas, hypotheses, and counterhypotheses. If the work is done in an industrial setting, the path to practice, process design, and manufacturing could be made shorter.

Structurally, the book goes from bioreactor design at smaller to larger scales. After an introductory chapter providing a historical and state-of-of-the-art perspective to bioreactor design, the book brings up issues around miniaturization and scale-down of bioreactor operation. Much focus is on screening of bioreactor performance, especially in relation to the requirements of the biological systems.

This is followed by the emerging use of the bioreactor for stem cells, for the production of materials for regenerative medicine and cell therapy, and for replacing human tissues and organs as is exemplified by artificial liver bioreactors.

The importance of scale for design and operation is highlighted in several of the chapters dealing with the study of effects related to dimensions of the bioreactor, how the experiments at smaller scale can predict operation at larger manufacturing scale, theoretically using simulation and fluid dynamics, or in scaled-down experimental models. In that respect, also alternatives of single-use equipment is discussed.

The last part of the book is devoted to methods that apply to the operation of bioreactors in order to control their performances. This includes the monitoring and control of bioreactors where measurements are extended with computational power mediated through mathematical modeling of process data, either derived online from sensors or from historical data. It includes the exploitation of powerful software and IT solution in real-time for data acquisition and interpretation in soft sensors. It includes statistical experimental design methods for improving the composition of the environment in the bioreactor, especially the composition of the culture media. And finally, it includes using the power of the computer to create virtual bioreactors in order to test, train, and try out how to use them on the operator level for reaching high-performance processing in the real world.

The authors of the chapters in the book represent a wide span of expertise and experience: from academia to industry, from cell biology to biophysics and engineering, from theoretical to practical labs and workshops. I thank all of them for their excellent contributions.

I also thank my colleagues Prof. Mats Björkman, Prof. Gunnar Hörnsten, and Prof. Johan Hyllner for most valuable discussion on the themes of this book, especially related to conceptual design, large-scale industrial development, and stem cell and cell therapy applications. My thanks also go to Prof. Anders Brundin for stimulating collaboration and contributions on experimental design and to Dr Inga Gerlach, Dr Jan Peter Axelsson, and Robert Gustavsson for valuable collaborations on bioprocess simulation and soft sensors. Finally, I also thank the staff of Wiley-VCH for their dedicated work in the production of this book.

*Carl-Fredrik Mandenius*

# List of Contributors

*Paula M. Alves*
Universidade Nova de Lisboa
Instituto de Tecnologia Química
e Biológica
Av. da República
Estação Agronómica Nacional
2780-157 Oeiras
Portugal

*and*

iBET
Instituto de Biologia
Experimental e Tecnológica
Av. República, Qta. do Marquês
Estação Agronómica Nacional
Edificio IBET/ITQB
2780-157 Oeiras
Portugal

*Birgit Andree*
Hannover Medical School
(MHH)
Leibniz Research Laboratories
for Biotechnology and Artificial
Organs (LEBAO)
REBIRTH - Center for
Regenerative Medicine
Department of Cardiothoracic
Transplantation and Vascular
Surgery
Carl Neuberg Str. 1
30625 Hannover
Germany

*Jochen Büchs*
AVT - Chair of Biochemical
Engineering
RWTH Aachen University
Worringer Weg 1
52074 Aachen
Germany

*Karen Coopman*
Loughborough University
Centre for Biological Engineering
Department of Biochemical
Engineering
Epinal Way
Leicestershire LE11 3TU
UK

*Pau Ferrer*
Autonomous University of
Barcelona
Department of Chemical,
Biological and Environmental
Engineering
Edifici Q - School of Engineering
Bellaterra (Cerdanyola del Vallés)
08193 Catalonia
Spain

***Jörg C. Gerlach***
University of Pittsburgh
McGowan Institute for
Regenerative Medicine
Department of Surgery,
Department of Bioengineering
3025 E. Carson Street
Suite 238
Pittsburgh PA 15203
USA

***Jarka Glassey***
CEAM, Merz Court
Newcastle University
Tyne and Wear
Newcastle upon Tyne NE1 7RU
UK

***Robert Gustavsson***
Division of Biotechnology/IFM
Linköping University
581 83 Linköping
Sweden

***Volker C. Hass***
Hochschule Furtwangen
University
Faculty of Medical and Life
Sciences
Campus Villingen-Schwenningen
Jakob-Kienzle-Str. 17
D-78056
Villingen-Schwenningen
Germany

***Thomas R.J. Heathman***
Loughborough University
Centre for Biological Engineering
Department of Biochemical
Engineering
Epinal Way
Leicestershire LE11 3TU
UK

***Christopher J. Hewitt***
Loughborough University
Centre for Biological Engineering
Department of Biochemical
Engineering
Epinal Way
Leicestershire LE11 3TU
UK

*and*

Aston University
Aston Medical Research Institute
School of Life and Health
Sciences
Aston Triangle
Birmingham B4 7ET
UK

***Stefan Junne***
Technische Universität Berlin
Department of Biotechnology
Chair of Bioprocess Engineering
Ackerstrasse 76 ACK24
13355 Berlin
Germany

***Lalita Kanwar Shekhawat***
Indian Institute of Technology
Department of Chemical
Engineering
Hauz Khas
110016 New Delhi
India

***Henning Kempf***
Hannover Medical School
(MHH)
Leibniz Research Laboratories
for Biotechnology and Artificial
Organs (LEBAO)
REBIRTH - Center for
Regenerative Medicine
Department of Cardiothoracic
Transplantation and Vascular
Surgery
Carl Neuberg Str. 1
30625 Hannover
Germany

***Christina Kropp***
Hannover Medical School
(MHH)
Leibniz Research Laboratories
for Biotechnology and Artificial
Organs (LEBAO)
REBIRTH - Center for
Regenerative Medicine
Department of Cardiothoracic
Transplantation and Vascular
Surgery
Carl Neuberg Str. 1
30625 Hannover
Germany

***Clemens Lattermann***
AVT - Chair of Biochemical
Engineering
RWTH Aachen University
Worringer Weg 1
52074 Aachen
Germany

***Varun Loomba***
Indian Institute of Technology
Department of Chemical
Engineering
Hauz Khas
New Delhi 110016
India

***Carl-Fredrik Mandenius***
Division of Biotechnology/IFM
Linköping University
581 83 Linköping
Sweden

***Peter Neubauer***
Technische Universität Berlin
Department of Biotechnology
Chair of Bioprocess Engineering
Ackerstrasse 76 ACK24
13355 Berlin
Germany

***Alvin W. Nienow***
Loughborough University
Centre for Biological Engineering
Department of Biochemical
Engineering
Epinal Way
Leicestershire LE11 3TU
UK

*and*

Aston University
Aston Medical Research Institute
School of Life and Health
Sciences
Aston Triangle
Birmingham B4 7ET
UK

***Nico M. G. Oosterhuis***
Celltainer Biotech BV
Bothoekweg 9
7115AK Winterswijk
The Netherlands

***Qasim A. Rafiq***
Loughborough University
Centre for Biological Engineering
Department of Biochemical
Engineering
Epinal Way
Leicestershire LE11 3TU
UK

*and*

Aston University
Aston Medical Research Institute
School of Life and Health Sciences
Aston Triangle
Birmingham B4 7ET
UK

***Anurag S. Rathore***
Indian Institute of Technology
Department of Chemical Engineering
Hauz Khas
New Delhi 110016
India

***Nigel Titchener-Hooker***
University College London
Department of Biochemical Engineering
Bernard Katz Building
Gordon Street
London WC1H 0AH
UK

***Rui Tostoes***
University College London
The Advanced Centre for Biochemical Engineering
Department of Biochemical Engineering
Gower Street
Torrington Place
London WC1E 7JE
UK

***Francisco Valero***
Autonomous University of Barcelona
Department of Chemical Biological and Environmental Engineering
Edifici Q – School of Engineering
Bellaterra (Cerdanyola del Vallés)
08193 Catalonia
Spain

***Danny van Noort***
Universidad de los Andes
Facultad de Ingeniería y Ciencias Aplicadas
Mons. Alvaro del Portillo 12455
Santiago 7620001
Chile

***Ajoy Velayudhan***
University College London
Department of Biochemical Engineering
Bernard Katz Building
Gordon Street
London WC1H 0AH
UK

***Katrin Zeilinger***
Charité - Universitätsmedizin Berlin
Campus Virchow-Klinikum
Bioreactor Group
Berlin-Brandenburg Center for Regenerative Therapies (BCRT)
Augustenburger Platz 1
13353 Berlin
Germany

***Robert Zweigerdt***
Hannover Medical School (MHH)
Leibniz Research Laboratories for Biotechnology and Artificial Organs (LEBAO)
REBIRTH – Center for Regenerative Medicine
Department of Cardiothoracic Transplantation and Vascular Surgery
Carl Neuberg Str. 1
30625 Hannover
Germany

# 1
# Challenges for Bioreactor Design and Operation

*Carl-Fredrik Mandenius*

## 1.1
## Introduction

As per definition, the bioreactor is the designed space where biological reactions take place. Hence, the bioreactor is essentially an engineering achievement and its design a challenge for bioengineers.

The bioreactor should create a biosphere that as profoundly and adequately as possible provides the ideal environment for the biological reaction.

The path for reaching, attaining, and maintaining this is the main task for bioreactor engineers to find. That task decomposes into several endeavors necessary to accomplish. One is to design the physical entity of the bioreactor itself – by that, ensuring favorable physical conditions for transport of gases and liquids and solids over time. Another is to ensure that the physical entity of the bioreactor is favorably adapted to the biological system that performs the bioreactions. Yet another is to ensure that the dynamic biophysical and biochemical events taking place are operable in an industrial environment.

In some of these design perspectives, bioreactor design is addressed at a process development stage where the performance of operations is independent of scale or biological system inside the bioreactor. Others address specific biological systems and the particular requirements of these. Others take the viewpoint at the holistic level: how to integrate the bioreactor and its design into an entire bioprocess with the constraints that this creates. Others concern provision of methodologies for observing the bioreactor at R&D as well as at operation stages in order to monitor and control and to optimize its performance from a variety of needs and purposes. Others provide better methods for supporting plant engineers and technicians to manage to operate the bioreactor processes under unpredictable industrial conditions where unexpected events, faults, and mishaps must be interpreted in short time and acted upon.

*Bioreactors: Design, Operation and Novel Applications*, First Edition. Edited by Carl-Fredrik Mandenius.

Importantly, all these aspects on design and operation may, and even must, be amalgamated into coherent design methodologies that are conceivable and practically achievable. It is the ambition of this book to provide a collection of design options where engineering principles and design tools are presented that facilitate to develop and apply good solutions to emerging needs in bioreactor design.

## 1.2
## Biotechnology Milestones with Implications on Bioreactor Design

The bioreactor is a historical apparatus known since ancient times. Old antique cultures were able to solve bioengineering design challenges for practical purposes such as wine and beer making from mere experience and observations. This paved the way for the evolvement of biotechnological processes, primarily for preparation and production of food products [1].

The notion that microscopic life is a huge industrial resource came gradually to man and with some resistance from the established scientific society itself. An array of fundamental scientific steps paved the way for the unfolding of industrial biotechnology. Growing understanding of the mechanisms of diseases and its interplay with cell biology supported the development.

In the early nineteenth century, scientists such as Lorenz Oken (1779–1851), Theodor Schwann (1810–1882), and others did stepwise begin to fathom the fundamental principles of the cell's behavior in the body and in culture [2]. Louis Pasteur (1822–1895) took these observations and conclusions further into a coherent description of the fermentation mechanisms [3]. Later, researchers such as Emile Roux (1853–1933) and Robert Koch (1843–1910) realized the implications to bacteriology and for spread of diseases. These consorted ascents in cell biology and medicine did synergistically create the necessary background for the exploitation of the industrial potential of cells. By that, also important prerequisites for a furthering of bioreactor design were set.

The microbiology research brought better insights into the up-till-then-hidden processes of the cell and, hence, to the development of bioengineering and to the widespread industrial biotechnology applications during the twentieth century. It is in this framework of bioindustrial activity and progress the bioreactors and their design have been shaped. Still, it is noteworthy that 100 years ago an industrial bioreactor facility did not look too different from today's industrial sites (Figure 1.1).

In the early twentieth century, large-scale fermentation processes were set up with impact onto the war-time industry of that period. Glycerol production for use in the manufacture of explosives, using yeast for conversion from glucose, was established. Another contemporary example is the large-scale production of butanol and acetone by butyric acid bacteria, as developed by Chaim Weizmann, used first for explosives and then for rubber manufacture in the emerging car industry [4]. However, these bioprocesses were soon abandoned for petroleum-based products that had better process economy.

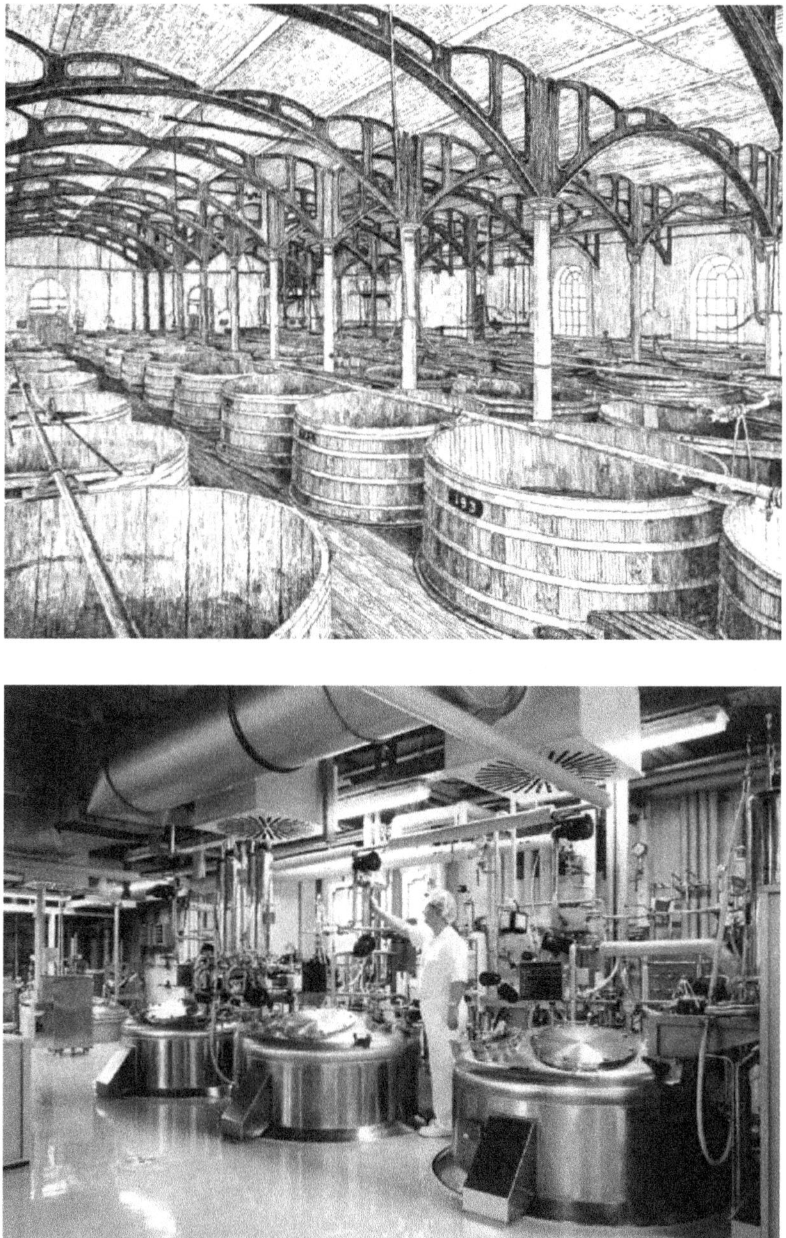

(a)

(b)

**Figure 1.1** (a) An old fermentation plant from the late nineteenth century. (b) A modern fermentation plant one century later. The gap in time between the plants reveals that some of the design features have undergone changes, while others are unchanged: the bioreactors are cylindrical vessels, the containment of the broth and concern about contamination were in former days less, piping are essential, many vessels are using the available plant space, and few plant operators are close to the process.

The story of the development of antibiotics is an impressive example of how microbiology and industrial biotechnology evolved over an extended period of time by consorted actions between academic research and industrial product development. The original discovery in 1929 by Alexander Fleming of the antibiotic effect of a *Penicillium* culture was in a series of steps for amplifying the yield and activity of cultures transferred into large-scale production [5]. And other renowned scientists such as Howard Florey, Ernest Chain, Norman Heatley, Marvin Johnson, and others in close collaboration with pharmaceutical companies managed to identify, stabilize, exploit, select strains, exploit genetics, mutational methods and, finally, establish large-scale bioproduction in bioreactors for meeting global medical needs for curing infections [6]. The latter did indeed challenge the engineering skills in understanding optimization in the design and operation of the bioreactor. It also gave ample examples of how knowledge and skills from one group of products could be transferred into others and, by that, pave way for other antibiotics such as cephalosporins, streptomycins, and aminoglycosides.

These endeavors and experiences contributed substantially to facilitate forthcoming bioprocess development of biotherapeutics. Undoubtedly, the concept of process intensification was driving the development although the term was not yet coined. The same was true for the transfer of the concept of continuous strain improvement of microbial strains and cell lines.

In parallel with the progress of developing antibiotics, other microbial primary and secondary products were realized. These included amino acids (e.g., glutamate and lysine) and organic acids (e.g., vitamins) used as food ingredients and commodity chemicals and reached considerable production volumes. Microbial polymers such as xanthan and polyhydroxyalkanoates are other examples of bioprocess unfolding during the mid-1950s [7].

Protein manufacture, especially industrial enzymes, became comparatively soon a part of the industrial biotechnology with large-scale production sites at a few specialized companies (e.g., Novo, Genencor, Tanabe). At these up-scaled processes, very important findings and experiences were reached concerning bioreactor design and operation. Although not yet exploiting gene transfer between species for these proteins, significant technology development for later use was accomplished [1].

Subsequently, the emerging industrial use of animal cells came about. Culturing at large scale, at lower cell densities than fungi and yeasts, and with much lower product titers posed a next challenge to bioreactor engineering [8].

With the ascents of Köhler and Milstein (1975) in expressing monoclonal antibodies in hybridoma cell culture and the ensuing setup of cell culture reactor systems for production, a new epoch came across which has impacted industrial biotechnology and bioengineering tremendously. It initiated a art of cultivation technology where conditions and procedures for the operation of a cell culture showed a number of constraints necessary to surpassed in order to make processing industrially feasible [10].

However, it was the genetic engineering and recombinant DNA technology that created a revolution in the field of industrial biotechnology with macromolecular products from cells, first in bacteria and yeast and subsequently in animal and human cells [11]. Industry was proactive and efficient in transforming science into business activity. In California, Cetus and Genentech were established in the early 1970s. In the years thereafter, Biogen, Amgen, Chiron, and Genzyme followed, all with successful biotherapeutic products in their pipelines – insulin, erythropoietin, interferons, growth hormones, blood coagulation factors, interleukins, and others reached the therapeutic market with relatively short development times, in spite of regulatory requirements and the multitude of novel production conditions spanning from clinical considerations to new manufacturing methodology. Especially, the latter embodied numerous challenges for bioprocess and bioreactor engineering to disentangle.

The latest steps in bioreactor engineering are related to cell production and applications with regenerative medicine products and pluripotent stem cells [12]. Certainly, this has had implication on bioreactor design in terms of new and diverse requirements of performing cellular transformation including cell differentiation, expansion, and maturation, and of longer process time compared with previous processing. The controllability demands of bioreactors for these purposes are higher due to more vulnerable cell types, more complicated growth behavior, and substantially different operations. This addresses again the critical issues of mass transfer and barriers of oxygen, nutrients, and sterility of the cultures.

Table 1.1 recapitulates the milestones of this industrial biotechnology evolution based on the events of modern biology and life science.

Furthermore, the industrial biotechnology development during the twentieth and early twenty-first centuries has been profoundly interconnected with a variety of specific challenges within biochemical engineering research [13]. These have, for example, regarded such issues as the lack of robustness of enzymes and microbial strains as compared with heterogeneous catalysts used in the chemical industry; the difficulties of redesigning cells and other biocatalysts with metabolic pathways adapted to the production of specific products; and challenges in the fermentation of complex raw material streams such as hydrolysates from lignocelluloses or other renewable resources. This has been an integral part of the development where bioreactors have been one of the enabling tools.

Solutions to these challenges are many: using combinations of biological and chemical reactions, genetically engineered crops with better properties for the actual production process, and systems biology methodologies applied on the production microorganisms or cells in order to engineer pathways for metabolic conversion, transcription, and expression.

The academic contributions to this research have been substantial and have led to important industrial improvements such as reduction in process development times and pretreatment, and the ability to handle renewable raw materials applicable to a wide variety of bioprocesses including large-scale biorefinery and waste treatment plants.

**Table 1.1** Milestones in industrial biotechnology.

| Major achievement | Discoverers | Year | Impact on industry applications | Implication on bioreactor design |
|---|---|---|---|---|
| Understanding of fermentation principles | L. Pasteur | 1857 | Initiating wider application of fermentation in industry | The needs for efficiency of design recognized |
| Anthrax bacteria | R. Koch | 1876 | Disease effect of bacteria and the uniqueness of a specific bacterium | |
| Use of antiseptics realized | I. Semmelweis | 1846 | Chemical control of infections | The success of large-scale microbial production and its dependency of sterility |
| The existence of biological catalysis | M. Traube | 1877 | The catalytic action of microorganisms | The optimization of the biocatalytic activity in a bioreactor |
| Glycerol from yeast cultures | Neuberg *et al.* | 1914–1918 | Need for glycerol in the war industry | Cultivation of yeast for other products than beer and wine |
| Acetone–butanol fermentation | C. Weizmann | 1914–1918 | Supply of bulk chemicals for explosives and car tires | Scale-up technology challenged to meet market demand |
| Penicillin discovered | A. Fleming | 1929 | Pharmaceutical biotechnology initiated | Strain improvement |
| Penicillin isolation | H.W. Florey and colleagues | 1939 | Product characterization | Yield improvements |
| Cephalosporin fermentation | Brotzu and Abraham | 1948 | Other microbial metabolites could act as antibiotics | Fed-batch operations |
| Antibiotic strain improvement | S. Waksman and others | 1940s–1950s | Higher yield per volume | Process intensification |

| | | | | |
|---|---|---|---|---|
| Amino acid fermentation | Kyowa Hakko Co. | 1957 | Metabolism in strains for amino acids is exploitable | Scale-up of microbial fermentations |
| Organic acid fermentation | Food industries | 1940 | | Large-scale fermenters |
| Vitamin fermentation | A. Guilliermond Reichstein | 1930s | Riboflavin (B2) (vitamin C) | Bioreactor processes including semisynthetic steps |
| Genetic engineering and recombinant DNA technique | P. Berg, D. Glaser | 1971 | Improved metabolism and expression in cells | Induction procedures in bioreactor |
| Recombinant insulin, growth hormone | H. Boyer and R. Swanson, Genentech | 1978 | The recombinant DNA-technique open for a biotherapeutic production | A new methodology of culturing recombinant microorganisms with induction protocols |
| Monoclonal antibodies in hybridoma cultures | G. Köhler, C. Milstein | 1975 | Diagnostics and therapeutics based on antibodies | Bioreactors to be developed for cell culture requirements, in particular, hybridoma |
| Recombinant DNA technique applied industrially in animal cell cultures | Pharmaceutical industries | 1990s | Recombinant products in mammalian cells human-like biotherapeutics (e.g., EPO, tPA, IFN, Factor VIII) | Bioreactors to be developed for other cell cultures such as CHO, HEK, and other cell lines |
| Pluripotent stem cells and derived cells | S. Yamanaka | 2006 | Cells from hES and iPS cells the potential to become new products for cell therapy | Bioreactors must be adapted to new cultivation conditions |

Also, the education and training of new bioengineers have evolved a thinking of biochemical engineering that has impacted industry with new perspectives where the biology scope was merged into the engineering framework of conceiving, designing, and operating industrial processes. It is, here, relevant to remind about the tremendous increases in biological knowledge with related technologies that have emerged, which must be integrated and constantly updated into bioengineering science.

## 1.3 General Features of Bioreactor Design

As Table 1.1 demonstrates, the ascents of the twentieth-century biotechnology have created a pull for advancing bioreactor design. Especially on large scales, the requirements of the cultivation system have dispersed into a variety of diverse technical issues that most of them have in common transfer of mass and energy [14]. In textbooks, a bioreactor is typically described as an apparatus shaped like a chamber for growing organisms such as bacteria or yeasts that can be used for the production of biomolecular metabolites or biopolymers or for the conversion of organic wastes. This very general bioreactor description clearly highlights the main purpose of the design efforts: to accomplish conditions where diverse cell types are able to grow efficiently and produce a variety of biological products with a wide range of molecular sizes in a single unit. This calls for profound adaption of the technical design of the bioreactor system and could expectedly result in many different design solutions. The diversity of the design mainly caused by the time factor; due to the fact that rates differ largely from one organism to another, in reproduction rates, in rates of molecular processing in the individual organisms, and transfer across biological barriers of the cellular systems.

The time factor also applies to the operational procedures. When cells grow, the design must adapt to compensate for the magnification of the dynamics due to higher cell numbers. This mostly concerns supply of nutrients and growth factors. However, it may also be about removal of mass and energy to avoid overloading the system with any of these. The operational procedures shall in combination with the design effectuate this.

A variety of conditions, operational procedures, and considerations are critical for the efficiency of the design (Table 1.2). Transfer rates of mass and energy are among the most critical issues [15].

Environmental factors in wider sense should be considered, as well as ambient temperature and moisture and occurrence of contaminants; all examples of factors may play a major role.

Sterilization is an operational procedure that differs only slightly depending on the organism but must be carefully adapted to the bioreactors' geometrical shape and construction materials. The prevalence of single-use units made in plastic materials highlights the actuality of this issue.

**Table 1.2** Bioreactor design criteria.

| Design issue | Purpose | Design means | Parameters |
|---|---|---|---|
| Gas transfer in submerged culture | Ensure high growth rate, avoiding oxygen starvation | Reactor geometry<br>Sparger design<br>Baffles<br>Overpressure<br>Impeller geometry | Aspect ratios<br>$K_L a$<br>*OTR*<br>*OUR*<br>*CER* |
| Mixing efficiency | Avoiding gradients of heat, nutrients and additives, stress<br>Reduce power | Impeller geometry<br>Baffles<br>Mixing analysis<br>CFD | Aspect ratios<br>Mixing time $t$<br>Power number |
| Nutrient supply and addition | Efficient transfer to bioreactor volume | Feeding regime<br>Multiple ports | Linear and exponential profile |
| Liquid–solid transfer | Enhance reaction rate<br>Reduce gradients | Flow distributors<br>Porous support | Thiele modulus |
| Heat transfer | Efficient removal of metabolic heat | Internal coils<br>Recycling of media<br>Jacket<br>Cooling media | Dimensionless numbers |
| Sterility | Ensure whole unit is devoid of foreign microorganisms to avoid infection | Sterilization procedure<br>Overpressure<br>Barriers<br>Containment<br>Microfilters | Sterilization time and temperature |
| Strain selection | Finding strain with properties adapted to media and reactor constraints | Microbial analysis<br>Omics | Specific rates ($\mu$, $q_P$, $q_S$)<br>Inhibition constants |
| Scale-up procedure | Ensuring same conditions at large scale | Design geometry of vessels and impellers<br>Range of mixing | Aspect ratio<br>Scale-up rule parameters<br>Dimensionless numbers |
| Rheology | | Additives affecting viscosity<br>CFD | Reynold's number<br>CFD data |
| Homogeneity of culture | Avoiding gradients for ideal reactor conditions | CFD | Zonal analysis data |
| Media composition | Balanced culture media | Factorial analysis<br>Omics methods | Model fit parameters |

Inoculation of cells is another consideration that relates to the size of the inoculum, the state of the cells to enter an exponential growth phase, their variability and sensitivity to microenvironmental conditions, and their purity.

Media composition is an example of yet another design issue comprising both the chemical aspects related to the nutritional value of the media as well as the biological significance of the components in relation to metabolic pathways involved in the growth and production of the cells.

Kinetic relationships in the bioreactor are described for a multitude of state and conditions. Table 1.2 summarizes most essential conditions and key parameters and how they influence design work and options. The access to these key parameters helps in understanding the frames of bioreactor design and why different design alternatives are needed as extensively discussed by many authors [16].

Based on these premises, a diversity of bioreactor design alternatives are have emerged (Figure 1.2). The stirred tank bioreactor is, with few exceptions, the predominantly used design for submerged cultures due to its versatility, operability, ability to cope with many of the aforementioned requirements, and manufacturability (Figure 1.2a).

The main shortcoming of the stirred bioreactor, its mechanical agitation, is solved in other bioreactor designs. In the bubble column bioreactor, the mechanical impeller is exchanged with raising bubbles, which, in the case of an aerobic fermentation will anyhow be required and to the benefit of fewer mechanical components and need for lesser electrical power (Figure 1.2b). But with a sometimes critical drawback, lower volumetric oxygen transfer, which may be a severe shortcoming for fast-growing organisms and high-density cultures.

The airlift bioreactor with a forced flow in an internal or external loop has the same advantage, although the design requires an addtional construction part, the down-comer tube (Figure 1.2c and d). These relatively minor design modifications appear to substantially limit the widespread use of these bioreactor types. Also, the fluidized bed reactor where cells are recycled by external pumping and soluble product harvested by overflow provides the same pros and cons, lesser mechanics, and lower oxygen transfer (Figure 1.2e). However, a density diversity between media and cells is needed, which makes aggregating cells such as flocculating yeast cells or possibly immobilized cells the ideal state for the biological component.

By that, the tank reactor design approaches the tubular reactor designs and solid-state fermentations. In the trickle-bed bioreactor, cells are grafted to a solid material while the medium is fluxed through a bed of biocatalyst (Figure 1.2f). This resembles the chemical engineering tubular reactor model where the catalyst is typically a transition metal catalyst. In contrast to the chemical reaction systems, the biocatalysed reactors harbour low-temperature aerobic processes with profoundly deviating kinetic regimes.

The solid-state bioreactor design can also follow the ancient Chinese tray reactor model as applied in *koji* fermentation. Collecting the culture on a support material, a kind of immobilization procedure, in trays placed in a container with controlled conditions is not optimal but convenient and well proven (Figure 1.2g). The tray bioreactor can be transformed into a static bed reactor, which provides

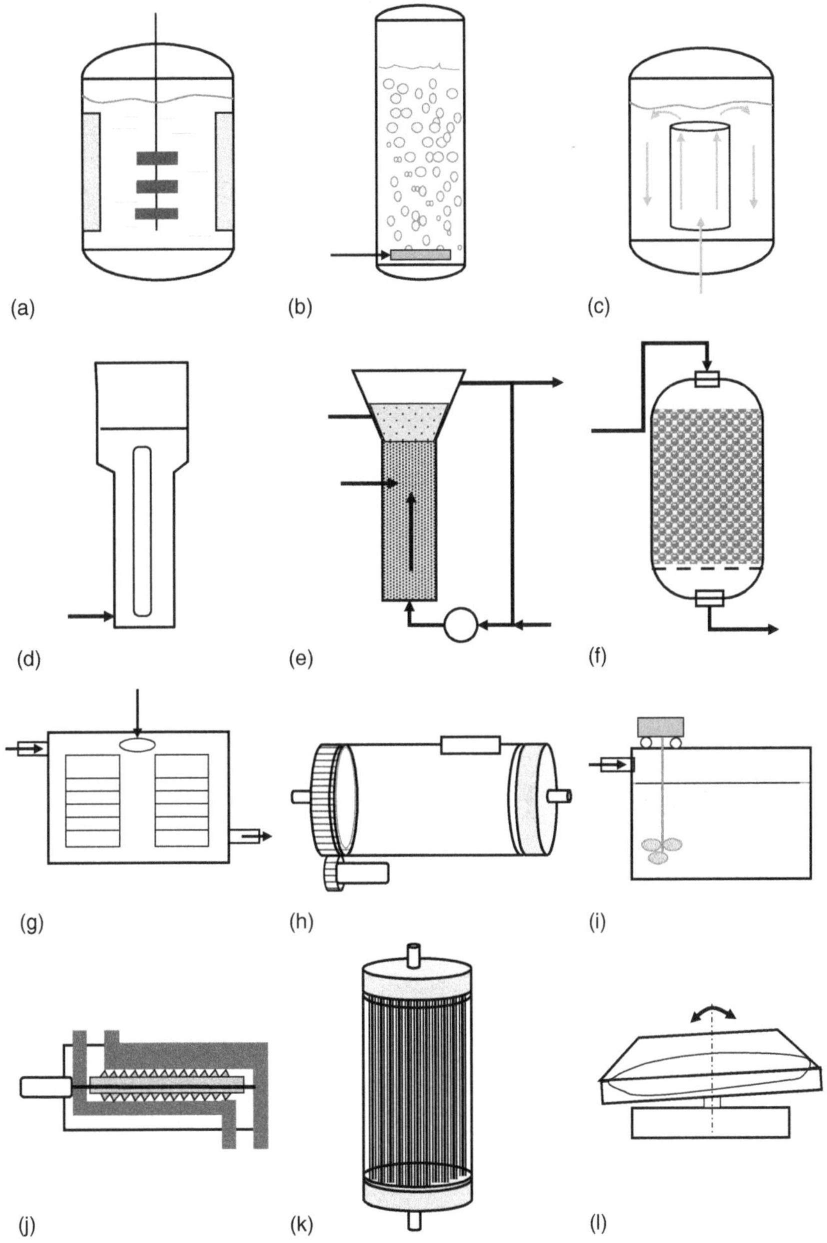


**Figure 1.2** Twelve examples of bioreactor designs: (a) stirred-tank reactor, (b) bubble reactor, (c) airlift reactor, (d) loop reactor, (e) reactor with immobilized cells, (f) fluidized reactor with recycling of cells, (g) solid-phase tray reactor, (h) rotary drum bioreactor, (i) agitated-tank reactor with movable impeller, (j) continuous screw bioreactor, (k) hollow-fiber reactor, and (l) wave bioreactor

better efficiency in contacting the reaction components in a flow-through manner. The static bed is a short version of a tunnel bioreactor that allows more catalyst to be contained at equivalent flow rate.

Rotating the bioreactor is another way to agitate the cells and reactants, where the gravity of the particles must be employed to cause the movement in the fluidium, either in a disk or a drum geometry (Figure 1.2h), or by moving a plastic bag (Figure 1.2i). Another technical solution is to move the impeller around in the bioreactor – back to where we started (Figure 1.2j). The design can of course here vary substantially. In waste water plant, for example, the reactor is a wide but low circular tank with the impeller on a rotating arm. As the example suggests, the approach is mostly a solution for the large scale. A continuous screw is a third alternative for creating movement in the bioreaction system, which is most appropriate if the liquid phase to be moved forward is viscous and where the energy input for driving the screw is negligible (Figure 1.2j). Cell cultivation conditions have generated a few additional design forms such as hollow-fiber reactors (Figure 1.2k) and wave bioreactors (Figure 1.2l).

The bioreactor designs quite often need to consider the forthcoming downstream steps. The volumetric rates and equipment size should cope in a realistic way. Often, a larger number of parallel units must be connected to a single reactor if the convenient operational volume of the subsequent step is too small to harbor the volumetric flow. However, the opposite can also apply; the capacity of a single high-speed centrifuge can suffice in a brewery with several fermenters.

## 1.4
## Recent Trends in Designing and Operating Bioreactors

The basic principles of bioreactor design outlined in the previous section have in more recent years unfolded into more elaborate methods due to several new premises. One is the access to computers for complex calculations. This has allowed computations previously too heavy and demanding to perform and too inconceivable to predict. By that, design work can be carried out in much reduced time and with much lesser brain efforts, in order to elucidate metabolic networks, fluid dynamics, and more complex kinetic models of the cell. However, the interaction between this newly created extended knowledge and tools and the problems that require solving have so far only been marginally exploited. Another is the manufacturing cost premises for equipment. Mass production of disposable units has become considerably less costly using new durable plastic construction materials and automated assembly of bioreactor equipment.

The sway from traditional bioreactors of the diverse designs as mentioned in the previous section to single-use reactors has changed the mind-set in the planning for plant engineering concerning estimates of equipment investments and operational costs. The emergence of the wave bioreactors with disposable bioreactor bags in sizes from 1 up to 2000 l took surprisingly long time despite the obvious

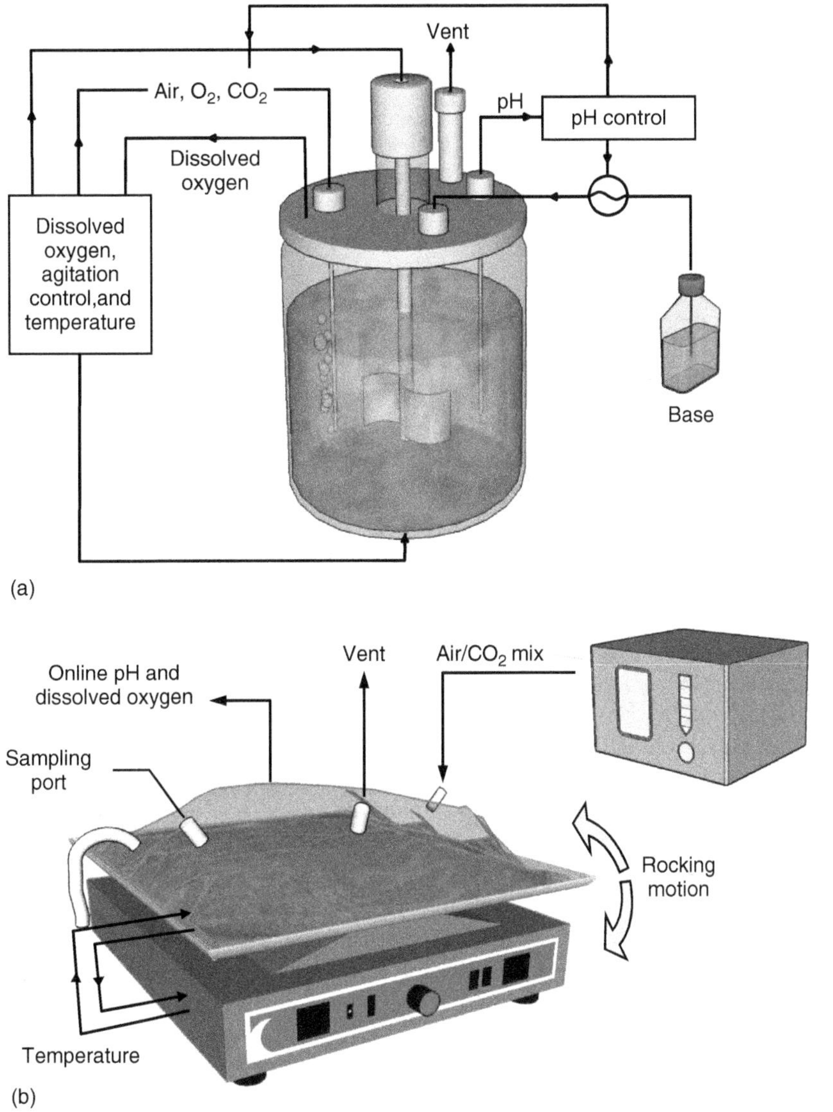


**Figure 1.3** An emerging new trend is the replacement of old stainless steel fermenter with single-use wave-bioreactors. It is a striking example of how smart designs based on fabrication technology use compatible low-cost materials and new conceptual thinking lead to a leap in design (from [17], with permission).

advantages they furnish for all involved – the operators, user-companies, and the supplies of units (Figure 1.3).

In principle, the same kind of progress in manufacturing technique, the capability of low-cost mass-production of polymeric materials, allows microfabrication of small devices, such as microbioreactors and microbiochips. This has in particular imposed implications on R&D processes by facilitating parallel testing of strains, cell lines, and media as well as basic cultivation parameters. Optimization of the bioreactor and the bioprocess can, using these tools, be accelerated significantly. The access to more compact and stable electronics, also due to the availability of mass-production and miniaturized fabrication techniques in the electronic and software industry, facilitates transfer of signals in the microelectronic environment of instrumentation of bioreactors. The capabilities of software programs and implementation of data processing have advanced considerably and could be anticipated to be substantially improved.

Better and more efficient education principles of bioengineers empower the biotech industry with the opportunity to improve process design and development. The conceive–design–implement–operate (CDIO) concept, founded at MIT in Boston and based on a revitalization of traditional engineering training and education with a new conceptual approach, also covers the key elements of bioreactor design and operations [18, 19]. The CDIO concept sees engineering work as a consecutive mental and practical process where ideas are formed and conceived due to eminence of understanding and analytical thinking, where these notions are applied in the design stage and further turned into implementation of systems, plants, and devices that are delivered for operational and perpetual usage (Figure 1.4). Although the CDIO concept is an initiative intended primarily for engineering education, it inevitably has the potential to influence industrial mindsets and work in a positive sense. If the CDIO principle is unanimously accepted and applied throughout the industry globally, it will endow plant engineering with a valuable leverage for efficiency.

Mathematical engineering models, in industrial practice, have been sometimes seen as something causing delays, difficulties, and inconveniences rather than being efficient design tools. Statistical methodology however, is an exception. Statistical multivariate data analysis and factorial analysis have become popular in industry and widely accepted as a tool to optimize or at least improve processes. This is much helped by user-friendly software that are easy to learn and apply. Another reason is, of course, that it works and leads to reduction of costs. These methods can definitely be applied in bioreactor design and operation in a multitude of ways.

Another trend close to this is the perhaps unexpected support to design from the regulatory bodies by enforcing the pharmaceutical industry to apply the principles of quality-by-design (QbD) and process analytical technology (PAT) [41, 44]. As the terms imply, quality should be associated with the design and, as repeatedly declared, be built into the design of the process. It is done by clearly defining the interdependency of critical parameters with the help of statistical measures in order to determine the parameter space where quality is maintained. Especially statistical factorial methodology, the so-called design of experiments (DoE), is an efficient tool for achieving QbD. The PAT is the enabling tool for QbD. Besides techniques such as DoE, other analytical methods and means provide relevant data for the critical parameter in the process and product. Moreover, once the

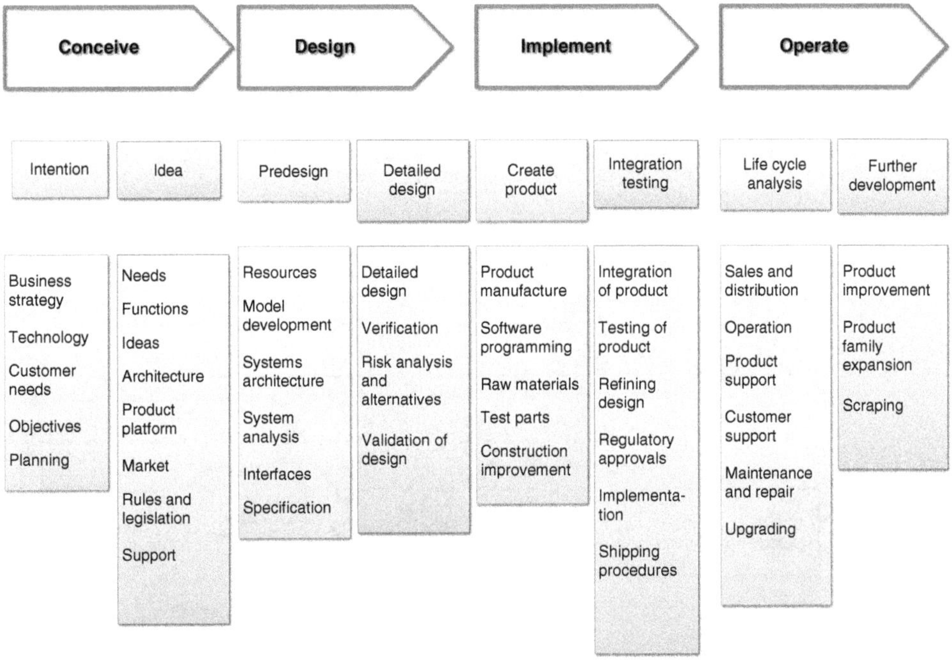


**Figure 1.4** The CDIO concept: the process of developing a new product or production system is considered as a consecutive activity spanning from conceiving the product and production concept, designing the product or production system, implementing it into full-scale production, and finally operating it continuously for regular production. It is advocated that The CDIO is applicable to all industrial development work and should, therefore, be the framework for all engineering activity – from training and education till operating a process (from [18]).

design and operational space are set, the product should be controlled dynamically to stay within it. This requires both adequate monitoring methods and reliable control approaches. For this, general methodologies are available and applied in industry. QbD has become an integral part of regulatory work and used by the development teams in the pharmaceutical industry. Although differently termed in other biotechnology applications, such as in food, biochemical, and enzyme and bioenergy production, the same quality aspects and design and development methods can be applied and are probably applied to a large extent. The same can be expected for a majority of manufacturing industries.

As pointed out in Section 1.1, the chapters of this book cover the majority of these recent trends. Evidently, although these new trends have already to a large extent spread over the entire industry, they do to a high degree apply to the challenges of bioreactor design and operations. This is, in particular, true for conceptual design methodologies. Viewpoints on bioprocess unit design, bioreactors in particular, with a conceptual mind-set have gradually unfolded in recent years.

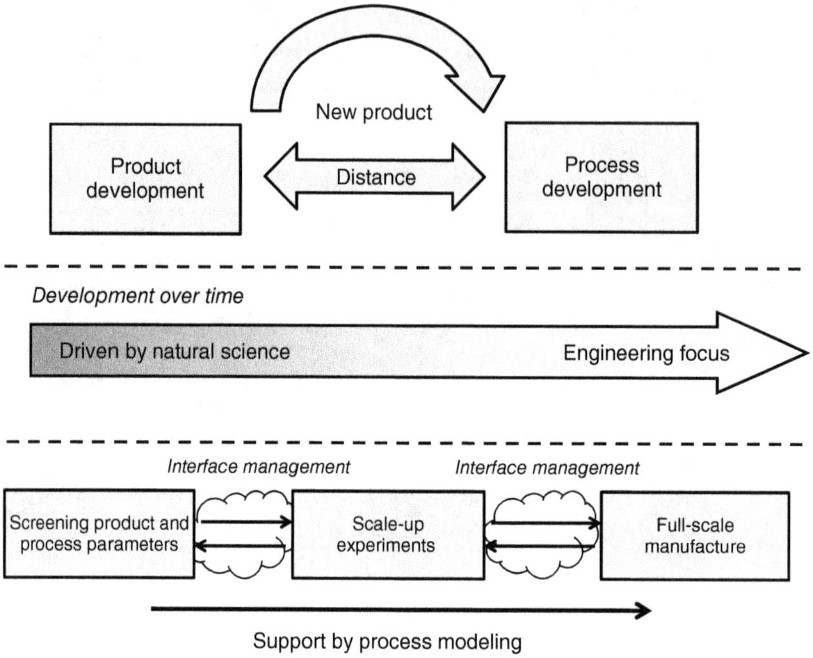


**Figure 1.5** Framework for the development of a new product and its production depicted from three critical hurdles: the transfer of the product concept into a bioprocess, the transcendence of development from natural science to engineering, and the interaction between screening, scale-up, and full-scale manufacture. In all, the bioreactor has a key role (from [20]).

This has been facilitated and probably inspired by the access of other technical means and tools.

However, it needs to be stated that a multitude of challenges encompass bioproduct and process development are more demanding and diverse than in other areas of industry, although bioproducts also share a number of similarites with other industrial products. This has been illuminated by Neubauer *et al.* [20] by combining basic biochemical and bioprocess engineering with management technology perspective. Basically, the biotechnology industry faces three dominating challenges: transferring the complex biological understanding of a production organism and its target product into the engineering domain, addressing key engineering objectives that are already in early development phase, and speeding up the path from lab- and pilot-scale to full-scale manufacture (Figure 1.5).

New tools are available for meeting these challenges that may have substantial impact on the product and process development. These are micro-multi-bioreactor systems for early screening experiments, statistical methodologies for optimization of production methods, scaled-down [38, 40, 43] process unit systems and new sensors, and other bioanalytical instruments. These tools need to be placed on a convenient development platform where they are used synergistically. Bioreactors, their design, and operation are key components in the approach, as high titers of target products are established thereby. However, the recovery of the target product should also be integrated into the platform approach to fully exploit the potential of the tools.

Undoubtedly, the majority of the trends discussed here are not restricted to biotechnology and bioreactors, which is of course a big advantage. Methodologies that are founded in industry in general are advancing more efficiently and faster. Biotechnology is small in comparison to other industrial activities and may benefit substantially from other adjacent technology areas.

## 1.5 The Systems Biology Approach

A most resourceful tool for the design and operation of bioprocesses, in particular for the bioreactor stage of a bioprocess, is systems biology, sometimes also termed systems biotechnology to highlight its utility for technical biological applications and distinguish it from biomedical or general applications in science. Probably no other current trends in modern biotechnology has the potential to impact bioprocessing as much, due to the process economic implications it could pave way for. The "omics" tools of systems biology have the power to substantially influence the designing of biological production systems, because they have, or at least have the potential, to measure minute events, rates, and metabolic flows within the cells that transcend other analytical means in consumed time and data volumes, and by that, reach parameters that can facilitate to effect better design and operation decisions of bioreactors. This concerns especially the biological functions that have a direct potential impact on the design of the bioreactor involving pathway engineering, transporter engineering, removal of negative regulation, and engineering of the regulatory network of the cell [50].

Objectives of systems biology typically aim at process intensification. If successful, it implies more demanding requirements of the transport of heat and energy, gases, and nutrients on the bioreactor system. Thus, efficiency of agitation and transfer becomes pronounced and calls for another round of optimization of engineering parameters. *In silico* simulations can be used to theoretically further predict, or verify, expected effects. The systems biology tools, or "omics" tools, are key in this, including genomic and transcriptomic arrays, proteomics tools for protein structure, metabolomics tools for pathway analysis supported by high-throughput instrumentation. This comes very close to the need of understanding of biological processes in the cell as is fundamental in medicine and for clinical purposes.

Park *et al.* [21] describe the systematic approach of using systems biology as a three-round procedure. Figure 1.6 illustrates the basic ideas behind the overall procedure of strain improvement by systems metabolic engineering. The factors to be considered are shown. In short, strain selection is followed by (i) the first round of metabolic engineering, which allows the development of a base strain.

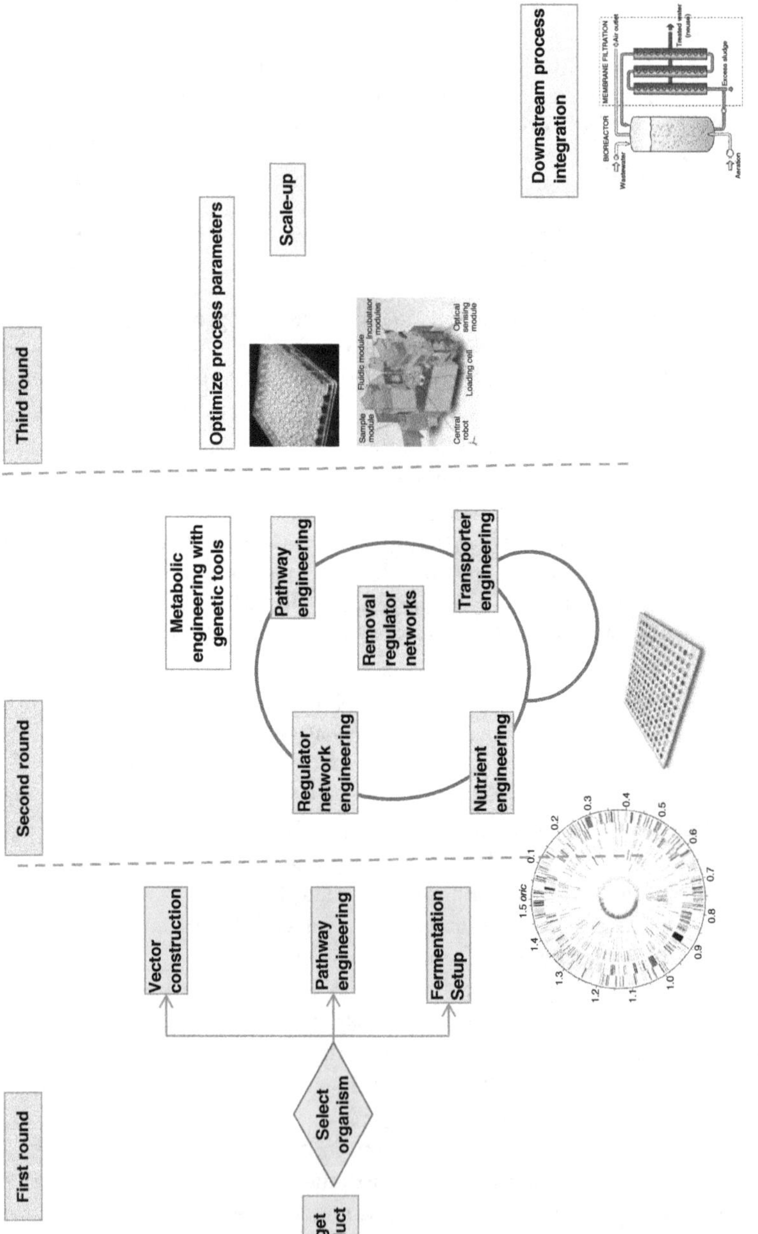


**Figure 1.6** A systems biology approach for bioreactor process development based on a three-round procedures.

The base strain is further engineered (ii) based on the results obtained from high-throughput genome-wide data and computational analyses. (iii) The performance of this preliminary production strain is then evaluated in an actual fermentation process. In this step, the downstream processes are also considered. The results are then fed back into further strain development until a superior strain showing desired performance is obtained.

Others put particular focus on the genomic step as being to keystone of successful bioprocessing. Herrgård *et al.* [22], for example, have convincingly shown how yield and expression rates of metabolite and protein products may result in manifold increases by optimal metabolic network identification (OMNI). By this method, potential changes in a metabolic model on genome scale are systematically identified by comparing model predictions of fluxes with experimental measurements. OMNI uses efficient algorithms to search through the space of potential metabolic model structures, thereby identifying bottleneck reactions and their associated genes. The OMNI method has been applied in the optimization of the metabolite production capacity of metabolically engineered strains [23, 24]. Thus, this method could unravel secretion pathways for desired byproducts and suggest ways for improving the strains. By that, a new tool is provided for efficient and flexible refining of metabolic network reconstructions using limited amounts of experimental data – this makes it a complementary resource for bioreactor bioprocess development.

As mentioned earlier, application to mammalian cells tend to dominate new industrial bioprocesses; in consequence, systems biology approaches must be able to deal with models of higher complexity for these cells to provide reliable predictions. The increased complexity of the systems biology task is apparent in the study performed by (Xu *et al.* 2011), where they present a map of the 2.5 GB genomic sequence of the CHO-K1 cell line comprising 24 400 genes located on 21 chromosomes, including genes involved in glycosylation, affecting therapeutic protein quality, and viral susceptibility genes, relevant to cell engineering and regulatory concerns. The huge data collection contributes to explain how expression and growth mechanisms may influence expression patterns related to human glycosylation-associated genes are present in the CHO genome. Again, conceiving systems biology data provide additional cues on the genome level that may facilitate the optimization of biopharmaceutical protein production in bioreactors.

One more key functionality of the biological system is the stability of the genetic material of the cell. The stability of a cloned cell line for recombinant protein expression is an essential function to maintain during a production batch as well as in a cell bank for repeated seeding of cultures. The sensitivity of production cell lines and the implications thereof have been addressed in a variety of studies [26, 27, 28].

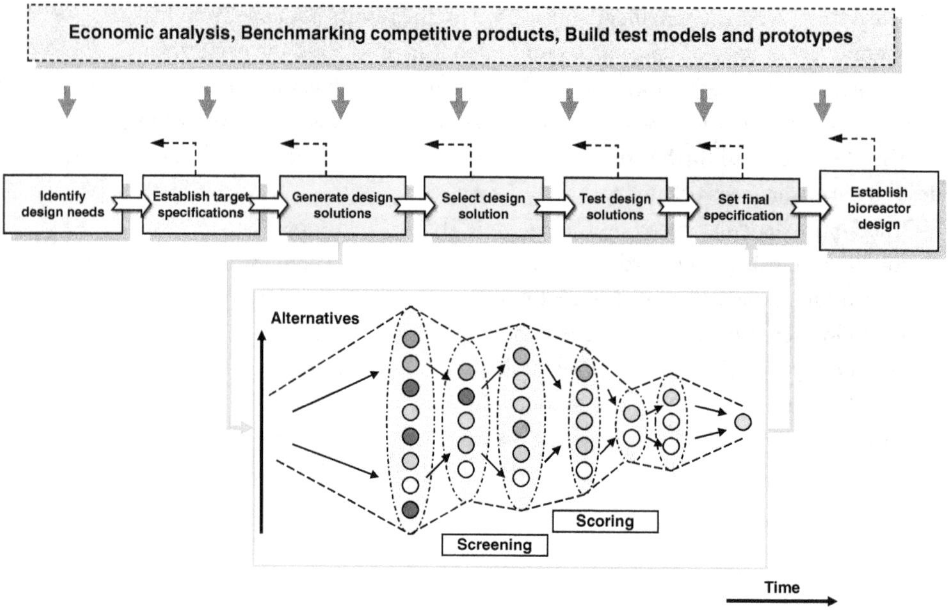


**Figure 1.7** Conceptual design principle according to Ulrich and Eppinger sequential design concept where product alternatives are screened versus customer needs.

## 1.6
## Using Conceptual Design Methodology

Another approach that can support bioreactor design and operation substantially is the conceptual design methodology. The basics of the methodology were established in mechanical engineering several decades ago and have since then gradually been refined [29–42]. The main intention was to systemize the design work in a development team in order to reach the best design architecture of a product (Figure 1.7). The approach is a typical top-down procedure: overview all alternative solutions and from that select the best constraints.

Recently, the concept was revived and demonstrated on applications in biotechnology, including bioreactors and bioprocesses [33]. The original methodology was expanded by bringing in the biological systems in the concept and showing how these in the best way could interact with mechanical and electronic systems in the product. Therefore, the methodology was termed *biomechatronics* as it merged complexities from three *per se* complex engineering disciplines: the bioengineering, the mechanics, and the electronics.

A key feature of the biomechatronics methodology is that it is user needs and functionality that guide the design toward the design targets.

Mandenius and Björkman [33, 34, 35] have in a number of examples shown how this can structure and improve the design work for typical biotechnology products and production systems, such as upstream and downstream equipments, biosensors, biochips, diagnostic devices, and bioprocesses, at the same time as the work process is facilitated and speeded up. [39, 45, 47–49]

Figure 1.8 recapitulates the cornerstones of the methodology: to precisely define and specify the needs and target metrics of the user or customer; to clearly define the expected transformation process (*Trp*) of the product or process and those systems that must interact with that process to carry it out efficiently; to consider all functional elements that must be present for this and to configure (or permute) these in a variety of more or less appealing alternatives; and, finally, to compare and assess these alternatives in order to screening out the ones that best cope with the original design and user targets.

As apparent in the figure, the methodology is based on a consecutive and iterative procedure where graphical and tabular tools support the design work. The flow of work depicted in the figure outlines the recommended steps in a sequential order. In the first step, the design mission is concisely stated. This is followed by identifying the needs of the users of the intended product or production process.

These needs are then further specified with target values. With the help of the specifications, an overview flow chart, the so-called Hubka–Eder map [30], is drawn, which shows the functions and systems required for accomplishing the specification.

The functions in this chart are represented by abstract functional components that are combined in as many realistic alternative permutations as imaginable. This is the key step in the design and is referred to as *concept generation*. The conceptual alternatives are screened and scored toward the original specification target values. This results in a ranking from which the best design alternatives are selected.

First, at this stage, actual physical, chemical, or biological objects are brought into the design work. These objects, the so-called anatomical components, are identified with concrete technical devices, instruments, or other technical gears, usually commercially available, or feasible to construct or prototype. After additional assessment, the anatomical objects form the final design structures of the product.

The conceptual approach is very useful in the design of bioreactors as well as for the layout of the operational procedures of bioreactors and integrated bioprocesses where the bioreactor is a part.

As mentioned earlier, the main purpose of a bioreactor is to control the biological transformations that take place in it.

One way of describing the *TrP* would be to follow the established biochemical engineering approach – to structure the transformation into the biological conversion steps based on metabolic maps and process flow diagrams [36, 37]. This would more or less automatically end up in a description with mass transport and rate constant-based kinetics. This would depend on the environmental state (e.g., temperature- and pressure-dependent constants) and supply of raw materials and media.

In the following example, the systems and subsystems necessary for carrying out the *TrP* in the bioreactor are instead described with the biomechatronic design

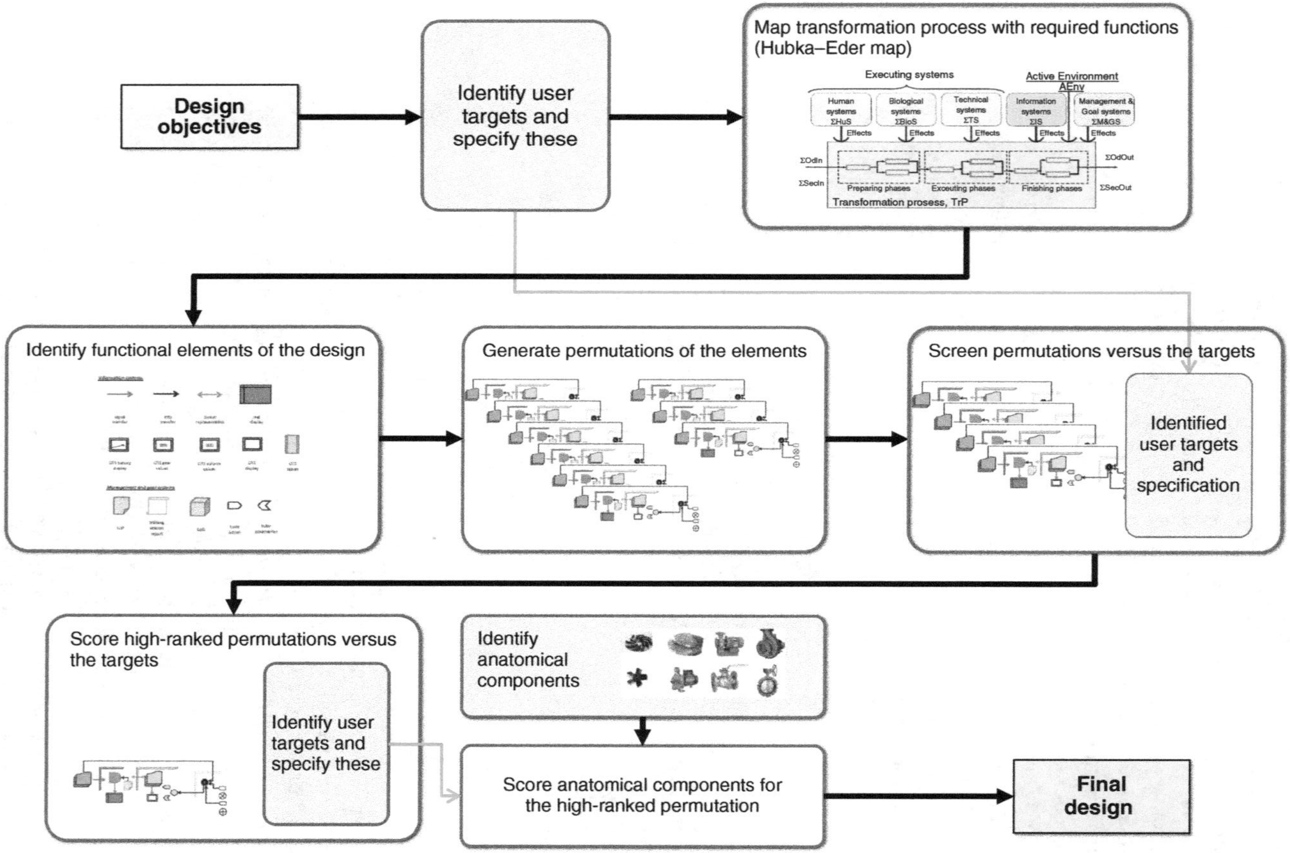


**Figure 1.8** The biomechatronic methodology according to Mandenius and Björkman where conceptual methods and tools are combined in a complementary methodology.

approach (Figure 1.9). The *TrP* in the example could be any bioconversion that is possible to realize in a submerged microbial or cell culture system, where nutrients are taken up by the cells and converted into metabolites or protein products. The Hubka–Eder mapping is now used to analyze the interactions between the systems in a generalized way. The biological ($\sum$ BioS) and technical system ($\sum$ TS) entities of the map are here described more thoroughly since these are, of course, pivotal in a bioreactor that performs biological conversions.

Also, as the figure illuminates, the *TrP* of the Hubka–Eder map has an inherent mass balance structure between the inputs and outputs. The map has defined phases (preparing, executing, and finishing), as a conventional process flow diagram has upstream and downstream sections, and in the Hubka–Eder map it is relatively easy to identify those phases where the biologically and kinetically controlled transformations take place. The Hubka–Eder map can be adapted to cover typical bioreactor processes such a recombinant protein expression, viral vector production, or stem cell differentiation.

Figure 1.9 and the zoom-in depiction in Figure 1.10 illustrate a well-known biotechnology application; protein production in a recombinant host cell line is exemplified. The biological systems have in the map been divided into four different biological systems: the culture media system (*BioS-1*), the transport system of the cells (*BioS-2*), the host cell metabolism (*BioS-3*), and the expression system (*BioS-4*). Also, a sub$^3$system and sub$^4$system can be included, preferably using a software tool to support the structuring of the information. It is noteworthy that at higher system levels only functions are described, whereas at the lower levels anatomical structures are introduced, such as a particular nutrient or biomolecules, for example, 30S ribosome and tRNA-amino acid. When the alternative anatomical units are identified the analysis is completed and an anatomical blueprint can be set up (cf. Figure 1.8).

The most essential functions needed in the $\sum$ TS and these functions' interactions with other systems in the Hubka–Eder map of the bioreactor are included in the descriptions shown in Figure 1.10. Here, the $\sum$ TS have been divided in subsystems for the functions of heat exchanging, agitation, pumping (transporting liquids and gases), containment, sterilization (partly overlapping with the previous), chemical state transformers, and pressure generation.

When possible we use the same groups of $\sum$ TS for different bioreactor types. Thus, the *TS-1* system concerns the function of nutrient handling (supply, store, and transport). On the next system level, this will result in pumping or injection, storage containment of nutrients, and means to contact the nutrients with the cells). For example, $CO_2$ supply through pH balancing of a buffer of gas head space are design alternatives to be ranked toward the cells' transport functions as described in *BioS-2*.

The functions of the *TS-2* system concern containment and agitation. The *TS-2* system should protect the culture from the environment and sometimes the opposite, protect the environment and the $\sum$ TS.

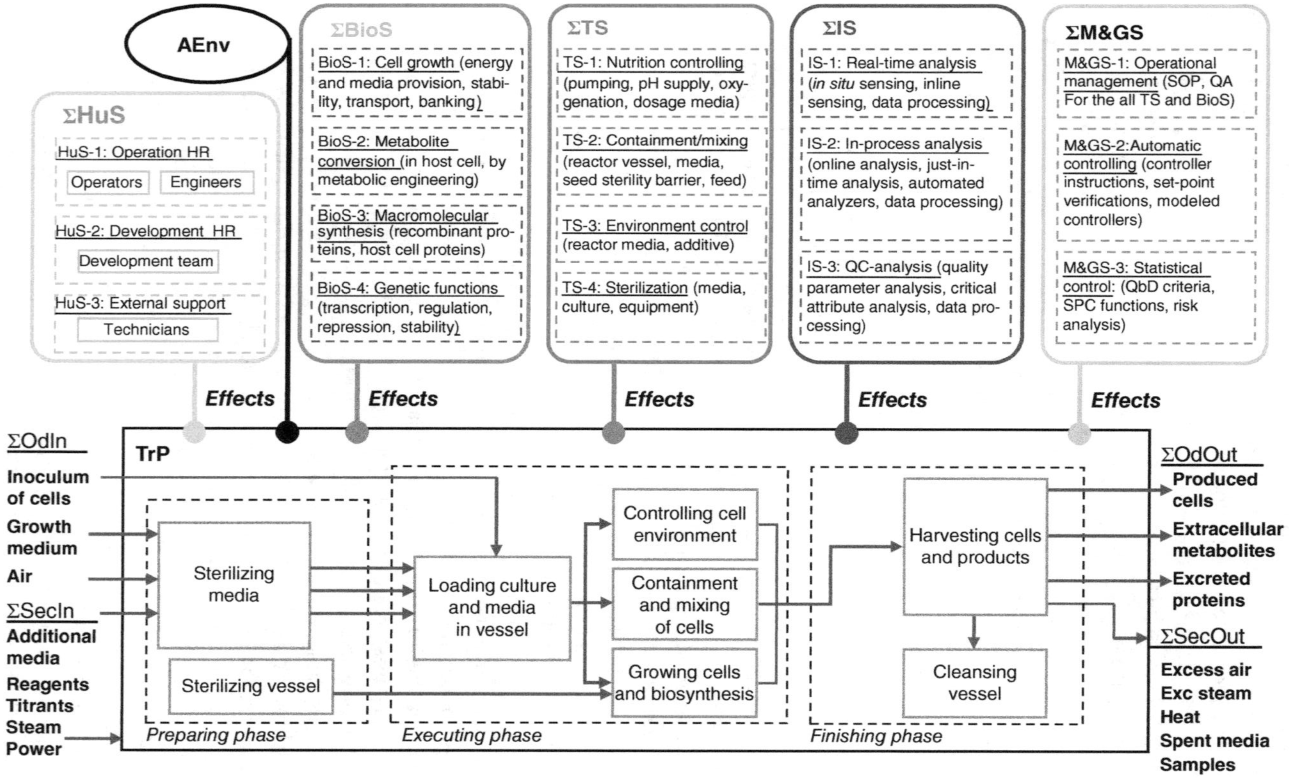


**Figure 1.9** Hubka–Eder map of a bioreactor for the microbial production of a recombinant protein. Overall Hubka–Eder map showing the transformation process and the systems and subsystems involved for performing the transformation.

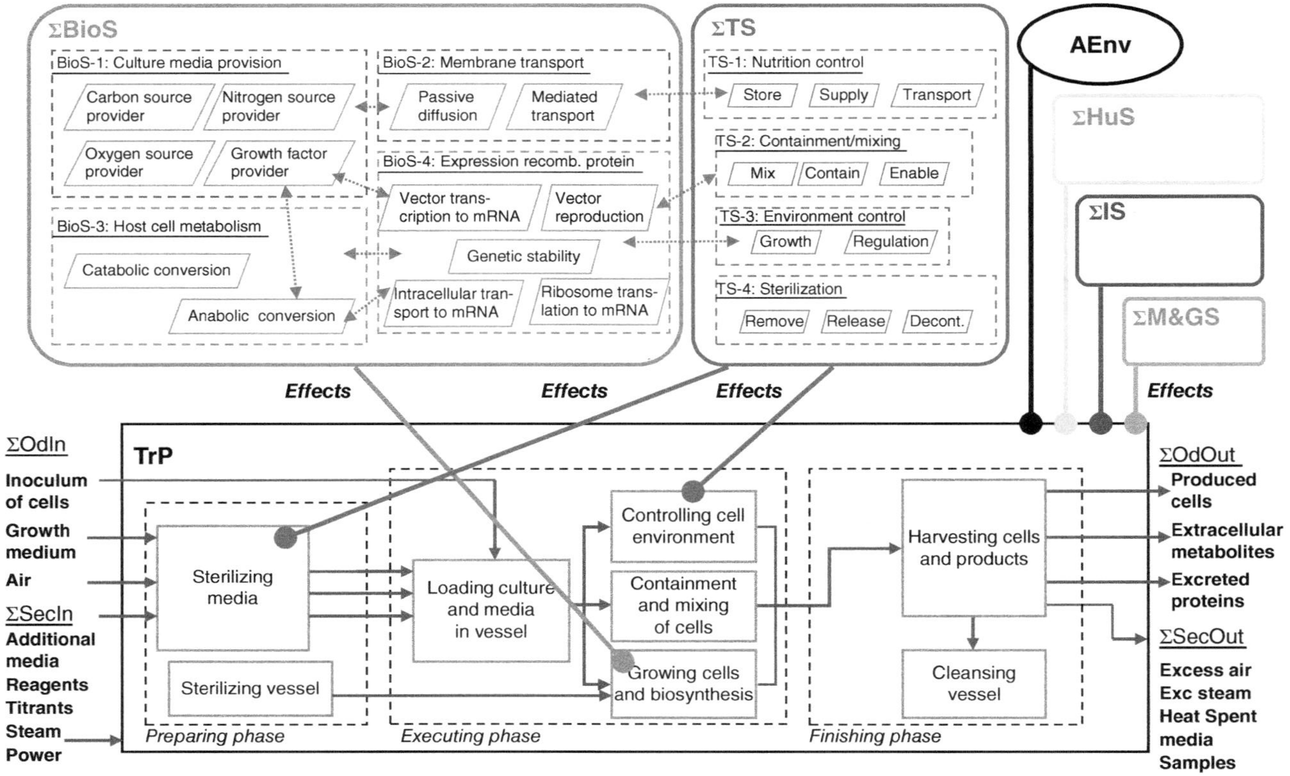


**Figure 1.10** A zoom-in of the Hubka–Eder map in Figure 1.9 showing the $\sum$ BioS and $\sum$ TS systems and the interaction of subsystems.

The mixing function is subdivided into agitation of cells, added media, and supplied gas. Anatomical parts for these subsystems could be turbines, draft tubes, and rotating vessels (cf. tissue culture application, discussed later).

Alternative ways of introducing oxygen (e.g., by spargers or silicone tubing) without provoking oxidative or shear stresses on the cells are considered at this stage.

The *TS-3* system provides the functions for control of the bioreactor environment. Subsystems involve heat transfer (e.g., by heat exchanging, pre-heated liquid media, reactor jacket), pH regulation, $pO_2$ regulation, $CO_2$ regulation, and pressure regulation and media additives (factors, shear force reducing polymers such as polyethylene glycol).

The function of the *TS-4* system is to provide sterility of the bioreactor. Common operations are *in situ* heating procedures, chemical treatment, and microfiltration. Here, it is also suitable to consider to bring up the less common alternatives such as radiation and toxicant treatment, or to introduce disposable bioreactor vessels that revolve the prerequisites for sterilization procedures significantly.

Table 1.3 resolves the map views of the $\sum$ TS in more detailed subsystems and functions, and gives examples of anatomical components.

For example, the heat exchange subsystem needs a subfunction for the removal of heat (produced by the culture), which could be cooling coils or a jacket. The heat exchanger subsystem also needs a function for heating up the reactor medium, which could be a heat cartridge, hot vapor perfusion, or, again, a heated coil.

Based on the identification and analysis of functional systems in the Hubka–Eder map, critical design elements are conceived and compiled (Figure 1.11a). Here, the most essential elements of the technical and biological systems are shown. Note that it is the functional capacity of the elements that are displayed, to avoid confusing the design work with physical objects at this state but to keep focus on what these objects shall achieve in the design solution.

The functional elements are subsequently combined in order to generate diverse conceptual alternatives (Figure 1.11b). The 12 functional elements are used with very small modifications to envisage combinations that will allow the *TrP* to be realized. The four configurations shown represent just a fraction of all combinations that are possible to generate. Especially, if additional elements were identified and introduced, a variety of other configuration alternatives could easily be generated. This would of course be the case in a large-scale design project (cf. [33]).

The generated alternatives are then screened and scored versus the user needs and specified targets identified in step 1. The total scores for each alternative are used to rank them and to assess which ones are preferable according to the users' targets. This results in a preferred conceptual design for which the functional elements are replaced with real physical objects, the so-called anatomical objects. The four configurations are shown. First, after this conceptual analysis and assessment, the prototyping of the bioreactor ensues. Figure 1.11b could be compared with the 12 bioreactor designs displayed in Figure 1.2.

**Table 1.3** The technical systems ($\sum$TS) and subsystems of a typical bioreactor.

| Technical system from functionality perspective | Technical subsystems and their functions | Examples of anatomical component for performing functions |
|---|---|---|
| Heat transfer system | To keep culture at optimal temperature level | Heat exchangers |
| | | External loops |
| | To sterilize the equipment | Steamers |
| Agitation/mixing system | Disperse air | Sparger, pressure valve |
| | | Bubbling device |
| | Mix liquid/air | Turbine impeller |
| | | Marine impeller |
| | | Anchor impeller |
| | | Toroid device |
| | | Baffles |
| Transport of media | To transport gaseous media | Pressure vessel |
| | | Gas flow system |
| | To transport liquid media | Displacement pump |
| | | Peristaltic pump |
| | | Syringe pump/device |
| | | Flask transfer device |
| | | Hydrostatic pressure system |
| Filtration of media | Particle removal | Mini-filtration |
| | Virus removal | Ultrafiltration |
| | Heat-labile molecule removal | Microfiltration |
| Containment | To contain batches repeatedly | Steel vessel |
| | | Glass vessel |
| | | Teflon vessel |
| | To contain one batch | Glass jar |
| | | Plastic bag |
| Sterilization of equipment and media | Sterilization of equipment and media together | *In situ* heat sterilization |
| | | *In situ* chemical treatment |
| | | *In situ* radiation sterilization |
| | | Microfiltration |
| | Gas media sterilization | Microfiltration |
| | | Flush sterilization |
| Pressure generation | Headspace pressure generation | Pressure valves/vents |
| | Air gas generation | In-house supply gas system |

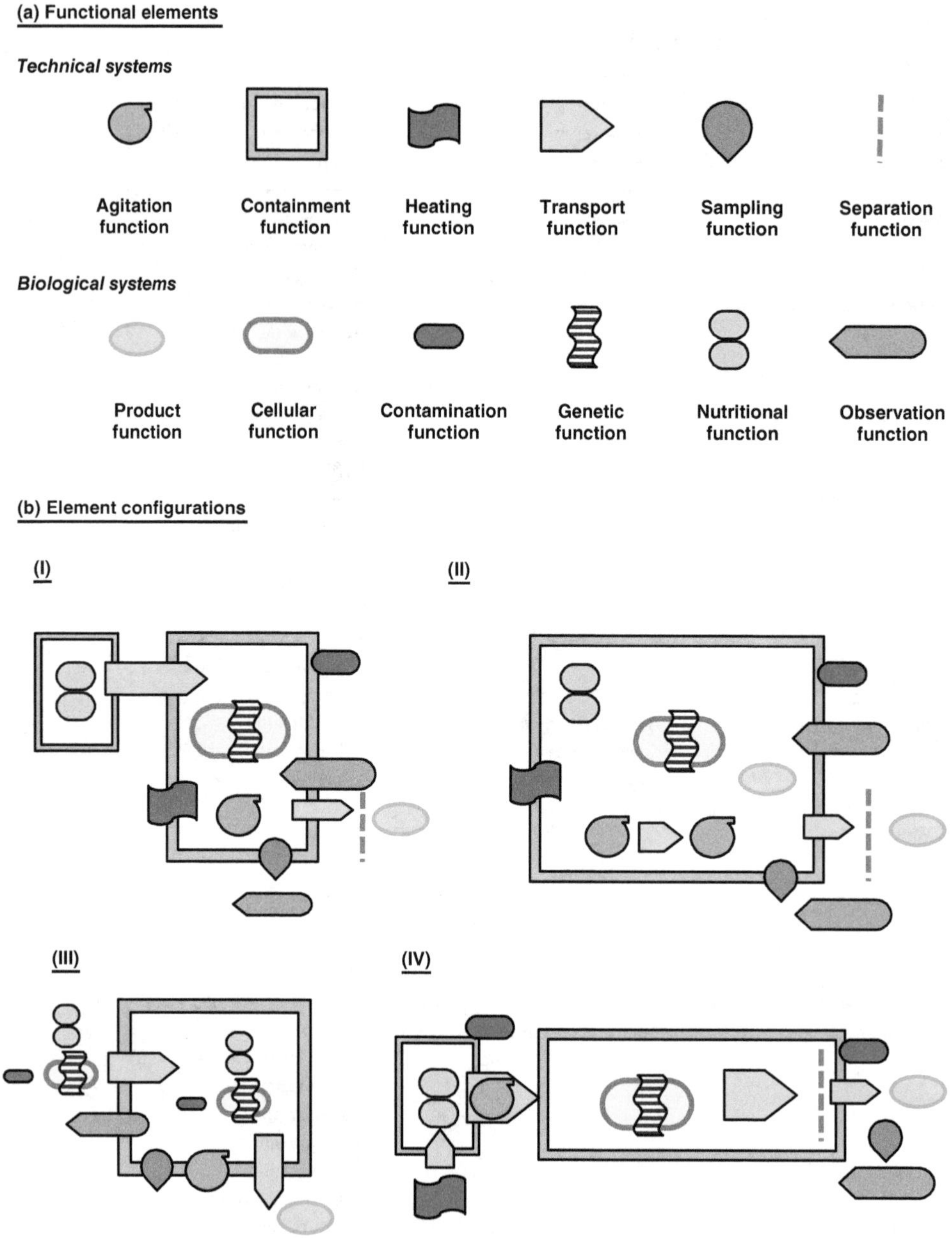


**Figure 1.11** (a) Functional elements derived from the systems in the Hubka–Eder map. In a real design, the number of elements may exceed 100. (b) The elements are combined in order to envisage various configurations. The four examples shown can be configured in a variety of other permutations.

## 1.7 An Outlook on Challenges for Bioreactor Design and Operation

Several of the design issues discussed in this chapter are further evolved in ensuing parts of this book. The challenging nature of the issues diverges. In Table 1.4, the character and potential impacts on bioreactors of these challenges are summarized with reference to where in this book these are further discussed.

Without doubt, a further exploitation of systems biology is one of the potential areas with substantial implications on bioreactors. To continue deriving information about production organisms and their behavior under relevant conditions is, however, a demanding task for future research work. High-throughput analytical machines able to carry out "omics" and data interpretation are currently employed. By that, the bioanalytical systems biology tools may facilitate and improve the conditions of design. The implementation of these data into the bioreactor and bioprocess scenarios versus the production engineering goals requires a synergistic mind-set that is not yet established in the industry. However, there are few reasons to believe that this will not happen in near future.

The combination of the systems biology view with microbial and cellular physiology and how this knowledge is transduced into design practice for more efficient processing is also a challenge required to be further pursued (see Chapter 8).

New biological production systems such as stem cells, tissues, and organs create their own challenges on the design of bioreactors where the intrinsic features and properties of these biological systems require careful consequence analyses for design and implementation (see Chapters 4–6). The early stage of development of these applications may today suffer from not being designed from typical bioengineering aspects, but from a cell biology perspective as suggested in the framework of Figure 1.3.

The use of novel inventive methods of immobilizing cells in order to improve their performance and stability in bioreactors fits well into the increased understanding of physiology of bioproduct-producing cells (see Chapter 7).

Still, traditional biological systems, such as microbial and cell cultures for metabolite and protein production, require the same kind of attention although this has historically been going on for a longer period.

Basic principles and implementation methods for scaling up and scaling out the production systems fit into production encompassing all cell types (see Chapters 4 and 11).

The access to reliable analytical platforms is necessary for good design work; this may include a variety of tools and methods, such as microbioreactors (see Chapter 2), single-use reactors (see Chapter 9), scale-down methods (see Chapter 11), and bioreactors-on-a-chip (see Chapter 3).

Moreover, the technical design of bioreactor equipment has also been supported by other resourceful tools such as DoE for optimization (see Chapter 15), better physical models, computational fluid dynamics, and scaled-down or miniaturized test platforms, which should offer better possibilities (see Chapter 10).

**Table 1.4** Challenges of the topics of the book chapters.

| Area of challenge | Character and potential of challenges | Chapter in book |
|---|---|---|
| Conceptual design | Approach bioreactor design conceptually and systematically; refining the design methodology for a user perspective | Chapter 1 |
| Exploiting systems biology and their tools | The basic principles for bioreactor kinetics, mass, and heat transfer are still applied but are also refined | Chapter 1 |
| The interface between cell physiology and bioreactors | Coping with cellular physiology in the bioreactor applying omics-derived understanding into biological reactions | Chapter 8 |
| Culture of stem cells at bioreactor scale | Adapting bioreactor systems to new cellular production requirements | Chapter 6 |
| | Adapting and scaling up and scaling out bioreactor systems to new cellular systems | Chapter 4 |
| Tissue and organ cell cultures in bioreactors | Adapting bioreactor systems to new cellular production requirements | Chapter 5 |
| Culture immobilized cells in bioreactor | Adapting bioreactor systems to new cellular production requirements | Chapter 7 |
| Down-scaling bioreactor processes | Providing tools representative for large-scale operation at the microbioreactor scale as a process development and optimization tool | Chapter 2 |
| | Providing tools representative for large-scale operation down to microfluidics dimensions | Chapter 3 |
| | Exploiting mass production and parallel process analysis | |
| Scale up/down methodology | Reducing gaps between scales | Chapter 11 |
| | Reducing gaps between scales | Chapter 10 |
| | Computational fluid dynamics for bioreactor design; understanding rheology of the bioreactor | |
| Single-use bioreactor design | Facilitating operation by convenience | Chapter 9 |
| Bioprocess integration | Integration of the bioreactor with the downstream process | Chapter 12 |
| Design of growth and production media for bioreactors | Accelerating media optimization by statistical factorial design methods | Chapter 15 |
| Efficient monitoring of bioreactors | Exploiting the information flow from the measurement with modeling Increasing observability by PAT approaches and multivariate data analysis | Chapter 13 |
| | Exploiting models for more information by using soft sensors | Mandenius (Chapter 14) |
| Training bioreactor operations | Training plant personnel in operating the complexity of bioreactor efficiently | Chapter 16 |

Although already widely used in bioengineering, it cannot be anticipated that information technology and computer applications will take design and operation of bioreactors several steps further allowing previous studies, methodologies, and existing know-how to be realized in industrial procedures. Examples are applications with multivariate data analysis and process monitoring and control (see Chapter 13) and use of factorial design and optimization of culture media and operation conditions (see Chapter 14).

Radically, new bioreactor designs have been accomplished that replaced old designs in favor of low-cost alternatives that are possible due to novel fabrication methods and materials as well as conditions of cost for operation and materials (see Chapter 9).

Further unfolding of statistics and data mining methods may be foreseen. Other engineering applications, for example, in chemical engineering, are ahead of bioengineering; DoE and related methodologies may be further advanced in the direction of coping with biological variation during extended process periods (see also Chapter 15).

The efforts of bioreactor design cannot be pursued efficiently without the integration of bioreactors into the entire bioprocess. This may essentially generate two gains: better process economics and processes of higher intensity. The implication of this may be huge (see also Chapter 12).

The increasing complexity of integrated bioprocess plants with bioreactors and digital communication requires qualified training procedure. This concerns especially the need for instantaneous decision-making by plant engineers and process operators. In pilot training, rescue training and clinical surgery virtual simulation is applied for accomplishing efficient and cost-effective training of new personnel. There is a challenge to adapt such simulators for bioprocess operator training where in particular variability and unpredictable events in the bioprocesses may be the focus of training (see also Chapter 16).

The CDIO engineering concept [18], referring to that all engineering should preferably be developed along a consecutive process of conceiving (C), designing (D), implementing (I), and operating (O) technical production systems, is indeed applicable to identify design and operation challenges. In Figure 1.12, an update of the earlier CDIO framework (Figure 1.4) is shown where the now-generalized CDIO activities are specified for bioreactor design and operation. The figures emphasize the consecutiveness of design and operation issues. And it provides a map of connectability of the challenges that are elaborately and with details discussed in this book content and placed into the frames of CDIO concept. However, it also reveals some gaps that need to be bridged by novel contributions.

So far, most of these progressing activities are still in the academic research environment. In a few cases, they emerge as new products from Small and Medium-sized Enterprises (SMEs).

Others are already in regular use at the process research and development units, especially at larger biotech companies.

Generation of knowledge and inventions may sometimes thrive best in the academic research supported by public resources, while sometimes it may best be

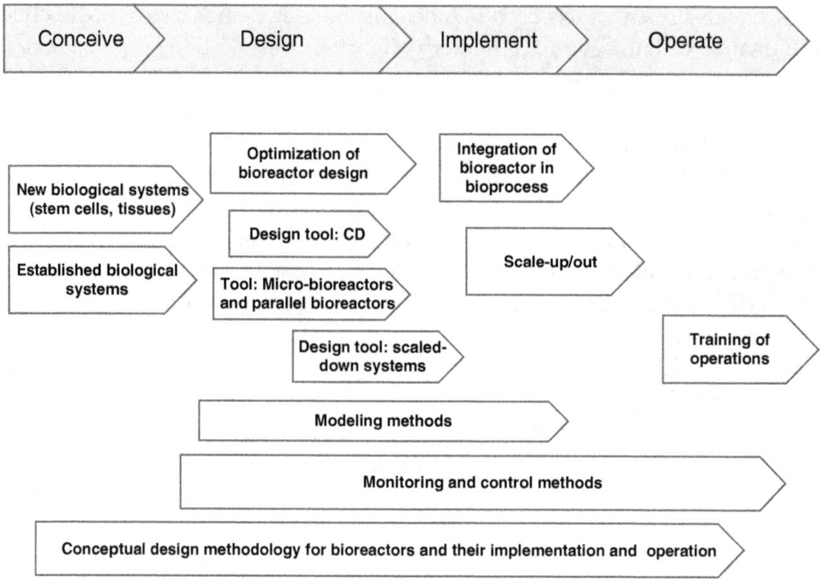


**Figure 1.12** The CDIO concept as defined in Figure 1.4, here adapted to bioreactor design and operation with several of the topics and challenges addressed in this book.

developed in-house by companies close to the applications and under knowledge protection.

This book, hopefully, contributes to overview the needs and possibilities and stimulate further progressing.

## References


1. Demain, A. (2010) History of industrial biotechnology, in *Industrial Biotechnology, Sustainable Growth and Economic Success* (eds W. Soetaert and E.J. Vandamme), Wiley-VCH Verlag GmbH, Weinheim.
2. Mason, S.F. (1964) *A History of Sciences*, Macmillan Publishing Company, New York.
3. Robbins, L. (2001) *Louis Pasteur and the Hidden World of Microbes*, Oxford University Press, New York.
4. Santangelo, J.D. and Durre, P. (1996) Microbial production of acetone and butanol. Can history be repeated. *Chem. Today*, **14**, 29–35.
5. Brown, K. (2005) *Alexander Fleming and the Antibiotic Revolution*, Sutton Publication Ltd.
6. Aminov, R. (2010) A brief history of the antibiotic era: lessons learned and challenges for the future. *Front. Microbiol.*, **1**, 134. doi: 10.3389/fmicb.2010.00134
7. Ratledge, C. and Kristiansen, B. (2006) *Basic Biotechnology*, Cambridge University Press.
8. Freshney, R.I. (2010) *Culture of Animal Cells: A Manual of Basic Techniques and Applications*, Wiley-Blackwell.
9. Köhler G and Milstein C (1975), Continuous cultures of fused cells secreting antibody of predefined specificity. Nature **256**, 495–497.
10. Shukla, A.A. and Thoemmes, J. (2010) Recent advances in large-scale production of monoclonal antibodies and related proteins. *Trends Biotechnol.*, **28**, 253–261.

11. Butler, M. and Meneses-Acosta, A. (2012) Recent advances in technology supporting biopharmaceutical production from mammalian cells. *Appl. Microbiol. Biotechnol.*, **96**, 885–894.
12. Serra, M., Brito, C., Correia, C., and Alves, P.M. (2012) Process engineering of human pluripotent stem cells for clinical applications. *Trends Biotechnol.*, **30**, 350–359.
13. Afeyan, N.B. and Cooney, C.L. (2006) Professor Daniel I.C. Wang: a legacy of education, innovation, publication, and leadership. *Biotechnol. Bioeng.*, **95**, 206–217.
14. Chisti, Y. (2010) Fermentation technology, in *Industrial Biotechnology, Sustainable Growth and Economic Success* (eds W. Soetaert and E.J. Vandamme), Wiley-VCH Verlag GmbH, Weinheim.
15. Kadic, E. and Heindel, T.J. (2014) *An Introduction to Bioreactor Hydrodynamics and Gas-Liquid Mass Transfer*, John Wiley & Sons, Inc.
16. Villadsen, J., Nielsen, J., and Lidén, G. (2011) *Bioreaction Engineering Principles*, Springer-Verlag, Berlin.
17. Derelöv, M., Detterfelt, J., Björkman, M., and Mandenius, C.-F. (2008) Engineering design methodology for bio-mechatronic products. *Biotechnol. Progr.*, **24**, 232–244.
18. Crawley, E.F., Malmqvist, J., Östlund, S., and Brodeur, D.R. (2007) *Rethinking Engineering Education: The CDIO Approach*, 1st edn, Springer-Verlag, Berlin.
19. Crawley, E.F., Malmqvist, J., Östlund, S., Brodeur, D.R., and Edström, K. (2014) *Rethinking Engineering Education: The CDIO Approach*, 2nd edn, Springer-Verlag, Berlin.
20. Neubauer, P., Cruz, N., Gluche, F., Junne, S., Knepper, A., and Raven, M. (2013) Consistent development of bioprocesses from microliter cultures to the industrial scale. *Eng. Life Sci.*, **13**, 224–238.
21. Park, J.H., Lee, S.Y., Kim, T.Y., and Kim, H.U. (2009) Application of systems biology for bioprocess development. *Trends Biotechnol.*, **26**, 404–412.
22. Herrgård, M.J., Fong, S.S., and Palsson, B.O. (2006) Identification of genome-scale metabolic network models using experimentally measured flux profiles. *PLoS Comput. Biol.*, **2** (7), e72.
23. Burgard, A.P., Pharkya, P., and Maranas, C.D. (2003) Optknock: a bilevel programming framework for identifying gene knockout strategies for microbial strain optimization. *Biotechnol. Bioeng.*, **84**, 647–657.
24. Pharkya, P., Burgard, A.P., and Maranas, C.D. (2004) OptStrain: a computational framework for redesign of microbial production systems. *Genome Res.*, **14**, 2367–2376.
25. Xu *et al.* (2011) The genomic sequence of the Chinese hamster ovary (CHO)-K1 cell line. *Nature Biotechnol.* **29**(8), 735–742.
26. Levy, S.F., Blundell, J.R., Venkataram, S., Petrov, D.A., Fisher, D.S., and Sherlock, G. (2015) Quantitative evolutionary dynamics using high-resolution lineage tracking. *Nature*, **519**, 181–186.
27. Pilbrough, W., Munro, T.P., and Grey, P. (2009) Intraclonal protein expression heterogeneity in recombinant CHO cells. *PLoS One*, **4** (12), e8432. doi: 10.1371/journal.pone.0008432
28. Oberbeck, A., Matasci, M., Hacker, D.L., Wurm, F.H. (2011) Generation of stable, high-producing CHO cell lines by lentil viral vector-mediated gene transfer in serum-free suspension culture. *Biotechnol. Bioeng*. **108**, 600–610.
29. Hansen, F. (1965) *Konstruktionssystematik*, VEB Verlag, Berlin.
30. Hubka, V. and Eder, W.E. (1988) *Theory of Technical Systems: A Total Concept Theory for Engineering Design*, Springer-Verlag, Berlin.
31. Pahl, G. and Beitz, W. (1996) *Engineering Design, A Systematic Approach*, Springer-Verlag, Berlin.
32. Ulrich, K.T. and Eppinger, S.D. (2007) *Product Design and Development*, 3rd edn, McGraw-Hill, New York.
33. Mandenius, C.F. and Björkman, M. (2011) Biomechatronic Design in Biotechnology: A Methodology for Development of Biotechnology Products, John Wiley & Sons, Inc., Hoboken, NJ.
34. Mandenius, C.F. and Björkman, M. (2010) Mechatronic design principles for biotechnology product development. *Trends Biotechnol.*, **28** (5), 230–236.

35. Mandenius, C.F. and Björkman, M. (2012) Scale-up of bioreactors using biomechantronic design methodology. *Biotechnol. J.*, 7 (8), 1026–1039.
36. Katoh, S., Horiuchi, J.I., and Yoshida, F. (2015) *Biochemical Engineering*, Wiley-VCH Verlag GmbH.
37. Mosberger, E. (2002) *Chemical Plant Design and Construction*, Wiley-VCH Verlag GmbH.
38. Carrondo, M.J.T., Alves, P.M., Carinhas, N., Glassey, J., Hesse, F., Merten, O.W., Micheletti, M., Noll, T., Oliveira, R., Reichl, U., Staby, A., Teixeira, A.P., Weichert, H., and Mandenius, C.F. (2012) How can measurement, monitoring, modeling and control advance cell culture in industrial biotechnology? *Biotechnol. J.*, 7 (11), 1522–1529.
39. Darkins, C.L. and Mandenius, C.F. (2014) Design of large-scale manufacturing of induced pluripotent stem cell derived cardiomyocytes. *Chem. Eng. Res. Des.*, **92**, 1142–1152.
40. Gernaey, K.V., Baganz, F., Franco-Lara, E., Kensy, F., Krühne, U., Luebberstedt, M., Marx, U., Palmquist, E., Schmid, A., Schubert, F., and Mandenius, C.F. (2012) Monitoring and control of microbioreactors: an expert opinion on development needs. *Biotechnol. J.*, 7 (10), 1308–1314.
41. Glassey, J., Gernaey, K.V., Oliveria, R., Striedner, G., Clemens, C., Schultz, T.V., and Mandenius, C.F. (2011) PAT for biopharmaceuticals. *Biotechnol. J.*, **6**, 369–377.
42. Hubka, V. and Eder, W.E. (1996) *Design Science*, Springer-Verlag, Berlin.
43. Luttmann, R., Bracewell, D.G., Cornelissen, G., Gernaey, K.V., Glassey, J., Hass, V.C., Kaiser, C., Preusse, C., Striedner, G., and Mandenius, C.F. (2012) Soft sensors in bioprocesses. Status report and recommendation. *Biotechnol. J.*, 7 (8), 1040–1047.
44. Mandenius, C.-F., Graumann, K., Schultz, T.W., Premsteller, A., Olsson, I.-M., Periot, E., Clemens, C., and Welin, M. (2009) Quality-by-Design (QbD) for biotechnology-related phamaceuticals. *Biotechnol. J.*, **4**, 11–20.
45. Mandenius, C.F. and Björkman, M. (2009) Process analytical technology (PAT) and Quality-by-Design (QbD) aspects on stem cell manufacture. *Eur. Pharm. Rev.*, **14** (1), 32–37.
46. Mandenius, C.-F. and Brundin, A. (2008) Bioprocess optimization using design-of-experiments methodology (DoE). *Biotechnol. Progr.*, **24**, 1191–1203.
47. Mandenius, C.-F., Derelöv, M., Detterfelt, J., and Björkman, M. (2007) Process analytical technology and design science. *Eur. Pharm. Rev.*, **12** (3), 74–80.
48. Mandenius, C.F. (2012) Biomechatronics for designing bioprocess monitoring and control systems: application to stem cell production. *J. Biotechnol.*, **162**, 430–440.
49. Mandenius, C.F. (2012) Design of monitoring and sensor systems for bioprocesses using biomechatronic principles. *Chem. Eng. Technol.*, **35** (8), 1412–1420.
50. Otero, J.M. and Nielsen, J. (2010) Industrial systems biology, in *Industrial Biotechnology, Sustainable Growth and Economic Success* (eds W. Soetaert and E.J. Vandamme), Wiley-VCH Verlag GmbH, Weinheim.

# 2
# Design and Operation of Microbioreactor Systems for Screening and Process Development

*Clemens Lattermann and Jochen Büchs*

## 2.1
## Introduction

Microbioreactors play an important role in screening and cultivation applications in the early phase of biotechnological process development. They are typically used for strain and media screening and for the evaluation and optimization of cultivation conditions. Low manufacturing costs, a high degree of parallelization, and their flexibility in application make microbioreactors a basic tool for efficient process development. However, they may differ significantly in size, design, and operating mode. For example, the size of microbioreactors ranges from microliter- to milliliter-range [1–7]. Their design typically varies in shape, material properties, and instrumentation, which directly influences the overall bioreactor performance. Today, both shaken and stirred microbioreactors are available, but also other systems, for example, bubble-aerated microbioreactors are used. A reliable transfer of the process from small scale to larger scale requires a broad knowledge of all relevant physical parameters and operating conditions. In other words, a profound understanding of the important engineering parameters and physical properties of the microbioreactors is necessary for an efficient process development.

In recent years, many efforts have been made to characterize microbioreactors in detail. As a consequence, fundamental knowledge on, for example, specific power input, gas–liquid mass transfer, and influence of ventilation in microbioreactors is available today. This knowledge is used to continuously optimize operating conditions and to improve the productivity of microbial cultivations. Simultaneously, new analytical methods and control strategies have been developed based on noninvasive online measurement techniques. As a result, precisely determined process parameters are available in small-scale bioreactors today. The latest trends of automated high-throughput applications as well as the feasibility of substrate feeding demonstrate that the full potential of microbioreactor development is not yet exploited. The

*Bioreactors: Design, Operation and Novel Applications*, First Edition. Edited by Carl-Fredrik Mandenius.

theoretical goal of this microbioreactor development process is an adaption of production-scale cultivation conditions in microbioreactors. However, microbioreactors are not able to imitate fully equipped stirred-tank bioreactors in every detail. The difference between the screening and production scale is in the order of $10^6 - 10^{10}$ (0.1 – 100 ml vs. 100 – 1000 $m^3$). This difference is simply too large to directly transfer a process between both scales. In addition, the high standard of technical equipment and analysis in stirred-tank bioreactors can often not be reproduced in microbioreactors with the same quality. Nevertheless, microbioreactors offer the possibility to obtain basic process parameters faster and, thus, may significantly accelerate biotechnological process development.

In this chapter, microbioreactors are divided into shaken systems, such as shaken microtiter plates and shake flasks, stirred and bubble column microbioreactors. Basic engineering parameters and fundamental subjects of bioreactor characterization are discussed. The realization of automated bioprocesses by use of robotic systems as well as fed-batch and continuous operation modes of microbioreactors are reported. Finally, issues regarding monitoring and control of microbioreactors are considered.

## 2.2
## Key Engineering Parameters and Properties in Microbioreactor Design and Operation

Key engineering parameters, such as specific power input, gas/liquid mass transfer, or liquid mixing, are crucial for the characterization of microbioreactors. The setting of appropriate operating conditions is basically determined by one or more of these parameters. In addition, surface properties of the reactor wall, liquid viscosity, and ventilation play an important role in ensuring a successful application of microbioreactors. This chapter focuses on the determination of the most important engineering parameters and their relevance for microbioreactor characterization. Shaken microbioreactors, such as glass shake flasks and microtiter plates, are of special interest due to their particular relevance in screening experiments. The advantage of shaken microbioreactors is their simple design, cheap operating costs, a high degree of parallelization, and well-defined mass transfer areas [1]. Therefore, shaken microbioreactors are often used for microbial cultivations during the early phase of process development. One of the first cultivation of microorganisms in shake flasks was reported in the early 1930s of the last century [8]. About 50 years later, the first microbial cultivations in microtiter plates were conducted [9]. In contrast, the systematic characterization of shaken microbioreactors and the development of control strategies started approximately 20 years ago and is still the topic of many research works. A selection of different shaken microbioreactor systems as well as some of the most important physicochemical and operating parameters is illustrated in Figure 2.1.

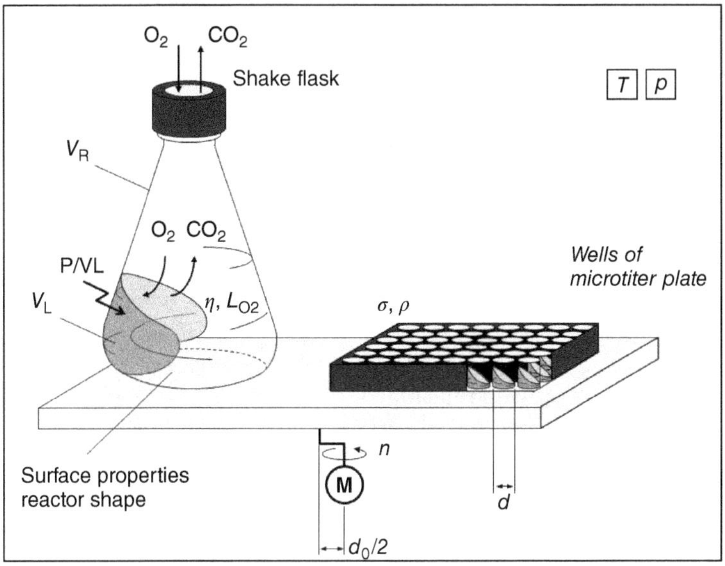


**Figure 2.1** Operating and physicochemical parameters influencing the reactor performance of shaken microbioreactor systems. Shake-flask and shaken-microtiter plates are illustrated schematically. Rotation of the bulk liquid inside the vessel is due to rotating centrifugal forces during the shaking process. The following parameters directly influence the performance of shaken microbioreactors: shaking frequency $n$, filling volume $V_L$, reactor volume $V_R$, vessel diameter $d$, shaking diameter $d_0$, viscosity of liquid $\eta$, oxygen solubility $L_{O_2}$, mass transfer resistance through the cover, mass transfer resistance at the gas–liquid interface, surface properties of the reactor wall, inner vessel design, temperature $T$, pressure $p$. Depending on the reactor size and properties of the bioreactor wall material, surface tension $\sigma$ of the liquid has an influence on liquid distribution.

### 2.2.1
### Specific Power Input

Specific power input is one of the basic engineering parameters for the characterization of bioreactors. Sumino *et al.* [10] mentioned specific power input as one of the crucial parameters for optimization and scale-up of biotechnological operating conditions from shaken to stirred bioreactors. It is defined as power input per liquid volume $P/V_L$ and describes the average power introduced into the liquid of a bioreactor. The specific power input directly influences the liquid distribution, fluid hydrodynamics, heat transfer and mixing as well as hydromechanical stress of the microorganisms. In addition, it is affected by the properties of the cultivation broth and increases with increasing viscosity. If the viscosity exceeds a critical value, the liquid in shaken bioreactors tends to be out-of-phase (see Section 2.2.2) and the specific power input will strongly be reduced [11]. As a consequence, oxygen transfer into the bulk liquid as well as mixing will also be reduced. Therefore, the specific power input directly influences the cultivation performance and represents one of the most important engineering parameters of bioreactors.

The power input into a bioreactor is caused by the transformation of electric power of the motor drive to a rotational or shaking movement. A mechanical element, that is, the stirrer in stirred bioreactors or, in the case of shaken bioreactors, that part of the vessel wall that is in contact with the rotating bulk liquid introduces the power into the liquid. As a consequence, fluid movement and mixing is induced. Finally, the energy of the liquid movement is dissipated as heat in the liquid. To determine the specific power input experimentally, measurements of the electric power consumption of the motor drive, mechanical torque measurements, or temperature measurement of the liquid can be applied [12]. While the determination of the electric power consumption is inaccurate due to an unknown energetic conversion efficiency of the electric drive, torque measurements are more accurate. For small liquid volumes, temperature measurements are inaccurate for the determination of the specific power input. To avoid elaborate experiments, a mathematical correlation to determine specific power input $P/V_L$ depending on the applied operating conditions is beneficial. In general, mechanical power $P$ of a rotating element can be calculated as the product of torque $M$ and angular velocity $\omega$. The torque can be calculated as product of a force $F$ and a lever arm $r$:

$$P = M\omega = Fr\omega \tag{2.1}$$

In shaken bioreactors, the power is introduced by friction between the rotating bulk liquid and the reactor wall. The force $F$ can be expressed as product of the wall shear stress $\tau_W$ and the friction area $A_W$. Then, Eq. (2.1) can be written as

$$P = Fr\omega = \tau_W A_W r\omega \tag{2.2}$$

The shear stress $\tau_W$ at the vessel wall is a function of the fluid flow and depends on the Reynolds number $Re$ [13]:

$$\tau_W = C_1 \rho r^2 \omega^2 f(Re) \tag{2.3}$$

$C_1$ is a constant and $\rho$, the liquid density. The Reynolds number $Re$ is defined as

$$Re = \frac{\rho n\, d^2}{\eta} \tag{2.4}$$

with dynamic viscosity of the liquid $\eta$, shaking frequency $n$, and diameter of the flask $d$ [12].

The friction area $A_W$ depends on the liquid distribution and has to be calculated numerically if all influencing parameters are considered. Büchs *et al.* [14] developed a model that enables the calculation of the contact area between liquid and flask wall in unbaffled shake flasks for liquids of low viscosity. If the influence of the shaking frequency and shaking diameter is neglected, the friction area $A_W$ in unbaffled shaken systems can be calculated with a simplified equation [12]:

$$A_W = 2\pi rh = 2\pi r V_L^{1/3} \tag{2.5}$$

The last term in Eq. (2.5) is derived if the liquid height $h$ is substituted by the characteristic length scale $V_L^{1/3}$. Considering the angular velocity $\omega = 2\pi n$ and the diameter of the shake flask $d = 2r$, and substituting Eqs. (2.3) and (2.5) in Eq. (2.2), the following equation is derived:

$$P = C_1 \pi^4 f(Re) \rho n^3 d^4 V_L^{1/3} \tag{2.6}$$

By introducing a modified power number $Ne'$

$$Ne' = \frac{P}{\rho n^3 d^4 V_L^{1/3}} \tag{2.7}$$

a dimensionless representation of the specific power input is obtained:

$$Ne' = \frac{P}{\rho n^3 d^4 V_L^{1/3}} = C_1 \pi^4 f(Re) \tag{2.8}$$

Eq. (2.8) is valid for unbaffled shaken bioreactors at in-phase operating conditions and turbulent flow behavior ($Re > 60\,000$ [15]): Turbulent flow behavior can typically be assumed for liquids with low viscosities [11]. For liquids with elevated viscosities, transitional or even laminar flow behavior is prevailing. Therefore, Eq. (2.8) is extended for different flow regimes [11]:

$$Ne' = \underbrace{C_{la} Re^{-1}}_{\text{laminar}} + \underbrace{C_{tr} Re^{-0.6}}_{\text{transitional}} + \underbrace{C_{tu} Re^{-0.2}}_{\text{turbulent}} \tag{2.9}$$

Eq. (2.9) is a dimensionless representation of specific power input in unbaffled shaken bioreactors with fitting parameters for laminar ($C_{la}$), transitional ($C_{tr}$), and turbulent ($C_{tu}$) flow at in-phase conditions. Büchs *et al.* [11] determined the fitting parameters of Eq. (2.9) for shake flasks [11]:

$$Ne' = 70 Re^{-1} + 25 Re^{-0.6} + 1.5 Re^{-0.2} \tag{2.10}$$

If commonly applied operating conditions are used, the order of specific power input values in shake flasks is of the same order of stirred-tank reactors ($1-10\,\text{kW}\,\text{m}^{-3}$) [12]. The specific power $P/V_L$ input increases with increasing shaking frequency $n$, decreasing filling volume $V_L$, increasing vessel diameter $d$ (only if $V_L^{1/3}/d = \text{constant}$), and increasing liquid viscosity $\eta$ (at in-phase-conditions). For shake flasks, the shaking diameter has no significant influence on the specific power input as tested for values of 2.5–10 cm as long as in-phase conditions are prevailing. The influence of the above-mentioned operating parameters on the specific power input is summarized in Table 2.1. It is worth to emphasize that the increase of $P/V_L$ for an increasing vessel diameter $d$ is in contrast to the maximum oxygen transfer capacity. The $OTR_{max}$ decreases for increasing vessel diameter $d$ (only if $V_L^{1/3}/d = \text{constant}$) [16]. This fact implies that the gas–liquid mass transfer in shake flasks is not directly related to specific power input, which is a basic difference to stirred-tank reactors.

Compared with unbaffled shaken bioreactors, the liquid distribution in baffled systems is not well defined. The liquid splashes randomly around on the flask bottom during shaking. Therefore, the friction area $A_W$ between the bulk liquid and the reactor wall cannot be calculated with Eq. (2.10). In shake-flask experiments, it has been observed that the specific power input $P/V_L$ in baffled

**Table 2.1** Influence of operating parameters on specific power input in shake flasks at in-phase shaking conditions.

| Change of specific power input | | Change of operating parameter | | Note |
|---|---|---|---|---|
| $P/V_L$ | ↑ | $n$ | ↑ | |
| $P/V_L$ | ↑ | $V_L$ | ↓ | |
| $P/V_L$ | ↑ | $d$ | ↑ | For $V_L/d^3 = \text{constant}$ |
| $P/V_L$ | ↑ | $\eta$ | ↑ | For in-phase conditions |

shake flasks is about one order of magnitude higher than that in unbaffled shake flasks for low liquid viscosities [10]. Furthermore, the modified power number $Ne'$ only depends on the filling volume $V_L$, but is no longer a function of the Reynolds number *Re* ($Ne \neq f(Re)$) [17]. This fact is already well known for baffled stirred-tank bioreactors.

Computational fluid dynamics (CFD) software represents an interesting tool to determine the specific power in bioreactors. The advantage of CFD analysis is the flexibility with respect to reactor geometry. The liquid distribution and the specific power input for nearly any reactor geometry can be calculated. In recent years, unbaffled shake flasks have been analyzed by means of CFD analysis [18–20]. Apart from the results of Zhang *et al.* [19], the calculated values of the specific power input agree satisfactorily with experimental results obtained by different measurement techniques [12, 17, 21]. Li *et al.* [22] investigated the hydrodynamics in baffled shake flask by means of CFD analysis and found reasonable results for the calculated specific power input. CFD analysis was used to determine the power input in unbaffled microtiter plates. For 24-well and 96-deep-well microtiter plates, average values for the specific power input were between $0.1-1\,\text{kW m}^{-3}$ for a shaking diameter of $d_0 = 3\,\text{mm}$ [23]. It is worth to mention that the reported specific power input for 96-well plates in that study is higher than that in 24-well plates. The authors explain this phenomenon with a region of high energy dissipation rates at the center of the well bottom, which was found to be existent in 96-deep-well plates. Barrett *et al.* [24] calculated the local maximum power input for 24-well microtiter plates and report values of up to $2\,\text{kW m}^{-3}$.

### 2.2.2 Out-of-Phase Phenomena

In shaken bioreactors, the shaking motion of the shaking table leads to a rotation of the bulk liquid inside the shaken vessel. At suitable shaking conditions, the bulk liquid exhibits a characteristic shape and follows the circular motion of the shaking table. The bulk liquid is oriented in the direction of the centrifugal force $F_c$ and rotates around the inner radius $r_i$ along the reactor wall with an

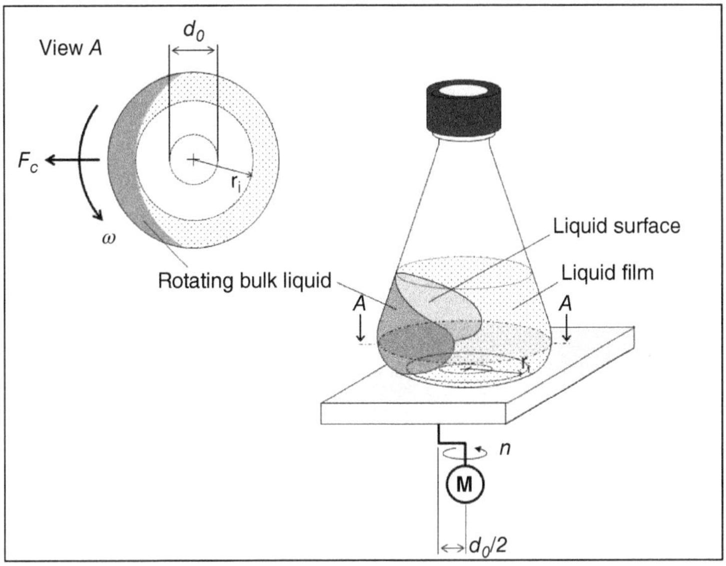


**Figure 2.2** Liquid distribution during shaking in shaken bioreactors at in-phase operating conditions. The liquid exhibits a characteristic shape and rotates inside the vessel around the inner radius $r_i$ with angular speed $\omega$. At in-phase conditions, the liquid is oriented in the direction of the centrifugal force $F_c$. A liquid film is formed on the reactor wall, which is not covered by the bulk liquid. Gas/liquid mass transfer occurs through the surface of the bulk liquid as well as through the liquid film.

angular speed $\omega$. These distinct operating conditions designate the so-called in-phase operating condition. In that case, the liquid shape is well defined, and the wall contact area $A_W$ and the gas/liquid interfacial area $A_S$ can be precisely calculated [14, 25]. In addition, a liquid film is formed on the reactor wall and bottom, which is not covered by the bulk liquid. This liquid film highly contributes to the gas/liquid mass transfer in shaken bioreactors (see Section 2.4). The shape of the liquid depends on the vessel geometry, the given shaking parameters and only to a little extent on the physicochemical properties of the liquid. In Figure 2.2, the rotating liquid and its rotational movement at in-phase condition are illustrated.

At unsuitable shaking conditions, the liquid neither exhibits a well-defined shape nor follows the rotational movement of the shaker anymore. In fact, a large part of the liquid sticks to the vessel bottom and shows no relative movement to the flask. As a result, the power input as well as the mass transfer and mixing is strongly reduced and the overall reactor performance and reproducibility are diminished. This operating condition is identified as out-of-phase operating condition [26–28]. Elevated viscosity promotes out-of-phase conditions. As a consequence, screening of microorganisms at out-of-phase conditions will possibly result in the selection of morphological mutants and low viscosity cultivation conditions due to the applied selection pressure [28]. At these conditions, it is unlikely that strains with optimal performance are identified. Therefore, out-of-phase conditions should be avoided for screening and other cultivation experiments. The probability of working at out-of-phase conditions is increased if baffled shaken bioreactors are used.

To determine out-of-phase conditions in unbaffled shake flasks, Büchs *et al.* [11] developed an equation based on the nondimensional phase number *Ph*:

$$Ph = \frac{d_0}{d}(1 + 3\log_{10}(Re_f)) > 1.26 \tag{2.11}$$

with the film Reynolds number $Re_f$:

$$Re_f = \left(\frac{\rho n d^2}{\eta}\right)\frac{\pi}{2}\left(1 - \sqrt{1 - \frac{4}{\pi}\left(\frac{V_L^{1/3}}{d}\right)^2}\right)^2 \tag{2.12}$$

Eq. (2.11) is only valid for an axial Froude number $Fr_a$ greater than 0.4:

$$Fr_a = \left(\frac{(2\pi n)^2 d_0}{2g}\right) > 0.4 \tag{2.13}$$

It is worth to mention that the out-of-phase phenomenon only occurs at elevated viscosity, for example, at filamentous growth of microorganisms or production of biopolymers if typical shaking conditions for shake flasks ($d_0 = 25-50$ mm, $V_L < 1$ l) are applied. Eqs. (2.11)–(2.13) are derived through the adaption of the fluid behavior in a partially filled rotating drum to shake flasks [11]. Considering this mechanistic background, the occurrence of out-of-phase conditions in other shaken bioreactors systems, for example, 48-well microtiter plates, can be roughly estimated with Eqs. (2.11)–(2.13) [6].

### 2.2.3
### Mixing in Microbioreactors

Mixing times in bioreactors directly depend on the specific power input introduced into the bioreactor. Consequently, the phase number *Ph* given in Eq. (2.11) directly influences the mixing time in shaken bioreactors. Besides the geometry of the shaken vessel and the filling volume $V_L$ also the liquid properties such as liquid density $\rho$ and viscosity $\eta$ influence the phase number. The liquid viscosity $\eta$ is of special interest for mixing times because it can change during cultivation of, for example, a filamentous fungi, if all other process parameters are kept constant. To determine mixing times experimentally, different methods can be applied. Easy to perform and technical simple methods are the measurement of the temperature, conductivity, and pH as well as colorimetric methods. Technical more challenging methods represent the measurement of the refraction index, fluorescence and radioactivity [29, 30]. Tan *et al.* [29] determined mixing times of 2 M and 10 M sulfuric acid (viscosities of 1 mPa s and 5 mPa s, respectively) in deionized water in shake-flask experiments with shaking frequency, varying flask size, and shaking diameter ($n = 100-350$ rpm, $V_R = 100-500$ ml, $d_0 = 25-50$ mm,

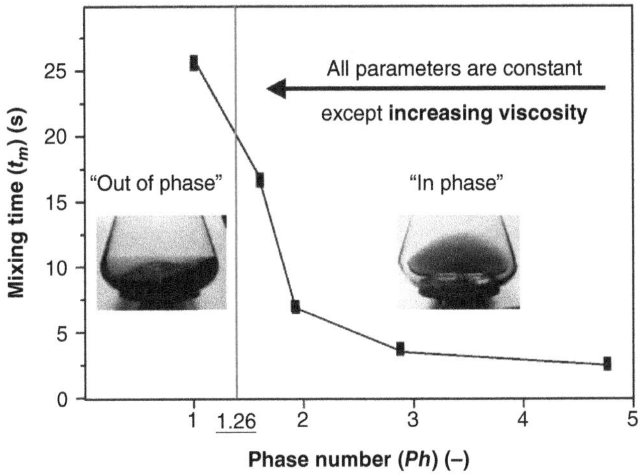


**Figure 2.3** Influence of the viscosity on mixing time in shake flasks. With increasing viscosity, the mixing time in shake flasks sharply increases at phase numbers of $Ph \leq 1.26$. Liquid sticking to the flask bottom during shaking indicates out-of-phase operating conditions. At in-phase operating conditions, the liquid follows the rotational movement of the shaking table. The following operating conditions were used: 250 ml flask, $n = 200$ rpm, $V_L = 25$ ml, $d_0 = 50$ mm.

10% relative filling volume). The authors report values between 1 and 10 s. Barrett *et al.* [24] determined mixing times in 24-well microtiter plates. For Reynolds numbers $Re > 1830$, mean mixing times of 1.7 s for 800 μl and 1000 μl filling volume are reported. Zhang *et al.* [23] used CFD analysis to investigate the mixing in microtiter plates. They show that the mixing intensity, that is, the strength of mixing of one or more phases, is higher in 96-deep-well microtiter plates compared with 24-well plates due to the appearance of an axial component of the fluid velocity during shaking.

Figure 2.3 illustrates the influence of the liquid viscosity on mixing times in shake-flask experiments. At low viscosities (i.e., high phase numbers), the mixing time is short. With increasing viscosity, that is, decreasing phase numbers, the mixing time increases. While the increase of the mixing time is moderate at higher phase numbers, a sharp increase can be observed at a phase number close to 1.26. As a consequence, the out-of-phase phenomenon (see Section 2.2) occurs, which strongly reduces the power input into the shaken bioreactor. If the phase number reaches values of ≤1.26, the bulk liquid does not follow the rotational movement of the shaking table any more. Consequently, the mixing of the liquid is drastically increased. The measurement of mixing times can be used to detect the presence of out-of-phase operating conditions. Thus, a relatively simple and cost-efficient method compared with oxygen transfer or power input measurements is available to determine this unsuitable operating condition. In general, insufficient mixing may cause unfavorable effects, such as substrate or product inhibition or oxygen limitation. Consequently, the overall bioreactor performance is directly influenced by mixing performance of a bioreactor. To ensure correct and reproducible results in cultivation experiments, a sufficient mixing of the bioreactor is essential.

### 2.2.4
### Gas–Liquid Mass Transfer

The oxygen transfer in shake flasks and microtiter plates was extensively investigated in recent years [5, 16, 31–35]. Gas–liquid mass transfer plays an important role in many microbial cultivations. The supply of gaseous educts as well as the removal of gaseous and volatile products is essential for a successful fermentation. In particular, sufficient oxygen supply in aerobic cultivations may become critical. Microbial response to oxygen limited cultivation conditions varies. Depending on the applied strain, deceleration of the metabolic activity, the formation of byproducts, or even the death of the microbial culture may occur. As a consequence, screening in microbioreactors at oxygen limited cultivation conditions will probably result in a distorted ranking of the investigated strains [36].

The discrepancy between low oxygen solubility in cultivation broths and high oxygen demand in aerobic cultivations is often a challenging task to overcome in process and bioreactor design. The oxygen solubility in salt solution and in culture media is in the range of $1-10\,\mathrm{mg\,l^{-1}}$ [37]. In shaken surface-aerated microbioreactors, such as shake flasks or microtiter plates, oxygen supply cannot easily be enhanced by increasing the aeration rate. The oxygen supply has to be adjusted by choosing appropriate operating parameters. The maximum oxygen transfer capacity in shaken bioreactors increases with increasing shaking frequency and shaking diameter and decreasing filling volume and vessel diameter (only if ${V_L}^{1/3}/d = \text{constant}$) [16]. Furthermore, the properties of the inner surface of the vessel as well as the liquid properties (viscosity, oxygen solubility, and diffusivity) influence the oxygen transfer considerably [16].

In shaken microbioreactors, oxygen transfer can be characterized by using the chemical sulfite system [16, 32, 38]. This chemical model system does not require any sterile handling, it is relatively simple to apply, and it is used to determine the maximum oxygen transfer capacity $\mathrm{OTR_{max}}$ or the volumetric mass transfer coefficient $k_L a$. In addition, the volumetric liquid surface area $a$ can be determined if the catalyst concentration is increased [38, 39]. In such cases, the reaction regime of the sulfite oxidation has to be in the enhanced regime ($Ha > 3$). Linek *et al.* [40] correctly pointed out that the sulfite oxidation method may lead to incorrect results if the liquid is not mix sufficiently. In that case, a drop of the local sulfite concentration near the liquid surface may occur, caused by insufficient sulfite ion transport from the inner bulk liquid to its surface. Another effect that might become critical is the change of the reaction kinetic at the end of the sulfite oxidation. If the sulfite concentration falls below a critical value of about 0.2 M, the reaction kinetic is decelerated. To avoid this effect and to ensure a correct determination of the volumetric surface area $a$, the time of the sulfite oxidation reaction has to be restricted to conditions before the pH sharply drops.

In the recent years, many research works focused on gas–liquid mass transfer and oxygen transfer measurements in bioreactors. Two excellent reviews summarize important aspects of mass transfer and established techniques for oxygen transfer measurements [41, 42]. The oxygen transfer rate (OTR) and the volumetric mass transfer coefficient $k_La$ are correlated by the following equation:

$$\mathrm{OTR} = k_La(c^*_{O_2} - c_{O_2,L}) = k_LaL_{O_2}p_{abs}\left(y^*_{O_2} - y_{O_2,L}\right) \quad (2.14)$$

Equation (2.14) results from an oxygen mass balance around the gas phase. In Eq. (2.14) $c^*_{O_2}$ represents the oxygen concentration in the liquid at the phase boundary, $c_{O_2,L}$ the oxygen concentration in the bulk liquid, $k_La$ the volumetric mass transfer coefficient, $L_{O_2}$ the oxygen solubility in the liquid, $p_{abs}$ the absolute pressure, $y^*_{O_2}$ the $O_2$-mole fraction in the liquid at the phase boundary, and $y_{O_2,L}$ the $O_2$-mole fraction in the bulk liquid. Equation (2.14) can be used to calculate either the OTR or the volumetric mass transfer coefficient $k_La$ in shaken microbioreactors. Assuming an oxygen concentration in the liquid phase close to zero ($c_{O_2,L} \approx 0$), the oxygen transfer becomes maximal ($\mathrm{OTR} = \mathrm{OTR}_{max}$). In shaken, surface-aerated bioreactors, the oxygen transfer into the bulk liquid occurs by diffusion of oxygen through the liquid surface area. Consequently, the oxygen mass transfer area, that is, the total liquid surface area, has a distinct influence on the oxygen transfer into the liquid. The total liquid surface area is composed of the bulk liquid surface area and the liquid film surface on the reactor wall and the bottom of the bioreactor, which highly contributes to the overall mass transfer [16]. Due to the rotation of the bulk liquid inside the vessel, the liquid film is periodically renewed. Büchs *et al.* [14] developed a model to calculate numerically the liquid distribution and the surface areas in unbaffled shake flasks. The liquid distribution is influenced by the shaking frequency, flask geometry, shaking diameter, and the filling volume. The model is only valid for water-like viscosities and if surface tension does not have to be considered. Results of this model have been visually compared with photographs of rotating liquids inside shake flasks. The calculated maximum liquid heights agreed within ±15% with the experimental values. Klöckner *et al.* [25] extended this model for larger shaken cylindrical bioreactors and included a calculation of the specific power input. In this extended model, only geometries with a vessel diameter larger than the shaking diameter can be applied. The simulation results agree well with CFD calculations for the liquid surface as well as experimental data for the total mass transfer area and specific power input.

As described earlier, the $\mathrm{OTR}_{max}$ of a shaken bioreactor can be experimentally determined by using the chemical sulfite system. Besides the calculation of the volumetric mass transfer coefficient $k_La$, several empirical equations are reported in literature to determine $k_La$ values in bioreactors [16]. These correlations are difficult to compare among each other because the applied measurement methods vary. For unbaffled shake flasks, Seletzky *et al.* [43] developed an empirical correlation to calculate the volumetric mass transfer of the chemical sulfite system:

$$k_La_{sulfite} = 6.67 \times 10^{-6} n^{1.16} V_L^{-0.83} d_0^{\,0.38} d^{1.92} \quad (2.15)$$

**Table 2.2** Geometry parameters for different microtiter plates to calculate the $k_L a$ value according to Eq. (2.16).

| Microtiter plate | Parameter *x* | Parameter *y* |
|---|---|---|
| 24-Well MTP, round, flat bottom | 0.86 | 0.03 |
| 96-Well MTP, round, flat bottom | 0.64 | 0.15 |
| 384-Well MTP, round, flat bottom | 0.51 | 0.18 |

In Eq. (2.15), the $k_L a$ is a function of the filling volume $V_L$, shaking frequency $n$, shaking diameter $d_0$, and flask diameter $d$. Calculated $k_L a$ values exhibit an accuracy of ±30% compared with experimental results, which is in the same accuracy range as mass transfer correlations for stirred-tank bioreactors [44]. For microtiter plates, Doig *et al.* [33] found a dimensionless correlation to predict $k_L a$ values:

$$k_L a_{\text{broth}} = 33.35 D a_i Re^{0.68} Sc^{0.36} Fr^x Bo^y \tag{2.16}$$

where $D$ is the diffusion coefficient of the diffusing species in a medium, $a_i$ the initial specific surface area, $Re$ the Reynolds number, $Sc$ the Schmidt number (ratio between momentum diffusivity (viscosity) and mass diffusivity), $Fr$ the Froude number (ratio between inertial and gravitational forces), and Bo the Bond number (ratio between gravitational and surface forces). Both, axial inertia forces (due to shaking) and surface forces (due to liquid properties) are considered in Eq. (2.16) by the Froude and the Bond number. The exponents $x$ and $y$ in Eq. (2.16) depend on the microtiter plate geometry, see Table 2.2. Calculated values from Eq. (2.16) have been correlated to measured $k_L a$ values of *Bacillus subtilis* cultivations in mineral medium and agrees within ±30% [33]. Furthermore, the results were compared with measured $k_L a$ values from Hermann *et al.*, which also agree well [39].

If the liquid distribution in microtiter plates is considered, surface tension of the liquid has to be taken into account. In small vessel sizes, the liquid surface tension hinders the rotation of the liquid. Duetz [9] states that the influence of the surface tension of aqueous solutions can be observed for small vessel sizes with a well diameter <12 mm. For well diameters of <4 mm, nearly no liquid movement was observed at a shaking frequency of 300 rpm and shaking diameter of 50 mm. If the surface tension is not negligible, a critical shaking frequency $n_{\text{crit}}$ has to be exceeded to initiate the liquid rotation inside the vessel. This critical shaking frequency can be calculated as follows [39]:

$$n_{\text{crit}} = \sqrt{\frac{\sigma d}{4\pi V_L \rho_L d_0}} \tag{2.17}$$

where $\sigma$ is the surface tension of the liquid, $d$ the well diameter, $V_L$ the filling volume, $\rho_L$ the liquid density, and $d_0$ the shaking diameter. If $n_{\text{crit}}$ is exceeded, the maximum oxygen transfer capacity $\text{OTR}_{\text{max}}$ increases with increasing shaking frequency. At shaking frequencies below $n_{\text{crit}}$, the $\text{OTR}_{\text{max}}$ is independent of the shaking frequency and reaches values at the level of the diffusive oxygen transfer into the stagnant nonshaken liquid. The surface tension $\sigma$ in Eq. (2.17) depends on the properties of the liquid. During cultivation, the liquid surface tension is reduced by cell-membrane fragments, proteins, biosurfactants, or extracellular glycolipids [9, 39].

Besides choosing appropriate operating conditions and bioreactor surface properties, the integration of baffles inside the microbioreactor represents an additional alternative to increase oxygen transfer in shaken bioreactors [45–47]. Even if the reproducibility of oxygen transfer between individual baffled bioreactors is poor [1, 46, 48], baffles are useful if high oxygen transfer is mandatory. Funke *et al.* [49] reported a five- to tenfold increase of the $\mathrm{OTR_{max}}$ in baffled microtiter plates. Based on these experimental data, Lattermann *et al.* [50] published an empirical correlation to calculate the maximum oxygen transfer capacities in baffled 48-well microtiter plates. In this correlation, the relative well perimeter Peri was introduced as key parameter. The oxygen transfer can be calculated as a function of the relative perimeter Peri, the shaking frequency $n$ and the filling volume $V_\mathrm{L}$:

$$\mathrm{OTR_{max,\ sulfite}} = 2.5 \times 10^{-7}\mathrm{Peri}^{6.0}n^{2.37}V_\mathrm{L}^{-0.64} + 3.3 \times 10^{-4} \tag{2.18}$$

Eq. (2.18) is valid for relative perimeters <1.1. The applicability of Eq. (2.18) for other shaken bioreactor systems, such as shake flask, still has to be verified. To calculate the mass transfer area $a$ in baffled bioreactors, the above-described models of Büchs *et al.* [14] and Klöckner *et al.* [25] cannot be applied. Due to the chaotic undefined liquid distribution, CDF simulations have to be applied to determine the liquid surface in baffled shaken bioreactors. Recently, Li *et al.* [22] conducted a CFD analysis to calculate mass transfer areas in baffled shake flasks for liquids of water-like viscosities.

The empirical $k_\mathrm{L}a$ correlations in Eqs. (2.15)–(2.18) consider experiments with different liquid properties, for example, the chemical sulfite system and cultivation broths. Therefore, the reported $k_\mathrm{L}a$ values cannot be directly compared among each other. In general, experimentally determined $OTR_\mathrm{max}$ values depend on the oxygen solubility in the liquid $L_{\mathrm{O_2}}$ and the oxygen diffusion coefficient $D$ in the liquid. As a consequence, the $k_\mathrm{L}a$ values that have been obtained for a specific liquid have to be corrected with a correction factor if other liquids are applied. In Table 2.3, upper limits of $k_\mathrm{L}a$ values reported in the literature are summarized.

#### 2.2.4.1 Influence of the Reactor Material

Commercially available shaken bioreactors have different technical specifications and properties. One important parameter is the bioreactor material. It has a strong influence on the overall performance of the bioreactor. Engineering parameters, such as the mass transfer coefficient, and operating parameters, such as the highest applicable shaking frequency, are influenced by the surface properties of a shaken bioreactor. Erlenmeyer shake flasks are traditionally made of borosilicate glass. Disposable flasks made of polycarbonate (PC) or polypropylene (PP) are commercially available today. Microtiter plates are mostly produced as disposables and are often made of polystyrene (PS). Also other raw materials, for example,

**Table 2.3** $k_L a$ Values for different shaken bioreactors reported in the literature.

| Shaken bioreactor system | | $k_L a$ Value ($h^{-1}$) | Experimental conditions | References |
|---|---|---|---|---|
| Shake flask | Unbaffled | <380 | 500 ml Flask $d_0$: 50 mm $n$: 500 rpm $V_L$: 20 ml 1 M sulfite system | [34] |
| Shake flask | Baffled | <115 | 500 ml Flask $d_0$: 30 mm $n$: 250 rpm $V_L$: 50–150 ml CDF analyses | [22] |
| 24-Well MTP | Unbaffled | <180 | $d_0$: 3 mm $n$: 800 rpm $V_L$: 1182 μl Cultivation broth | [33] |
| 48-Well MTP | Unbaffled | <330 | $d_0$: 3 mm $n$: 1100 rpm $V_L$: 300 μl 0.5 M Sulfite system | [6] |
| 48-Well MTP | Unbaffled | <290 | $d_0$: 50 mm $n$: 350 rpm $V_L$: 300 μl 0.5 M Sulfite system | [6] |
| 48-Well MTP (Flowerplate) | Baffled | <600 | $d_0$: 3 mm $n$: 1000 rpm $V_L$: 500 μl 0.5 M Sulfite system | [49] |
| 96-Well MTP | Unbaffled | <90 | $d_0$: 3 mm $n$: 1000 rpm $V_L$: 200 μl 0.5 M Sulfite system | [39] |
| 96-Well MTP | Unbaffled | <160 | $d_0$: 50 mm $n$: 350 rpm $V_L$: 200 μl 0.5 M Sulfite system | [39] |
| 96-Well MTP | Unbaffled | <190 | $d_0$: 3 mm $n$: 1100 rpm $V_L$: 200 μl Cultivation broth | [33] |
| 384-Well MTP | Unbaffled | <115 | $d_0$: 3 mm $n$: 1400 rpm $V_L$: 65 μl Cultivation broth | [33] |
| Bench-scale STR | | Up to 400 | $n$: 2000 rpm $q$: 2 vvm $V_L$: 100 ml NaCl in deionized water | [51] |
| Bench-scale STR, 1 l | | Up to 300 | $n$: 500 rpm $V_L$: 600 ml Water + 10% (w/w) glycerol | [52] |

polytetrafluoroethylene (PTFE, Teflon), polypropylene, polymethylmethacrylate (PMMA), or glass, can be used for microtiter plate manufacturing [33]. The properties of the materials vary considerably. The surface properties can be hydrophilic or hydrophobic. Specific surface treatments may be used to modify surface properties. This approach was chosen to compare oxygen transfer in shaken small-scale bioreactors. It was found that the $OTR_{max}$ decreases by at least 50% in unbaffled hydrophobic shake flasks compared with hydrophilic flasks; see Figure 2.4 [16]. This decrease in hydrophobic flasks is caused by a reduced formation of the liquid film on the flask wall. Thus, nearly no mass transfer area on the flask wall is formed. A similar behavior is observed in microtiter plates. In plates with a hydrophilic well surface, an increase of the $OTR_{max}$ compared with hydrophobic plates could be observed at the same shaking frequencies [39]. However, the highest applicable shaking frequency at which still no liquid flows out of the wells is reduced in hydrophilic plates. As a consequence, the overall $OTR_{max}$ which can be reached

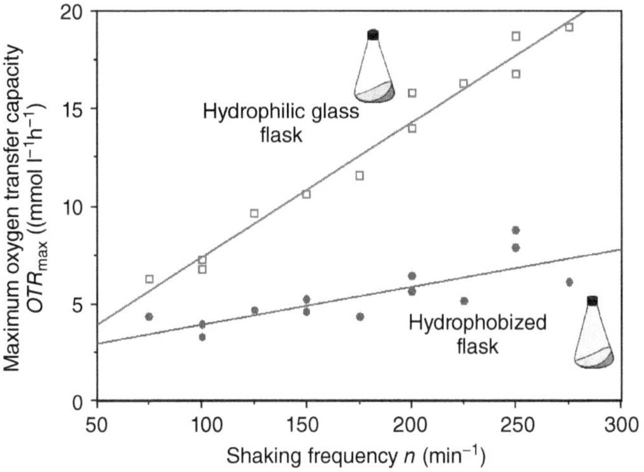


**Figure 2.4** Influence of the reactor surface properties on the maximum oxygen transfer capacity in shaken bioreactors. Maximum oxygen transfer capacity ($OTR_{max}$) was determined in hydrophilic and hydrophobic shake flask. Experimental condition: 1 M sulfite system, 250 ml flask, $d_0 = 50$ mm, $V_L = 26$ ml. (Figure adapted from [16].)

in both plate types is roughly the same [39]. Besides the surface properties of the shaken vessel, the liquid property itself influences the wetting characteristic of the vessel wall. Doig *et al.* [33] have verified this effect for different material/liquid combinations. An up to twofold increase of the total mass transfer area was found if a liquid with reduced surface tension was used.

#### 2.2.4.2
#### Influence of the Viscosity

The characterization of the mass transfer in shaken bioreactors in literature is commonly based on the assumption of liquids of water-like viscosities. Published correlations about the volumetric mass transfer coefficient $k_L a$ (Eqs. (2.15) and (2.16)) and the maximum oxygen transfer capacity $OTR_{max}$ (Eq. (2.18)) cannot be applied for liquids of elevated viscosities. In biotechnological processes, however, the viscosity of the cultivation broths may increase due to the formation of specific products, such as, for example, alginate [53], $\gamma$-polyglutamic acid [54], or xanthan [27]. Elevated viscosities of up to 110 mPa s in shake flasks are reported in the literature (summarized in [55]). To improve the mixing performance and the oxygen transfer in shaken bioreactors at elevated viscosities, baffles can be introduced. Baffles, however, increase the risk to work at out-of-phase operating conditions. Consequently, a suitable bioreactor configuration has to be carefully selected.

In unbaffled shake flasks, a counterintuitive increase of the $OTR_{max}$ could be observed for liquids of elevated viscosities of up to 10 mPa s [55]. This $OTR_{max}$ increase is explained by an increase of the film thickness on the flask wall at elevated viscosities. The increase of the film thickness elevates the driving concentration gradient of oxygen diffusion into the liquid film. It is shown that the

$OTR_{max}$ and the highest driving concentration gradient is reached at viscosities of 10 mPa s. At even higher viscosities (>20 mPa s), the $OTR_{max}$ decreases again due to a reduced oxygen diffusivity in the liquid. However, Giese *et al.* [55]showed for viscosities up to 80 mPa s that the $OTR_{max}$ does not undermatch the initial $OTR_{max}$ values of water-like viscosity (1 mPa s) again. In contrast, in stirred-tank bioreactors, a decrease of the $OTR_{max}$ of up to 95% of the initial $OTR_{max}$ value at water-like viscosities is observed at a viscosity of 80 mPa s [55]. This strong decrease of the $OTR_{max}$ is explained by a reduction of the mass transfer area due to a reduced bubble breakup and an increased coalescence of the gas bubbles in bubble-aerated bioreactors. In addition, the diffusion coefficient decreases with increasing viscosity. In shaken bioreactors, the power input and, thus, the mass transfer coefficient $k_L$ increases with increasing viscosity (at in-phase operating conditions). The increase of the $k_L$ value compensates the reduction of the diffusion coefficient. As a consequence, the oxygen transfer in shaken bioreactors is only slightly reduced at elevated viscosities.

### 2.2.5
### Influence of Shear Rates

A comprehensive characterization of relevant operating parameters is mandatory for a successful process development and scale-up. An important parameter, which might strongly influence the cultivation conditions of microorganisms, is the effective shear rate $\dot{\gamma}_{eff}$. The effective shear rate influences the apparent viscosity, and subsequently, the mixing and heat transfer in shaken bioreactors [26, 55]. While liquids of low viscosities typically show Newtonian flow behavior, that is, constant viscosity over shear rate, fermentation broths of elevated viscosity generally show a non-Newtonian flow behavior, that is, changing viscosity over shear rate. The risk of adjusting unfavorable operating conditions, that is, the out-of-phase operating conditions (see Section 2.2), increases if the effective shear rate of the cultivation broth is unknown [28]. Giese *et al.* [56] systematically investigated the effective shear rate in shake flasks for a wide range of operating conditions. By using dimensional analysis, the following equation was obtained:

$$\dot{\gamma}_{eff} = L^{1/m+1}\left(\frac{P/V_L}{K}\right)^{1/m+1}\left(\frac{V_L^{1/3}}{d}\right)^{x/m+1} \tag{2.19}$$

In agreement to newer findings for stirred-tank reactors [57], it could be shown that the shear rate depends on the specific power input and not on the stirring speed or the shaking frequency. By using Eq. (2.19), the effective shear rate $\dot{\gamma}_{eff}$ in shake flasks can be calculated as a function of the specific power input $P/V_L$, the filling volume $V_L$, the flask diameter $d$, the shaking diameter $d_0$, the flow behavior index $m$, the flow consistency index $K$, the proportional factor $L$ and the exponent $x$. The correlation is valid for shake flasks with nominal volumes ranging from 50 ml to 1000 ml and for in-phase operating conditions. Values for $L$ and $x$ were found to be $L = 2.06$ and $x = -0.331$. These values are proven for the following shaking conditions: $V_R = 50-1000$ ml, $V_L = 10-100$ ml, $d_0 = 25-100$ mm, specific power input: 0.71 – 15.1 kW $m^{-3}$, consistency index $K$: 49 – 23 233 mPa $s^m$, flow behavior index $m$: 0.107 – 0.867. Giese *et al.* [56] report that, depending on the broth's flow behavior index, the effective shear rate in shake flasks is at least 1.55 times higher than that in stirred-tank bioreactors compared with the same specific power input. This increased value might lead to a 50% lower apparent viscosity in shake flasks compared with stirred-tank bioreactors.

### 2.2.6 Ventilation in Shaken Microbioreactors

Gas exchange between the liquid phase and the surrounding gas phase is determined by two mass transfer resistances in shaken bioreactors. Besides the resistance of the gas–liquid interface area, there is the resistance of the sterile closure at the neck of the shaken bioreactor. The resistance of the sterile closure is determined by the cross-section area of the flask neck, the height of the sterile closure and the material properties. Cotton plugs are typically used as sterile closure in shake flasks. The plug resistance can be neglected if moderate oxygen transfer rates are prevailing. However, for high oxygen transfer rates, for example, in baffled shake flasks, the resistance of the cotton plug may become the limiting step [58]. In aerobic cultivations, also the removal of carbon dioxide out of the liquid phase is important to avoid inhibitory effects during the cultivation. In addition, the evaporation of water has to be minimized to ensure uniform and reproducible cultivation conditions. Four methods are known to determine experimentally the resistance of sterile closures. Besides the classic dynamic method, which determines the increase of oxygen in the flask headspace after nitrogen flushing, a steady-state method is reported by Henzler *et al.* [59] that compares different headspace composition of microbial cultivations in sealed and unsealed shake flasks. Another method is based on the steady-state assumption but additionally considers water evaporation [58]. Finally, a low-cost method is established in which the water loss from two equivalent shaken bioreactors is measured [60]. One flask is filled with distilled water and the second flask is filled with a saturated salt solution. Out of the different evaporation rates in both flasks, the diffusion coefficients of oxygen and carbon dioxide in the sterile closure and the relative humidity in the environment are calculated. To reduce the experimental effort, a mathematical model was developed to determine the resistance of sterile closures in shake flasks [61]. This model allows the prediction of oxygen mass transfer coefficients of the gas phase in the shake flask headspace. Similar to shake flasks, the oxygen transfer into microtiter plates is determined by adhesive sealing foils on top of the microtiter plate wells. In addition, water loss out of the wells due to evaporation plays a dominant role, compared with shake flasks, due to the higher surface-to-volume ratio. To ensure sufficient oxygen supply and simultaneously minimize evaporation, an appropriate microtiter plate sealing has to be chosen. Up to now, no comprehensive investigation about microtiter plate sterile closures is available. Zimmermann *et al.* [62] investigated different commercially available membrane sealings. The authors specify the required sealing properties and state that an optimal product does not exist that combines high oxygen permeability and little water evaporation. Duetz *et al.* [63] developed a special lid for 96-well plates. It consists of a perforated steel lid in which a sandwich layer consisting of cotton wool and perforated silicone is placed. Schlepütz and Büchs [64] presented a special lid for 48-well microtiter plates that reduces the evaporation of ethanol during vinegar production. In the same work, a diffusion model was used to calculate the dimensions of a gassing hole in the lid to guarantee a suitable oxygen supply during cultivation.

### 2.2.7
### Hydromechanical Stress

The power input of a bioreactor is not uniformly introduced into the bulk liquid. At some spots of the bioreactors higher local power input occurs, whereas at other spots reduced local power input is observed. The maximum local power input $(P/V_L)_{max}$ is a parameter that indicates the level of hydromechanical stress the microorganisms are exposed to. In general, cell damage is not a precisely defined phenomenon. It includes the alteration of aggregate sizes, release of intracellular compounds, variation of specific respiration rates, loss of viability, reduced biomass production, and finally changes in the cell wall composition [65]. To avoid these effects, the quantity of hydromechanical stress in bioreactors has to be evaluated. In particular, if a second liquid phase of an organic carbon source, such as plant oil, alkanes, or water insoluble chemicals, are present, the level and the spatial distribution of hydromechanical stress has to be known to sufficiently characterize a bioprocess [65, 66]. While bacteria are in general quite robust, animal cells or filamentous growing organisms such as fungi might be more sensitive to hydromechanical stress. The level of hydromechanical stress inside the bioreactor depends on the prevailing operating conditions, the bioreactor design, and the liquid properties, such as the viscosity. The microbial response to hydromechanical stress varies, depending on the particular cell properties such as size and wall rigidity of the cells. Animal cells are generally sensitive to hydromechanical stress due to their relatively large size and the lack of a protecting cell wall. While shaken bioreactors show relatively low local power input, the local power input in stirred-tank bioreactors at the same specific power input is approximately by a factor of 10–20 higher compared with the shake flasks [66]. It should be noted that the average power input $(P/V_L)_ø$ in shaken and stirred bioreactors at usual operating conditions is in a similar range. Consequently, the power input is introduced more evenly in shaken bioreactors than in stirred-tank bioreactors. This finding corresponds to the observation during cultivations of hydromechanically sensitive cells in both reactor types, in which different morphologies have been observed [15]. In addition to local power input in the bulk liquid generated by the stirrer, bubble burst on the liquid surface highly contributes to the hydromechanical stress of mammalian cells in aerated bioreactors. If bubbles burst on the liquid surface, small droplets are torn out of the surface and high energy densities are generated. In particular, sensitive animal cells might be damaged due to bubble burst [67].

This effect is completely absent in surface-aerated unbaffled shake flasks where no bubbles are generated.

The influence of aeration in stirred-tank bioreactors on the hydromechanical stress of microorganisms was investigated recently [68]. It was found that the maximum local energy dissipation rate is reduced by more than 50% in the presence of air compared with unaerated conditions and equal specific power input. The exact reduction factor depends on the specific geometry of the applied 6-bladed Rushton impeller. Interestingly, similar maximum power input is observed in unbaffled and baffled shake flask if the same average specific power input is adjusted. Peter *et al.* [15] presented a correlation to calculate the maximum local energy dissipation rate $\varepsilon_{max}$ in unbaffled shake flasks depending on the flow regime and operating conditions. The authors determined the transition from nonturbulent to turbulent flow regime in unbaffled shake flasks to a Reynolds number of $Re > 60\,000$. To better understand the effect of hydromechanical stress on microorganisms and to obtain a better comparability between shake flask and stirred-tank bioreactors, further investigations have to be carried out in future.

## 2.3 Design of Novel Stirred and Bubble Aerated Microbioreactors

In recent years, different miniaturized bioreactors have been developed and published in literature. Besides shaken systems, such as shake flasks or microtiter plates, other systems, such as stirred and bubble-aerated systems, have been investigated. A broad and comprehensive overview of the latest developments is given in several reviews and research articles [2, 5, 7, 69–71]. In this chapter, a summary of relevant stirred, bubble aerated, or shaken miniaturized bioreactors with volumes <100 ml are presented.

Stirred microbioreactor systems have some general advantages. They provide a homogeneous environment, which offers the possibility to comprehensively monitor important process parameters online during cultivation [7]. Furthermore, stirred bioreactors provide several options to increase mixing and mass transfer. Based on the closed design reproducibility and quality of measuring, results are typically high in stirred microbioreactor systems. However, stirred microbioreactors are often expensive and poorly suited for high-throughput experimentation. One of the first miniaturized stirred bioreactors was reported in 2001 [72]. This low-cost microbioreactor consists of a 4 ml polystyrene cuvette with 2 ml working volume. Optodes for dissolved oxygen tension (*DOT*) and pH measurements, an OD measurement, and magnetic stir bar were integrated. Stirring speeds of up to 300 rpm and $k_La$ values up to $21\,h^{-1}$ in *Escherichia coli* cultivation have been achieved. In 2003, Lamping *et al.* published a 10 ml Rushton turbine sparged bioreactor [73]. The system is mechanically driven by a microfabricated electric motor and aeration is achieved with a single tube sparger. By means of fiber-optics, DOT, pH, and OD measurements were realized. $k_La$ values up to $\sim 360\,h^{-1}$ in water–air system were achieved, but, for *E. coli* cultivation values of up to 128 $h^{-1}$ are reported. This microbioreactor system was later comprehensively characterized in terms of overall volumetric oxygen transfer coefficient and mixing time over a wide range of impeller speeds [74]. In addition, the power input was determined by means of electric power input measurements. Yang *et al.* [75] used a miniaturized stirred bioreactor with 12 ml filling volume for metabolic flux analyses. A perfluoroalkoxyalkane membrane inlet interface was applied to connect the bioreactor to a mass spectrometer for $^{18}O$ tracer studies to determine respiratory activity and metabolic fluxes during cultivation of *Corynebacterium glutamicum*. A pH probe was installed as well to determine the pH. Puskeiler *et al.* [76] presented a miniaturized sparged and agitated polystyrene bioreactor system, which allows up to 48 parallel experiments. At a filling volume of 8 ml, $k_La$ values of about 1600 $h^{-1}$ were reached at agitation speed of 2300 rpm. By integration into a robotic fluid handling system, pH control, OD measurements, and fed-batch operation were realized. Later, a DOT sensor was integrated in these systems [77]. Betts and Baganz report about a prototype of a 18 ml working volume miniaturized stirred-tank bioreactor [2]. Equipped with DOT and pH measurement, a $k_La$ value of 480 $h^{-1}$ is reached at 7000 rpm. Klein *et al.* [78] presented a small-scale system of eight parallel Hungate tubes with a working volume of 10 ml. The system is operated as a chemostat and allows $CO_2$ monitoring and *DOT* measurement. $k_La$ of 48.5 $h^{-1}$ was reached at 2000 rpm.

If only a moderate oxygen supply is required, an interesting alternative to stirred microbioreactors are bubble column or shaken microbioreactors. Both bioreactor types are technically simple and cost-efficient. Doig *et al.* developed a miniature bubble column reactor with 2 ml working volume that reached $k_La$ values of 220 $h^{-1}$ [79]. The system enables optical DOT and pH measurements. In 2006, the Micro 24 system ($\mu$24) was introduced, which is a shaken miniaturized bubble column bioreactor based on a microtiter plate with integrated optical pH and DOT sensor and thermal conductor temperature sensor [80]. Individual control of pH, DOT, and temperature in each well is possible. The commercially available microtiter plate-based system with DOT and pH measurement and temperature control was tested by Isett *et al.* with cultivations of *E. coli*, *Saccharomyces cerevisiae*, and *Pichia pastoris*. $k_La$ of 56 $h^{-1}$ at 800 rpm was reached [81].

## 2.4
## Robotics for Microbioreactors

In recent years, highly automated screening experiments gained more and more importance in research laboratories and industry. Quantitative determination of process parameters is a fundamental task for optimizing strains and cultivation conditions during bioprocess development. However, the number of possible parameter combinations exponentially grows with every additional parameter [82]. Automation of microbioreactor handling can highly reduce work effort and widens the range of process variables to be examined [83]. A large number of parallel experiments can be performed at minimized experimental error rate and high accuracy by combining robotic systems with high-throughput microbioreactor cultivation devices. Sample taking, analytical assays, downstream processing, or media preparation is easily performed by the robotic system. To determine important process parameters and to reduce costs, design of experiments (DoE) is applied [84]. Latest research on automation of microbioreactors deals with the replication of the whole process chain in small-scale bioprocesses [82].

Several research on microbioreactor automation was published in recent years. Puskeiler *et al.* [76] integrated a stirred miniaturized bioreactor and a microtiter plate reader into a liquid handling system. Samples are taken automatically and pipetted into a sample microtiter plate on a cooling unit. A robotic arm transfers the plate to the reader for at-line analyses and places it back for reuse after washing. pH control is realized by addition of titration solution. Zimmermann and Rieth [85] presented a fully automated cultivation system encased in a temperature and humidity controlled chamber that allows the cultivation of 768 experiments in parallel. This system covers all steps from preparation, cultivation, monitoring by means of fluorescence measurements in an external plate reader, downstream processing (centrifugation), and storage (deep freezer). Even sterile sealing of microtiter plates is done automatically by the robotic system. A fully automated system with integrated autoclaving and stock solution storage is presented by Wollerton *et al.* (Piccolo™) [86]. To avoid interruptions of the cultivation process during sample taking, Huber *et al.* [87] developed an automated cultivation system that combines a liquid-handling system, a laminar airflow cabinet and a BioLector device for online scattered light and fluorescence measurements in microtiter plate cultures. A great advantage of this system is the continuous cultivation without interruption during optical measurements. Thus, the data quality is increased and a large number of process parameters can be obtained in short time. This system was used to perform automated biomass concentration-based induction of different *E. coli* cultures to ensure an identical growth status of all microbial cultures [87]. Thus, even at different growth kinetics of the microorganisms, the product formation at the end of the cultivation was comparable among each other. The deviation of product formation was reduced to ±7%. Induction profiles with varying inducer concentration and induction times were used to improve the productivity of the microorganisms. In addition, different growth kinetics of precultures can be compensated by calculating individually the required preculture volumes for inoculation. Rohe *et al.* [82] combined a liquid handling system with a BioLector device but extended the system by integration of a cooling rack, plate reader, and microtiter plate centrifuge. The whole setup was enclosed in a laminar airflow cabinet. The automated cultivation system was used for optimization of the secretory production of a cutinase by C. *glutamicum*. Even though many process parameters can be determined today in small scale, the full functionality to mimic large-scale process conditions, such as for example, pH measurement over the full range, very high $k_L a$ values ($>1500\,h^{-1}$) or feeding and control of any substance, is not given up to now [70].

## 2.5
## Fed-Batch and Continuous Operation of Microbioreactors

Industrial bioprocesses are often carried out in fed-batch or in continuous operating mode. Both operating modes are used to overcome substrate inhibition, catabolite repression, or overflow metabolism [88]. Limited substrate supply is suitable to avoid oxygen limitation at high cell densities. Furthermore, optimal conditions for both growth and product formation can be adjusted by appropriate application of fed-batch operating mode. Finally, an often underestimated problem in screening is the nonparallelized growth of batch cultures due to a fixed time protocol [89]. Fed-batch operating modes is suitable to equalize the growth in precultures and, subsequently, to ensure the same initial cultivation condition for the microorganisms in main cultures. In contrast to fed-batch conditions in production processes, screening experiments in microbioreactors are commonly performed in batch mode. The fundamental difference between both operating modes may lead to a wrong selection of strains [90]. As a consequence, process modifications might become necessary at a later time and the process development will become less efficient. To overcome this issue, fed-batch screening systems are currently developed and investigated.

### 2.5.1
### Diffusion-Controlled Feeding of the Microbioreactor

A technically simple and cost-efficient way to provide fed-batch cultivation conditions in small scale is the diffusion-controlled release of substrates into the cultivation broth. The advantage of diffusion-driven systems is that no additional technical equipment, such as pumps and valves, is needed. This approach is applicable at any pH value. Several approaches of diffusive fed-batch screening systems are published. Jeude *et al.* [90] presented a parallel screening technique for microbioreactors based on elastomer disks (FeedBeads®) that were loaded with glucose. Solid glucose is embedded into a silicone matrix and the polymer disk is added to the cultivation broth. If water diffuses into the silicone matrix, dissolved glucose diffuses out of the silicone elastomer and the substrate is released into the cultivation broth. The substrate release is adjusted by the substrate concentration in the elastomer, the properties of the silicone and the disk geometry. This system was successfully applied in shake flasks and microtiter plates [90, 91]. It was shown that overflow metabolism of *Hansenula polymorpha* could be reduced and an increase of the biomass yield of 85% was reached. In addition, GFP formation was increased 35-fold in Syn6-MES mineral medium and even 420-fold in YNB mineral medium compared with batch cultivations with the same amount of totally metabolized glucose [90]. A slightly different design of this feeding system is used in 96-deep-well plates, in which the silicone matrix is fixed on the bottom of each well (FeedPlate®). Successful fed-batch cultivations of *H. polymorpha* have been conducted with that system [92]. Instead of glucose Scheidle *et al.* [93] embedded sodium carbonate $Na_2CO_3$ in the silicone matrix to control the

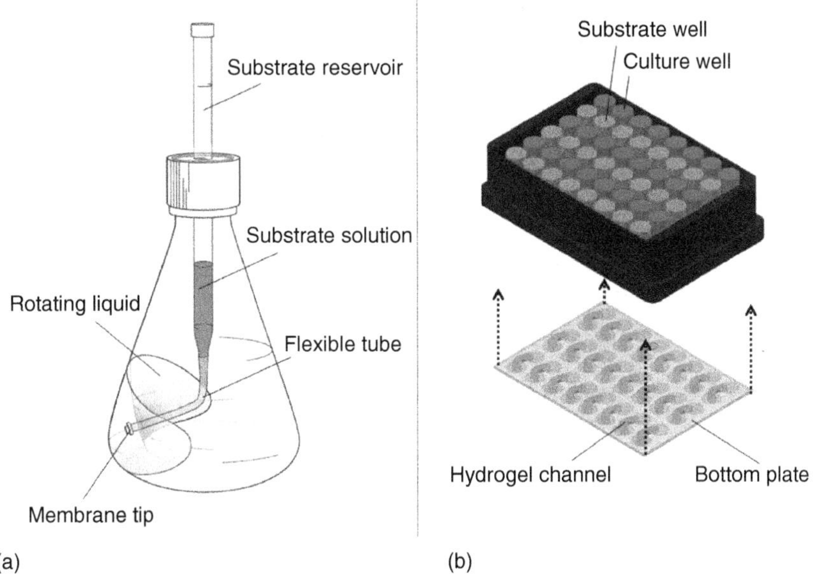


**Figure 2.5** Schematic representation of a fed-batch microtiter plate and dialysis shake flask. (a) Dialysis shake flask. A substrate reservoir is installed inside the flask. Through a membrane, feed solution diffuses into the bulk liquid. (b) Fed-batch MTP. Two wells are connected through a hydrogel channel in the bottom of the plate. Substrate diffuses from substrate well to culture well, depending on gel properties, channel geometry, and substrate feed concentration.

pH value in *E. coli* cultivations in shake flasks. Also, other solid substrates can be embedded in the elastomer matrix. Liquids, such as for example, glycerol, can, however, not be fed. A technical similar approach for substrate feeding was used by Hegde *et al.* [94]. Hydrogel disks, made of 2-hydroxyethyl methacrylate (HEMA) hydrogel, were loaded with glucose and protein hydrolysate for fed-batch cultivation of CHO cells in shake flasks.

A technical different approach for diffusion-controlled substrate release is presented by Bähr *et al.* [95]. This system consists of a dialysis shake flask (see Figure 2.5), made up of a conventional Erlenmeyer flask and a central tubular substrate reservoir. The substrate reservoir is fitted in the center of the screw cap. At the end, the substrate reservoir is sealed with a membrane fixed on a tip. The membrane rotates synchronously with the liquid inside the flask during shaking, and thus, permanent contact between the membrane and the liquid is ensured. Feeding of liquids and solid dissolved substrates is possible with that system. The substrate diffusion rate can be adjusted by the substrate feed concentration and the geometry and properties of the membrane.

Another diffusion-driven feeding systems is the recently developed fed-batch microtiter plate, which is illustrated in Figure 2.5. In that microtiter plate, half of the wells serve as substrate reservoirs, while the other half serve as cultivation wells. A substrate reservoir is connected to a cultivation well via a hydrogel channel. A special bottom plate with channels is used, in which the hydrogel is filled (see Figure 2.5). The substrate diffuses from the substrate reservoir into the hydrogel channel and further into the culture well. The diffusion rate depends on the substrate feed concentration, the channel geometry, and the properties of the hydrogel. An advantage of this system is that the time, which is needed by the substrate molecules to pass the hydrogel, can be can be adjusted by appropriate choice of the channel length. Thus, the time at which substrate feeding of the microorganisms begins can be defined. This feeding system, applicable for almost any substrates, was successfully tested in cultivations of *E. coli* and *H. polymorpha* [96].

### 2.5.2
### Enzyme Controlled Feeding of the Microbioreactor

A technical simple, cost-efficient, and easy-to-apply method to feed glucose in microbioreactors is the enzymatic approach [97]. In this approach, starch and hydrolytic enzymes, such as amylase and maltase, are added to the cultivation medium. Amylase breaks down the starch into glucose and maltose. Maltose is converted into glucose by the maltase. At sufficiently high starch concentrations, the glucose release rate depends on the amount of active enzymes in the cultivation broth. It can easily be adjusted by changing the enzyme concentration. This enzymatic substrate release method was applied in a commercially available substrate autodelivery system [98]. In this system, a two-phase gel is used. The first agar layer contains starch. Above the first layer, a second starch-free agar layer is placed to improve the release control if the starch diffuses out of the agar. The two-phase agar gel can be fixed at the bottom of shake flasks or microtiter plates. In the presence of glucoamylases, the starch is degraded to glucose, which is metabolized by the microorganisms. A drawback of the enzymatic substrate release system is that it is limited to glucose as substrate. This system cannot be used for amylase-producing microorganisms because in the presence of further amylases the glucose release will become undefined. Furthermore, if proteases are present during cultivation, the glucosidase can be degraded and, subsequently, the substrate release is inhibited. It should be noted that optima of temperature and pH may limit the applicability of an enzymatic substrate release system significantly. An enhanced version of this enzymatic glucose release system was developed. Instead of using a gel phase, a polymer is used that is completely soluble in aqueous solutions [99]. Fed-batch cultivations of *P. pastoris* with a methanol inducible AOX1 promoter were successfully conducted using this enhanced release system [100]. By applying the enzymatic glucose feeding, carbon starvation phases between the commonly applied methanol pulses have been avoided. As a result, increased cell densities and a threefold higher product activity (fungal lipase) were reached. A enzymatic controlled glucose release system was also used for fed-batch cultivation of *P. pastoris* [101]. In this application, a cultivation medium based on a Syn6 medium was used, which contains a soluble glucose-polymer. After addition of glucosidase to the cultivation medium, glucose release is initiated.

### 2.5.3
### Feeding of Continuous Microbioreactors by Pumps

A continuous cultivation of microorganisms has some advantages compared with batch or fed-batch operating mode of bioreactors. Continuously operated bioreactors) provide constant cultivation conditions (steady-state) over long times. Thus, they are suitable to investigate and quantify the specific properties of microbial strains due to the fact that some microorganisms react highly sensitive to changes of the environment [102]. For conventional continuous bioprocesses, mainly stirred-tank bioreactors are used. To ensure similar cultivation condition in production and screening, the application of stirred microbioreactors (see Chapter 3) is highly suitable. However, stirred microbioreactors have relatively high investment costs, and they are scarcely suitable for high-throughput applications. The time to reach equilibrium and steady-state operating conditions is too long for efficient screening of microorganisms [103]. In contrast, shaken continuous microbioreactor systems offer the possibility of fast and cost-efficient high-throughput experimentation [54, 102, 104]. Parallel continuous cultivations can be conducted in a continuous parallel shaken bioreactor system (CosBios) presented by Akgün *et al.* [102]. For the feeding a multichannel peristaltic pump is applied in this system. A large number of experimental data are obtained in one experiment if different operating conditions in the parallel shaken cultivation vessels are adjusted. It was shown that the results of *S. cerevisiae* cultivations in this system agree well with the results obtained in a continuously operated 1 l stirred-tank bioreactor. In Figure 2.6, the biomass concentration of a continuous

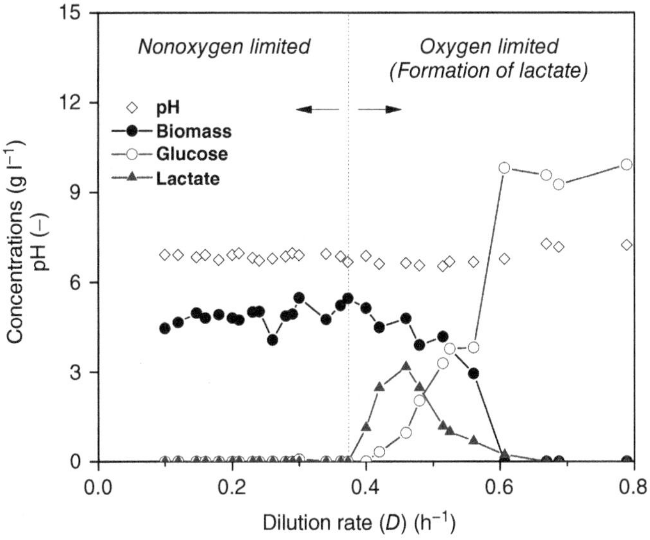


**Figure 2.6** X-D diagram of a continuous cultivation of *C. glutamicum* in the CosBios bioreactor. At dilution rate higher than $0.4\,h^{-1}$, oxygen limitation occurs. Cultivation conditions: $10\,g\,l^{-1}$ glucose in the feed stock, $n = 275$ rpm, resulting in a filling volume of 27 ml, $d_0 = 50$ mm, $q = 2$ vvm, $T = 30$°C. (Figure adapted from [102].)

cultivation of *C. glutamicum* in the CosBios bioreactor system is depicted over the dilution rate. Up to a dilution rate of approximately 0.38 $h^{-1}$, the value of the biomass concentration is constant at 5 $g\,l^{-1}$ and the glucose concentration is close to 0 $g\,l^{-1}$. At higher dilution rates, lactate is formed due to oxygen limitation. Simultaneously, the biomass concentration decreases and the glucose concentration increases due to the reduced microbial growth at oxygen limitation. Recently, the continuous parallel shaken bioreactor system was extended and successfully applied in repeated batch cultivations of bacteria for vinegar production [105]. It was demonstrated that the systems enabled a cost-efficient fully automated cultivation of eight parallel repeated batch cultivations.

## 2.6 Monitoring and Control of Microbioreactors

Monitoring of biological parameters is essential for a systematic process analysis and reasonable screening results. Relevant parameters, such as temperature, pH, DOT, OTR, and signals for biomass and product formation, are required for process optimization and scale-up. To comprehensively characterize a bioprocess, different operating conditions and measurements have to be applied. The operating parameters need to be applied in different combinations to cover a broad range of cultivation conditions. This variation of operating parameters implicates laborious and time-intensive experiments. To reduce the experimental effort and to save time, DoE may be used [84]. The amount of obtained process data is increased and, simultaneously, experimental effort is minimized if DoE is applied.

Today, different physicochemical sensors and various measurement techniques for microbioreactor systems are available [3, 4, 71, 106]. Physicochemical sensors can generally be applied in bench-scale bioreactors without major restrictions, whereas the application in small-scale bioreactors is more challenging due to small vessel volumes and little space inside the bioreactor. Due to these challenges, optical measurement techniques are often applied in microbioreactors. The feasibility of noninvasive measurements, miniaturized design, and fast response times makes these techniques qualified for applications in small-scale bioreactors. Furthermore, integrated optical measurement allows a continued process design without interrupting the microbial cultivation during measurements in shaken microbioreactors. For example, scattered light and fluorescence measurements can be performed without interrupting the shaking process during a microbial cultivation in microbioreactors [107]. Thus, optimal cultivation conditions are provided during the entire cultivation time. Another drawback of invasive measurement technique, for example, pH measurement with electrochemical electrodes, is the alteration of the hydrodynamics of the liquid inside the bioreactors. As a consequence, the cultivated culture may be negatively affected [108]. In particular, at small liquid volumes, the influence of the sensors on the liquid hydrodynamic might become critical. Therefore, scattered light and fluorescence measurements are preferred to determine biomass and product formation in microtiter plates [107]. Specific fluorescent proteins are required to optically monitor the product formation. The fluorescent proteins are linked as fusion tags to the desired target molecules in genetically modified microorganisms. Conventional fusion tags are the green fluorescent protein (GFP), and its derivatives, that is, the yellow fluorescent protein (YFP), flavin mononucleotide–based fluorescent proteins (FbFP), and tryptophan tags [109, 110]. Fluorescence measurements are also used to determine pH, $O_2$, $CO_2$, and $NH_4$. The application of these measurements is comprehensively described in the literature [111]. A cultivation device for microtiter plates that is based on optical scattered light and fluorescence measurements was developed by Samorski *et al.* [107]. In this device, the microtiter plate is placed on a shaking table in a tempered cultivation chamber. A fiber-optic cable is moved via a $x-y$-drive below the microtiter plate and performs automated optical measurements of individual wells of the microtiter plate. This BioLector, called measurement technology, was extensively validated and is commercially available from the company m2p-labs GmbH, Baesweiler, Germany [112]. A prototype of this device is used at the Chair of Biochemical Engineering at RWTH Aachen University. With that device, the parallel cultivation of up to four microtiter plates is possible. If 96-well plates are used, in total 384 cultivations monitored online via scattered light and fluorescence measurements can be run in parallel. Figure 2.7 shows a picture of the prototype.

Fluorescence measurements can be used to determine the qualitative temperature profiles in microbioreactors [113]. However, accurate absolute temperature measurement is more challenging. In stirred microbioreactors or microfluidic

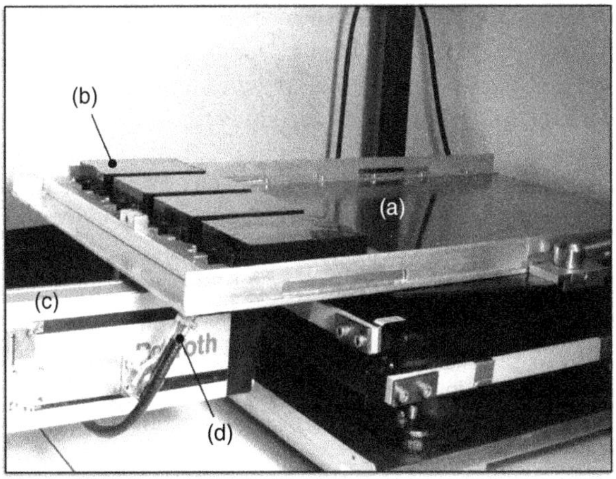


**Figure 2.7** Prototype of an optical online measuring device based on the BioLector technology. Up to four microtiter plates can be operated in parallel. The wells are monitored by scattered light or fluorescence measurements subsequently. The device is placed in a climate chamber with constant temperature. (a) Shaking table, (b) microtiter plates, (c) $x-y$ drive to move the fiber-optic cable, (d) fiber-optic cable below the microtiter plates.

chips, the temperature is typically measured with thermistors, resistance temperature detectors, or thermocouples [4]. In shaken systems, however, temperature measurement and control are commonly integrated into the incubation chamber in which the microbioreactor is placed. An individual control of the single wells in shaken microtiter plate is complex and expensive. So far, a well-specific temperature control is only known for one commercial system for shaken microtiter plates (micro-Matrix (former: $\mu$24), Applikon, Delft, the Netherlands).

### 2.6.1 DOT and pH Measurement

The DOT and pH of a cultivation broth represent two of the most important process parameters in biological cultivations. The parameters are indispensable for a fundamental characterization of bioprocesses. For shake flasks, autoclavable DOT sensor spots for optical fluorescence measurements have been reported [47, 114]. These sensors contain an oxygen-sensitive luminescence dye. The oxygen concentration is determined by lifetime measurements of the luminophore. One challenge of DOT measurement in shake flask is that the spots have to be continuously covered by cultivation medium to avoid measurement errors [108]. In particular, at high shaking frequencies and low filling volumes, there is no location in the flask that is continuously covered by the rotating bulk liquid [14, 108]. Similar sensor spots for DOT measurements in microtiter plates exist. The fluorescence signal is detected noninvasively by means of a fiber-optic cable. In microtiter plates, the liquid commonly covers the total bottom during shaking. Another sensor for oxygen measurement in living cells was developed by Potzkei *et al.* [115]. This sensor is based on the genetically encoded Förster resonance energy transfer (FRET). It consists of a yellow fluorescent and a flavin mononucleotide–based fluorescent protein (YFP-FbFP). If oxygen is present, YFP is formed and matured. Consequently, the measured FbFP fluorescence intensity decreases due to a nonradiative energy transfer between the chromophores of FbFP and YFP. The sensor was successfully applied for continuous real-time monitoring of temporal changes of $O_2$ levels in the cytoplasm of *E. coli* cells during a batch cultivation.

For pH measurements, similar sensor spots for shake flasks and microtiter plates are available. The spots are based on fluorescence measurement and contain a pH-sensitive fluorescent dye. The range of optical pH measurement is limited to a range of pH = 4.5–8.5. A commercial device for online measurements of DOT and pH in shake flasks is described in the literature [116]. Badugu *et al.* developed a ratiometric optical pH sensor film for bioprocesses, which can easily be miniaturized. It is applicable in microwell plates and microbioreactors [117]. The sensor is based on an allylhydroxyquinolinium (AHQ) derivative copolymerized with polyethylene glycol (PEG). It has two excitation maxima at $\lambda_{ex,1} = 375\,nm$ and $\lambda_{ex,2} = 425\,nm$ with a single emission peak at $\lambda_{em} = 520\,nm$ and can be used to determine the pH values in the range of pH = 5–8. The pH response of this sensor is not sensitive to the ionic strength and temperature of the medium. A technically simple method to determine the pH in aqueous solutions is the use of the fluorescent dye HPTS (8-hydroxypyrene-1,3,6-trisulfonic acid trisodium salt). HPTS is water soluble and can be added to an aqueous solution. Depending on the prevailing pH the fluorescence intensity of HPTS changes. A successful application of this dye to determine the pH in microtiter wells is reported in the literature [118]. The authors conducted a ratiometric measurement. At two excitation wavelengths of $\lambda_{ex,1} = 405$ nm and $\lambda_{ex,2} = 450$ nm, the fluorescence emission intensity is measured at $\lambda_{em} = 520$ nm, and the ratio of both emission intensities is calculated. This ratiometric measurement strongly reduces measurement errors due to the fact that possible alterations of the light source, optical detectors, and slight changes of the light transmission through the optical fibers are nearly completely compensated. Besides ratiometric measurement, the dual lifetime referencing (DLF) is a highly accurate method to conduct fluorescence measurement. In that case, the phase angle of two periodically stimulated fluorescence signals is measured. While one dye is a pH-insensitive reference dye, the second one is pH sensitive. If the pH changes, the phase angle of the pH-sensitive dye changes as well while the phase angle of the reference dye remains constant. Finally, the pH can be determined from the change of the measured overall phase angle. This method is highly accurate, and possible errors such as signal drifts of the measurement device or variation of the dye concentration, for example, bleaching, are minimized. Just recently, Kunze *et al.* [119] pointed out that the measurement of pH and *DOT* may become incorrect if the fluorescence signals of the optical sensors are influenced by fluorescent proteins in the cultivation broth (see Figure 2.8). If the excitation spectra of the fluorescent proteins in the cultivation broth, and the pH and DOT optodes are similar and, in addition, the fluorescence emission wavelength of the proteins is similar to the emission wavelength of the optodes, DOT, and pH signal might be incorrect.

### 2.6.2 Respiratory Activity

The respiratory activity of microorganisms is characterized by the OTR and carbon dioxide transfer rate (CTR). The measurement of both parameters is a powerful tool to obtain quantitative information about the current performance of a microbial culture. Information about substrate limitation, substrate or product inhibition, and pH inhibition can be obtained by these parameters. The ratio of CTR *and* OTR, the respiratory quotient RQ, provides additional information about activated metabolic pathways in microorganisms. Conventional off-gas analyses are can be applied to determine CTR and OTR. However, this technique is cost-intensive and not applicable for application in small-scale high-throughput cultivations in microbioreactors. Anderlei and Büchs presented a device to determine the OTR in shake flasks [120]. It is based on partial oxygen pressure measurements. It was later extended by a total pressure sensor to determine CTR and *RQ*. This respiratory activity monitoring system (RAMOS) is used to determine online OTR, CTR, and RQ in eight parallel shaken RAMOS flasks [31]. Each flask has an air inlet and outlet. Air inlet and outlet can be

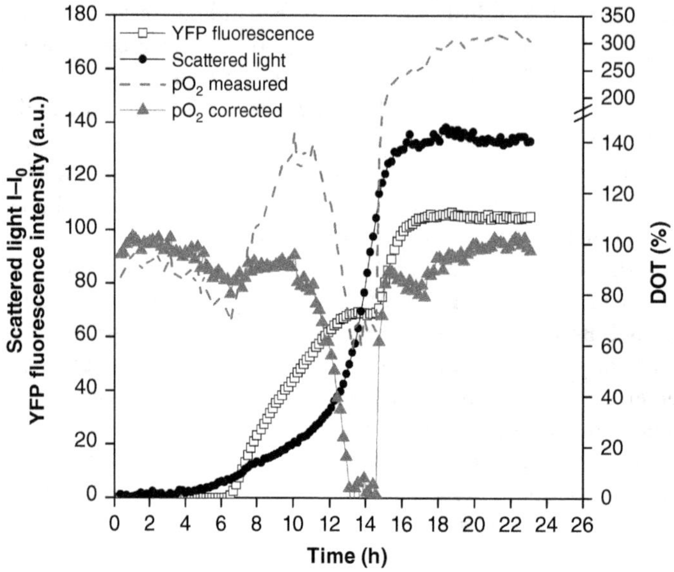


**Figure 2.8** Influence of fluorescent proteins on online monitoring in microtiter plates. *In vitro* characterization and correction of the influence of YFP fluorescent protein on the optical DOT signal is shown. The DOT signal is corrected for the cultivation of *E. coli* expressing YFP by using fluorescence-dependent calibration curves. The measured original DOT signal shows final values of 300%. By using an appropriate calibration curve, the signal is corrected. Cultivation conditions: 48-well FlowerPlate with optodes for DOT and pH measurement, $V_L = 800$ µl, $n = 1100$ rpm, $d_0 = 3$ mm, $T = 37$°C, Wilms-MOPS medium with 20 g $l^{-1}$ glucose, induction with 0.1 mM IPTG after 6 h. (Figure adapted from [119].)

opened or closed by valves. For CTR and OTR measurement, a measuring phase is started in which the inlet and outlet of a flask are closed. The respiration activity of aerobic microorganisms leads to a decrease of the oxygen partial pressure in the flask. This decrease is measured by the partial oxygen pressure sensor, and the OTR is calculated. Simultaneously, carbon dioxide is formed, which increases carbon dioxide partial pressure. The total pressure is influenced by the decrease of the oxygen partial pressure and increase of the carbon dioxide partial pressure. Consequently, by measuring the total pressure and the oxygen partial pressure, the *RQ* and CTR can be calculated. After this measuring phase, the inlet and outlet valves are opened and the head space of the flask is flushed by fresh air. During this rinsing phase, a specific air flow rate is adjusted to compensate the oxygen decrease during the measuring phase. Afterward, the valves close again and the measuring phase is started again. The RAMOS technique was extensively used in several studies and is widely described in the literature [36, 53, 121–123]. Furthermore, the RAMOS device was successfully applied in scale-up studies from shake-flask to stirred-tank bioreactor [43]. To improve the information content, a combination of OTR measurements and an optical pH measurement was presented [124]. Recently, a new evaluation algorithm for the

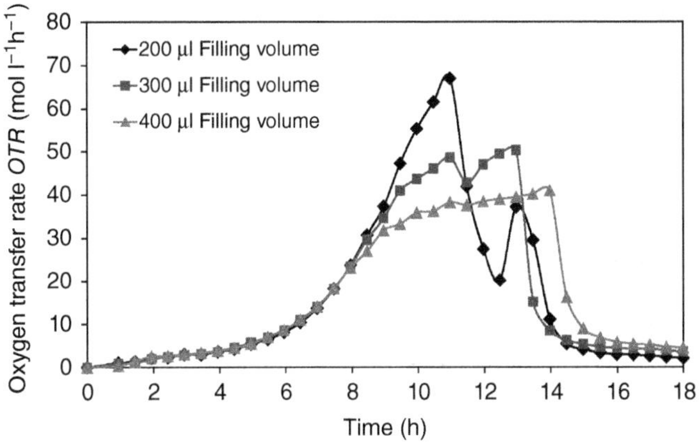


**Figure 2.9** RAMOS measurement of *H. polymorpha* RB11 FMD-GFP cultivation in 96-deepwell microtiter plate. Oxygen transfer rate was determined online in three different microtiter plates with different filling volumes. Cultivation parameters: Syn6-MES medium, $c_{Glucose} = 20\,g\,l^{-1}$, $n = 350$ rpm, $d_0 = 50$ mm, $T = 37\,°C$.

evaluation of the raw data was presented, which enables a higher resolution of the respiration activity measurements [125]. In particular, for fast-growing organisms, this new software offers better evaluation performance due to higher data density.

The RAMOS technique was also adapted to microtiter plates [6]. Figure 2.9 shows three exemplary OTR curves of a *H. polymorpha* cultivation in 96-deepwell microtiter plate with the RAMOS technique. The cultivation was run in Syn6-MES-medium with $20\,g\,l^{-1}$ glucose and three different filling volumes of 200, 300, and 400 µl. The oxygen transfer rate increases exponentially for all filling volumes. At this time, the organisms start to metabolize glucose. For 200 and 300 µl filling volume, a first maximum of the oxygen transfer rate is reached after 11 h ($OTR_{max}$ of $67\,mmol\,l^{-1}h^{-1}$ and $50\,mmol\,l^{-1}h^{-1}$, respectively). Subsequently, both curves decrease before they increase a second time. This second increase indicates that the organisms metabolize a second substrate. For the filling volume of 400 µl, a maximum of the oxygen transfer rate of $41\,mmol\,l^{-1}h^{-1}$ is reached after 14 h. The plateaus of the OTR curves of 300 and 400 µl filling volume clearly indicate an oxygen limitation. If the microorganisms are oxygen limited, the cultivation time is increased due to reduced consumption of the main substrate (glucose). It is worth mentioning that the total oxygen consumption, that is, the area below the OTR curves, is the same for all three filling volumes. It should be emphasized that the OTR curves in Figure 2.9 do not represent oxygen transfer of single wells of a microtiter plate but rather the sum of all wells of the microtiter plate. Consequently, each well needs to be filled with the same liquid at identical volumes. Even though the high-throughput advantage is lost, respiratory measurements in microtiter plates help to illuminate in general microbial behavior in these bioreactors and, in particular, to design small-scale screening systems

[6]. This example illustrates that online respiratory activity measurements provide important information about physiological conditions of a microbial cultivation and may help to improve cultivation conditions and scale-up processes to larger scale.

Another online monitoring system for OTR, CTR, and RQ in shake flasks with a minimal nominal flask volume of 500 ml was developed by the company Blue-Sens gas sensor GmbH, Herten, Germany. This device measures the oxygen and carbon dioxide concentration in the headspace of a shake flask with zirconium dioxide sensors or infrared sensors, respectively. Ge and Rao [126] used noninvasive, disposable fluorescence sensor spots to determine pH, DOT, and $CO_2$ in the liquid phase and $O_2$ and $CO_2$ in the headspace in shake flask. The sensors spots are positioned on the flask bottom and on the flask wall in the headspace. In the same study, the headspace and the off-gas is analyzed by means of the optical sensor spots. Van Leeuwen *et al.* [127] developed a system for accurate online monitoring of the gas consumption or production rates in microbioreactors if only one single gas is involved. The system is based on a highly sensitive pressure sensor that controls the gas pumping into the reaction chamber. Depending on the pressure signal, gas is pumped into the reaction chamber. The amount of supplied oxygen is used to determine the oxygen consumption over time. This system can be applied for measurements of gas consumptions or production rates in the nanomole per minute range. The system was validated for the enzymatic oxidation of glucose to gluconic acid in a microbioreactor with a working volume of 100 µl.

## 2.7 Conclusion

Microbioreactors play an important role in biotechnological screening and process development. They offer a high degree of parallelization, require small liquid volumes, and allow high-throughput screening of microorganisms. In addition, they offer a high flexibility regarding experimental design and investigation of operating parameters. The characterization of important process parameters, such as the oxygen transfer coefficient $k_L a$, power input $P/V_L$, and mixing time, is essential to optimize and scale-up cultivation conditions. For shaken microbioreactors, several empirical correlations are published. Unfavorable out-of-phase operating conditions (Eq. (2.11)), the volumetric mass transfer coefficient in unbaffled shaken bioreactors (Eqs. (2.15) and (2.16)), the modified power number in unbaffled shake flasks (Eq. (2.10)), the maximum oxygen transfer capacity in baffled microtiter plates (Eq. (2.18)), and the effective shear rate in unbaffled shake flasks (Eq. (2.19)) can be determined by applying these correlations. In addition, simulation models and CFD analyses might be useful to determine mass transfer areas, mixing times, and specific power input in bioreactors.

Besides a comprehensive characterization, monitoring and control of microbioreactors is fundamental for successful screening. To increase the available measurement data per time, and due to spatial limitation inside the vessels, noninvasive online measurement techniques are usually preferred to monitor bioprocesses in microbioreactors. Several parameters, such as biomass, product formation, pH, and DOT, can be determined by means of scattered light and fluorescence measurements. To determine product formation in a cultivation broth, specific fluorescent proteins, such as GFP, YFP, FbFP, and tryptophan tags, have to be linked to the molecule of interest. The respiration activity of microorganisms is determined online by means of electrochemical measurements of the oxygen partial and the total pressure in the head space of shaken bioreactors. To reduce the total amount of data and to reduce experimental effort, DoE may be applied. Automation of bioprocesses provides a further possibility to increase the efficiency of process development. In recent years, several studies with robotic systems have been published. This trend of automation is also driven by the aim to mimic large-scale bioprocess conditions in small scale. Fully automated small-scale bioprocesses including preparation steps, autoclaving, product separation, or deep freezing of samples are presented in the literature. Large-scale production processes are often performed at fed-batch conditions to increase the product yield and to avoid, for example, oxygen limitation or substrate inhibition. In contrast, screening is typically performed at batch conditions. To close this gap, fed-batch screening technologies have been developed and successfully applied in recent years. Several studies demonstrate that biomass concentration and product formation is increased at fed-batch operating conditions, compared with batch conditions. In future, novel research fields and new applications will be established, if operating conditions in small and large scales are further assimilated. This assimilation will increase the efficiency of process development, although small-scale bioreactors will never be identical to large-scale bioreactors in all detail. Finally, new measurement techniques will contribute to advance microbioreactor technology in future.

## Terms

| | |
|---|---|
| $a$ | Volume-specific surface area ($m^{-1}$) |
| $a_i$ | Initial specific area ($m^{-1}$) |
| $A_W$ | Wall area of the shake flask ($m^2$) |
| $A_S$ | Surface area gas/liquid ($m^2$) |
| $C_1$, $C_{la}$, $C_{tr}$, $C_{tu}$ | Constants (-) |
| $c$ | Concentration ($kg\,m^{-3}$) |
| $c^*_{O_2}$ | Oxygen concentration in the liquid at the phase boundary ($kg\,m^{-3}$) |
| $c_{O_2,L}$ | Oxygen concentration in the bulk liquid ($kg\,m^{-3}$) |
| CTR | Carbon dioxide transfer rate ($mol\,l^{-1}\,h^{-1}$) |
| $d$ | Flask/vessel diameter (m) |
| $D$ | Diffusion coefficient ($m^2\,s^{-1}$) |
| $d_0$ | Shaking diameter (m) |

| | |
|---|---|
| DOT | Dissolved oxygen tension (%) |
| $F$ | Force (N) |
| $F_c$ | Centrifugal force (N) |
| $g$ | Gravitational acceleration ($m\,s^{-2}$) |
| $h$ | Height (m) |
| $K$ | Flow consistency index ($Pa\,s^m$) |
| $k_L a$ | Volumetric mass transfer coefficient ($h^{-1}$) |
| $L$ | Proportional factor (–) |
| $L_{O_2}$ | Oxygen solubility in the liquid ($mol\,l^{-1}\,bar^{-1}$) |
| $m$ | Flow behavior index (–) |
| $M$ | Torque (N m) |
| $n$ | Shaking frequency ($s^{-1}$) |
| $n_{crit}$ | Critical shaking frequency ($s^{-1}$) |
| OD | Optical density (–) |
| OTR | Oxygen transfer rate ($mol\,l^{-1}\,h^{-1}$) |
| $OTR_{max}$ | Maximum oxygen transfer capacity ($mol\,l^{-1}\,h^{-1}$) |
| $p$, $p_{abs}$ | Absolute pressure (bar) |
| $P$ | Mechanical power (W) |
| $P/V_L$ | Specific power input ($W\,m^{-3}$) |
| $(P/V_L)_{max}$ | Maximum local specific power input ($W\,m^{-3}$) |
| $(P/V_L)_ø$ | Average specific power input ($W\,m^{-3}$) |
| $q$ | Aeration rate (vvm) |
| *Peri* | Relative perimeter (–) |
| $r$ | Radius, lever arm (m) |
| *RQ* | Respiratory quotient (–) |
| $T$ | Temperature (°C) |
| $t_m$ | Mixing time (s) |
| $V_L$ | Liquid volume, filling volume, working volume ($m^3$) |
| $V_R$ | Reactor volume ($m^3$) |
| $x$ | Exponent of geometric numbers (–) |
| $y_{O_2,L}$ | Mole fraction of oxygen in the liquid (–) |
| $y^*_{O_2}$ | Mole fraction of oxygen in the liquid at the phase boundary (–) |

**Greek Letters**

| | |
|---|---|
| $\varepsilon_{max}$ | Maximum local energy dissipation ($W\,kg^{-1}$) |
| $\dot{\gamma}_{eff}$ | Effective shear rate ($s^{-1}$) |
| $\eta$ | Dynamic viscosity of the liquid (Pa s) |
| $\rho$ | Density ($kg\,m^{-3}$) |
| $\rho_L$ | Liquid density ($kg\,m^{-3}$) |
| $\sigma$ | Surface tension of the liquid ($N\,m^{-1}$) |
| $\tau_W$ | Shear stress in the wall ($N\,m^{-2}$) |
| $\omega$ | Angular velocity ($s^{-1}$) |

### Dimensionless Numbers

| | |
|---|---|
| *Bo* | Bond number |
| *Fr* | Froude number |
| $Fr_a$ | Axial Froude number |
| *Ha* | Hatta number |
| *Ne'* | Modified power number |
| *Ph* | Phase number |
| *Re* | Reynolds number |
| $Re_f$ | Film Reynolds number |
| *Sc* | Schmidt number |

### List of Abbreviations

| | |
|---|---|
| AHQ | Allylhydroxyquinolinium |
| CFD | Computational fluid dynamics |
| CHO | Chinese hamster ovary |
| CosBios | Continuous parallel shaken bioreactor system |
| DoE | Design of experiments |
| DLF | Dual lifetime referencing |
| FbFP | Flavin mononucleotide–based fluorescent protein |
| FRET | Förster resonance energy transfer |
| GFP | Green fluorescent protein |
| HEMA | 2-Hydroxyethyl methacrylate |
| HPTS | 8-Hydroxypyrene-1,3,6-trisulfonic acid trisodium salt |
| IPTG | Isopropyl-$\beta$-d-thiogalactopyranoside |
| MTP | Microtiter plate |
| PC | Polycarbonate |
| PMMA | Polymethylmethacrylate |
| PP | Polypropylene |
| PS | Polystyrene |
| PTFE | Polytetrafluoroethylene |
| PEG | Polyethylene glycol |
| PDMS | Polydimethylsiloxane |
| RAMOS | Respiration activity monitoring system |
| YFP | Yellow fluorescent protein |
| YNB | Yeast nitrogen base |

### References

1. Büchs, J. (2001) Introduction to advantages and problems of shaken cultures. *Biochem. Eng. J.*, 7, 91–98.
2. Betts, J.I. and Baganz, F. (2006) Miniature bioreactors: current practices and future opportunities. *Microb. Cell Fact.*, **5**, 21.

3. Fernandes, P. and Cabral, J.M.S. (2006) Microlitre/millilitre shaken bioreactors in fermentative and biotransformation processes – a review. *Biocatal. Biotransform.*, **24**, 237–252.
4. Schäpper, D., Alam, M.N., Szita, N., Eliasson Lantz, A., and Gernaey, K.V. (2009) Application of microbioreactors in fermentation process development: a review. *Anal. Bioanal. Chem.*, **395**, 679–695.
5. Kirk, T.V. and Szita, N. (2013) Oxygen transfer characteristics of miniaturized bioreactor systems. *Biotechnol. Bioeng.*, **110**, 1005–1019.
6. Kensy, F., Zimmermann, H.F., Knabben, I., Anderlei, T., Trauthwein, H., Dingerdissen, U., and Büchs, J. (2005) Oxygen transfer phenomena in 48-well microtiter plates: determination by optical monitoring of sulfite oxidation and verification by real-time measurement during microbial growth. *Biotechnol. Bioeng.*, **89**, 698–708.
7. Kumar, S., Wittmann, C., and Heinzle, E. (2004) Minibioreactors. *Biotechnol. Lett.*, **26**, 1–10.
8. Calam, C. (1969) in *Methods of Microbiology*, vol. **1** (eds J.R. Norris and D.W. Ribbons), Academic Press, London, pp. 225–336.
9. Duetz, W.A. (2007) Microtiter plates as mini-bioreactors: miniaturization of fermentation methods. *Trends Microbiol.*, **15**, 469–475.
10. Sumino, Y., Akiyama, S., and Fukada, H. (1972) Performance of the shaking flask. I. Power consumption. *J. Ferment. Technol.*, **50**, 203–208.
11. Büchs, J., Maier, U., Milbradt, C., and Zoels, B. (2000) Power consumption in shaking flasks on rotary shaking machines: II. Nondimensional description of specific power consumption and flow regimes in unbaffled flasks at elevated liquid viscosity. *Biotechnol. Bioeng.*, **68**, 594–601.
12. Büchs, J., Maier, U., Milbradt, C., and Zoels, B. (2000) Power consumption in shaking flasks on rotary shaking machines: I. Power consumption measurement in unbaffled flasks at low liquid viscosity. *Biotechnol. Bioeng.*, **68**, 589–593.
13. Schlichting, H. (1979) *Boundary Layer Theory*, 7th edn, McGraw-Hill.
14. Büchs, J., Maier, U., Lotter, S., and Peter, C.P. (2007) Calculating liquid distribution in shake flasks on rotary shakers at waterlike viscosities. *Biochem. Eng. J.*, **34**, 200–208.
15. Peter, C.P., Suzuki, Y., and Büchs, J. (2006) Hydromechanical stress in shake flasks: correlation for the maximum local energy dissipation rate. *Biotechnol. Bioeng.*, **93**, 1164–1176.
16. Maier, U. and Büchs, J. (2001) Characterisation of the gas-liquid mass transfer in shaking bioreactors. *Biochem. Eng. J.*, **7**, 99–106.
17. Peter, C.P., Suzuki, Y., Rachinskiy, K., Lotter, S., and Büchs, J. (2006) Volumetric power consumption in baffled shake flasks. *Chem. Eng. Sci.*, **61**, 3771–3779.
18. Mehmood, N., Olmos, E., Marchal, P., Goergen, J.L., and Delaunay, S. (2010) Relation between pristinamycins production by *Streptomyces pristinaespiralis*, power dissipation and volumetric gas-liquid mass transfer coefficient, $k_La$. *Process Biochem.*, **45**, 1779–1786.
19. Zhang, H., Williams-Dalson, W., Keshavarz-Moore, E., and Shamlou, P.A. (2005) Computational-fluid-dynamics (CFD) analysis of mixing and gas-liquid mass transfer in shake flasks. *Biotechnol. Appl. Biochem.*, **41**, 1–8.
20. Ottow, W., Kümmel, A., and Buchs, J. (2004) Shaking flask: flow simulation and validation. Conference on Transport Phenomena with Moving Boundaries, Berlin, Germany, October 9–10, 2004.
21. Raval, K., Kato, Y., and Büchs, J. (2007) Comparison of torque method and temperature method for determination of power consumption in disposable shaken bioreactors. *Biochem. Eng. J.*, **34**, 224–227.
22. Li, C., Xia, J.Y., Chu, J., Wang, Y.H., Zhuang, Y.P., and Zhang, S.L. (2013) CFD analysis of the turbulent flow in baffled shake flasks. *Biochem. Eng. J.*, **70**, 140–150.

23. Zhang, H., Lamping, S.R., Pickering, S.C.R., Lye, G.J., and Shamlou, P.A. (2008) Engineering characterisation of a single well from 24-well and 96-well microtitre plates. *Biochem. Eng. J.*, **40**, 138–149.
24. Barrett, T.A., Wu, A., Zhang, H., Levy, M.S., and Lye, G.J. (2010) Microwell engineering characterization for mammalian cell culture process development. *Biotechnol. Bioeng.*, **105**, 260–275.
25. Klöckner, W., Lattermann, C., Pursche, F., Werner, S., Eibl, D., and Büchs, J. (2014) Time efficient way to calculate oxygen transfer areas and power input in cylindrical disposable shaken bioreactor. *Biotechnol. Progr.*, **30**, 1441–1456.
26. Büchs, J., Lotter, S., and Milbradt, C. (2001) Out-of-phase operating conditions, a hitherto unknown phenomenon in shaking bioreactors. *Biochem. Eng. J.*, **7**, 135–141.
27. Lotter, S. and Büchs, J. (2004) Utilization of specific power input measurements for optimization of culture conditions in shaking flasks. *Biochem. Eng. J.*, **17**, 195–203.
28. Peter, C.P., Lotter, S., Maier, U., and Büchs, J. (2004) Impact of out-of-phase conditions on screening results in shaking flask experiments. *Biochem. Eng. J.*, **17**, 205–215.
29. Tan, R.K., Eberhard, W., and Büchs, J. (2011) Measurement and characterization of mixing time in shake flasks. *Chem. Eng. Sci.*, **66**, 440–447.
30. Weiss, S., John, G.T., Klimant, I., and Heinzle, E. (2002) Modeling of mixing in 96-well microplates observed with fluorescence indicators. *Biotechnol. Progr.*, **18**, 821–830.
31. Anderlei, T., Zang, W., Papaspyrou, M., and Büchs, J. (2004) Online respiration activity measurement (OTR, CTR, RQ) in shake flasks. *Biochem. Eng. J.*, **17**, 187–194.
32. Hermann, R., Walther, N., Maier, U., and Büchs, J. (2001) Optical method for the determination of the oxygen-transfer capacity of small bioreactors based on sulfite oxidation. *Biotechnol. Bioeng.*, **74**, 355–363.
33. Doig, S.D., Pickering, S.C.R., Lye, G.J., and Baganz, F. (2005) Modelling surface aeration rates in shaken microtitre plates using dimensionless groups. *Chem. Eng. Sci.*, **60**, 2741–2750.
34. Maier, U., Losen, M., and Büchs, J. (2004) Advances in understanding and modeling the gas-liquid mass transfer in shake flasks. *Biochem. Eng. J.*, **17**, 155–167.
35. Henzler, H.-J. and Schedel, M. (1991) Suitability of the shaking flask for oxygen supply to microbial cultures. *Bioprocess. Eng.*, **7**, 123–131.
36. Zimmermann, H.F., Anderlei, T., Büchs, J., and Binder, M. (2006) Oxygen limitation is a pitfall during screening for industrial strains. *Appl. Microbiol. Biotechnol.*, **72**, 1157–1160.
37. Weisenberger, S. and Schumpe, A. (1996) Estimation of gas solubilities in salt solutions at temperatures from 273 K to 363 K. *AIChE J.*, **42**, 298–300.
38. Linek, V. and Vacek, V. (1981) Chemical engineering use of catalyzed sulfite oxidation kinetics for the determination of mass transfer characteristics of gas-liquid contactors. *Chem. Eng. Sci.*, **36**, 1747–1768.
39. Hermann, R., Lehmann, M., and Büchs, J. (2003) Characterization of gas-liquid mass transfer phenomena in microtiter plates. *Biotechnol. Bioeng.*, **81**, 178–186.
40. Linek, V., Kordač, M., and Moucha, T. (2006) Evaluation of the optical sulfite oxidation method for the determination of the interfacial mass transfer area in small-scale bioreactors. *Biochem. Eng. J.*, **27**, 264–268.
41. Garcia-Ochoa, F. and Gomez, E. (2009) Bioreactor scale-up and oxygen transfer rate in microbial processes: an overview. *Biotechnol. Adv.*, **27**, 153–176.
42. Suresh, S., Srivastava, V.C., and Mishra, I.M. (2009) Techniques for oxygen transfer measurement in bioreactors: a review. *J. Chem. Technol. Biotechnol.*, **84**, 1091–1103.
43. Seletzky, J.M., Noak, U., Fricke, J., Welk, E., Eberhard, W., Knocke, C., and Büchs, J. (2007) Scale-up from shake flasks to fermenters in batch and continuous mode with *Corynebacterium*

*glutamicum* in lactic acid based on oxygen transfer and pH. *Biotechnol. Bioeng.*, **98**, 800–811.
44. Van't Riet, K. (1979) Review of measuring methods and results in nonviscous gas–liquid mass transfer in stirred vessels. *Ind. Eng. Chem. Process Des. Dev.*, **18**, 357–364.
45. Gaden, E.L. (1962) Improved shaken flask performance. *Biotechnol. Bioeng.*, **4**, 99–103.
46. McDaniel, L.E., Bailey, E.G., and Zimmerli, A. (1965) Effect of oxygen supply rates on growth of *Escherichia coli* – I. Studies in unbaffled and baffled shake flasks. *Appl. Microbiol.*, **13**, 109–114.
47. Gupta, A. and Rao, G. (2003) A study of oxygen transfer in shake flasks using a non-invasive oxygen sensor. *Biotechnol. Bioeng.*, **84**, 351–358.
48. Delgado, G., Topete, M., and Galindo, E. (1989) Interaction of cultural conditions and end-product distribution in *Bacillus subtilis* grown in shake flasks. *Appl. Microbiol. Biotechnol.*, **31**, 288–292.
49. Funke, M., Diederichs, S., Kensy, F., Müller, C., and Büchs, J. (2009) The baffled microtiter plate: increased oxygen transfer and improved online monitoring in small scale fermentations. *Biotechnol. Bioeng.*, **103**, 1118–1128.
50. Lattermann, C., Funke, M., Hansen, S., Diederichs, S., and Büchs, J. (2014) Cross-section perimeter is a suitable parameter to describe the effects of different baffle geometries in shaken microtiter plates. *J. Biol. Eng.*, **8**, 18.
51. Gill, N.K., Appleton, M., Baganz, F., and Lye, G.J. (2008) Design and characterisation of a miniature stirred bioreactor system for parallel microbial fermentations. *Biochem. Eng. J.*, **39**, 164–176.
52. Özbek, B. and Gayik, S. (2001) The studies on the oxygen mass transfer coefficient in a bioreactor. *Process Biochem.*, **36**, 729–741.
53. Pena, C., Galindo, E., and Büchs, J. (2011) The viscosifying power, degree of acetylation and molecular mass of the alginate produced by *Azotobacter vinelandii* in shake flasks are determined by the oxygen transfer rate. *Process Biochem.*, **46**, 290–297.
54. Wilming, A., Begemann, J., Kuhne, S., Regestein, L., Bongaerts, J., Evers, S., Maurer, K.H., and Büchs, J. (2013) Metabolic studies of gamma-polyglutamic acid production in *Bacillus licheniformis* by small-scale continuous cultivations. *Biochem. Eng. J.*, **73**, 29–37.
55. Giese, H., Azizan, A., Kümmel, A., Liao, A.P., Peter, C.P., Fonseca, J.A., Hermann, R., Duarte, T.M., and Büchs, J. (2014) Liquid films on shake flask walls explain increasing maximum oxygen transfer capacities with elevating viscosity. *Biotechnol. Bioeng.*, **111**, 295–308.
56. Giese, H., Klöckner, W., Peña, C., Galindo, E., Lotter, S., Wetzel, K., Meissner, L., Peter, C.P., and Büchs, J. (2014) Effective shear rates in shake flasks. *Chem. Eng. Sci.*, **118**, 102–113.
57. Henzler, H.J. and Kauling, J. (1985) Scale-up of mass transfer in highly viscous liquids. In *Deutsche Vereinigung für Chemie- und Verfahrenstechnik*, editors. Papers presented at the 5th European Conference on Mixing Würzburg/Germany. BHRA, Cranfield, pp. 303–312.
58. Mrotzek, C., Anderlei, T., Henzler, H.J., and Büchs, J. (2001) Mass transfer resistance of sterile plugs in shaking bioreactors. *Biochem. Eng. J.*, **7**, 107–112.
59. Henzler, H.-J., Schedel, M., and Müller, P.F. (1986) Non-equimolar diffusion in shaking flasks - method for determination of respiratory activity of biological cultures. *Chem. Ing. Tech.*, **58**, 234–235.
60. Anderlei, T., Mrotzek, C., Bartsch, S., Amoabediny, G., Peter, C.P., and Büchs, J. (2007) New method to determine the mass transfer resistance of sterile closures for shaken bioreactors. *Biotechnol. Bioeng.*, **98**, 999–1007.
61. Nikakhtari, H. and Hill, G.A. (2006) Closure effects on oxygen transfer and aerobic growth in shake flasks. *Biotechnol. Bioeng.*, **95**, 15–21.

62. Zimmermann, H.F., John, G.T., Trauthwein, H., Dingerdissen, U., and Huthmacher, K. (2003) Rapid evaluation of oxygen and water permeation through microplate sealing tapes. *Biotechnol. Progr.*, **19**, 1061–1063.
63. Duetz, W.A., Kühner, M., and Lohser, R. (2006) Microbial and cell growth in microtiter plates. *Genet. Eng. News*, **26**, 44.
64. Schlepütz, T. and Büchs, J. (2014) Scale-down of vinegar production into microtiter plates using a custom-made lid. *J. Biosci. Bioeng.*, **117**, 485–496.
65. Suresh, S., Srivastava, V.C., and Mishra, I.M. (2009) Critical analysis of engineering aspects of shaken flask bioreactors. *Crit. Rev. Biotechnol.*, **29**, 255–278.
66. Büchs, J. and Zoels, B. (2001) Evaluation of maximum to specific power consumption ratio in shaking bioreactors. *J. Chem. Eng. Jpn.*, **34**, 647–653.
67. Henzler, H.-J. (2000) in *Influence of Stress on Cell Growth and Product Formation*, Advances in Biochemical Engineering/Biotechnology, vol. **67** (eds K. Schügerl, G. Kretzmer, H.J. Henzler, P.M. Kieran, P.E. MacLoughlin, D.M. Malone, W. Schumann, P.A. Shamlou, and S.S. Yim), Springer, Berlin, Heidelberg, pp. 35–82.
68. Daub, A., Böhm, M., Delueg, S., Mühlmann, M., Schneider, G., and Büchs, J. (2014) Maximum stable drop size measurements indicate turbulence attenuation by aeration in a 3 $m^3$ aerated stirred tank. *Biochem. Eng. J.*, **86**, 24–32.
69. Fernandes, P., Carvalho, F., and Marques, M.P.C. (2011) Miniaturization in biotechnology: speeding up the development of bioprocesses. *Recent Pat. Biotechnol.*, **5**, 160–173.
70. Bareither, R. and Pollard, D. (2011) A review of advanced small-scale parallel bioreactor technology for accelerated process development: current state and future need. *Biotechnol. Progr.*, **27**, 2–14.
71. Marques, M.P.C., Cabral, J.M.S., and Fernandes, P. (2009) High throughput in biotechnology: from shake-flasks to fully instrumented microfermentors. *Recent Pat. Biotechnol.*, **3**, 124–140.
72. Kostov, Y., Harms, P., Randers-Eichhorn, L., and Rao, G. (2001) Low-cost microbioreactor for high-throughput bioprocessing. *Biotechnol. Bioeng.*, **72**, 346–352.
73. Lamping, S.R., Zhang, H., Allen, B., and Shamlou, P.A. (2003) Design of a prototype miniature bioreactor for high throughput automated bioprocessing. *Chem. Eng. Sci.*, **58**, 747–758.
74. Betts, J.I., Doig, S.D., and Baganz, F. (2006) Characterization and application of a miniature 10 mL stirred-tank bioreactor, showing scale-down equivalence with a conventional 7 L reactor. *Biotechnol. Progr.*, **22**, 681–688.
75. Yang, T.H., Wittmann, C., and Heinzle, E. (2004) Membrane inlet mass spectrometry for the on-line measurement of metabolic fluxes in the case of lysine-production by *Corynebacterium glutamicum*. *Eng. Life Sci.*, **4**, 252–257.
76. Puskeiler, R., Kaufmann, K., and Weuster-Botz, D. (2005) Development, parallelization, and automation of a gas-inducing milliliter-scale bioreactor for high-throughput bioprocess design (HTBD). *Biotechnol. Bioeng.*, **89**, 512–523.
77. Weuster-Botz, D., Puskeiler, R., Kusterer, A., Kaufmann, K., John, G., and Arnold, M. (2005) Methods and milliliter scale devices for high-throughput bioprocess design. *Bioprocess Biosyst. Eng.*, **28**, 109–119.
78. Klein, T., Schneider, K., and Heinzle, E. (2013) A system of miniaturized stirred bioreactors for parallel continuous cultivation of yeast with online measurement of dissolved oxygen and off-gas. *Biotechnol. Bioeng.*, **110**, 535–542.
79. Doig, S.D., Diep, A., and Baganz, F. (2005) Characterisation of a novel miniaturised bubble column bioreactor for high throughput cell cultivation. *Biochem. Eng. J.*, **23**, 97–105.
80. Tang, Y.J.J., Laidlaw, D., Gani, K., and Keasling, J.D. (2006) Evaluation of the effects of various culture conditions on Cr(VI) reduction by *Shewanella*

*oneidensis* MR-1 in a novel high-throughput mini-bioreactor. *Biotechnol. Bioeng.*, **95**, 176–184.

81. Isett, K., George, H., Herber, W., and Amanullah, A. (2007) Twenty-four-well plate miniature bioreactor high-throughput system: assessment for microbial cultivations. *Biotechnol. Bioeng.*, **98**, 1017–1028.
82. Rohe, P., Venkanna, D., Kleine, B., Freudl, R., and Oldiges, M. (2012) An automated workflow for enhancing microbial bioprocess optimization on a novel microbioreactor platform. *Microb. Cell Fact.*, **11**, 144.
83. Lye, G.J., Ayazi-Shamlou, P., Baganz, F., Dalby, P.A., and Woodley, J.M. (2003) Accelerated design of bioconversion processes using automated microscale processing techniques. *Trends Biotechnol.*, **21**, 29–37.
84. Mandenius, C.-F. and Brundin, A. (2008) Bioprocess optimization using design-of-experiments methodology. *Biotechnol. Progr.*, **24**, 1191–1203.
85. Zimmermann, H.F. and Rieth, J. (2006) A fully automated robotic system for high throughput fermentation. *J. Assoc. Lab. Autom.*, **11**, 134–137.
86. Wollerton, M.C., Wales, R., Bullock, J.A., Hudson, I.R., and Beggs, M. (2006) Automation and optimization of protein expression and purification on a novel robotic platform. *J. Assoc. Lab. Autom.*, **11**, 291–303.
87. Huber, R., Ritter, D., Hering, T., Hillmer, A.-K., Kensy, F., Müller, C., Wang, L., and Büchs, J. (2009) Robo-Lector – a novel platform for automated high-throughput cultivations in microtiter plates with high information content. *Microb. Cell Fact.*, **8**, 42.
88. Weuster-Botz, D., Altenbach-Rehm, J., and Arnold, M. (2001) Parallel substrate feeding and pH-control in shaking-flasks. *Biochem. Eng. J.*, **7**, 163–170.
89. Huber, R., Scheidle, M., Dittrich, B., Klee, D., and Büchs, J. (2009) Equalizing growth in high-throughput small scale cultivations via precultures operated in fed-batch mode. *Biotechnol. Bioeng.*, **103**, 1095–1102.
90. Jeude, M., Dittrich, B., Niederschulte, H., Anderlei, T., Knocke, C., Klee, D., and Büchs, J. (2006) Fed-batch mode in shake flasks by slow-release technique. *Biotechnol. Bioeng.*, **95**, 433–445.
91. Scheidle, M., Jeude, M., Dittrich, B., Denter, S., Kensy, F., Suckow, M., Klee, D., and Büchs, J. (2010) High-throughput screening of *Hansenula polymorpha* clones in the batch compared with the controlled-release fed-batch mode on a small scale. *FEMS Yeast Res.*, **10**, 83–92.
92. Stöckmann, C., Scheidle, M., Dittrich, B., Merckelbach, A., Hehmann, G., Melmer, G., Klee, D., Büchs, J., Kang, H.A., and Gellissen, G. (2009) Process development in *Hansenula polymorpha* and *Arxula adeninivorans*, a re-assessment. *Microb. Cell Fact.*, **8**, 22.
93. Scheidle, M., Dittrich, B., Klinger, J., Ikeda, H., Klee, D., and Büchs, J. (2011) Controlling pH in shake flasks using polymer-based controlled-release discs with pre-determined release kinetics. *BMC Biotechnol.*, **11**, 25.
94. Hegde, S., Pant, T., Pradhan, K., Badiger, M., and Gadgil, M. (2012) Controlled release of nutrients to mammalian cells cultured in shake flasks. *Biotechnol. Progr.*, **28**, 188–195.
95. Bähr, C., Leuchtle, B., Lehmann, C., Becker, J., Jeude, M., Peinemann, F., Arbter, R., and Büchs, J. (2012) Dialysis shake flask for effective screening in fed-batch mode. *Biochem. Eng. J.*, **69**, 182–195.
96. Wilming, A., Bähr, C., Kamerke, C., and Büchs, J. (2014) Fed-batch operation in special microtiter plates: a new method for screening under production conditions. *J. Ind. Microbiol. Biotechnol.*, **41**, 513–525.
97. Green, H. and Rheinwald, J.G. (1975) Method of controllably releasing glucose to a cell culture medium. US Patent 3,926,723 A.
98. Panula-Perälä, J., Siurkus, J., Vasala, A., Wilmanowski, R., Casteleijn, M.G., and Neubauer, P. (2008) Enzyme controlled glucose auto-delivery for high cell density cultivations in microplates

and shake flasks. *Microb. Cell Fact.*, **7**, 31.

99. Krause, M., Ukkonen, K., Haataja, T., Ruottinen, M., Glumoff, T., Neubauer, A., Neubauer, P., and Vasala, A. (2010) A novel fed-batch based cultivation method provides high cell-density and improves yield of soluble recombinant proteins in shaken cultures. *Microb. Cell Fact.*, **9**, 11.
100. Panula-Perälä, J., Vasala, A., Karhunen, J., Ojamo, H., Neubauer, P., and Mursula, A. (2014) Small-scale slow glucose feed cultivation of *Pichia pastoris* without repression of AOX1 promoter: towards high throughput cultivations. *Bioprocess Biosyst. Eng.*, **37**, 1261–1269.
101. Hemmerich, J., Adelantado, N., Barrigon, J., Ponte, X., Hormann, A., Ferrer, P., Kensy, F., and Valero, F. (2014) Comprehensive clone screening and evaluation of fed-batch strategies in a microbioreactor and lab scale stirred tank bioreactor system: application on *Pichia pastoris* producing *Rhizopus oryzae* lipase. *Microb. Cell Fact.*, **13**, 36.
102. Akgün, A., Müller, C., Engmann, R., and Büchs, J. (2008) Application of an improved continuous parallel shaken bioreactor system for three microbial model systems. *Bioprocess Biosyst. Eng.*, **31**, 193–205.
103. Büchs, J. (1994) in *Process Computations in Biotechnology* (ed T. Ghose), McGraw-Hill, New Delhi, pp. 194–237.
104. Akgün, A., Maier, B., Preis, D., Roth, B., Klingelhöfer, R., and Büchs, J. (2004) A novel parallel shaken bioreactor system for continuous operation. *Biotechnol. Progr.*, **20**, 1718–1724.
105. Schlepütz, T. and Büchs, J. (2013) Investigation of vinegar production using a novel shaken repeated batch culture system. *Biotechnol. Progr.*, **29**, 1158–1168.
106. Harms, P., Kostov, Y., and Rao, G. (2002) Bioprocess monitoring. *Curr. Opin. Biotechnol.*, **13**, 124–127.
107. Samorski, M., Müller-Newen, G., and Büchs, J. (2005) Quasi-continuous combined scattered light and fluorescence measurements: a novel measurement technique for shaken microtiter plates. *Biotechnol. Bioeng.*, **92**, 61–68.
108. Hansen, S., Kensy, F., Käser, A., and Büchs, J. (2011) Potential errors in conventional DOT measurement techniques in shake flasks and verification using a rotating flexitube optical sensor. *BMC Biotechnol.*, **11**, 49.
109. Siepert, E.-M., Gartz, E., Tur, M.K., Delbrück, H., Barth, S., and Büchs, J. (2012) Short-chain fluorescent tryptophan tags for on-line detection of functional recombinant proteins. *BMC Biotechnol.*, **12**, 65.
110. Drepper, T., Huber, R., Heck, A., Circolone, F., Hillmer, A.K., Büchs, J., and Jaeger, K.E. (2010) Flavin mononucleotide-based fluorescent reporter proteins outperform green fluorescent protein-like proteins as quantitative in vivo real-time reporters. *Appl. Environ. Microbiol.*, **76**, 5990–5994.
111. Wolfbeis, O.S. (ed) (1997) *Chemical Sensing Using Indicator Dyes*, Artech House.
112. Kensy, F., Zang, E., Faulhammer, C., Tan, R.K., and Büchs, J. (2009) Validation of a high-throughput fermentation system based on online monitoring of biomass and fluorescence in continuously shaken microtiter plates. *Microb. Cell Fact.*, **8**, 31.
113. Kunze, M., Lattermann, C., Diederichs, S., Kroutil, W., and Büchs, J. (2014) Minireactor-based high-throughput temperature profiling for the optimization of microbial and enzymatic processes. *J. Biol. Eng.*, **8**, 22.
114. Tolosa, L., Kostov, Y., Harms, P., and Rao, G. (2002) Noninvasive measurement of dissolved oxygen in shake flasks. *Biotechnol. Bioeng.*, **80**, 594–597.
115. Potzkei, J., Kunze, M., Drepper, T., Gensch, T., Jaeger, K.E., and Büchs, J. (2012) Real-time determination of intracellular oxygen in bacteria using a genetically encoded FRET-based biosensor. *BMC Biol.*, **10**, 28.
116. Schneider, K., Schütz, V., John, G.T., and Heinzle, E. (2010) Optical device for parallel online measurement of dissolved oxygen and pH in shake flask

cultures. *Bioprocess Biosyst. Eng.*, **33**, 541–547.

117. Badugu, R., Kostov, Y., Rao, G., and Tolosa, L. (2008) Development and application of an excitation ratiometric optical pH sensor for bioprocess monitoring. *Biotechnol. Progr.*, **24**, 1393–1401.
118. Rachinskiy, K., Schultze, H., Boy, M., Bornscheuer, U., and Büchs, J. (2009) "Enzyme test bench," a high-throughput enzyme characterization technique including the long-term stability. *Biotechnol. Bioeng.*, **103**, 305–322.
119. Kunze, M., Roth, S., Gartz, E., and Büchs, J. (2014) Pitfalls in optical on-line monitoring for high-throughput screening of microbial systems. *Microb. Cell Fact.*, **13**, 53.
120. Anderlei, T. and Büchs, J. (2001) Device for sterile online measurement of the oxygen transfer rate in shaking flasks. *Biochem. Eng. J.*, **7**, 157–162.
121. Guez, J.-S., Müller, C.H., Danze, P.M., Büchs, J., and Jacques, P. (2008) Respiration activity monitoring system (RAMOS), an efficient tool to study the influence of the oxygen transfer rate on the synthesis of lipopeptide by *Bacillus subtilis ATCC6633*. *J. Biotechnol.*, **134**, 121–126.
122. Stöckmann, C., Maier, U., Anderlei, T., Knocke, C., Gellissen, G., and Büchs, J. (2003) The oxygen transfer rate as key parameter for the characterization of *Hansenula polymorpha* screening cultures. *J. Ind. Microbiol. Biotechnol.*, **30**, 613–622.
123. Losen, M., Frolich, B., Pohl, M., and Büchs, J. (2004) Effect of oxygen limitation and medium composition on *Escherichia coli* fermentation in shake-flask cultures. *Biotechnol. Progr.*, **20**, 1062–1068.
124. Scheidle, M., Klinger, J., and Büchs, J. (2007) Combination of on-line pH and oxygen transfer rate measurement in shake flasks by fiber optical technique and Respiration Activity Monitoring System (RAMOS). *Sensors*, **7**, 3472–3480.
125. Hansen, S., Hariskos, I., Luchterhand, B., and Büchs, J. (2012) Development of a modified Respiration Activity Monitoring System for accurate and highly resolved measurement of respiration activity in shake flask fermentations. *J. Biol. Eng.*, **6**, 11.
126. Ge, X.D. and Rao, G. (2012) Real-time monitoring of shake flask fermentation and off gas using triple disposable noninvasive optical sensors. *Biotechnol. Progr.*, **28**, 872–877.
127. van Leeuwen, M., Heijnen, J.J., Gardeniers, H., Oudshoorn, A., Noorman, H., Visser, J., van der Wielen, L.A.M., and van Gulik, W.M. (2009) A system for accurate on-line measurement of total gas consumption or production rates in microbioreactors. *Chem. Eng. Sci.*, **64**, 455–458.

# 3
# Bioreactors on a Chip

*Danny van Noort*

## 3.1
## Introduction

Understanding the behavior of cells and their functions during growth and production, and the relation to nutrient supply is essential in the field of biotechnology. Most conventional cell culture systems operate in batch-mode, in which a fixed amount of nutrients and oxygen is provided for the initial cell culture, supporting growth until there is a lack of nutrients and oxygen. This means that due to the constant change in the environment, batch cultures are not ideal to characterize and bring better understanding of the cellular processes. Instead, a more constant and precisely controlled environment is necessary. The chemostat principle provides this by its continuous supply of nutrients and oxygen [1]. Unlike batch cultures, continuous chemostat cultures can run for weeks at steady state. However, the consumption and cost of growth media, up to 500 l per chemostat culture run for a bench-scale reactor, is not realistic, and keeping the system sterile during medium addition or sampling is another problem.

Microfluidics can offer a way to address these difficulties. A microfluidic-based bioreactor system will consume in the order of 10 ml to 1 l, rather than 500 l. The integration possibility of optical or electronic sensors to measure and control various environmental conditions, as well as the integration of fluidic handling, such as pumps and valves and the possibility to massive parallelism, can give microfluidic-based bioreactors an advantage that conventional bioreactors do not have.

In Chapter 2, the concept and possibilities of microbioreactors were introduced. Microbioreactors have a reactor chamber with volumes of around 50–800 μl, while bench-scale bioreactors have a typical volume of 0.5–5 l [2]. Some advantages of microbioreactors are the reduced amount of substrate and utilities needed per experiment and reduced space requirements for parallel operation. As these devices are disposable, they also reduce the efforts to prepare for the experiment, as they do not have to be cleaned and sterilized after use. Also, automation is made easier.

*Bioreactors: Design, Operation and Novel Applications*, First Edition. Edited by Carl-Fredrik Mandenius.

A variety of bioproducts are obtained through microbial fermentation process. Products can include primary metabolites, secondary metabolites, enzymes, therapeutic proteins, vaccines, and gums [3]. The development of these products begins with a screening phase in which many potential bacterial strains are screened toward largest product yield at certain growth conditions [4]. When a microbial candidate has been identified, the strain is transferred to the development phase where the physiology of the strain is studied in more detail, as are the growth conditions in bench-scale bioreactors with volumes of 0.5–10 l. The final stage is a scale-up of the bioreactor volume until production scale is reached.

The time and labor bottleneck in this development work is the screening phase. Typically, the experiments are carried out using a combination of multiwell plates, Petri dishes, and shake flasks. These conventional means permit only limited control of the microenvironment, and only endpoint data of the cells performance can be obtained.

A key issue is that cells in bioreactors need to be stirred to obtain a more homogeneous culture environment, by reducing spatial gradients of soluble factors. This holds true for large-scale suspension bioreactors, as well as for microbioreactors in the microliter range. In the latter, stirring can be achieved by using magnetic stirrer bars [5] or buoyancy-driven thermoconvection [6]. Magnetic stirrer bars can be useful when using single bioreactors, but when using many microbioreactors in parallel this method will be harder to control. The thermoconvection method has been applied to a 96-well plate format. The convection is achieved by placing a microfabricated heater under each well and creating a temperature difference up to 4 °C. It is claimed that only 5% of the medium evaporation was detected after 7 days. However, when a culture is perfused with media in a microfluidic cell culture system, there is no need for a stirrer. The diffusion distances inside a true microbioreactor are very short, up to 100 μm. As such, there will be cell culture condition as is found in *in vivo* tissues. The question is, of course – what the objective of the cell culture is: mimic the *in vivo* physiology or optimize a cell culture for the highest yield of products, totally disregarding *in vivo* similarities? The answer is as follows: if drugs are tested for human consumption, the former situation is essential; if drugs are produced by means of microbial cell cultures, the latter situation is preferable.

There is a need to develop novel microbial cell cultivation technologies in high-throughput fashion to enhance our understanding of the functions of microorganisms in different environmental conditions. Industrial and bench-scale bioreactors have the ability to control and measure the temperature, pH, dissolved oxygen (DO), $CO_2$, and cell densities (OD). However, these bioreactors are expensive and also time consuming to run in parallel batches. The increasing number of genetic and process permutations and optimizations needed to screen for new products will need faster ways to collect all that data. Smaller reactor volumes combined with integrated sensors with a real-time flow of data would facilitate high-throughput and cost-effective screening. Although bench-scale offers some real-time parameters, such as DO and pH, biomass is still measured off-line, which increases contamination and reduces volume during the experiments. This leads to altered process conditions. Tubes and shake flasks, which count for more than 90% of all the cell culture experiments in biotechnology [7], can be operated in parallel with smaller volumes, but data are only collected at endpoint. Scaling down microbioreactors even further could lead to systems such as microfermenter arrays or microbiological assay kits [8].

By using microfabrication techniques, the volumes of the bioreactor chambers can be reduced down to picoliters. The question that immediately arises is what would be the advantage if bioreactors are further scaled down. One answer is that the dimensions of the microbioreactors are then similar to those in *in vivo* systems. As is highlighted later, there is already research performed on single-cell chemostats. It is known that the physiological conditions within a microfluidic chamber are not the same as within industrial-sized bioreactors. This will present problems when trying to scale up the process from picoliters to full-sized bioreactors. However, it might be possible to scale up the process by creating a large array of BRoCs.

It must be stated that the border between microbioreactors and BRoCs is vague. There are a number of microbioreactors using microfluidics to address the reactor chamber (e.g., Ref. [9]), as well as some reactor chambers using microtechnology to fabricate cell culture platforms imbedded in a 1.35 ml microreactor [10].

Another advantage of microtechnology is the ability to integrate sensors into the system. There is a wide range of sensors necessary to monitor and control the bioreactor process. These include foremost pH, $O_2$, and $CO_2$ sensors, as well as monitoring the cell density. As microfluidic devices are usually transparent, it is possible to examine the reactors optically and observe, for instance, fluorescence signals from transgenic systems such as monitoring the production of proteins that are tagged with GFP reporter strands.

Microfabricated bioreactors with integrated sensors can be operated in parallel and are thus better suited to acquire high-throughput process data. The materials used to fabricate microbioreactors, such as PMMA and PDMS, are disposable, thus reducing assembly, cleaning, and sterilization time. They also need fewer resources and have less waste, which is of benefit when using expensive enzymes.

Microfabrication allows for control of the microenvironment, especially the extracellular matrix (ECM), cell–cell interaction, the interaction of the cells with the ECM, soluble factors, and mechanical forces, such as shear stress [11–13]. Micropatterning techniques allow for a further control of cell–cell and cell–matrix interactions [14–16].

## 3.2
## Advantages of Microsystems

Biological systems are on the scale of micrometers if we are looking at the fluidic and diffusive transport properties. As such, microfabricated bioreactors make a lot of sense, as they can mimic these properties *in vitro*. Devices in the micro-range need less space, less reagents, and less energy, while the response times and, therefore, the reaction rates go up. There is a gain on information over space, compared with standard equipments, which would also reduce the costs. One major advantage of microfluidic systems is the possibility of high-throughput screening (HTS) [17], which is required to have a large number of system runs to obtain statistically significant data. This especially applies to microbial perfusion microbioreactors where there has been a significant progress in implementing high-throughput [18, 19].

Another advantage of microsystems is the possibility to construct cell-specific microenvironments due to the precise control of microscale structures. The ability to control, for example, the flow pattern, gives control over the transport of growth factors, reagents, oxygen, and the amount of hydrodynamic stress on the cells [20]. The size of the bioreactor chamber will affect the behavior of the cells, altering transport phenomena, due to diffusion distances and cell–cell interactions. The flow rate can have an effect on the cell growth dynamics [21]. For example, fibroblasts were grown under various flow conditions in a microbioreactor: from static (no flow) conditions to high flow rates [21]. The cells displayed little growth in either no-flow or in high-flow ($0.3\,\text{ml}\,\text{h}^{-1}$) conditions, but displayed optimal growth at $0.2\,\text{ml}\,\text{h}^{-1}$. Embryonic stem cells showed the same behavior toward flow rates [22]. Not only cell growth is affected, but viability as well [23]. This shows that the microenvironment does impact cell functions and, therefore, the toxicity response toward drugs. Culturing cells in a microfluidic environment in a continuous perfused system offer the ability to control cell-media interaction by producing a steady-state chemical gradient, compared with standard culture system where the chemical composition changes over time [22]. Under static conditions, diffusion is the dominant mass transport method.

Cells can be in direct contact with a flow, or shielded by a microfabricated barrier [23, 24], micro-groove substrate [25], internal membrane oxygenators [26], or hydrogels [27]. When shielded, the effect of shear stress is still present but in lower magnitude than when the cells are in direct contact with the flow. Cell cultures need to be perfused to refresh the culture medium, but cells should also be incubated in secreted soluble factors, such as feeder cells for human embryonic stem cell (hESC) cultures. A flow would wash out these factors. A method that facilitates long-term culture (>7 days) in direct flow is using a "flow-stop" perfusion system: a short temporal exposure followed by long static incubation periods [28]. Short pulses of shears would allow shear-sensitive cells (such as hESCs) to withstand medium renewal flows, while the long static incubation periods would allow for secreted factors (i.e., growth factors) to be accumulated locally. The "flow-stop" method might be suitable for culturing different cell culture with different reactions to shear stress in the same microbioreactor, when exposed directly to the flow. As such, the "low-stop" protocol should be designed according to the cell culture requirements. Under these conditions, the cells resemble conventional culture dish characteristics [28].

There has also been research conducted to use microbioreactors on space missions as biological sentinels or remote monitoring tools [29]. Microbioreactors would greatly reduce the payload. It also shows that microfluidic cell culture systems can be autonomous.

### 3.2.1 Concentration Gradients

An important advantage of microfluidic systems is the ability to create concentration gradients of chemicals inside the microbioreactor using laminar flows [30] (Figure 3.1a). Basically, a gradient generator is a series of channels with different fluidic contents connected to a chamber where the gradient is formed. Due to the laminar character of the flows inside the microchannels, mixing only occurs by diffusion. A simple way to form a concentration gradient is to have two solutions flowing in one microchannel (Figure 3.1b). This creates a parallel flow, which will slowly mix down the channel. Other, more complex, types of concentration gradients have also been developed, including linear, sigmoidal, and logarithmic distributions [32–34] (Figure 3.1c). The formation of concentration gradients in multiwell systems is not straight forward as doses have to be created in different wells. Another application of gradient generator is the creation of combinatorial drug dosages, to achieve a more synergistic effects and establishing the optimal ratio between drugs. An example is to test the effect of various concentrations

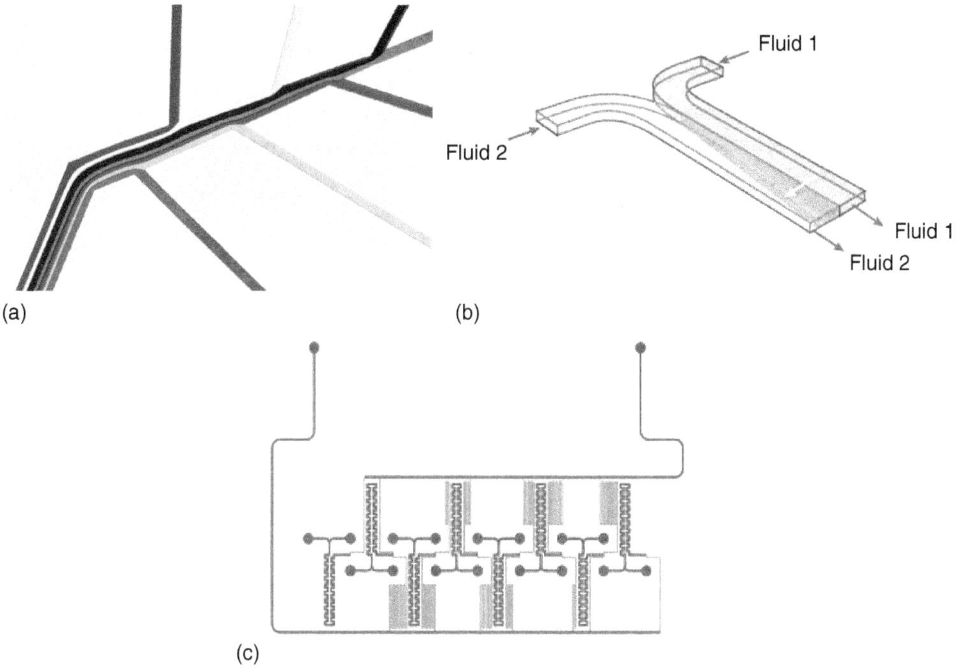


**Figure 3.1** (a) Multiphase laminar flow patterning (from [31]) (b) Mixing of two laminar flows by diffusion. (c) A linear concentration generator with eight different concentrations at the outputs (from [32]).

of anesthetics, bupivacaine, and lidocaine on myoblasts [35]. Such a control over reagent deliveries at defined concentrations shows the potential of BRoCs not easily achievable with conventional microwell platforms.

An oxygen gradient across a microbioreactor can also be achieved by controlling the flow rate, to expose cells to heterogeneous oxygen environment. As oxygen is consumed at the beginning of the microreactor where the medium enters, there will be less oxygen left at the other end of the microreactor at the outlet. The oxygen gradient in the microreactor thus mimics physiological conditions in tissues, for example, liver, where localized functions in the liver control the local oxygen concentration [36].

## 3.3 Scaling Down the Bioreactor to the Microfluidic Format

The overall drive to scale down the bioreactor is to use them for process optimization. Due to their small size, they reduce the cost, increase the throughput, and reduce the manual labor involved [37]. Also, reaction and process times are shorter due to shorter diffusion times of soluble factors. Another driving factor is that with smaller devices it is easier to optimize processes in parallel and that microfluidic devices lend themselves to be automated. However, they are not up to the same level of automation as robotic liquid-handling systems. There are actual examples where microreactors perform better than conventional systems. An example is the efficiency of digestion of enzyme immobilized on the walls of a microtube, which had a higher performance compared with in-solution digestion systems [38]. This is because microfluidics have a much higher surface-to-volume ratio than reactor vessels.

Many of the described reactors in this chapter have been compared with bench-scale reactors showing similar results. The question whether the microbioreactor can be scaled up to industrial-sized bioreactors remains.

## 3.4 Microfabrication Methods for Bioreactors-On-A-Chip

There are various ways to fabricate a microsystem. Here, we describe a number of fabrication techniques of which some can be easily implemented in standard labs, while others are more costly techniques. Also, processes that can be made in transparent material are described. One of these processes is soft lithography, a technique that is one of the most widely used methods for microfluidic fabrication. Also, fine-mechanical fabrication technique is an option, as holes can be bored down to 100 μm in diameter. These days, laser-machining is more readily accessible, but to reach high accuracy a high-precision laser is needed, which will drive the price up.

To perform microfabrication, in most cases, a clean room is needed to prevent dust settling on the device. Depending on the fabrication method used, a different class of clean room is needed. Class, in this case, means the number of dust particles per cubic meter. To get an idea, for soft lithography a cleanliness of 1000–10 000 particle $m^{-3}$ (ISO class 4–5) is needed. On the other side of the scale is the fabrication of microelectronics, from 1 to10 particle $m^{-3}$ (ISO class 1–2), when defining a particle size of 0.3 μm.

### 3.4.1
### Etching of Silicon/Glass

Although glass might be the most desired material to use in biotech applications because of its optical properties and biological compatibility, it is a difficult material to process due to its brittleness. There are a few methods to process glass: powder blasting, deep reactive-ion etching (DRIE), wet etching [39], and laser machining. Of all these methods, wet etching with HF gives good results, even though it is an isotropic process, but it is the most hazardous of methods, which is not recommended for a standard lab, as special training and safety regulations are necessary. Furthermore, not every lab allows HF on their premises. DRIE is a very slow etching method, and the equipment is very expensive, which is not always available in clean rooms. However, the process can be extremely accurate for small features (0.5 μm). Powder blasting will also give significant results except that the walls are tapered at an angle of about 12–15°, but the minimum feature size is around 50 μm. Laser machining is discussed later. It is always possible to outsource the fabrication of glass devices. There are a number of companies specialized in processing glass.

### 3.4.2
### Soft Lithography

Soft lithography was introduced by Whitesides [30, 40] and has since then caused microfluidics to take off. It is a relative cheap replica molding method with a high possible turnover (Figure 3.2). However, obtaining a mold would require the use of a clean room equipped with lithography, which is basically a mask aligner (or another UV light source; [42]), spin coater, hot plates, and a solvent bench. The most commonly used photoresist is SU8 (from MicroChem Corp.), which is available in different viscosities resulting in different feature heights. Depending on the feature size, there are different mask options. For everything larger than 20 μm, foil masks can be printed (around $40, depending on the country). For features between 5 and 20 μm, soda-lime masks are needed (~$500–1000, depending on the number of features on the design and on the country). For features between 0.5 and 5 μm, quartz masks are needed (>$1500). For smaller feature sizes, the process becomes slightly more complicated by using a stepper, which basically reduces the feature sizes on the mask by optically transferring the pattern via a lens.

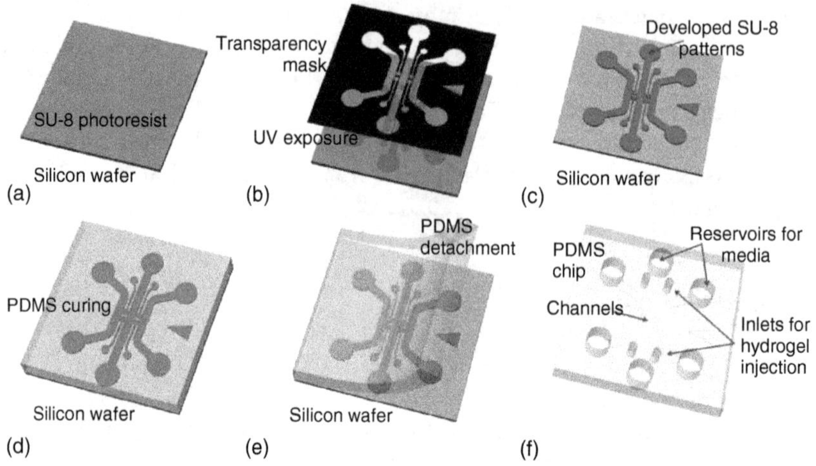


**Figure 3.2** Schematics of the photolithography (a–c) and soft lithography (d–f) procedures. (a) SU-8 is spin-coated and prebaked on a bare wafer. (b) With a transparency photomask (black), UV light is exposed on the SU-8. (c) Exposed SU-8 is then baked after exposure and developed to define channel patterns. (d) PDMS mixed solution is poured on the wafer and cured. (e) Cured PDMS is then peeled from the wafer. (f) The device is trimmed, punched, and autoclaved ready for assembly (from [41]).

Once the master is fabricated, the mold is cast against PDMS, a transparent two-component elastomer (provided by companies such as Dow Corning (Sylgard-184) or General Electric (RTV-615)). By punching holes through the PDMS, to create inlets and outlets, the device can be readily connected to syringe pumps. Finally, the PDMS device is oxygen plasma bonded to a microscope slide or coverslip.

### 3.4.3 Hot Embossing

To scale up production, imprinting could be considered. This method uses plastics, such as PMMA (or acrylic) to make chips in big quantities by using a silicon or metal mold to press under pressure in heated plastic. Feature sizes in the submicron range can be replicated. This method is only advisable for large-scale production, not for rapid development processes.

### 3.4.4 Mechanical Fabrication Technique (Or Poor Man's Microfluidics)

Computer numerical control (CNC) milling and microdrilling can create holes down to 100 μm (http://www.datron.com/applications/small-hole-drilling.php), while features can be made using balls, squares, drills, and keyseat end mills of various sizes. Milling and drilling leaves a rough surface, which can be smoothed

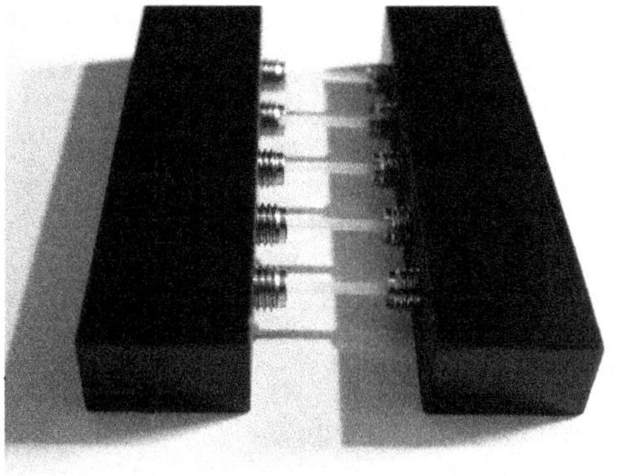

**Figure 3.3** A set of two micromachined Delrin holders with short Teflon microtubes. There are frits with filters at both sides of the microtubes. The diameter of the tubes is 150 μm (from [44]).

by polishing with methylene chloride vapor [43] or a laser-assisted surface treatment [10]. Prewarming followed by melting the surface up to 50 μm deep with a defocused $CO_2$-laser give extreme smooth surfaces, eliminating the mechanical roughness of the milling process.

With off-the-shelve tubing, ferules, nuts, and filters, it is possible to set up a relatively cheap microfluidic system, one which I like to call *poor man's microfluidics*. Off-the-shelve components such as tubing are available with bore sizes down to 50 μm. By using other components, such as unions, T-junctions, and filters, microfluidic systems can be devised. One could even incorporate solenoid valves to control the directions of the flow, thus creating a more complex system. A macroscale plumbing system would be the analogy. In this manner, by using a two-micromachined pieces of Delrin with channel diameter of 150 μm were connected to each other by 50 μm bore Teflon tubing, capped with frits (filters) with a pore size of 2 μm to be able to load DNA single stranded functionalized beads, DNA hybridization experiments have been performed, while the event was recorded by using a intercalating dye and a standard fluorescence microscope (Figure 3.3) [44]. Another example is the use of a 500 μm inner diameter PTFE tube to perform rapid proteolysis for proteomic analysis [38].

### 3.4.5 Laser Machining

With the reduction in prices of laser machining, this might be another viable method for smaller labs. Depending on the quality of the laser, submicrons and any material can be processed. Standard laser cutters using an infrared laser can only cut a limited number of materials, such as paper, plastic, wood, and leather. In general, materials that do not conduct heat very well cannot be used, such as glass. Plastic, or PMMA (known as acrylic), is a good candidate for microfluidics and especially BRoCs, as they sometimes have a reactor chamber in the millimeter range. However, for most commercial laser cutters, the smallest channel one can make might be only around 400 μm wide. By using an optional beam focuser, slightly smaller features might be possible, as this reduces the spot size.

To fabricate smaller features in a wider range of materials, other laser types are necessary, such as an excimer laser or femtosecond laser. The downside is that these lasers are quite expensive, somewhere between $50K and $250K.

### 3.4.6
### Thin Metal Layers

When wanting to fabricate on-chip sensors or heaters, there is a need of a piece of equipment that can deposit thin metal layers (usually Au, Pt, Ti, Cr) on to the surfaces. There are two types of methods: sputtering and evaporation. The difference is the way the metal is heated. Examples are plasma-induced molecular beam epitaxy (PIMBE) and metal–organic chemical vapor deposition (MOCVD).

## 3.5
## Fabrication Materials

The material from which the microbioreactor is fabricated will influence its functions. To realize these certain functions will mean attention has to be given to the material and the properties, such as hard versus soft materials, or the transparency or thermo properties of the material. Following are some examples of materials used in microfluidic devices. The materials can be used by itself, or a combination of these materials can used, to form hybrid structures, such as sandwiching a soft material between two hard ones [45], or combining materials to increase permeability or elasticity in certain regions [46].

### 3.5.1
### Inorganic Materials

Before microfluidics were popularized, it was already extensively used, for example, glass and quartz were used for capillaries for gas chromatography and capillary electrophoresis (CE), while flow reactors were micromachined in metal. Microfabrication technology developed in the semiconductor industry meant that the first generation of microfluidic devices were fabricated in silica or glass [47, 48]. Glass is optically transparent and acts as an electrical insulator, while silica is transparent to infrared light, while both are solvent resistant and biocompatible. One problem is the high cost of fabrication for each chip and another is the use of dangerous chemicals, such as HF. These materials are not really suitable for long-term cell cultures, as neither glass nor silicon is gas permeable.

## 3.5.2 Elastomers and Plastics

There is a vast variety of different polymers with specific properties [49, 50]. Polymers are easy to access and relatively cheap and, therefore, the preferred material for microfluidic devices. Polymers can be classified into three groups: elastomers, thermosets, and thermoplastics.

### 3.5.2.1 Elastomers

Elastomers consist of cross-linked polymer chains and are compressible when applying external force, after which they return to their original shape when the force is removed. The most widely used elastomer is PMDS [51], provided by companies such as Dow Corning (Sylgard-184) or General Electric (RTV-615). PDMS is a two-component liquid, which is curable between 40 and 80 °C, and it can be cast at nanometer resolution against a photoresist pattern on a silicon wafer. PDMS can be reversibly bonded to glass or another piece of PDMS, or irreversibly bonded after an oxygen plasma treatment. PDMS is elastic, which allows it to be used as pneumatic valves [52]. A thin layer of PDMS (40 μm) is placed between two channel layers. By applying a pressure on the top channel, the membrane is pushed into the other channel, effectively blocking it. In this manner, high-density integration of valves ($10^6$ valves $cm^{-2}$) can be realized, as well as pneumatic pumps, which is based on three valves [53, 54]. PDMS is also gas permeable, which makes it an excellent material for long-term cell culture systems. However, since PDMS is a porous matrix at molecular level, it is incompatible with organic solvents as well, as it adsorbs small hydrophobic molecules and biomolecules.

### 3.5.2.2 Thermosets

Negative photoresists, such as SU-8 (a photosensitive epoxy resin) and polyimide, are mainly used for microfabrication molds on silicon wafers. However, lately, they have also been used to create microchannels. When heated and radiated, the thermosets crosslink to form rigid networks, which are stable at high temperatures and resistant against most solvents, while being optically transparent. It also allows for vertical sidewalls with high aspect ratios. With proper bonding, microfluidic devices can be entirely fabricated in thermosets [55]. A good bonding between SU8 and PDMS can be achieved by using a gas-phase silanization step of SU8, oxygen plasma treatment of PDMS, and heating under a slight pressure [56].

### 3.5.2.3 Thermoplastics

Unlike thermosets, thermoplastics can be reshaped after being cured. They can be reshaped multiple times by reheating around their glass transition temperature.

Typical thermoplastic used in microfluidics are PMMA, polycarbonate (PC), polystyrene (PS), polyethylene terephthalate (PET), and polyvinylchloride (PVC) [55]. Thermoplastics are normally sold solid as sheets of plastic, which can be laser cut or thermo-molded (or nanoimprinted). With nanoimprinting, it is easy to quickly make thousands of replicas at low cost but it requires a template of metal (depending on the feature size, it can be micromachined or fabricated using X-ray lithography) or silicon. This method is not very economical for prototyping. Thermoplastics will have to be thermobonded to other thermoplastics. Similar to PDMS, their surface can be covalently modified, but are more stable than PDMS. For example, an oxygen plasma-treated surface can maintain its hydrophilicity for a few years. Other perfluorinated polymers of interest are TeflonPFA (perfluoroalkoxy) and TeflonFEP (fluorinated ethylenepropylene) as they are extremely inert to chemicals and solvents, nonsticky, and antifouling. Furthermore, they are optically transparent, soft enough to make vales and moderately gas permeable. They can be thermo-molded at a high temperature (over 280 °C) [57].

### 3.5.3 Hydrogels

Hydrogels resemble the extra cellular matrix (ECM) in which cells are encapsulated and are thus widely used to embed cells [58]. Microfluidic channels can be fabricated in the hydrogels to deliver solutions, cells, or other substances [58, 59]. Hydrogels are a 3D network of hydrophilic polymer chains with more than 99% water content. They are highly porous allowing small molecules or particles to diffuse through. As such, hydrogels offer a function similar to natural vasculatures, allowing bulk 3D cell cultures [60]. Because of their low density, hydrogels only allow microscale features. There are two strategies of structuring hydrogels [61]: one is using direct laser writing method, and the other is the gelation of the hydrogel from a moving nozzle. When cells are added to the hydrogel, the latter is called bioprinting.

### 3.5.4 Paper

Lately, there has been an increasing interest in paper-based microfluidics. Paper is a highly porous matrix made of cellulose with excellent capillarity. By rendering certain areas of the paper hydrophobic, aqueous solution can be guided through the hydrophilic regions by capillary effect [62]. To obtain the hydrophilic regions, lithographic methods can apply a polymer solution to the paper [63]. Another method is to cut out the channels in parafilm, or other wax films, and press the pattern into a piece of paper with a hot press. Paper has a few advantages: it is cheap; it acts as a passive pump; paper can be stacked to make multilayered microfluidic systems [64]; and paper can filter out particles, such as remove blood cells from the blood. It also has a few disadvantages: the evaporation problem of liquid from open channels; high-density integration is hard because the minimum channel width is 200 μm; and the fact that only liquids with high-surface tensions can be used.

## 3.6 Integrated Sensors for Key Bioreactor Parameters

Sensing and control are crucial to obtain meaningful results in bioreactors. For example, the growth and productivity of microorganisms depend on the pH and temperature, while other factors such as dissolved oxygen and $CO_2$ concentration also play a significant role, at least in standard bioreactors.

Control and measurement on smaller scale become exceedingly more difficult due to the low concentration and volumes of products, be it metabolic or enzymatic products. A typical HPLC requires a sample of 50 μl, but the volume of a BRoC might be 100 nl. Thus, an essential requirement for microfluidic-based bioreactors is online sensors to measure the reaction parameters. The advantage of microtechnology is that it allows to integrate sensors on the chip. Various sensor techniques to measure the different parameters include electrochemical sensors, fluorescence sensor, quorum sensing, and bioluminescence sensors. Chapter 2.6 has described in more detail the working of some of the following sensors based on optodes. The problem with microbioreactors and optical sensors is the size. Filters might have to be used to separate the excitation and emission wavelengths from the different wavelengths needed to excite the different optodes or optical sensors. Optical fibers can be used to guide the light in small spaces [65–67].

### 3.6.1 Temperature

Due to the size of the BRoC, maintaining a constant temperature is not so difficult, if there is a big enough heat sink, compared with conventional systems, which can take a long time to raise the temperature. It takes nearly 10 min to raise the temperature in a 16 l vessel [68], but only a few minutes in a microbioreactor to raise the temperature by 10 °C [2]. A stable hotplate [27], water bath [5, 69], foil heater [70], on–off heater [71], Peltier heater with PID controller [29, 72], or placing the device in an incubator are some ways to keep a stable temperature. Also, heaters can be integrated on chip using microfabricated heaters [73, 74]. It should be noted that, due to its size heat, transfer is large and very rapid. Thus, the system should be placed in an area with no convection, such as a closed biosafety cabinet.

The most common way to measure the temperature on a chip is to use integrated thermistors or resistance temperature detectors (RTDs) [71]. In RTDs, the resistance of the sensor varies according to the temperature. They are usually made of platinum (e.g., Pt100 or Pt1000 sensors) but can also be made of gold. The measurement and control of the temperature are done off-chip and can be computer controlled.

### 3.6.2
### pH

Standard pH sensors cannot be used on chip, due to their size. Other methods that allow for pH measurements on chip include optical sensors on fluorescence spots (optodes) [5, 65, 67, 69, 75] or pH-sensitive metal ions, and solid-state ion-sensitive field effect transistors (ISFETs) [67, 73, 74]. However, optical sensors are preferred due to their noninvasive nature, ease of integration, and price, which lend itself well for disposable BRoCs. The optical pH sensors (such as the HP2A from PreSens) are excited with a square wave-modulated blue LED (465 nm) and measured by a lock-in amplifier to determine the phase shift, which correlates to the pH. The latest development is in nanosensors based on aluminum gallium nitrides (AlGaN/GaN) [76]. Compared with an ISFET, it exhibits superior chemical stability and biocompatibility, while having a favorable transparency in the visible spectrum.

### 3.6.3
### $O_2$

Dissolved oxygen concentration can also be optically measured using optodes [5, 65, 67, 69, 77, 78], which are based on the quenching of fluorescence by oxygen [79]. The optimal sensitivity of these optodes is at low concentrations of oxygen, which is relevant for a range of bioreactor processes. Sensor spots can also be fabricated by embedding platinum(II) octaethylporphyrine-ketone (PtOEPK) in polystyrene and immobilizing it on glass disks [79]. Another method is to place an oxygen sensor spot (such as the PSt3 from PreSens) at the bottom of a transparent microbioreactor and excite it with a square wave-modulated blue–green LED (505 nm). [65]. Again, with the aid of a lock-in amplifier, the phase shift between excitation and emission can be measured and correlated to the dissolved oxygen concentration. Other sensors include amperometric sensors, which can measure the oxygen concentration based on the electrochemical reduction of oxygen [73]. Another method for oxygen and pH sensing is by using a fluorescent PEG hydrogel microarray sensor using BCECF-dextran [80].

### 3.6.4
### $CO_2$

Also, $CO_2$ can be measured with an optical sensor based on a fluorescent dye on a silicone membrane [81].

### 3.6.5
### Cell Concentration (OD)

In bioreactors, there is a need to monitor the cell mass in real time, usually by means of optical methods such as Beer–Lambert law [82] or near infrared (NIR) [83]. Optical density can be calculated from a transmission measurement using an orange LED (600 nm), a collimating lens, and a photodetector [65]. Another method utilizes impedance spectroscopy [73, 84]. This method applies an alternate current across the bioreactor and measures the cell conductivity as a function of the frequency. Because only the membranes of living cells can be polarized, dead cells are excluded from the measurement. Methods to measure the cell density of *Escherichia coli* in microfluidic systems includes quorum sensing to observe the dynamics of the cell culture [53].

### 3.6.6 Humidity and Environment Stability

BRoCs fabricated in PDMS can have some problem with evaporation of the media when operating the device as a static culture. To avoid evaporation, the device should be placed in a high humidity environment. One option is to place the device (and all necessary pumps, control, and detection equipment in an incubator along with a few dishes of water. An incubator can also control the gas composition and the temperature needed. Another, much easier, way is to enclose the device in a small, airtight aluminum box with windows at the top and bottom for optical interrogation [65]. Again, placing dishes with water will ensure high humidity. As the interior volume of the box is large compared with the bioreactor chamber, gasses can be flowed in to control gas composition above the cell culture. Also, the aluminum box provides a large thermal mass to keep the temperature stable at the desired set point. The temperature can be controlled by using a water bath, to flow the water through the base of the box.

### 3.6.7 Oxygenation

Media should contain enough oxygen to sustain a cell culture in microfluidics. There are a few ways to accomplish this. One way not to execute this is by bubbling oxygen through the media before injecting it in or flowing this into the microbioreactor. This will cause bubbles inside the BRoCs, which will block the channel and lead to cell death. BRoCs fabricated in PDMS have some advantages with regard to gas transfer. PDMS has a high permeability to gasses, including oxygen and carbon dioxide [85]. Thus, oxygen can easily transfer an aeration membrane with a thickness of 100 μm between the bioreactor and gas chamber [29, 65]. In this instance, the membrane was placed directly above the cell culture chamber. Another way for aeration is through gas-permeable silicon tuning [67].

## 3.7 Model Organisms Applied to BRoCs

Various microbial cells are used in bioreactors, but not all have made it to BRoCs yet. Typical models include fermentation processes involving *E. coli* (Gram-negative bacteria) and *Saccharomyces cerevisiae* (yeast), which are used for protein production [86–88] and genomic studies [89, 90], and *Cyclotella cryptica* (algae) as a potential source for biofuel [91, 92]. Other models are based on mammalian cells, with the most important being the liver model. However, other organs are being rapidly included in an area called "organ-on-a-chip" (Table 3.1). Recently, there has been another development utilizing hESC cultures under steady-state and dynamic conditions [127]. A special case is the platelet BRoC [128]. The main focus here is to reproduce the bone marrow microenvironment to allow human-induced pluripotent stem cells-derived megakaryocytes to produce platelets. These platelets should generate clinically sufficient numbers of functional human platelets to offset risks associated with donated platelets procurement and storage.

But in recent years, what constitutes as bioreactor has been blurred. In general, bioreactors can be considered as devices that involve designed or programmed flow, where the flow is used to enhance molecular transport, provide mechanical stimulation, control addition of drugs and biological regulators, or otherwise to influence cell cultures that are otherwise not possible in static cultures [9].

## 3.8 Applications of Microfluidic Bioreactor Chip

In the following, we describe some examples of microfluidic-based bioreactors. Fermentation process can be run in different modes, such as batch modes, fed-batch, and continuous cultivations. Other applications include mammalian cells in microfluidic devices. The motivation in most of the applications is either to investigate the response of the cell culture for process development purposes or to capture the physiology and pathophysiology effects *in vitro* to satisfy therapeutic needs.

### 3.8.1 A Chemostat BRoC

A chemostat, or continuous bioreactor, is a bioreactor that is supplied with continues fresh medium, while culture liquid is constantly removed, keeping the volume of liquid in the bioreactor constant while including media mixing and online measurements of DO, pH, and OD (Figure 3.4a). Also, the cell biomass and product concentrations stay constant. By changing the flow rate of the medium into the reactor, the growth rate of the microorganism can be controlled. In a macro-bioreactor, both the inflow and outflow have to be equally controlled. However, in a microfluidic system, a chemostat would be a perfusion system in which the biomass and product concentrations stay constant. Due to the laminar behavior of the flow, anything that is flowed in is automatically flowed out of the microbioreactor, as no turbulent flow is possible. This also implies that the microbioreactor volume is constant. The only variable, apart from the media composition, will

**Table 3.1** Current development of organ-on-a-chip.

| Organ | Achievements | References |
|---|---|---|
| Brain | Modeling of the blood–brain barrier, using microslits or a membrane as barrier Measuring the tightness of the cell junctions with TEER | [93–95] |
| Breast | Migration of mammalian tumors into a 3D matrix | [96] |
| Eye | Retinal cells seeded on PLGA scaffolds | [97] |
| Heart | A pump powered by cardiomyocytes Real-time measurement of the reactive oxygen species and cardiac contraction | [98–101] |
| Intestine | Cultivating Caco-2 cells on a porous membrane exposed from two sides with media Emulating peristaltic bowel movement | [102–105] |
| Kidney | Shear stress and controlled topology *in vitro* result in a tube-like environment Renal cells cultured on a porous membrane with a fluidic compartment on both sides | [106–109] |
| Liver | Liver cells co-cultured with fibroblasts or microvascular endothelial cells Liver cells in chamber and a barrier between the flow and the hepatocyte culture | [27, 110–114] |
| Lung | Cyclic stretching with cultured human alveolar epithelial cells and human pulmonary microvascular endothelial cells on opposite sides of a porous PDMS membrane Fabricating lungs in microfluidics consisting of a multilayer of branched microvascular network Alveoli-like 3D scaffolds to study cancer cell migration | [46, 115–118] |
| Neuron | Spatial control that enables the control of a single domain in the neuron network Electrodes to measure the signals from single neurons | [119–124] |
| Pancreas | Isle of Langerhans cultured in device and exposed to glucose to measure the insulin response | [125, 126] |
| Skin | None | http://m.technologyreview.com/biomedicine/24384/ |

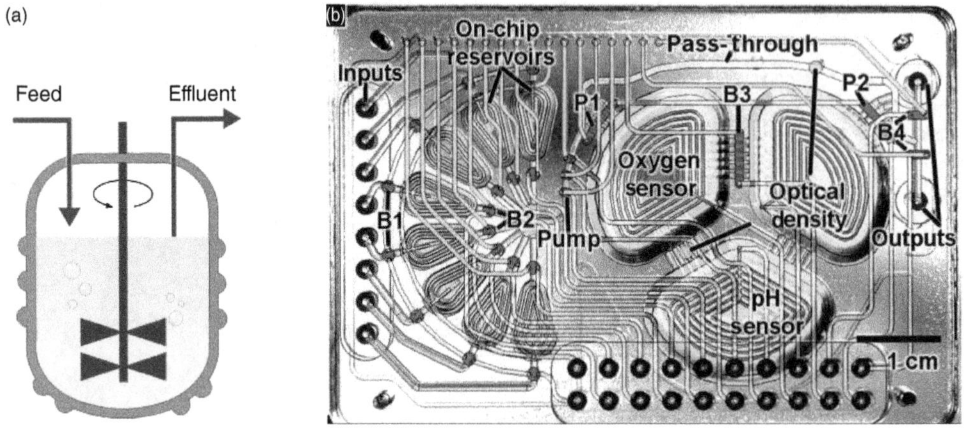


**Figure 3.4** (a) A schematic of a chemostat. It is a continuous-flow reactor, keeping the amount of medium inside the reactor constant. (b) False-color photograph of the continuous culture device with device components. B and P are individual on-chip blocking valves (from [72]).

be the flow rate of the medium, which in turn would determine the amount of nutrient directed to the reactor.

A microchemostat requires a balance between media inflow and cell growth, while avoiding wash-out or overpopulation of cells in the microbioreactor. Another problem in realizing a chemostat is the avoidance of cell growth on the reactor walls and avoiding chemotaxis of bacteria into nutrient-rich microfluidic channels. Cell growth on the walls of the reactor is a problem since these cells have physiological characteristics that are different from those growing in suspension [129]. This problem is cause by the high surface-to-volume ratio of microscale devices [130]. The reactors can be periodically cleaned by using a lysis buffer [53] or the reactor walls can be coated with poly(ethylene glycol)-grafted poly(acrylic acid) (PEG-PAA) copolymer films on PDMS and PMMA surfaces, effectively reducing cell adhesion and cell wall growth [67]. Chemotaxis can be prevented by having physical obstacles in the channel such as weirs or gratings, or by applying local heating at the input and output channel [67] as cells have the tendency to adverse chemotaxis toward high temperature [131, 132].

Various microchemostats have been proposed. Balagaddé *et al.* [53] demonstrated a microchemostat made of PDMS incorporating peristaltic pumps and valves. Here, about $10^4$ *E. coli* cells were cultured in chambers of 15 nl. Solutions were mixed in loops by using three PDMS valves to pump around the solution for a few seconds. Another perfused microfluidic device involves physically trapped cells in chambers of a few nanoliters [133]. Although, technically, this was a batch process, due to the absence of active mixing, constant growth conditions were obtained and were measured by monitoring the expression of fluorescent proteins [134]. Zhang *et al.* [67] introduced a PDMS-based membrane-aerated microchemostat with integrated optical sensors for OD, pH, and DO measurements. Active mixing was obtained by using a mini magnetic stir bar. The volume of 150 μl was determined to be ideal to be used for offline analysis such as HPLC and gene expression analysis [135]. Also, a nanoliter-scale turbidostat, a chemostat with feedback between turbidity and dilution rate, was developed [136]. However, this device had no control over the environmental parameters as no sensor were integrated in the small reactor. Furthermore, the cell removal was performed by diffusion with no direct connection between the growth chamber and the waste channels.

One system demonstrated an automated continuous culture device with a 1 ml working volume. It incorporated fast mixing (2 s), accurate flow control, a closed-loop temperature control, and integrated OD, DO, and pH sensors (Figure 3.4b) [72]. The integrated peristaltic pumps and valves control the input concentrations and allow the systems to be configured in different modes, such as batch, chemostat, and turbidostat continuous cultures.

### 3.8.2
### Using a BRoC as a Single-Cell Chemostat

Chemostats have also been used to measure the dynamics in single bacteria, observing the cell growth and GFP expression while maintaining a bacterial population of the same age [137, 138]. The main advantage for cell isolation is the ability to study single cells in the absence of communication. However, there is a need to have continuous nutrient supply to the isolated cells and excess cells should be removed from the system to maintain a steady-state population density. The cells were cultured in grooves such that the parental cells are removed from the channel at both ends by the flowing media. These single-cell chemostats can be fabricated in agarose casted against an intermediary PDMS mold (soft lithography) with submicron grooves [139]. Agarose has numerous benefits: less mechanical stress on the cells, a more uniform nutrient environment, and a better exchange of small molecules between the cells, compared with a PDMS substrate. A system such as the single-cell chemostat could be used to optimize the bacterial culture by selecting those cells with the best properties for the task at hand. Other single-cell systems include microcavities allowing for one cell per cavity [140], however, physical barriers might affect the cell behavior, and optical tweezers that allow to study, for example, the chemotactic motion of a single *E. coli* [141], however, a focused laser beam raises some concerns about local heating and irradiative photodamage to the cellular structure [142]. Growth analysis has shown that the doubling time in single-cell systems is around 42 min, while in bulk this is about 60 min for *E. coli* as single cells are not affected by population-level signaling or cell crowding [137].

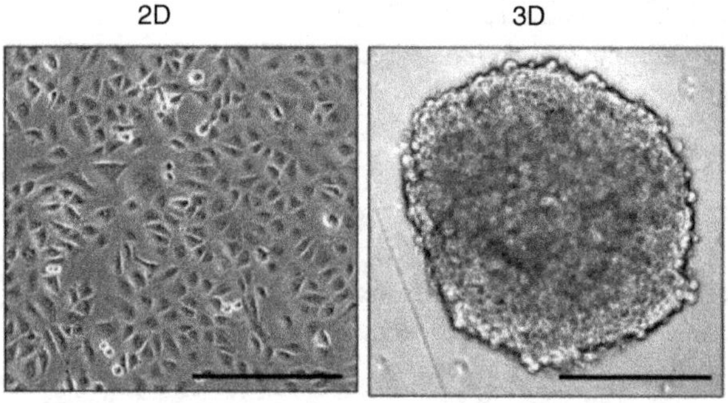


**Figure 3.5** An *ex vivo* tumor spheroid model. Microscopic images of monolayers and spheroids of human mesothelioma cell line NCI-H226. 2D, monolayers; 3D, spheroids. Scale: 400 μm. Microarray analysis revealed that 142 probe sets were differentially expressed between tumor spheroids and monolayers (from [145]).

### 3.8.3
### Mammalian Cells in the Bioreactor on a Chip

Microfluidic devices lend themselves well for mammalian cell cultures. The rise of microfluidic devices with mammalian cell cultures will serve as an invaluable link between *in vitro* and *in vivo* models, and thus reducing the number of animal studies and costs to develop new drugs.

Only one out of ten drugs entering the clinical trials reaches the approval stage [143, 144]. The main reason for this failure is the unforeseen lack of efficacy and toxicity [144]. As such, there is a need to improve *in vitro* assays to predict the efficacy and toxicity of drug candidates. The "golden standard" is still to evaluate cell-based drug toxicity and efficacy in microwell plates. However, the predictability of such assays is not satisfactory as they do not represent the human body in any aspect; only a single cell type is cultured at a time, without considering the typical cell–cell interaction found in tissues. In plates, cells have a 2D configuration, whereas in tissues, the cells are surrounded by an ECM and other supporting cells. Also, the microenvironment surrounding the plated 2D cell cultures lacks nutrient transport, shear stress, chemical, and mechanical signaling found in tissues. This 2D morphology does not reflect the *in vivo* state of the cells in 3D, having a different expression over 142 genes, altering the cell's function and response (Figure 3.5) [145]. This means that 2D cell cultures will intrinsically behave differently than the 3D cell cultures, which would also explain the fact of drug failure in human clinical trials. 3D cell cultures thus offer a more *in vivo*-like environment compared with 2D, both for mammalian and microbial cells. Microfluidic technology allows the implementation of a three-dimensional (3D) culture, which can mimic the *in vivo* state of cells [146]. Another method to create a 3D environment is to encapsulate the cells in hydrogels [147].

One aspect of 3D cell cultures, which should not be overlooked, is the limited transport of nutrients in and waste out of the culture. After a few days (3 days) of culture, the 3D cell culture is a dense structure of cells and ECM. Diffusive transport is predominant inside the 3D cell matrix, where often nutrients and oxygen are not able to reach the center of the culture. This causes cell necrosis at the center of the 3D cell culture. To overcome this problem, one solution is to limit the size of the 3D cell culture. A culture with a 100 μm distance from the edge of the culture to the center of it can maintain itself. The core concept here is to maintain in a tissue-like state by using microfluidic mass transport that mimics the tissue microvasculature [24]. This is where microfabrication can assist. The key feature is to have a narrow (<200 μm wide) cell chamber, which is exposed to the medium via porous walls. An array of micropillars has been fabricated, flanked by a medium-feeding channel on both sides (Figure 3.6a) [27]. Another method was to fabricate a microporous microfluidic barrier that formed a sieve-pocket to collect the cells during seeding [24]. A medium flow was maintained around the cell pocket (Figure 3.6b). These 3D cell culture devices are high-density cell culture (>2000 cells $mm^{-2}$), as opposed to cell suspensions. Using microfluidic traps can also control the 3D cultures. A trap facilitates cell capture and as such aggregation while offering efficient nutrient and gas exchange [148]. The size of the trap determines the size of the formed spheroid. Other ways to control the size of the spheroid include a centrifugal forced aggregation [149] or the use of microfabricated adhesive stencils [150].

Maintaining or even increasing the cell functions is obtained by 3D cell cultures and the appropriate soluble factors. Another approach to increase cell functions is by co-cultures. For example, when primary hepatocytes are co-cultured with non-parenchymal cells, there is a significant increase in hepatocyte functions compared with a monoculture of hepatocytes [151, 152]. Also, in co-cultures, a perfused culture has significantly higher metabolic production rates compared with static co-cultures [112]. In the case of a co-culture of human hepatocytes and non-parenchymal cells, the hepatocytes expressed bile canaliculi after the fourth day of culture [112].

The integration of conventional cell culture techniques with microfabrication technology resulted in "cells-on-a-chip" [153, 154]. With microfabrication, it is possible to reproduce multiorgan systems with blood circulation, the so-called "body-on-a-chip" (BoaC; see Section 3.8.4) [155]. Taking into account 3D cell cultures, such systems can mimic *in vivo* systems more realistically. As such, there has been an increase in 3D cell culture microfluidic systems [156, 157] and drug toxicity screening, especially involving hepatocytes [27, 158, 159]. By using gradient generators, it is possible to obtain a value for the $IC_{50}$ in one run of the system [32]. Also, by using arrays of $20 \times 50$ with hydrogel encapsulated cells could produce toxicity assay results in a high-throughput fashion [160] with $IC_{50}$ values comparable to conventional 96-well technology.

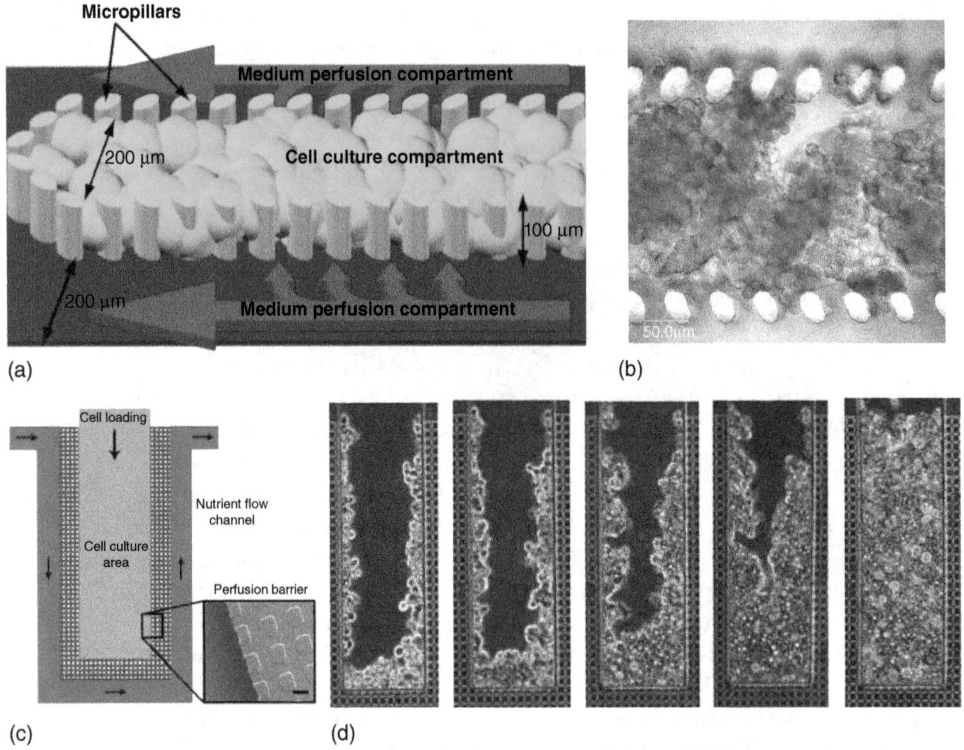


**Figure 3.6** The principle of a 3D cell culture chip. (a) A pillar array 200 μm with a 20 μm gap between the pillars retains the cells. On both sides of the pillar array, a medium is perfused (from [27]). (b) Scanning electron micrograph of pillar array. (c) Microfluidic culture unit design. The microfluidic unit consisted of three parts: a 150 μm wide by 440 μm long cell culture area (blue), a microfluidic perfusion barrier (gray), and a medium-flow channel (red). Cells were introduced from the top port and localized into the cell culture area. The perfusion barrier consisted of a grid of channels 5 μm wide and 2 μm long, serving to prevent cells from passing through, while enabling nutrient exchange from the flow channel on the opposite side. Inset shows scanning electron micrograph of the perfusion channels. Scale bar represents 5 μm. (d) HepG2/C3A human hepatoma cell growth inside the microfluidic cell culture array. Scale bar represents 100 μm (from [24]).

Other various 3D culture systems include spheroid cultures [161, 162], scaffolds [25, 163], hollow fibers [164], sandwich cultures [165, 166], and hydrogels [167].

### 3.8.4 Body-on-a-Chip Bioreactors

To achieve a model for multiorgan interaction *in vitro* requires seeding different cell types in one device with a precise control of the different microenvironments. The organs in our body work together to maintain homeostasis. Extracellular fluid, which circulates throughout the body, is a key component in maintaining a constant internal environment. Similar to our body, microorgans in separate compartments of a chip should be systematically connected. These systems are known as "body-on-a-chip," and depending on cell types used, "animal-on-a-chip" or "human-on-a-chip". BoaC consists of microorgan compartments linked through a microfluidic circulatory system mimicking the extracellular fluid[168]. Michael Shuler's group performed a groundbreaking pioneer work toward the BoaC concept by showing the possibility of a human surrogate to predict human response in clinical trials using a three-chamber (lung–liver–other organ) microscale cell culture analog ($\mu$CCA) device with a PBPK model[168]. Most recently, the same group demonstrated that a $\mu$CCA device comprising liver–tumor–marrow compartments can be used to predict the toxicity of an anticancer drug, 5-fluorouracil, with the aid of pharmacokinetics–pharmacodynamics (PK–PD) modeling[169].

An elegant way of fabricating a multiorgan chip is by the use of a direct 3D cell writing [170]. The bioprinter used a syringe to deposit 3D cell cultures in alginate layer by layer, which was later enclosed in a PDMS housing. Another multiple cell device has been developed by having small inner wells with various cell types embedded inside a larger well; termed "wells-inside-a-well" [171, 172]. This provides a simple, yet rudimentary interaction between different cell types without them being in contact. The different cell types shared a common cell culture medium.

However, a shared common culture medium is not favorable. Each cell type will need specific additions of soluble factors, such as growth factors, to the base medium. While some growth factors increase cell functionality, others inhibited certain cell types [173]. A way to create local microenvironments is to load gelatin microspheres with growth factors and seed them with the specific cell type in their specific microbioreactor. This was shown on a chip with four interconnected microreactors containing liver (C3A), lung (A549), kidney (HK-2), and fat cells (HDA) (Figure 3.7). It has been shown that there is little cross-talk between the different bioreactors, as the growth factors leached from the microspheres have the most effect in the specific microbioreactor in which they are located, as the concentration of the growth factor in the total media volume is very small [174].

### 3.8.5 Organ-on-a-Chip Bioreactor-Like Applications

When is a bioreactor with mammalian cells an organ-on-a-chip? The difference is the objective of the cell culture, although both try to optimize the functionality of the cells. Bioreactors tend to be used, historical, for fermentation or protein production (some kind of production), while organ-on-a-chip strives to mimic the structures of organs that than can be used as disease or toxicity models for drug screening. While bioreactors produce drugs, organ-on-a-chip tests drugs. However, since the borders have slightly blurred between these two systems, a short overview is given about the achievements on organs-on-a-chip.

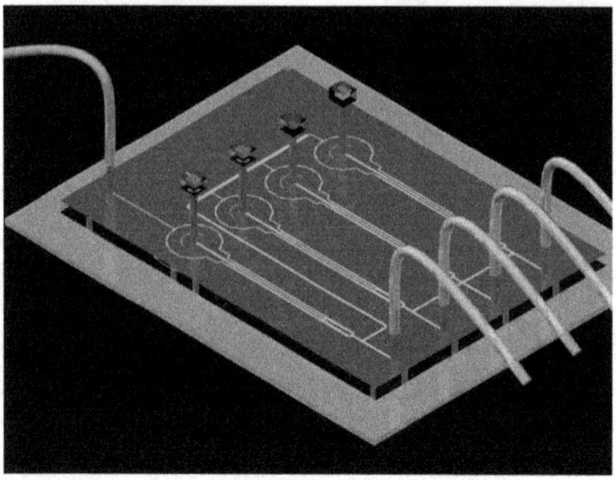

Figure 3.7 A human-on-a-chip with four-cell chambers used to culture multiple cell types with a base medium and cell-specific microenvironments created by gelatin microspheres loaded with soluble factors (from [173]).

An organ is a system with complex interactions between different cell types. A micropatterned co-culture with hepatocytes and fibroblasts could maintain the functionality of the hepatocytes for more than 4 weeks. The continuously diminishing functionality of pure hepatocyte cultures lasted only 2 weeks [110], as did a culture in a collagen/matrigel sandwich culture [114]. In another example, the pancreas is a system of at least five cell types, of which the $\alpha$- and $\beta$-cells are the most important [175]. Together, as an ensemble, they regulate a part of the glucose household in the blood. And again, the physical location of the cells plays an important role, as some cell signals have spatiotemporal limits. The distance between the two different cell types within a co-culture determines the fate of the soluble signal. This has been elegantly demonstrated by culturing hepatocytes and stroma cells in a micromechanical reconfigurable culture [176].

In anatomical terms, an organ is a collection of tissues joined together to perform a function [177]. Strictly speaking, a system should have at least two tissue types to qualify as an organ-on-a-chip. The combination of 3D cell cultures with microtechnology and the integration of chemical and mechanical cues have led to the development of more complex devices mimicking some essential traits of organs. Table 3.1 gives a summary on the work carried out so far on organs-on-a-chip. The microfluidic devices may not necessarily mimic the entire organ, but at least of some organ-specific functions.

## 3.9
## Scale Up

Now, the important question is whether it is possible scale up the processes with the parameters obtained from microfluidic-based bioreactors. It should be appreciated that cell cultures in industrial-sized stirred bioreactors will have a different behavior and cell functions than 3D cell cultures in microfluidics. Most probably cell cultures in large bioreactors have not reached their optimal cell functions, as cell–cell interaction is at a different level than that found in the microfluidic-based 3D cell cultures. Having said that, studies have been carried out to compare microscale bioreactors with bench-top bioreactors up to 10 l, which show that the behavior of bacteria is the same between microbioreactors and a 500 ml bench-scale bioreactor [65], at least when looking at factors such as the growth kinetics, dissolved oxygen, and pH over time, as well as the final number of cells and cell morphology. The question is whether this still holds true in production-scale bioreactors from 100 000 to 300 000 l.

To scale up microfluidic-based production, another route could be followed. We could create a batch of BRoCs in such parallel fashion so as to facilitate production on a larger scale, while preserving the optimal culture conditions as found in single BRoCs. This has been shown by constructing a high-density cell culture device, by stacking 10 layers of PMDS-based cell cultures of the liver cells HepG2 to achieve a cell density of $4 \times 10^7$ cells $cm^{-2}$. This is roughly the same density as in a macroscale bioreactor [178]. By measuring the glucose consumption and albumin production, it was shown the metabolic activities of the cells were maintained.

## 3.10
## Conclusion

At this time, there are some companies offering bench-scale systems with multiple stirred bioreactors with the possibility to add sensors. The “Multifors 2” operates six reactors in parallel (Infors HT, Switzerland), the “DASGIP® Parallel Bioreactor System for Microbial Applications” has up to eight culture vessels (DASGIP AG, Germany, now under Eppendorf). In the milliliter range, there are several companies offering bioreactors, such as Dasbox (Eppendorf), which can stack 24 or more bioreactors with volumes from 60 to 250 ml; the “MiniBio” with volumes of 250–1000 ml, as well as “micro-Matrix” with volumes from 0.2 to 5 ml and the ability to control 24 microbioreactors (Applikon Biotechnology, the Netherlands), or the 0.8–2 ml BioLector (M2P Labs, Germany), which controls 48 microbioreactors in parallel. Other parallel systems would be microtiter plates, which offer a great number of reactor wells and reduced working volumes, while integration of DO [179, 180] and pH [181] sensors have been shown. However, mixing is accomplished by shaking the whole plate, which complicates optical and fluidic integration to control fermentation parameters and medium evaporation.

The use of microfluidics for microbioreactors to culture mammalian or microbial cells is very promising. In particular, with the integration of analytical tools, which are now performed off-line, such as HPLC, more reliable systems will emerge. However, this is one of the obstacles that should be overcome. Miniaturization results in smaller sample sizes, down to picoliters, which makes it impossible to analyze using conventional analytical systems. OD, DO, and pH cell viability can readily be measured on chip, but metabolic products are not being measured on chip. For example, the pancreatic $\beta$-cell produces 80 ng of insulin per one million cells [182]. However, only 5000–10 000 cells are typically used in a chip.

Another challenge is the operation of microfluidic systems, especially when involving cell culture. It takes skill to run experiments on microfluidic devices, and every microfluidic chip has its own specificities. Automation will be key factor in producing user-friendly devices, which can be used by anyone. As microfluidic devices become more complicated, a fully controlled system is desired. A good example for a fully automated system is the microfluidic large scale integration (mLSI), a system that is comparable with integrated circuits in electronics [183]. Although on-chip integration is progressing, it has not reached the same level of control as a robotic liquid-handling system, which had similar issues before [184].

As a final point, automation and integration of the chip will result in better repeatability, as it does not depend on the user of the chip. To seamlessly connect with existing automated tools, microfluidics could be integrated into a robotic liquid-handling system [185].

## Abbreviations

| | |
|---|---|
| BCECF | 2′,7′-Bis-(2-carboxyethyl)-5-(and-6)-carboxyfluorescein |
| BoaC | body-on-a-chip |
| BRoC | bioreactor-on-a-chip |
| CE | capillary electrophoresis |
| DO | dissolved oxygen |
| DRIE | deep reactive ion etching |
| ECM | extra cellular matrix |
| FEP | fluorinated ethylenepropylene |
| GFP | green fluorescent protein |
| hESC | human embryonic stem cell |
| HF | hydrofluoric acid |
| HTS | high-throughput screening |
| OD | optical density |
| LED | light emitting diode |
| PAA | poly(acrylic acid) |
| PC | polycarbonate |
| PDMS | polydimethylsiloxane |
| PEG | poly(ethylene glycol) |
| PET | polyethylene terephthalate |
| PFA | perfluoroalkoxy |
| pH | measure of the acidity or basicity of an aqueous solution |
| PMMA | polymethylmethacrylate |
| PS | polystyrene |

PtOEPK platinum(II) octaethylporphine-ketone
PVC polyvinylchloride

## References


1. Novick, A. and Szilard, L. (1950) Description of the chemostat. *Science*, **112**, 715–716.
2. Schapper, D., Zainal Alam, M.N.H., Szita, N., Eliasson Lantz, A., and Gernaey, K.V. (2009) Application of microbioreactors in fermentation process development: a review. *Anal. Bioanal. Chem.*, **395**, 679–695.
3. Schmid, R.D. and Hammelehle, R. (2003) *Pocket Guide to Biotechnology and Genetic Engineering*, Wiley-Blackwell.
4. Shanks, J.V. and Stephanopoulos, G. (2000) Biochemical engineering—bridging the gap between gene and product. *Curr. Opin. Biotechnol.*, **11**, 169–170.
5. Boccazzi, P., Zhang, Z., Kurosawa, K., Szita, N., Bhattacharya, S., Jensen, K.F., and Sinskey, A.J. (2006) Differential gene expression profiles and real-time measurements of growth parameters in Saccharomyces cerevisiae grown in microliter-scale bioreactors equipped with internal stirring. *Biotechnol. Progr.*, **22**, 710–717.
6. Luni, C., Feldman, H.C., Pozzobon, M., De Coppi, P., Meinhart, C.D., and Elvassore, N. (2010) Microliter-bioreactor array with buoyancy-driven stirring for human hematopoietic stem cell culture. *Biomicrofluidics*, **4**, 034105.
7. Buchs, J. (2001) Introduction to advantages and problems of shaken cultures. *Biochem. Eng. J.*, **7**, 91–98.
8. Hegab, H.M., ElMekawy, A., and Stakenborg, T. (2013) Review of microfluidic microbioreactor technology for high-throughput submerged microbiological cultivation. *Biomicrofluidics*, **7**, 021502.
9. Ebrahimkhani, M.R., Neiman, J.A., Raredon, M.S., Hughes, D.J., and Griffith, L.G. (2014) Bioreactor technologies to support liver function in vitro. *Adv. Drug Delivery Rev.*, **69–70**, 132–157.
10. Fernekorn, U., Hampl, J., Weise, F., Augspurger, C., Hildmann, C., Klett, M., Läffert, A., Gebinoga, M., Weibezahn, K.-F., Schlingloff, G., Worgull, M., Schneider, M., and Schober, A. (2011) Microbioreactor design for 3-D cell cultivation to create a pharmacological screening system. *Eng. Life Sci.*, **11** (2), 133–139.
11. Koh, W.G. and Pishko, M.V. (2006) Fabrication of cell-containing hydrogel microstructures inside microfluidic devices that can be used as cell-based biosensors. *Anal. Bioanal. Chem.*, **385**, 1389–1397.
12. Kou, S., Pan, L., van Noort, D., Meng, G., Wu, X., Sun, H., Xu, J., and Lee, I. (2011) A multishear microfluidic device for quantitative analysis of calcium dynamics in osteoblasts. *Biochem. Biophys. Res. Commun.*, **408** (2), 350–355.
13. Torisawa, Y., Shiku, H., Yasukawa, T., Nishizawa, M., and Matsue, T. (2005) Multi-channel 3-D cell culture device integrated on a silicon chip for anticancer drug sensitivity test. *Biomaterials*, **26**, 2165–2172.
14. Albrecht, D.R., Underhill, G.H., Wassermann, T.B., Sah, R.L., and Bhatia, S.N. (2006) Probing the role of multicellular organization in three-dimensional microenvironments. *Nat. Methods*, **3**, 369–375.
15. Lu, H., Koo, L.Y., Wang, W.C.M., Lauffenburger, D.A., Griffith, L.G., and Jensen, K.F. (2004) Microfluidic shear devices for quantitative analysis of cell adhesion. *Anal. Chem.*, **76**, 5257–5264.
16. Takayama, S., McDonald, J.C., Ostuni, E., Liang, M.N., Kenis, P.J.A., Ismagilov, R.F., and Whitesides, G.M. (1999) Patterning cells and their environments using multiple laminar fluid flows in capillary networks. *Proc. Natl. Acad. Sci. U.S.A.*, **96**, 5545–5548.
17. Yang, S.T., Zhang, X., and Wen, Y. (2008) Microbioreactors for high

throughput cytotoxicity assays. *Curr. Opin. Drug Discovery Dev.*, **11**, 111–127.

18. Kim, B.S., Lee, S.C., Lee, S.Y., Chang, Y.K., and Chang, H.N. (2004) High cell density fed-batch cultivation of *Escherichia coli* using exponential feeding combined with pH-stat. *Bioprocess Biosyst. Eng.*, **26**, 147–150.
19. Lee, P.C., Lee, S.Y., and Chang, H.N. (2008) Cell recycled culture of succinic acid-producing *Anaerobiospirillum succiniciproducens* using an internal membrane filtration system. *J. Microbiol. Biotechnol.*, **18**, 1252–1256.
20. Walker, G.M., Zeringue, H.C., and Beebe, D.J. (2004) Microenvironment design considerations for cellular scale studies. *Lab Chip*, **4**, 91–97.
21. Korin, N., Bransky, A., Dinnar, U., and Levenberg, S. (2007) A parametric study of human fibroblasts culture in a microchannel bioreactor. *Lab Chip*, **7**, 611–617.
22. Kim, L., Vahey, M.D., Lee, H.Y., and Voldman, J. (2006) Microfluidic arrays for logarithmically perfused embryonic stem cell culture. *Lab Chip*, **6**, 394–406.
23. Ong, S.M., Zhang, C., Toh, Y.C., Kim, S.H., Foo, H.L., Tan, C.H., van Noort, D., Park, S., and Yu, H. (2008) A gel-free 3D microfluidic cell culture system. *Biomaterials*, **29**, 3237–3244.
24. Zhang, M.Y., Lee, P.J., Hung, P.J., Johnson, T., Lee, L.P., and Mofrad, M.R.K. (2008) Microfluidic environment for high density hepatocyte culture. *Biomed. Microdevices*, **10**, 117–121.
25. Park, J., Berthiaume, F., Toner, M., Yarmush, M.L., and Tilles, A.W. (2005) Microfabricated grooved substrates as platforms for bioartificial liver reactors. *Biotechnol. Bioeng.*, **90**, 632–644.
26. Roy, P., Baskaran, H., Tilles, A.W., Yarmush, M.L., and Toner, M. (2001) Analysis of oxygen transport to hepatocytes in a flat-plate microchannel bioreactor. *Ann. Biomed. Eng.*, **29**, 947–955.
27. Toh, Y.-C., Zhang, C., Zhang, J., Khong, Y.M., Chang, S., Samper, V.D., van Noort, D., Hutmacher, D.W., and Yu, H. (2007) A novel 3D mammalian cell perfusion-culture system in microfluidic channels. *Lab Chip*, **7**, 302–309.
28. Korin, N., Bransky, A., Dinnar, U., and Levenberg, S. (2009) Periodic "flow-stop" perfusion microchannel bioreactors for mammalian and human embryonic stem cell long-term culture. *Biomed. Microdevices*, **11**, 87–94.
29. Moore, S.K. and Kleis, S.J. (2008) Characterization of a novel miniature cell device. *Acta Astronaut.*, **62**, 632–638.
30. McDonald, J.C., Duffy, D.C., Anderson, J.R., Chiu, D.T., Wu, H., Schueller, O.J., and Whitesides, G.M. (2000) Fabrication of microfluidic systems in poly(dimethylsiloxane). *Electrophoresis*, **21** (1), 27–40.
31. Kenis, P.J.A. *et al.* (1999) Microfabrication inside capillaries using multiphase laminar flow patterning. *Science*, **285**, 83–85.
32. Toh, Y.-C., Lim, T.C., Tai, D., Xiao, G., van Noort, D., and Yu, H. (2009) A microfluidic 3D hepatocyte chip for drug toxicity testing. *Lab Chip*, **9**, 2026–2035.
33. Keenan, T.M. and Folch, A. (2008) Biomolecular gradients in cell culture systems. *Lab Chip*, **8**, 34–57.
34. Walker, G.M., Monteiro-Riviere, N., Rouse, J., and O'Neill, A.T. (2007) A linear dilution microfluidic device for cytotoxicity assays. *Lab Chip*, **7**, 226–232.
35. Tirella, A., Marano, M., Vozzi, F., and Ahluwalia, A. (2008) A microfluidic gradient maker for toxicity testing of bupivacaine and lidocaine. *Toxicol. In Vitro*, **22**, 1957–1964.
36. Camp, J.P. and Capitano, A.T. (2007) Induction of zone-like liver function gradients in HepG2 cells by varying culture medium height. *Biotechnol. Progr.*, **23**, 1485–1491.
37. Au, S.H., Shih, S.C., and Wheeler, A.R. (2011) Wheeler Integrated microbioreactor for culture and analysis of bacteria, algae and yeast. *Biomed. Microdevices*, **13**, 41–50.
38. Yamaguchi, H., Miyazaki, M., Honda, T., Briones-Nagata, M.P., Arima, K., and Maeda, H. (2009) Rapid and efficient proteolysis for proteomic analysis by

protease-immobilized microreactor. *Electrophoresis*, **30**, 3257–3264.

39. Jacobson, S.C., Hergenröder, R., Koutny, L.B., and Ramsey, J.M. (1994) Open channel electro-chromatography on a microchip. *Anal. Chem.*, **66**, 4184–4189.
40. Anderson, J.R., Chiu, D.T., McDonald, J.C., Jackman, R.J., Cherniavskaya, O., Wu, H., Whitesides, S., and Whitesides, G.M. (2000) Fabrication of topologically complex three-dimensional microfluidic systems in PDMS by rapid prototyping. *Anal. Chem.*, **72**, 3158–3164.
41. Shin, Y., Han, S., Jeon, J.S., Yamamoto, K., Zervantonakis, I.K., Sudo, R., Kamm, R.D., and Chung, S. (2012) Microfluidic assay for simultaneous culture of multiple cell types on surfaces or within hydrogels. *Nat. Protoc.*, **7**, 1247–1259.
42. Guijt, R.M. and Breadmore, M.C. (2008) Maskless photolithography using UV LEDs. *Lab Chip*, **23** (8), 1402–1404.
43. Lee, K.S., Lee, H.L., and Ram, R.J. (2007) Polymer waveguide backplanes for optical sensor interfaces in microfluidics. *Lab Chip*, **7**, 1539–1545.
44. van Noort, D. (2006) A poor man's microfluidic DNA computer, in *DNA Computing*, Lecture Notes in Computer Science, vol. **3892**, Springer, Berlin, Heidelberg, pp. 380–386.
45. Mu, X., Liang, Q.L., Hu, P., Ren, K.N., Wang, Y.M., and Luo, G.A. (2009) Laminar flow used as "liquid etch mask" in wet chemical etching to generate glass microstructures with an improved aspect ratio. *Lab Chip*, **9**, 1994–1996.
46. Huh, D., Matthews, B.D., Mammoto, A., Montoya-Zavala, M., Hsin, H.Y., and Ingber, D.E. (2010) Reconstituting organ-level lung functions on a chip. *Science*, **328**, 1662–1668.
47. Reyes, D.R., Iossifidis, D., Auroux, P.A., and Manz, A. (2002) Micro total analysis systems. 1. Introduction, theory, and technology. *Anal. Chem.*, **74**, 2623–2636.
48. Whitesides, G.M. (2006) The origins and the future of microfluidics. *Nature*, **442**, 368–373.
49. Berthier, E., Young, E.W.K., and Beebe, D. (2012) Engineers are from PDMS-land, biologists are from polystyrenia. *Lab Chip*, **12**, 1224–1237.
50. Sollier, E., Murray, C., Maoddi, P., and Di Carlo, D. (2011) Rapid prototyping polymers for microfluidic devices and high pressure injections. *Lab Chip*, **11**, 3752–3765.
51. Stroock, A.D. and Whitesides, G.M. (2003) Controlling flows in microchannels with patterned surface charge and topography. *Acc. Chem. Res.*, **36**, 597–604.
52. Unger, M.A., Chou, H.P., Thorsen, T., Scherer, A., and Quake, S.R. (2000) Monolithic microfabricated valves and pumps by multilayer soft lithography. *Science*, **288**, 113–116.
53. Balagadde, F.K., You, L., Hansen, C.L. *et al.* (2005) Long-term monitoring of bacteria undergoing programmed population control in a microchemostat. *Science*, **309**, 137–140.
54. Huang, B., Wu, H.K., Bhaya, D., Grossman, A., Granier, S., Kobilka, B.K., and Zare, R.N. (2007) Counting low-copy number proteins in a single cell. *Science*, **315**, 81–84.
55. Becker, H. and Gäertner, C. (2007) Polymer microfabrication technologies for microfluidic systems. *Anal. Bioanal. Chem.*, **390**, 89–111.
56. Talaei, S., Frey, O., van der Wal, P.D., de Rooij, N.F., and Koudelka-Hep, M. (2009) Hybrid microfluidic cartridge formed by irreversible bonding of SU-8 and PDMS for multi-layer flow applications. *Procedia Chem.*, **1**, 381–384.
57. Ren, K.N., Zheng, Y.Z., Dai, W., Ryan, D., Fung, C.Y., and Wu, H.K. (2010) Soft-lithography-based high temperature molding method to fabricate whole Teflon microfluidic chips, in 14th International Conference on Miniaturized Systems for Chemistry and Life Sciences, *Groningen, The Netherlands, October 3–7, 2010*, RSC Publishing, London, pp. 554–556.
58. Ghaemmaghami, A.M., Hancock, M.J., Harrington, H., Kaji, H., and Khademhosseini, A. (2012) Biomimetic

tissues on a chip for drug discovery. *Drug Discovery Today*, **17**, 173–181.

59. Domachuk, P., Tsioris, K., Omenetto, F.G., and Kaplan, D.L. (2010) Biomicrofluidics: biomaterials and biomimetic designs. *Adv. Mater.*, **22**, 249–260.
60. Choi, N.W., Cabodi, M., Held, B., Gleghorn, J.P., Bonassar, L.J., and Stroock, A.D. (2007) Microfluidic scaffolds for tissue engineering. *Nat. Mater.*, **6**, 908–915.
61. Huang, G.Y., Zhou, L.H., Zhang, Q.C., Chen, Y.M., Sun, W., Xu, F., and Lu, T.J. (2011) Microfluidic hydrogels for tissue engineering. *Biofabrication*, **3**, 012001.
62. Martinez, A.W., Phillips, S.T., Butte, M.J., and Whitesides, G.M. (2007) Patterned paper as a platform for inexpensive, low-volume, portable bioassays. *Angew. Chem. Int. Ed.*, **46**, 1318–1320.
63. Li, X., Ballerini, D.R., and Shen, W. (2012) A perspective on paper-based microfluidics: current status and future trends. *Biomicrofluidics*, **6**, 011301–1130113.
64. Derda, R., Tang, S.K.Y., Laromaine, A., Mosadegh, B., Hong, E., Mwangi, M., Mammoto, A., Ingber, D.E., and Whitesides, G.M. (2011) Multizone paper platform for 3D cell cultures. *PLoS One*, **6**, e18940.
65. Zanzotto, A., Szita, N., Boccazzi, P., Lessard, P., Sinskey, A.J., and Jensen, K.F. (2004) Membrane-aerated microbioreactor for high-throughput bioprocessing. *Biotechnol. Bioeng.*, **87** (2), 243–254.
66. Zanzotto, A., Boccazzi, P., Gorret, N., Van Dyk, T.K., Sinskey, A.J., and Jensen, K.F. (2006) In situ measurement of bioluminescence and fluorescence in an integrated microbioreactor. *Biotechnol. Bioeng.*, **93** (1), 40–47.
67. Zhang, Z., Boccazzi, P., Choi, H.-G., Perozziello, G., Sinskey, A.J., and Jensen, K.F. (2006) Microchemostat—microbial continuous culture in a polymer-based, instrumented microbioreactor. *Lab Chip*, **6**, 906–913.
68. Hua, M., Xuan, F., Tu, S.-T., Xia, C., Zhu, H., and Shao, H. (2011) Study of an efficient temperature measurement for an industrial bioreactor. *Measurement*, **44**, 875–880.
69. Szita, N., Bocazzi, P., Zhang, Z., Boyle, P., Sinskey, A.J., and Jensen, K.F. (2005) Development of a multiplexed microbioreactor system for high-throughput bioprocessing. *Lab Chip*, **5**, 819–826.
70. Lee, H.L., Bocazzi, P., Ram, R.J., and Sinskey, A.J. (2006) Microbioreactor arrays with integrated mixers and fluid injectors for high throughput experimentation with pH and dissolved oxygen control. *Lab Chip*, **6**, 1229–1235.
71. Muhd, N. H., Zainal, A., Amir, A., Amiri, M., and Abbas. K. (2014) Establishment of temperature control scheme for microbioreactor operation using integrated microheater. *Microsyst. Technol.*, doi: 10.1007/s00542-014-2088-9.
72. Lee, K.S., Boccazzi, P., Sinskey, A.J., and Ram, R.J. (2011) Microfluidic chemostat and turbidostat with flow rate, oxygen, and temperature control for dynamic continuous culture. *Lab Chip*, **11**, 1730–1739.
73. Krommenhoek, E.E., van Leeuwen, M., Gardeniers, H., van Gulik, W.M., van den Berg, A., Li, X., Ottens, M., van der Wielen, L.A.M., and Heijnen, J.J. (2008) Lab-scale fermentation tests of microchip with integrated electrochemical sensors for pH, temperature, dissolved oxygen and viable biomass concentration. *Biotechnol. Bioeng.*, **99** (4), 884–892.
74. Maharbiz, M.M., Holtz, W.J., Howe, R.T., and Keasling, J.D. (2004) Microbioreactor arrays with parametric control for high-throughput experimentation. *Biotechnol. Bioeng.*, **85**, 376–381.
75. Maiti, T.K. (2006) A novel lead-wire-resistance compensation technique using two-wire. resistance temperature detector. *IEEE Sens. J.*, **6** (6), 1454–1458.
76. Schober, A., Augspurger, C., Fernekorn, U., Weibezahn, K.-F., Schlingloff, G., Gebinoga, M., Worgull, M., Schneider, M., Hildmann, C., Weise, F., Hampl, J., Silveira, L., Cimalla, I., and Lübbers, B. (2010) Microfluidics and biosensors as

tools for NanoBioSystems research with applications in the "Life Science". *Mater. Sci. Eng., B*, **169**, 174–181.

77. van Leeuwen, M., Heijnen, J.J., Gardeniers, H., Oudshoorn, A., Noorman, H., Visser, J., van der Wielen, L.A.M., and van Gulik, W.M. (2009) A system for accurate on-line measurement of total gas consumption or production rates in microbioreactors. *Chem. Eng. Sci.*, **64**, 455–458.
78. Zhang, Z., Perozziello, G., Boccazzi, P., Sinskey, A.J., Geschke, O., and Jensen, K.F. (2007) Microbioreactors for bioprocess development. *J. Assoc. Lab. Autom.*, **12** (3), 143–151.
79. Papkovsky, D.B., Ponomarev, G.V., Trettnak, W., and O'Leary, P. (1995) Phosphorescent complexes of porphyrin ketones: optical properties and application to oxygen sensing. *Anal. Chem.*, **67**, 4112–4117.
80. Lee, S. *et al.* (2008) Measurement of pH and dissolved oxygen within cell culture media using a hydrogel microarray sensor. *Sens. Actuators B*, **128**, 388–398.
81. Liebsch, G., Klimant, I., Frank, B., Holst, G., and Wolfbeis, O.S. (2000) Luminescence lifetime imaging of oxygen, pH, and carbon dioxide distribution using optical sensors. *Appl. Spectrosc.*, **54** (4), 548–559.
82. Jensen, K.H., Alam, M.N., Scherer, B., Lambrecht, A., and Mortensen, N.A. (2008) Slow-light enhanced light-matter interactions with applications to gas sensing. *Opt. Commun.*, **281**, 5335–5339.
83. Cervera, A.E., Petersen, N., Eliasson Lantz, A., Larsen, A., and Gernaey, K.V. (2009) Application of near-infrared spectroscopy for monitoring and control of cell culture and fermentation. *Biotechnol. Progr.*, **25** (6), 1561–1581, doi: 10.1021/bp.280.
84. Krommenhoek, E.E., Gardeniers, J.G.E., Bomer, J.G., Li, X., Ottens, M., van Dedem, G.W.K., van Leeuwen, M., van Gulik, W.M., van der Wielen, L.A.M., Heijnen, J.J., and van den Berg, A. (2007) Integrated electrochemical sensor array for on-line monitoring of yeast fermentations. *Anal. Chem.*, **79** (15), 5567–5573.
85. Merkel, T.C., Bondar, V.I., Nagai, K., Freeman, B.D., and Pinnau, I. (2000) Gas sorption, diffusion, and permeation in poly(dimethylsiloxane). *J. Polym. Sci., Part B: Polym. Phys.*, **38** (3), 415–434.
86. Cereghino, G.P.L. and Cregg, J.M. (1999) Applications of yeast in biotechnology: protein production and genetic analysis. *Curr. Opin. Biotechnol.*, **10**, 422–427.
87. Demain, A.L. and Adrio, J.L. (2008) Contributions of microorganisms to industrial biology. *Mol. Biotechnol.*, **38**, 41–55.
88. Swartz, J.R. (2001) Advances in *Escherichia coli* production of therapeutic proteins. *Curr. Opin. Biotechnol.*, **12**, 195–201.
89. Koonin, E.V. and Galperin, M.Y. (1997) Prokaryotic genomes: the emerging paradigm of genome-based microbiology. *Curr. Opin. Genet. Dev.*, **7**, 757–763.
90. Piškur, J. and Langkjær, R.B. (2004) Yeast genome sequencing: the power of comparative genomics. *Mol. Microbiol.*, **53**, 381–389.
91. Chisti, Y. (2008) Biodiesel from microalgae beats bioethanol. *Trends Biotechnol.*, **26**, 126–131.
92. Yu, E., Zendejas, F., Lane, P., Gaucher, S., Simmons, B., and Lane, T. (2009) Triacylglycerol accumulation and profiling in the model diatoms *Thalassiosira pseudonana* and *Phaeodactylum tricornutum* (Baccilariophyceae) during starvation. *J. Appl. Phycol.*, **21**, 669–681.
93. Griep, L.M., Wolbers, F., de Wagenaar, B., ter Braak, P.M., Weksler, B.B., Romero, I.A., Couraud, P.O., Vermes, I., van der Meer, A.D., and van den Berg, A. (2013) BBB on chip: microfluidic platform to mechanically and biochemically modulate blood-brain barrier function. *Biomed. Microdevices*, **15**, 145–150.
94. Prabhakarpandian, B., Shen, M.-C., Nichols, J.B., Mills, I.R., Sidoryk-Wegrzynowicz, M., Aschner, M., and Pant, K. (2013) SyM-BBB: a

microfluidic Blood Brain Barrier model. *Lab Chip*, **13**, 1093–1101.

95. Yeon, J.H., Na, D., Choi, K., Ryu, S.-W., Choi, C., and Park, J.-K. (2012) Reliable permeability assay system in a microfluidic device mimicking cerebral vasculatures. *Biomed. Microdevices*, **14**, 1141–1148.
96. Liu, T., Li, C., Li, H., Zeng, S., Qin, J., and Lin, B. (2009) A microfluidic device for characterizing the invasion of cancer cells in 3-D matrix. *Electrophoresis*, **30**, 4285–4291.
97. McUsic, A.C., Lamba, D.A., and Reh, T.A. (2012) Guiding the morphogenesis of dissociated newborn mouse retinal cells and hES cell-derived retinal cells by soft lithography-patterned microchannel PLGA scaffolds. *Biomaterials*, **33**, 1396–1405.
98. Agarwal, A., Goss, J.A., Cho, A., McCain, M.L., and Parker, K.K. (2013) Microfluidic heart on a chip for higher throughput pharmacological studies. *Lab Chip*, **13**, 3599–3608.
99. Cheah, L.-T., Dou, Y.-H., Seymour, A.-M.L., Dyer, C.E., Haswell, S.J., Wadhawan, J.D., and Greenman, J. (2010) Microfluidic perfusion system for maintaining viable heart tissue with real-time electrochemical monitoring of reactive oxygen species. *Lab Chip*, **10**, 2720–2726.
100. Park, J., Kim, I.C., Baek, J., Cha, M., Kim, J., Park, S., Lee, J., and Kim, B. (2007) Micro pumping with cardiomyocyte–polymer hybrid. *Lab Chip*, **7**, 1367–1370.
101. Tanaka, Y., Sato, K., Shimizu, T., Yamato, M., Okano, T., and Kitamori, T. (2007) A micro-spherical heart pump powered by cultured cardiomyocytes. *Lab Chip*, **7**, 207–212.
102. Esch, M.B., Sung, J.H., Yang, J., Yu, C., Yu, J., March, J.C., and Shuler, M.L. (2012) On chip porous polymer membranes for integration of gastrointestinal tract epithelium with microfluidic 'body-on-a-chip' devices. *Biomed. Microdevices*, **14**, 895–906.
103. Imura, Y., Asano, Y., Sato, K., and Yoshimura, E. (2009) A microfluidic system to evaluate intestinal absorption. *Anal. Sci.*, **25**, 1403–1407.
104. Kim, H.J., Huh, D., Hamilton, G., and Ingber Human, D.E. (2012) Human gut-on-a-chip inhabited by microbial flora that experiences intestinal peristalsis-like motions and flow. *Lab Chip*, **12**, 2165–2174.
105. Kimura, H., Yamamoto, T.I., Sakai, H., Sakai, Y., and Fujii, T. (2008) An integrated microfluidic system for long-term perfusion culture and on-line monitoring of intestinal tissue models. *Lab Chip*, **8**, 741–746.
106. Frohlich, E.M., Zhang, X., and Charest, J.L. (2012) The use of controlled surface topography and flow-induced shear stress to influence renal epithelial cell function. *Integr. Biol.*, **4**, 75–83.
107. Gao, X., Tanaka, Y., Sugii, Y., Mawatari, K., and Kitamori, T. (2011) Basic structure and cell culture condition of a bioartificial renal tubule on chip towards a cell-based separation microdevice. *Anal. Sci.*, **27**, 907–912.
108. Jang, K.-J. and Suh, K.-Y. (2010) A multi-layer microfluidic device for efficient culture and analysis of renal tubular cells. *Lab Chip*, **10**, 36–42.
109. Sciancalepore, A.G., Sallustio, F., Girardo, S., Passione, L.G., Camposeo, A., Mele, E., Di Lorenzo, M., Costantino, V., Schena, F.P., and Pisignano, D. (2014) A bioartificial renal tubule device embedding human renal stem/progenitor cells. *PLoS One*, **9**, e87496.
110. Khetani, S.R. and Bhatia, S.N. (2008) Microscale culture of human liver cells for drug development. *Nat. Biotechnol.*, **26**, 120–126.
111. Lee, P.J., Hung, P.J., and Lee, L.P. (2007) An artificial liver sinusoid with a microfluidic endothelial-like barrier for primary hepatocyte culture. *Biotechnol. Bioeng.*, **97**, 1340–1346.
112. Novik, E., Maguire, T.J., Chao, P., Cheng, K.C., and Yarmush, M.L. (2010) A microfluidic hepatic coculture platform for cell-based drug metabolism studies. *Biochem. Pharmacol.*, **79**, 1036–1044.
113. Sudo, R., Chung, S., Zervantonakis, I.K., Vickerman, V., Toshimitsu, Y., Griffith, L.G., and Kamm, R.D. (2009) Transport-mediated angiogenesis in

3D epithelial coculture. *FASEB J.*, **23**, 2155–2164.

114. Ukairo, O., Kanchagar, C., Moore, A., Shi, J., Gaffney, J., Aoyama, S., Rose, K., Krzyzewski, S., McGeehan, J., Andersen, M.E., Khetani, S.R., and Lecluyse, E.L. (2013) Long-term stability of primary rat hepatocytes in micropatterned cocultures. *J. Biochem. Mol. Toxicol.*, **11** (3), 204–212.

115. Huh, D., Leslie, D.C., Matthews, B.D., Fraser, J.P., Jurek, S., Hamilton, G.A., Thorneloe, K.S., McAlexander, M.A., and Ingber, D.E. (2012) A human disease model of drug toxicity-induced pulmonary edema in a lung-on-a-chip microdevice. *Sci. Transl. Med.*, **4**, 159ra147.

116. Kniazeva, T., Hsiao, J.C., Charest, J.L., and Borenstein, J.T. (2011) A microfluidic respiratory assist device with high gas permeance for artificial lung applications. *Biomed. Microdevices*, **13**, 315–323.

117. Kniazeva, T., Epshteyn, A.A., Hsiao, J.C., Kim, E.S., Kolachalama, V.B., Charest, J.L., and Borenstein, J.T. (2012) Performance and scaling effects in a multilayer microfluidic extracorporeal lung oxygenation device. *Lab Chip*, **12**, 1686–1695.

118. Sun, Y.S., Peng, S.-W., Lin, K.H., and Cheng, J.Y. (2012) Electrotaxis of lung cancer cells in ordered three-dimensional scaffolds. *Biomicrofluidics*, **6** (1), 014102–14.

119. Gebinoga, M.l., Mai, P., Donahue, M., Kittler, M., Cimalla, I.A., Lübbers, B., Klett, M., Lebedev, V., Silveira, L., Singh, S., and Schober, A. (2012) Nerve cell response to inhibitors recorded with an aluminum-galliumnitride/galliumnitride field-effect transistor. *J. Neurosci. Methods*, **206**, 195–199.

120. Moraes, C., Labuz, J.M., Leung, B.M., Inoue, M., Chun, T.H., and Takayama, S. (2013) On being the right size: scaling effects in designing a human-on-a-chip. *Integr. Biol.*, **5**, 1149–1161.

121. Reyes, D.R., Perruccio, E.M., Becerra, S.P., Locascio, L.E., and Gaitan, M. (2004) Micropatterning neuronal cells on polyelectrolyte multilayers. *Langmuir*, **20**, 8805–8811.

122. Takayama, Y., Kotake, N., Haga, T., Suzuki, T., and Mabuchi, K. (2012) Formation of one-way-structured cultured neuronal networks in microfluidic devices combining with micropatterning techniques. *J. Biosci. Bioeng.*, **114**, 92–95.

123. Tang, K.C., Reboud, J., Kwok, Y.L., Peng, S.L., and Yobas, L. (2010) Lateral patch-clamping in a standard 1536-well microplate format. *Lab Chip*, **10**, 1044–1050.

124. Zeck, G. and Fromherz, P. (2001) Noninvasive neuroelectronic interfacing with synaptically connected snail neurons immobilized on a semiconductor chip. *Proc. Natl. Acad. Sci. U.S.A.*, **28**, 10457–10462.

125. Lee, D., Wang, Y., Mendoza-Elias, J.E., Adewola, A.F., Harvat, T.A., Kinzer, K., Gutierrez, D., Qi, M., Eddington, D.T., and Oberholzer, J. (2012) Dual microfluidic perfusion networks for concurrent islet perfusion and optical imaging. *Biomed. Microdevices*, **14**, 7–16.

126. Wang, Y., Lee, D., Zhang, L., Jeon, H., Mendoza-Elias, J.E., Harvat, T.A., Hassan, S.Z., Zhou, A., Eddington, D.T., and Oberholzer, J. (2012) Systematic prevention of bubble formation and accumulation for long-term culture of pancreatic islet cells in microfluidic device. *Biomed. Microdevices*, **14** (2), 419–426.

127. Cimetta, E., Figallo, E., Cannizzaro, C., Elvassore, N., and Vunjak-Novakovic, G. (2009) Micro-bioreactor arrays for controlling cellular environments: design principles for human embryonic stem cell applications. *Methods*, **47**, 81–89.

128. Thon, J.N., Mazutis, L., Wu, S., Sylman, J.L., Ehrlicher, A., Machlus, K.R., Feng, Q., Lu, S., Lanza, R., Neeves, K.B., Weitz, D.A., and Italiano, J.E. Jr., (2014) Platelet bioreactor-on-a-chip. *Blood*, **124** (12), 1857–1867.

129. Pilyugin, S.S. and Waltman, P. (1999) The simple chemostat with wall growth. *SIAM J. Appl. Math.*, **59**, 1552–1572.

130. Boccazzi, P., Zanzotto, A., Szita, N., Bhattacharya, S., Jensen, K.F., and Sinskey, A.J. (2005) Gene expression analysis of *Escherichia coli* grown in miniaturized bioreactor platforms for high-throughput analysis of growth and genomic data. *Appl. Microbiol. Biotechnol.*, **68**, 518–532.
131. Adler, J. (1976) The sensing of chemicals by bacteria. *Sci. Am.*, **234**, 40–47.
132. Maeda, K., Imae, Y., Shioi, J.I., and Oosawa, F. (1976) Effect of temperature on motility and chemotaxis of *Escherichia coli*. *J. Bacteriol.*, **127**, 1039–1046.
133. Groisman, A., Lobo, C., Cho, H., Campbell, J.K., Dufour, Y.S., Stevens, A.M., and Levchenko, A. (2005) A microfluidic chemostat for experiments with bacterial and yeast cells. *Nat. Methods*, **2**, 685–689.
134. Balaban, N.Q. (2005) Szilard's dream. *Nat. Methods*, **2**, 648–649.
135. Lidstrom, M.E. and Meldrum, D.R. (2003) Life-on-a-chip. *Nat. Rev. Microbiol.*, **1**, 158–164.
136. Luo, X., Shen, K., Luo, C., Ji, H., Ouyang, Q., and Chen, Y. (2010) An automatic microturbidostat for bacterial culture at constant density. *Biomed. Microdevices*, **12**, 499–503.
137. Johnson-Chavarria, E.M., Agrawal, U., Tanyeri, M., Kuhlman, T.E., and Schroeder, C.M. (2014) Automated single cell microbioreactor for monitoring intracellular dynamics and cell growth in free solution. *Lab Chip.*, **14** (15), 2688–2697, doi: 10.1039/c4lc00057a
138. Long, Z., Nugent, E., Javer, A., Cicuta, P., Sclavi, B., Cosentino Lagomarsino, M., and Dorfman, K.D. (2013) Microfluidic chemostat for measuring single cell dynamics in bacteria. *Lab Chip*, **13** (5), 947–954.
139. Moffitt, J.R., Lee, J.B., and Cluzel, P. (2012) The single-cell chemostat: an agarose-based, microfluidic device for high-throughput, single-cell studies of bacteria and bacterial communities. *Lab Chip*, **12** (8), 1487–1494.
140. Meier, B., Zielinski, A., Weber, C., Arcizet, D., Youssef, S., Franosch, T., Rädler, J.O., and Heinrich, D. (2011) Chemotactic cell trapping in controlled alternating gradient fields. *Proc. Natl. Acad. Sci. U.S.A.*, **108**, 11417–11422.
141. Min, T.L., Mears, P.J., Chubiz, L.M., Rao, C.V., Golding, I., and Chemla, Y.R. (2009) High-resolution, long-term characterization of bacterial motility using optical tweezers. *Nat. Methods*, **6**, 831–835.
142. Neuman, K.C., Chadd, E.H., Liou, G.F., Bergman, K., and Block, S.M. (1999) Characterization of photodamage to *Escherichia coli* in optical traps. *Biophys. J.*, **77**, 2856–2863.
143. Dingemanse, J. and Appel-Dingemanse, S. (2007) Integrated pharmacokinetics and pharmacodynamics in drug development. *Clin. Pharmacokinet.*, **46**, 713–737.
144. Kola, I. and Landis, J. (2004) Can the pharmaceutical industry reduce attrition rates? *Nat. Rev. Drug Discovery*, **3**, 711–715.
145. Kim, H., Phung, Y., and Ho, M. (2012) Changes in global gene expression associated with 3D structure of tumors: an ex vivo matrix-free mesothelioma spheroid model. *PLoS One*, **7** (6), e39556.
146. Abbott, A. (2003) Cell culture: biology's new dimension. *Nature*, **424**, 870–872.
147. McGuigan, A.P., Bruzewicz, D.A., Glavan, A., Butte, M.J., and Whitesides, G.M. (2008) Cell encapsulation in sub-mm sized gel modules using replica molding. *PLoS One*, **3**, e2258.
148. Khoury, M., Bransky, A., Korin, N., Konak, L.C., Enikolopov, G., Tzchori, I., and Levenberg, S. (2010) A microfluidic traps system supporting prolonged culture of human embryonic stem cells aggregates. *Biomed. Microdevices*, **12**, 1001–1008.
149. Ungrin, M.D., Joshi, C., Nica, A., Bauwens, C., and Zandstra, P.W. (2008) Reproducible, ultra high-throughput formation of multicellular organization from single cell suspension-derived human embryonic stem cell aggregates. *PLoS One*, **3** (2), e1565.
150. Park, J., Cho, C.H., Parashurama, N., Li, Y., Berthiaume, F., Toner, M. *et al.* (2007) Microfabrication-based modulation of embryonic

stem cell differentiation. *Lab Chip*, 7 (8), 1018–1028.

151. Bhatia, S.N., Balis, U.J., Yarmush, M.L., and Toner, M. (1998) Microfabrication of hepatocyte/fibroblast co-cultures: role of homotypic cell interactions. *Biotechnol. Progr.*, **14**, 378–387.

152. Zeilinger, K., Sauer, I.M., Pless, G., Strobel, C., Rudzitis, J., Wang, A. *et al.* (2002) Three dimensional co-culture of primary human liver cells in bioreactors for in vitro drug studies: effects of the initial cell quality on the long-term maintenance of hepatocyte-specific functions. *Altern. Lab. Anim.*, **30**, 525–538.

153. El-Ali, J., Sorger, P.K., and Jensen, K.F. (2006) Cells on chips. *Nature*, **442**, 403–411.

154. Park, T.H. and Shuler, M.L. (2003) Integration of cell culture and microfabrication technology. *Biotechnol. Progr.*, **19**, 243–253.

155. Khamsi, R. (2005) Labs on a chip: meet the stripped down rat. *Nature*, **435**, 12–13.

156. Kim, M.S., Yeon, J.H., and Park, J.K. (2007) A microfluidic platform for 3-dimensional cell culture and cell-based assays. *Biomed. Microdevices*, **9**, 25–34.

157. Kim, M.S., Lee, W., Kim, Y.C., and Park, J.K. (2008) Microvalve-assisted patterning platform for measuring cellular dynamics based on 3D cell culture. *Biotechnol. Bioeng.*, **101**, 1005–1013.

158. Gomez-Lechon, M.J., Donato, M.T., Castell, J.V., and Jover, R. (2003) Human hepatocytes as a tool for studying toxicity and drug metabolism. *Curr. Drug Metab.*, **4**, 292–312.

159. Vignati, L., Turlizzi, E., Monaci, S., Grossi, P., Kanter, R., and Monshouwer, M. (2005) An in vitro approach to detect metabolite toxicity due to CYP3A4-dependent bioactivation of xenobiotics. *Toxicology*, **216**, 154–167.

160. Lee, M.Y., Kumar, R.A., Sukumaran, S.M., Hogg, M.G., Clark, D.S., and Dordick, J.S. (2008) Three-dimensional cellular microarray for high-throughput toxicology assays. *Proc. Natl. Acad. Sci. U.S.A.*, **105**, 59–63.

161. Kunz-Schughart, L.A., Freyer, J.P., Hofstaedter, F., and Ebner, R. (2004) The use of 3-D cultures for high-throughput screening: the multicellular spheroid model. *J. Biomol. Screen.*, **9**, 273–285.

162. Ong, S.M., Zhao, Z., Arooz, T., Zhao, D., Zhang, S., Du, T., Wasser, M., van Noort, D., and Yu, H. (2010) Engineering a scaffold-free 3D in vitro tumor model for drug penetration studies. *Biomaterials*, **31**, 1180–1190.

163. Gottwald, E., Giselbrecht, S., Augspurger, C., and Lahni, B. (2007) A chip-based platform for the in vitro generation of tissues in three dimensional organization. *Lab Chip*, **7**, 777–785.

164. Miyoshi, H., Ehashi, T., Kawai, H., Ohshima, N., and Suzuki, S. (2010) Three-dimensional perfusion cultures of mouse and pig fetal liver cells in a packed-bed reactor: effect of medium flow rate on cell numbers and hepatic functions. *J. Biotechnol.*, **148**, 226–232.

165. Xia, L., Ng, S., Han, R., Tuo, X. *et al.* (2009) Laminar-flow immediate-overlay hepatocyte sandwich perfusion system for drug hepatotoxicity testing. *Biomaterials*, **30**, 5927–5936.

166. Zhang, S., Xia, L., Kang, C.H., Xiao, G., Ong, S.M., Toh, Y.C., Leo, H.L., van Noort, D., Kan, S.H., and Yu, H. (2008) Microfabricated silicon nitride membranes for hepatocyte sandwich culture. *Biomaterials*, **29**, 3993–4002.

167. Nicodemus, G.D. and Bryant, S.J. (2008) Cell encapsulation in biodegradable hydrogels for tissue engineering applications. *Tissue Eng. Part B Rev.*, **14**, 149–165.

168. Viravaidya, K., Sin, A., and Shuler, M.L. (2004) Development of a microscale cell culture analog to probe naphthalene toxicity. *Biotechnol. Progr.*, **20**, 316–323.

169. Sung, J.H., Kam, C., and Shuler, M.L. (2010) A microfluidic device for a pharmacokinetic-pharmacodynamic (PK-PD) model on a chip. *Lab Chip*, **10**, 446–455.

170. Chang, R., Nam, J., and Sun, W. (2008) Direct cell writing of 3D microorgan for in vitro pharmacokinetic

model. *Tissue Eng. Part C Methods*, **14**, 157–166.
171. Li, A.P., Bode, C., and Sakai, Y. (2004) A novel in vitro system, the integrated discrete multiple organ cell culture (IdMOC) system, for the evaluation of human drug toxicity: comparative cytotoxicity of tamoxifen towards normal human cells from five major organs and MCF-7 adenocarcinoma breast cancer cells. *Chem. Biol. Interact.*, **150**, 129–136.
172. Li, A.P. (2008) In vitro evaluation of human xenobiotic toxicity: scientific concepts and the novel integrated discrete multiple cell co-culture (IdMOC) technology. *ALTEX*, **25**, 43–49.
173. Zhang, C., Chia, S.M., Ong, S.M., Zhang, S., Toh, Y.-C., van Noort, D., and Yu, H. (2009) The controlled presentation of TGF-beta1 to hepatocytes in a 3D-microfluidic cell culture system. *Biomaterials*, **30**, 3847–3853.
174. Zhang, C., Zhao, Z., Abdul Rahim, N.A., van Noort, D., and Yu, H. (2009) Towards a human-on-chip: culturing multiple cell types on a chip with compartmentalized microenvironments. *Lab Chip*, **9**, 3185–3192.
175. Elayat, A.A., el-Naggar, M.M., and Tahir, M. (1995) An immunocytochemical and morphometric study of the rat pancreatic islets. *J. Anat.*, **186**, 629–637.
176. Hui, E.E. and Bhatia, S.N. (2007) Micromechanical control of cell-cell interactions. *Proc. Natl. Acad. Sci. U.S.A.*, **104**, 5722–5726.
177. Widmaier, E.P., Raff, H., and Strang, K.T. (2003) *Vander's Human Physiology*, 11th edn, McGraw-Hill.
178. Leclerc, E., Sakai, Y., and Fujii, T. (2004) Microfluidic PDMS (polydimethylsiloxane) bioreactor for large-scale culture of hepatocytes. *Biotechnol. Progr.*, **20**, 750–755.
179. John, G.T., Klimant, I., Wittmann, C., and Heinzle, E. (2003) Integrated optical sensing of dissolved oxygen in microtiter plates: a novel tool for microbial cultivation. *Biotechnol. Bioeng.*, **81**, 829–836.
180. Stitt, D.T., Nagar, M.S., Haq, T.A., and Timmins, M.R. (2002) Determination of growth rate of microorganisms in broth from oxygen-sensitive fluorescence plate reader measurements. *Biotechniques*, **32**, 684–689.
181. John, G., Goelling, D., Klimant, I., Schneider, H., and Heinzle, E. (2003) PH-sensing 96-well microtitre plates for the characterization of acid production by dairy starter cultures. *J. Dairy Res.*, **70**, 327–333.
182. Asfari, M., Janjic, D., Meda, P., Li, G., Halban, P.A., and Wollheim, C.B. (1992) Establishment of 2-mercaptoethanol-dependent differentiated insulin-secreting cell lines. *Endocrinology*, **130**, 167–178.
183. Melin, J. and Quake, S.R. (2007) Microfluidic large-scale integration: the evolution of design rules for biological automation. *Annu. Rev. Biophys. Biomol. Struct.*, **36**, 213–231.
184. Chapman, T. (2003) Lab automation and robotics: automation on the move. *Nature*, **421**, 661.
185. Meyvantsson, I., Warrick, J.W., Hayes, S., Skoien, A., and Beebe, D.J. (2008) Automated cell culture in high density tubeless microfluidic device arrays. *Lab Chip*, **8**, 717–724.

# 4
# Scalable Manufacture for Cell Therapy Needs

*Qasim A. Rafiq, Thomas R.J. Heathman, Karen Coopman, Alvin W. Nienow, and Christopher J. Hewitt*

## 4.1
## Introduction

Cell therapies, where whole cells form the basis of the therapeutic, represent a scientifically, commercially, and clinically important treatment option. With the potential to address previously unmet clinical conditions and reduce the social, economic, and clinical burden of chronic and age-related conditions, cell therapies are emerging as a key therapeutic treatment. Among the various cell types currently being considered as effective cell therapies, the therapeutic potential of stem cells makes them a primary cell candidate, and as such, have been the focus of significant research and development activities. Their unique ability to self-renew and differentiate, as well as the immunomodulatory, proangiogenic, and antifibrotic properties exhibited by some stem cells (e.g., mesenchymal stem cells), provide manufacturers and clinicians with an array of new treatment options. An example of this is the engineering of toxin-resistant stem cells to produce and secrete *Pseudomonas* endotoxin to target glioblastomas (GBMs), representing a novel treatment strategy for GBM therapy [1].

If the cell therapy industry comes to fruition, it is predicted that the industry could replicate, or even exceed, the success of the biopharmaceutical industry [2]. Global revenues generated for the cell therapy industry from both the public and private sector exceeded US$1 billion for the first time in 2012, with more than 160 000 patients receiving cell therapy treatments [2, 3]. With more than 150 companies worldwide and a predicted annual growth rate of 29% in market size from 2013 to 2018 [4], it is predicted that the global revenues for the regenerative medicine industry (of which cell therapy plays a significant role) will increase to US$35 billion by 2019 [5].

Although there have been marginal successes with cell therapies, for example, the market approval of Prochymal®, Mesoblast's (USA) human mesenchymal stem cell (hMSC) therapy for treating pediatric acute graft versus host disease

*Bioreactors: Design, Operation and Novel Applications*, First Edition. Edited by Carl-Fredrik Mandenius.

(aGvHD) [6], the next phase of growth for the cell therapy industry will require a coordinated effort on the part of the biomanufacturing community to develop novel, cost-effective, and efficacious therapies. With more than 200 clinical trials involving hMSCs alone, the majority of which are currently in phases I and II [7, 8], it is becoming increasingly clear that many cell therapy companies require a shift in their manufacturing process to meet the increasing demand for clinical-grade quality cells for increasing numbers of patients, a demand that cannot be supplied by the current production methods. It has been estimated that lot size requirements will exceed billions, or potentially trillions of cells, when therapies targeting larger clinical indications achieve market approval [9], and as such, biomanufacturing has become a key focus for the cell therapy industry. The difficulties associated with the translation of efficacious cell therapies is not limited to developing our biological understanding of the cell, but rather encompasses engineering, biomanufacturing, regulatory, commercial, and clinical challenges that must be addressed in parallel.

The focus of this chapter, therefore, is to highlight the key engineering and biomanufacturing challenges associated with the large-scale production of cells intended for patient administration. The primary challenges (discussed in detail later) relate to the production of clinical-grade cells at sufficient scale using robust and consistent processes. Given the diversity of source material for cell therapies (with respect to both cell candidates and variability in source material), the different treatment paradigms (autologous, allogeneic, haplobank), and the need to produce complex biological products (living cells that will undergo little or no purification postexpansion), it will be necessary to develop customizable, controlled, flexible manufacturing platforms to accommodate the specific requirements of cellular therapies.

It is worth noting from the outset that although the terms regenerative medicine and cell therapy are often used synonymously, they are in fact different. Mason *et al.* [2] highlight this distinction between the two and suggest that regenerative medicine is a medical speciality (such as pathology or endocrinology), which may or may not employ cell therapies, whereas cell therapy is a platform technology that utilizes living cells irrespective of the intended clinical indication, thereby representing a larger collective market opportunity than regenerative medicine. This is because the regeneration of tissue can occur from the application of therapeutics, which would not be considered to be cell based, for example, the administration of recombinant human erythropoietin to stimulate erythropoiesis, resulting in the regeneration of a patient's red blood cells. Cell therapies, meanwhile, possess a variety of mechanisms of action, which may or may not include regeneration. For example, the aforementioned hMSC therapy, Prochymal®, is an immunomodulatory treatment for aGvHD, while Dendreon's (USA) Provenge treatment is an autologous cell-based prostate cancer vaccine. As such, throughout the course of this chapter, the term cell therapy refers to the concept of cell therapies being a platform technology and not limited to the regeneration of tissue alone.

## 4.2 Requirements for Cell Therapy

Many of the challenges associated with traditional biopharmaceutical production, such as scalability, consistent and controlled manufacture, and obtaining sufficient yield to achieve economically viable processes, also apply to the production of cell therapies. However, cell therapies also pose additional manufacturing constraints, which are not necessarily applicable to biopharmaceuticals. As such, although there is much to be gained by learning from the biopharmaceutical sector, there will be some specific cell therapy production–related challenges that must be addressed, thereby adding a level of complexity to cell therapy manufacture [10]; these are outlined and summarized in Table 4.1.

### 4.2.1 Quality

Developing therapeutic treatments intended for patient administration that are comprised wholly of, or in part (e.g., as part of a tissue-engineered based construct), living cells requires stringent focus on product quality and characterization right from collection of source material all the way until delivery into the patient. As the living cell forms the product of interest, it is paramount that cell quality, as defined by the critical quality attributes (CQAs), identity, purity, potency, and safety, is retained throughout the manufacturing process, irrespective of the manufacturing paradigm or platform adopted [11, 12]. For the biopharmaceutical and vaccine industry, the cell is used as a factory and the product of interest; for example, a therapeutic protein or live, attenuated virus, is secreted by the cell and the fate of the cell is irrelevant to the process postexpansion and purification.

Characterization in traditional bioprocesses focuses on viable cell concentration and product titer, with measurement and possibly the control of a number of process parameters (e.g., pH, dissolved oxygen, temperature, off-gas, and metabolite concentrations) possible [13]. However, for cell therapies, the focus is much broader where the product needs to be characterized to ensure the cells retain their potency while also guaranteeing that the product meets the necessary release criteria with respect to identity, purity, and safety. Thus, manufacturing processes need to be designed to ensure that the CQAs are not affected, particularly when the process is scaled-up or scaled-out (these terms are defined and discussed in detail in Section 4.5). This also requires the development of suitable assays for cell characterization, particularly those that can be integrated with the expansion system for online or at-line measurement [11]. As such, unlike traditional biopharmaceutical production, the quality of the cell will always be given precedence over the quantity of the cells as it would be of no value to produce cells that do not retain their CQAs.

It must be stated that while assay development for product release testing is pivotal, there should arguably be a greater focus on ensuring that quality

**Table 4.1** Key considerations for the manufacture of cell therapies.

| Consideration | Implications | Perspective |
|---|---|---|
| Measuring cellular quality | • Basis of comparability, allows for process changes, scale-up, and multiple site manufacture during clinical development.<br>• Basis of release tests to demonstrate product function.<br>• Provides a metric for process control to maintain product consistency. | • Understanding the *in vivo* product mechanism of action is critical.<br>• Development of biological assays based on this understanding.<br>• Bespoke assays will be required for each product and clinical indication. |
| High donor variability | • Difficulty in maintaining a consistent manufacturing process.<br>• Increased risk that product batches will not meet quality specification. | • Identify the effect of critical donor characteristics (e.g., age) on input quality.<br>• Set minimum thresholds on process input based on this.<br>• Group patient populations to stratify the process input.<br>• Set strict controls on the isolation of donor material. |
| Autologous quality assurance (QA) | • Validation of multiple smaller batches greatly increases costs.<br>• Limited material is available for QA testing.<br>• There is risk of patient going untreated if product batch fails QA tests.<br>• Short shelf-life reduces time available for QA testing. | • Development of rapid and nondestructive QA tests will reduce this burden.<br>• Pooling of multiple lots from one patient could increase the material available for QA. |
| Immunological complications | • Acute or chronic rejection of the cell therapy.<br>• Use of immunosuppressive drugs can lead to further complications. | • Development of iPS or other autologous cell therapy.<br>• Use of cells with the potential for relative immune suppression. |
| Preservation | • Long-term storage methods have the potential to impact product quality.<br>• Short-term preservation adds complexity to the product distribution and delivery. | • The effect of long-term storage on product quality should be assessed.<br>• Increased understanding of cryopreservation process is required.<br>• Local manufacturing facilities can reduce complexity of supply chain. |

Table 4.1 (Continued)

| Consideration | Implications | Perspective |
|---|---|---|
| Harvest and pooling | • Many cell types are adherent and must be removed from a culture surface.<br>• This process has the potential to impact product quality.<br>• Allowable pooling time will limit the scale of the process. | • Harvesting protocol should be developed that are based on quality.<br>• Effective process scale-up will reduce the pooling time of the product. |
| Regulation | • Given the complexity of cell therapies there is little regulatory precedence.<br>• There is uncertainty around how much data are required to establish product profiles. | • Regulatory pathway is likely to be streamlined as more products receive approval. |
| Delivery | • Bespoke delivery devices will be required to fit the vast array of cell therapy products. | • Delivery devices should be designed early on in clinical development.<br>• These devices must fit in with the clinical setting and effective training must be provided. |

standards are built into the process and manufacturing facility to ensure that the cell therapy is produced in line with good manufacturing practice (GMP) protocols. The focus on quality becomes even more significant when, for example, a multisite manufacturing approach is adopted or when a change to the process is made [14]. With the need to establish and maintain product reproducibility and comparability to satisfy the expectations of the regulator in such instances, product and process understanding are a necessary prerequisite. This understanding can only be acquired via stringent quality assurance and control systems. For cell therapies, it is likely that the regulator will take the position that product comparability will require the inclusion of measures of biological activity as part of the testing regime [12].

### 4.2.2 Number of Cells Required

Given that there are relatively few cell therapy treatments on the market, in particular stem cell treatments, it is difficult to determine the exact number of cells required per dose or number of doses per lot. As mentioned previously, however, it is envisaged that lot sizes would range between $10^9$ and $10^{12}$ cells, although for each treatment, the quantity of cells required will vary depending on the clinical indication being targeted. Approximate quantities of cells can be ascertained from preclinical and clinical studies, as well as from approved therapies.

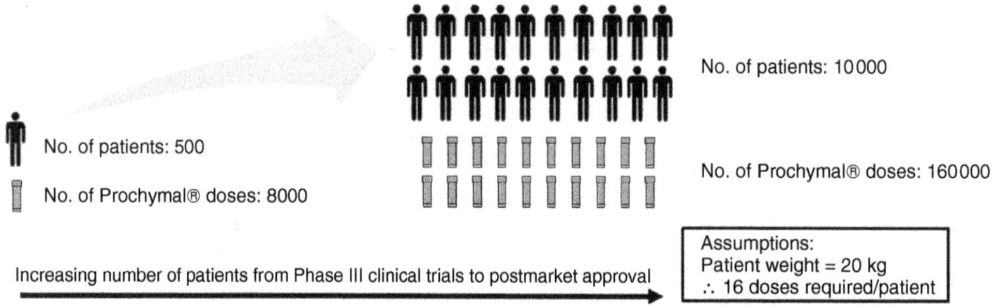


**Figure 4.1** Overview of approximate number of Prochymal® doses required for increasing numbers of patients from Phase III clinical trials to post-market approval (example scale).

Prochymal® was approved by Health Canada in 2012 and as part of the "Summary Basis of Decision" for the therapy [6], it states that the recommended dose of Prochymal® is based on the weight of the patient and should be $2 \times 10^6$ hMSCs kg$^{-1}$, which is administered by intravenous (IV) infusion twice per week for four consecutive weeks (eight IV infusions in total). However, to account for potential post-thaw viability loss, each Prochymal® dose is prepared at approximately $25 \times 10^6$ cells. As Prochymal® will be used to treat pediatric aGvHD, assuming that a patient weighing 20 kg requires treatment, this would mean that $40 \times 10^6$ hMSCs would be required per administration (8 in total), therefore requiring $320 \times 10^6$ hMSCs. However, given that each Prochymal® dose is prepared at $25 \times 10^6$ hMSCs, this will mean 2 doses will be required per treatment, which implies that 16 doses will be required to treat the patient (total of $400 \times 10^6$ hMSCs).

Acute GvHD has a relatively low incidence, but assuming that 10 000 patients per year require the treatment, this would, therefore, require an annual production of $4 \times 10^{12}$ hMSCs (Figure 4.1). Similarly, as part of their phase I clinical study for spinal cord injury, Geron (USA) provided a dose GRNOPC1, human embryonic stem cell (hESCs)-derived cells, at $2 \times 10^6$ cells per patient [15]. Moreover, as part of a clinical study, it has also been reported that for the treatment of acute myocardial infarction, a dose ranging between $0.5 \times 10^6$ and $5 \times 10^6$ hMSC kg$^{-1}$ per patient was administered [16].

What is apparent is that if a clinical indication with a significant patient population was to be treated, the number of cells per lot would range likely between $10^9$ and $10^{12}$ cells, and given that the culture of stem cells is still predominantly attained via small sized, static, planar expansion systems, which will not yield this number of cells, other platforms must be considered for the larger scale culture of stem cells.

### 4.2.3 Anchorage-Dependent Cells

Unlike common expression systems used in traditional bioprocessing such as *Escherichia coli* and suspension-adapted Chinese hamster ovary (CHO) cells, most cell candidates currently being investigated for their use as cell therapies are anchorage-dependent, meaning they require a surface by which to adhere and proliferate. Although this is not a phenomenon associated only with cells intended for cell therapy, for example, Vero cells are commonly used for vaccine production and are also anchorage-dependent [17, 18], it has greater implications for cells intended for cell therapy because reiterating what was mentioned earlier about cell quality, it is of paramount importance that the cells are not compromised in any way as the product of interest is the living cell. This is not to say that the quality of cells used as an expression is not important in the production of vaccines or therapeutic proteins, but ultimately the product of interest is what the cell expresses and not the cell itself. To examine this further, if a cell used as an expression system were to become compromised with respect to karyotype, as the established CHO cell line has become, for example [19], this may not necessarily affect the quality of the protein of interest, yet for cell therapies, it is unlikely that this would be acceptable to the regulator. It is also worth noting that the cell types commonly used for biopharmaceutical production are immortalized cell lines. However for cell therapies, the cells have a finite period for expansion and will senesce during prolonged *in vitro* culture. Wagner and colleagues identified that within 43–77 days of *in vitro* culture, hMSCs demonstrated morphological abnormalities, enlargement, attenuated expression of cell-surface markers, and cellular senescence [20]. Moreover, it has been reported that karyotypic abnormalities have arisen with prolonged hESC *in vitro* culture, whereby cells have acquired chromosome 17q and chromosome 12 [21]. Particular consideration needs to be given, therefore, to the manufacturing conditions, particularly where cells undergo multiple population doublings to achieve the desired cell number.

## 4.3 Stem Cell Types and Products

Stem cells are defined as cells that have the ability to self-renew and differentiate into mature cells of a particular tissue. It is this ability to self-perpetuate and produce large quantities of therapeutically active cells that have generated significant interest around the use of stem cells to not only treat symptoms, but potentially cure the underlying condition.

Stem cell therapies have been around for a number of years with the first successful allogeneic bone marrow transplant taking place in 1968 [22]. The bone marrow contains a number of stem cell populations, particularly related to the production of blood cells. Hematopoietic stem cells (HSCs) form the vast majority of cell therapy clinical trials and are being used to treat hematological malignancies as well as other oncology indications [23]. These HSCs are typically transplanted from one patient to another without any cell expansion taking place; however, systems are available to expand these cells in suspension [24].

In addition to HSCs, the bone marrow also contains a population of adherent and self-renewing cells, termed mesenchymal stem or stromal cells.

Ever since the term mesenchymal stem cell was first introduced by Arnold Caplan in the early 1990s [25], much excitement has been generated around the potential for MSCs to treat and in some cases cure diseases, such as diabetes, stroke, and myocardial infarction. This interest has been mainly due to their ease of isolation as well as their ability to proliferate *ex vivo* under appropriate culture conditions. Human MSCs have been reportedly isolated from many tissue types such as bone marrow [26–28], adipose tissue [29, 30], umbilical cord [31], and cord blood [32, 33] with their putative mechanism of action thought to be a combination of: modulating the patients' own immune system, release of trophic factors to stimulate native tissue regeneration, and reducing the inflammatory response of the patient's native tissue. Part of the defining criteria for an MSC is the ability to differentiate into different cell types including osteoblasts, chondrocytes, and adipocytes. Along with the desirable characteristics described, MSCs have also received a lot of attention due to their ability for donor mismatched transplantation without ectopic formation and their nonengraftment following infusion, achieving their mode of action via a "hit-and-run" mechanism [34]. It is these unique properties that have made MSCs a popular choice for companies developing stem cell products, with Athersys Inc. (USA) and Mesoblast Ltd. (Australia) in late-stage clinical development of products targeting immune disorders, cardiovascular disease, and neurological conditions.

Since the discovery of embryonic stem cells in 1998 [35], pluripotent stem cells have been investigated as a promising source of stem cells due to their ability to self-renew and form all the tissues of the body. In 2010, Geron Co. (USA) went into clinical development with an allogeneic hESC product for the treatment of spinal cord injury. Despite no adverse safety issues with this hESC-derived therapy, Geron Co. halted the clinical trial citing financial and regulatory issues [36]. Mature human cells, such as dermal fibroblasts, can also be reprogrammed to become pluripotent using a technique involving transduction of defined transcription factors [37, 38]. This novel technique allows for the development of patient-specific pluripotent stem cells, termed induced pluripotent stem (iPS) cells [37, 38]. It seems that as the field develops, iPS cells are increasingly taking the place of ESCs as the pluripotent cell of choice with reduced ethical constraints associated with the destruction and manipulation of human embryos. Human iPS cells also have the potential to be patient specific, which reduces the potential for immunological rejection and opens up opportunities for their use as a personalized cell therapy.

## 4.4 Paradigms in Cell Therapy Manufacture

Cellular therapies represent a distinctive bioprocess challenge, owing to the different paradigms that exist in the field to treat both acute and chronic disease by the application of cells (e.g., MSCs) or tissues (e.g., skin). The development of cell therapy products will be largely driven by whether the cells are universal (allogeneic), donor matched (haplobank), or patient specific (autologous), each carrying their own set of manufacturing challenges and commercial drivers.

### 4.4.1
### Haplobank

To limit the complications associated with immunological rejection of cell therapies, it has been suggested that cell banks can be constructed, which contain a sufficient number of human leukocyte antigen (HLA)-matched samples to cover a large patient population [39]. This concept has been driven by the use of pluripotent cell types as a cell therapy as the progeny of differentiated cells will still contain HLAs with the potential to cause rejection of the transplant. It has been shown that 200 embryonic cell lines from randomly donated oocytes can provide beneficial matching for around 80% of the Japanese population [40]. Sufficient donor matching in this way would allow for the production of a cell therapy product for multiple patients, although a haplobank would not be sufficient to target an entire patient population.

### 4.4.2
### Autologous Products

Cell therapies where the donor and the recipient are the same individual are known as autologous products [41] (Figure 4.2). These products are patient specific and their manufacture must be scaled-out to ensure that patient material is segregated and cross-contamination of material is avoided. Scale-out of autologous therapies will likely necessitate multiple manufacturing facilities, creating the need for local automation and the demonstration of comparability between these sites [14]. Personalized healthcare creates a service-based business model where the complex logistics surrounding manufacture and delivery of live cells creates significant opportunity. However, these logistics must be carefully considered as there are significant operating costs associated with implementing this supply network for a product with a short shelf life.

The main advantage of autologous cell therapies is the lack of immune rejection associated with donor transplant material, eliminating the need for immunosuppressive medication that would add significant cost to the treatment. Besides immune complications, the use of autologous material also greatly reduces the risk of transmitting infectious diseases between patients, which typically forms part of safety testing during the manufacturing process [42]. Autologous therapies provide the potential for point-of-care devices, where functional closed devices can be used to manufacture cell therapies at the bedside. These typically involve the isolation and enrichment of cells directly from the patient and are returned on-site as “minimally manipulated,” which have the advantage of not being regulated as a medicinal product [43]. Examples of this bedside

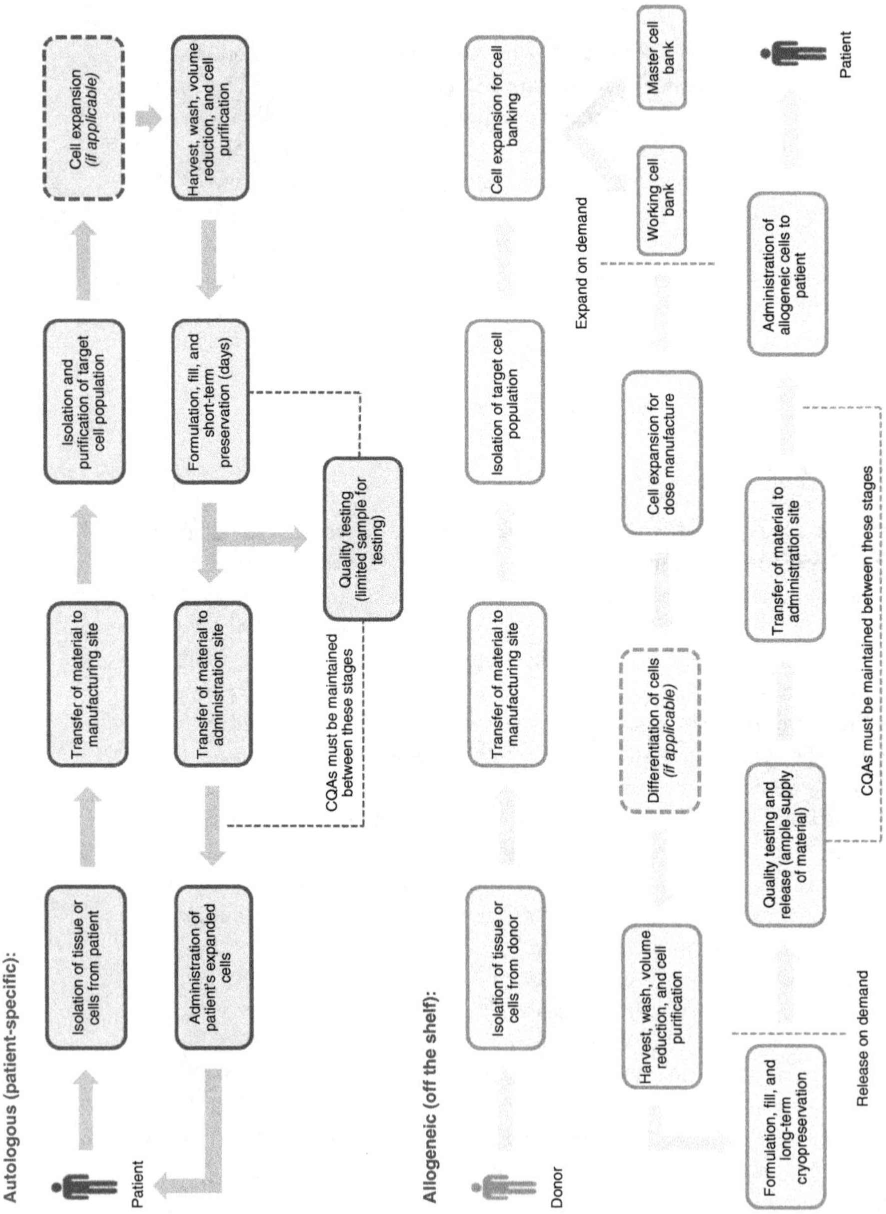


**Figure 4.2** An overview of the two main manufacturing paradigms for cell therapy; process flow diagrams for autologous (patient specific) manufacture and allogeneic (off the shelf) manufacture.

manufacturing paradigm include the CliniMACS® Prodigy (Miltenyi Biotech) and Celution® (Cytori Therapeutics Inc.). Autologous therapies also have an economic advantage as they generally incur reduced capital costs for start-up, and unit operations can be scaled-out to match market demand. This scale-out approach also reduces the number of product population doublings, which can lead to chromosomal instability [44] and reduced product quality [45], if the product undergoes multiple population doublings.

Despite these advantages, many challenges remain in the development of autologous cell therapies. Extracting cells from each patient requires a biopsy, which carries inherent risk to the patient and requires extra time in the clinic. It is also possible that the route of the target disease might be with the patient's own cells and it would, therefore, be better to avoid using them. The inherent variability between donors is also high, which must be controlled during manufacture, but can be improved by stratifying patient populations into groups based on characteristics such as age or gender.

Issues surrounding the quality test burden and logistics add to the complexity for production and delivery of a cost-effective autologous cell-based therapy [46]. This aspect is largely due to the reduced economies of scale with scaling-out a process and the reduced market demand due to the lower patient numbers [47]. The delivery of autologous cell-based products can be complex due to the short-term preservation methods used for transport, introducing significant logistical costs associated with their supply. A key issue with this approach is the risk that a manufacturing lot will not meet specification and the patient does not receive their treatment. This problem could have huge implications for patient health and must be carefully considered.

### 4.4.3 Allogeneic Products

Regenerative therapies where the donor and the recipient are different individuals are known as allogeneic products [41]. This creates an off-the-shelf business model, which is far more akin to current biopharmaceuticals, representing an attractive commercial opportunity as cell products can be stored for a long term and are always available for delivery (Figure 4.2). This business model has been adopted by a number of regenerative medicine companies, such as Mesoblast Ltd. (Australia) and Athersys Inc. (USA) who are developing off-the-shelf platform products, to treat a range of clinical indications.

In contrast to autologous therapies, allogeneic products have the potential to be scaled-up, benefitting from the economies of scale experienced by traditional bioprocesses. There is a lot of expansion technology available for the scale-up of cellular products, with substantial legacy data supporting their large-scale design and operation [48]. Manufacturing a large number of product doses in this manner allows companies to target clinical indications with high patient numbers, creating the prospect of increased financial reward. The larger lot size for allogeneic processes also reduces the burden of quality assurance, which must be applied to every patient lot for an autologous therapy. In addition, an allogeneic therapy reduces the impact of donor variability on the process, as the cell banks are expanded to treat multiple patients. The value of process consistency should not be underestimated, as the increased cost in developing and validating a variable process could be substantial. Allogeneic therapies also introduce the prospect of utilizing pluripotent stem cell types such as embryonic stem cells, which offer huge potential in treating a wide range of clinical indications.

The use of pluripotent stem cells does, however, introduce safety challenges within downstream purification and the elimination of cells that have the potential to form teratomas in the patient [42]. Introducing high levels of process and product characterization will likely increase direct process costs, from forming a master cell bank to final product testing. There are also potential issues if the cell therapy is foreign to the immune system of the recipient, although some cell types have displayed relative immune privilege post-delivery [49]. This issue has the potential to limit the cell types amenable to large-scale allogeneic processes, although it has been suggested that multiple cell lines can be derived to provide a reasonable match for most patients [39]. The development of an allogeneic product also necessitates a master cell bank, from which the product can be manufactured [50]. The cost of developing a master cell bank is typically high and cannot be recovered until the product receives market approval, adding to the funding gap in development. The quality and consistency of the master bank is critical, because once it has been created it cannot be changed. This requires a high level of product understanding coupled with rigorous safety and quality testing to ensure that the product in the master cell bank is suitable for the manufacture of the cell therapy product. Any changes to the manufacturing process during clinical development will require validation of the process to ensure that it remains comparable before and after the change. This aspect will require significant time and resource, as well as the development of functional assays to demonstrate that there has been no change to the safety or function of the cell therapy product.

## 4.5 Cell Therapy Manufacturing Platforms

With multiple cell-based therapies in late-stage clinical development and moving toward commercial production, it is clear that selecting suitable manufacturing technology is becoming increasingly important [23]. Given the diversity of cell types, disease indications, and business models, it is likely that multiple manufacturing platforms will be utilized in the production of cell therapies. Careful consideration must be given to the selection of manufacturing platforms before clinical development, to produce scalable and cost-effective cell therapy products.

With respect to cell therapy production, we have used the terms scale-out (scaling horizontally) and scale-up (scaling vertically) in accordance with the Cell Therapy and Regenerative Medicine Glossary PAS84, where scale-out is defined as increasing production by an increase in the number of units, and scale-up is defined as increasing the size of the process rather than increase production by an increase in the number of units [51].

### 4.5.1 Scale-Out Technology

Adherent cell culture has traditionally taken place in tissue culture flasks, which are cultured in incubators under static conditions. Production can then be scaled out by increasing the number of flasks, with double the number of cells requiring double the number of flasks. For an autologous therapy, this approach requires a lot of clean room space, as each patient lot must be segregated and manufacture becomes costly for high patient numbers. This method of scaling-out tissue culture flasks is highly labor intensive and due to the static conditions, sampling and control is not possible. Some of these limitations can be resolved by automating the culture of flasks and a number of robotic platforms now exist, such as TAP Biosystems (now a Sartorius Stedim company, Germany) CompacT SelecT (Figure 4.3). Such platforms are capable of handling upward of 100 flasks with a combined surface area of 17 500 $cm^2$ and have been shown to reduce variability in the process [52]. The pooling time required during a scale-out manufacturing process has the potential to negatively impact product characteristics [53] and will, therefore, limit the scale of these techniques. There has been a push to increase the surface area contained in tissue culture flasks, and companies such as Corning Inc. (USA) have developed systems with multiple layers that offer up to 25 440 $cm^2$ of surface area per flask. One of the concerns with using multi-layer flasks or cell stacks is the development of a heterogenous environment within the system due to issues with mass and heat transfer. Therefore, there have also been efforts to produce closed multiple layer technology with perfused medium so that process parameters can be monitored and controlled during culture. Pall Corporation (USA) have recently acquired the Integrity® Xpansion™, which contains 122 400 $cm^2$ of surface area in a disposable, closed, and controlled multiplate bioreactor system.

Rotating flasks, commonly referred to as roller bottles, have been widely employed in traditional bioprocessing for the production of vaccines and proteins from adherent cell types. These systems utilize the full inner surface area of the flask and require a much lower volume of culture medium per surface area compared with static culture flasks. The rotation also provides a level of mixing, which improves gaseous exchange and temperature control and allows for sampling and monitoring of the culture medium. As with static flasks, rotating flasks are labor intensive and also incur the penalty of high pooling times, which can limit the potential scale of the manufacturing process. A further constraint of a roller bottle process for cell therapy manufacture is the limitation on the control of $O_2$ and $CO_2$ composition in both the gas and the liquid phase of the culture. Rotating vessels have traditionally been cultured and manipulated by operators in large temperature controlled warehouses and therefore operation

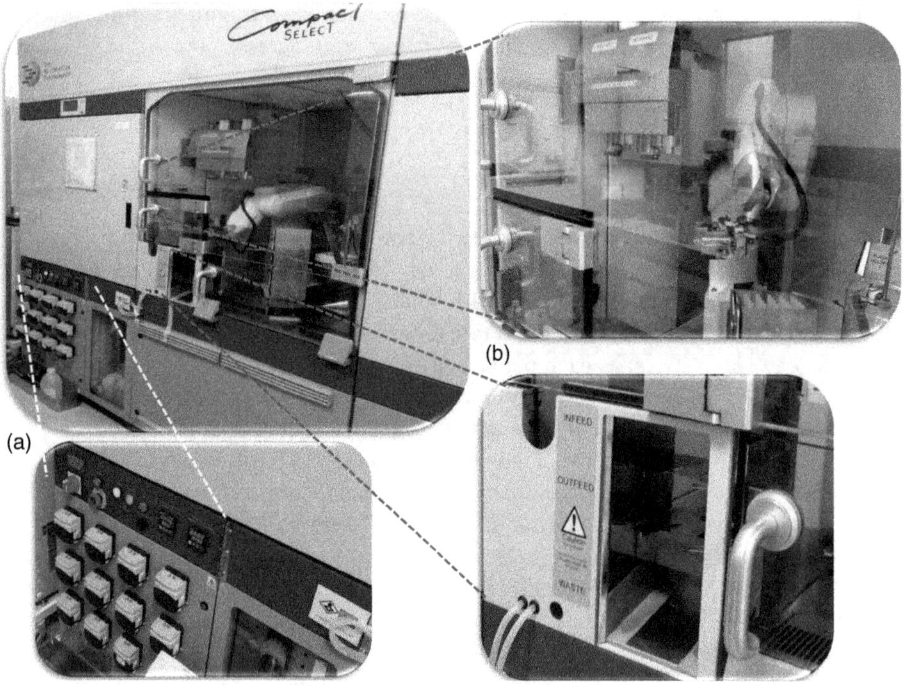


**Figure 4.3** (a) A scale-out expansion platform for stem cells; an automated robotic tissue culture flask handler, the CompacT SelecT, (b) Overview of the main internal components within the CompacT SelecT including robotic arm, stripette holder, decapper and flask holder, (c) The barcoded T-flask infeed and outfeed ports for traceability and monitoring, (d) Inlet, waste, and reservoir pumps and automated Cedex mammalian cell counter.

at $O_2$ concentrations lower than atmospheric, which is commonly used for cell therapy culture, will not be possible.

Hollow fiber bioreactors can be used to provide high surface areas for cell expansion, for example, the Quantum® Cell Expansion System by Terumo BCT (USA), which is currently being evaluated by Athersys Inc. (USA) for the manufacture of Athersys' multipotent adult progenitor cells. This functionally closed system provides 21 000 $cm^2$ per bioreactor, is GMP compliant, and multiple units can be operated and harvested by a single operator [54]. Medium is perfused through the fibers meaning that automated monitoring and control of metabolite concentrations can take place, which is important in maintaining process consistency. Due to the nature of the hollow fiber system, there is the potential for the formation of longitudinal concentration gradients as culture medium or dissociation reagent flows down the fibers. Furthermore, large-scale expansion using multiple hollow fiber units requires interprocess cryopreservation steps as the number of units required increases exponentially and would quickly become cost prohibitive.

### 4.5.2
### Scale-Up Technology

The scale-up of a bulk allogeneic cell therapy provides an attractive business model, with large product batches manufactured in bioreactors using a similar process to those adopted by the current manufacturers of biopharmaceuticals using free suspension cell culture. There are a number of types of bioreactor available, each having different advantages and disadvantages. These different types are discussed in the following.

Rocking-motion bioreactors are used as a noninvasive, single-use technology for the culture of cells in suspension. Mixing is achieved by the rocking-motion and, therefore, does not require any internal moving parts, making it highly amenable to disposable manufacturing. Control is achieved by noninvasive monitoring of process parameters such as temperature and dissolved oxygen concentration ($dO_2$) (usually expressed as % with respect to saturation with air (100%)), in a functionally closed process, which is used for GMP applications. Anchorage-dependent cell types can be cultured on microcarriers in suspension though there are some issues associated with microcarriers becoming stranded on the base during the rocking cycle, In addition, due to the gentle mixing from rocking, the cell harvest from the microcarrier surface may have to take place in a separate bioreactor vessel (as discussed later). The current market leader for rocking-motion technology is the WAVE Bioreactor from GE Healthcare (USA), which supplies rocking-motion bioreactors ranging from 300 ml up to 1000 l, though there are many others on the market.

Pneumatically driven bioreactors have been developed for free suspension cell culture to provide a so-called "low-shear" mixing environment, driven by air sparging into a vertical wheel-based impeller. The motion of the wheel mixes the contents of the bioreactor and aids oxygen and $CO_2$ transfer allowing sampling and control of the process. The motion is also able to suspend microcarriers (discussed in detail later), which provide a surface for the cells to grown on. The bubbling required to drive the mixer is an important consideration for the design and operation of such bioreactors as bursting bubbles are known to be detrimental to cell viability [55]. This problem has meant that the inclusion of protective agents, such as Pluronic™ F68, has become necessary during free suspension culture to limit cell damage as well as reduce foaming within the bioreactor. Considering that bubbling gas is fundamental to the operation of pneumatically driven bioreactors, these implications must be carefully considered. Due to the vertical impeller, the operation of these bioreactors at reduced volumes may be somewhat impaired, which could also impact on operability. An example of this is the cell harvesting step, which typically requires *in situ* volume reduction before cell detachment as discussed later. Combined with the low energy input from the slowly rotating wheel, this requirement could make the detachment and harvest of cell therapy products difficult in pneumatically driven bioreactors.

Packed and fluidized beds have been employed for cell therapy production due to the high cell densities achievable during expansion. In these systems, adherent cells are grown on packing material, while culture medium is perfused through the bioreactor, to reach cell densities approaching $10^8$ cells $ml^{-1}$ [56]. Perfusion of the culture medium allows for monitoring and control of the process conditions, which, as mentioned previously, is critically important in the development of a reproducible manufacturing process. Despite the advantages of high cell densities, this structure does introduce a risk of fouling and the formation of axial and radial concentration gradients especially at the large-scale. Furthermore, the harvesting of cells from these systems can be problematic due to the high cell densities present and the difficulty of introducing the detachment fluids effectively into the beds. Harvesting is a key consideration that needs as much consideration as cell culture as the cellular quality must be maintained throughout the expansion and harvest process. These systems are currently being developed alongside clinical programs by companies such as Pluristem Therapeutics Inc (Israel), with clinical trials using expanded placental cells for the treatment of peripheral artery disease, pulmonary hypertension, and muscle injury.

Stirred tank bioreactors are by far the most common platform used for the culture of mammalian cells whether free suspension or microcarrier based in current bioprocesses (Figure 4.4a). There are well-established conventions for scale-up of operation [48, 55]. Many companies are applying this current understanding to the development of cell therapy products and bioreactors are routinely operated at 50 l by companies such as Celgene Corporation (USA), EMD Millipore (USA), and Lonza (USA). These systems provide a homogenized environment for cell culture, with existing technology available for process monitoring and control. The culture of stem cells in stirred tanks has been demonstrated in suspension for ESCs [57, 58], iPSCs [59, 60], MSCs [61], and HSCs [24, 62]. The expansion and harvest of adherent cells have also been demonstrated for the production of MSCs in stirred-tank bioreactors [63, 64], although the cell densities achieved to date are relatively low. The cell densities in stirred bioreactors are, however, constantly increasing and will likely experience similar improvements in product yield as have been experienced in traditional cell culture bioprocesses. Considering the strong legacy data that exist and the decades of experience supporting their large-scale operation, it is likely that a number of bulk allogeneic cell therapies will be produced in stirred-tank bioreactors.

## 4.6
## Microcarriers and Stirred-Tank Bioreactors

Research into microcarrier and stirred-tank bioreactor for cell therapy applications has rapidly increased since the mid-2000s. This is primarily due to the realization that other cell therapy manufacturing platforms are unlikely to be able to generate the quantity of cells for clinical indications, which require $>10^9$ cells per lot. Numerous studies using stirred-tank bioreactors and microcarriers have been performed within both the academia and industry, and although scalability has been the main driver in moving to a stirred-tank bioreactor system, there are a number of other advantages. These include the capability to monitor and

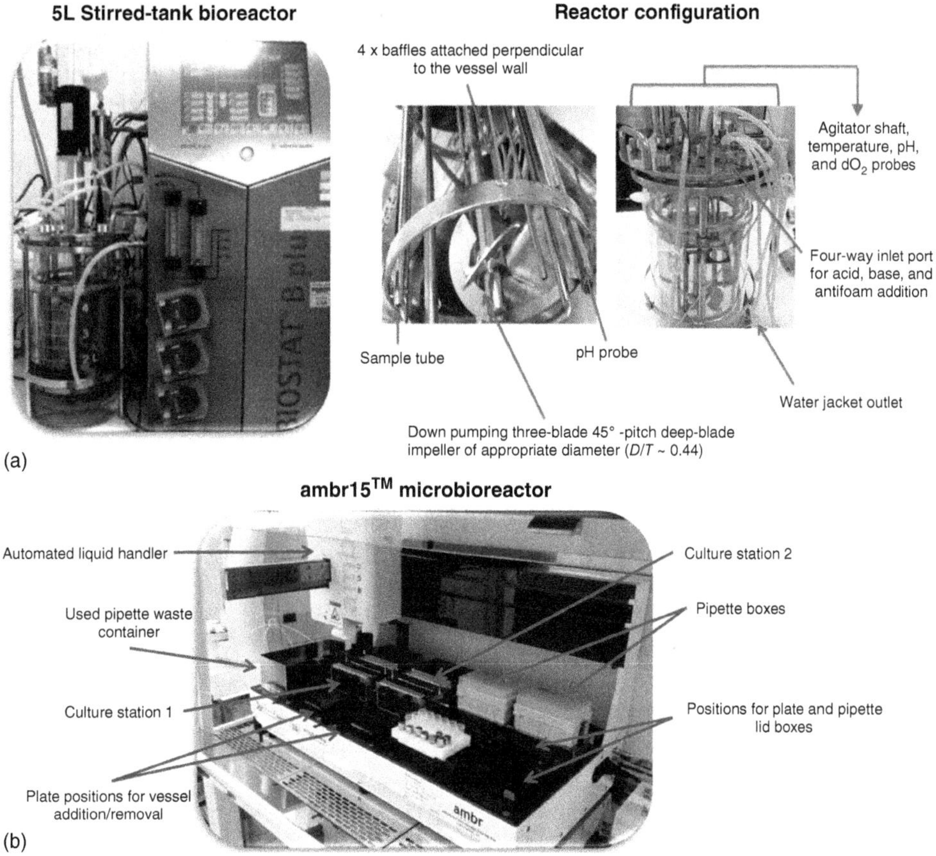


**Figure 4.4** (a) An overview of the 5 l stirred-tank bioreactor configuration used in our laboratories. The microcarriers (with cells attached) are kept in suspension with a downward pumping three-blade 45° pitch-blade impeller ($D/T \sim 0.44$) agitated at the minimum speed required to keep the microcarriers in suspension, (b) The high-throughput, automated ambr15™ microbioreactor that allows for the parallel culture of 24 microbioreactor vessels (15 ml working volume), which we have recently demonstrated is amenable for hMSC microcarrier culture.

control key bioprocess parameters, the flexibility to operate via different modes of operation (i.e., batch, fed-batch, perfusion) under a completely closed system and the ability to achieve a well-mixed system, thereby creating a homogenous culture environment. Moreover, we have shown that they also offer the possibility of harvesting cells *in situ* [64].

An overview of the stirred-tank bioreactor configuration used in our laboratories is illustrated in Figure 4.4a. No specific adaptations were required to successfully grow the hMSCs in the stirred-tank bioreactor. The system comprised four baffles attached perpendicular to the vessel wall with a downward pumping three-blade 45° pitch-blade impeller, which was agitated at the minimum speed required to keep the microcarriers in suspension. It is worth noting that this pumping direction, combined with an impeller diameter ($D$) of ~0.4 of the bioreactor tank diameter ($T$) (here $D/T = 0.44$) enables suspension of microcarriers with the lowest agitation intensity.

Stirred-tank bioreactors used in conjunction with microcarriers have the advantage that they are well-understood and well-characterized systems, proving to be both cost-effective and reliable for biopharmaceutical production [65]. In addition to this, with the need to be sufficiently customizable to accommodate multiple cell candidates for therapies, microcarrier systems are advantageous in that specific microcarriers are currently available or can be developed with varying charge, surface coating, and shape to accommodate specific cell types. Moreover, with respect to customization and flexibility, stirred-tank bioreactors allow for different modes of operation such as batch, fed-batch, or perfusion culture, which again can be tailored for each process depending on the specific requirements of the cell.

### 4.6.1 Overview of Studies Using a Stirred-Tank Bioreactor and Microcarrier System

Studies have investigated the culture of multiple cell types on microcarriers including human and mouse ESCs [66, 67], hiPSCs [68], and hMSCs [63, 69, 70] (Table 4.2). However, the majority of the literature on microcarrier expansion of stem cells has been conducted on hMSCs, with a range of commercially available microcarriers being used.

Although originally much of the research on MSC and microcarrier expansion focused on tissue engineering applications [84, 85], in 2007, Frauenschuh *et al.* [73] published the first study of MSC expansion on microcarriers with the sole aim of cell production. Many of the studies conducted so far have demonstrated the ability to grow hMSCs on microcarriers, some even at the liter scale [63], and over the years, there has been a steady increase in the reported cell densities attainable via these systems [77]. Even at the current cell density, we demonstrated that one 5 l bioreactor (Figure 4.4a) containing 2.5 l culture medium can produce as many cells as ~65 fully confluent T-175 flasks being cultivated in a robotically controlled scale-out system [63], equivalent to ~2/3 of the CompacT SelecT (Figure 4.3) capacity.

Figure 4.5 provides data from work we have performed in our laboratories including phase-contrast and fluorescent images of identical microcarriers with hMSCs attached (Figure 4.5a and b), hMSCs that have been successfully harvested from microcarriers (Figure 4.5c), hMSCs that have differentiated toward osteogenic (Figure 4.5d), chondrogenic (Figure 4.5e), and adipogenic (Figure 4.5f), and multiparameter flow cytometry data of hMSCs that have been successfully harvested (Figure 4.5g) displaying > 99% expression of the expected positive markers (CD105, CD73, and CD90) and <2% expression of the negative markers (HLA-DR and CD34).

**Table 4.2** Overview of bioreactor studies for different stem cell types.

| Stem cell type | Species | Source | Method of expansion | Purpose of investigation | References |
|---|---|---|---|---|---|
| HSC | Human | Umbilical cord blood | Suspension culture | Comparison of T-flask and bioreactor expansion for HSCs | [71] |
| HSC | Human | Umbilical cord blood | Suspension culture | Comparative gene expression of HSCs grown in static and stirred culture systems | [72] |
| MSC | Porcine | Bone-marrow | Microcarrier (Cytodex-1) | Investigate multiple microcarriers and cell densities for MSC culture | [73] |
| MSC | Human | Bone-marrow | Microcarrier (Cytodex-1) | Investigation of optimal culture conditions for hMSC microcarrier cultures | [69] |
| MSC | Porcine | Bone-marrow | Microcarrier (Cytodex-1) | Identify effect of agitation rate and cell density on MSC culture | [74] |
| MSC | Human | Placenta | Microcarrier (Cytodex-3) | Investigation of different impeller geometries, microcarrier concentrations and cell density for hMSC culture | [75] |
| MSC | Human | Bone-marrow | Microcarrier (Cultispher-S) | Development of new protocols for improved hMSC microcarrier expansion based on agitation, removal of FBS, and pH adjustments | [76] |
| MSC | Human | Bone-marrow | Microcarrier (Plastic P102-L) | Expansion of hMSCs on microcarriers under xeno-free conditions | [77] |
| MSC | Human | Bone-marrow | Microcarrier (Cultispher-S) | Establishing an optimal expansion strategy using a low-serum containing medium | [78] |
| MSC | Human | Bone-marrow | Microcarrier (Plastic P102-L) | Demonstration of expansion of hMSC microcarrier culture at the liter scale in a stirred-tank bioreactor | [63] |

(*continued overleaf*)

**Table 4.2** (*Continued*)

| Stem cell type | Species | Source | Method of expansion | Purpose of investigation | References |
|---|---|---|---|---|---|
| MSC | Human | Bone-marrow | Microcarrier (Plastic P102-L) | Development of a novel, scalable harvesting method for hMSCs on microcarriers | [64] |
| MSC | Human | Fetal | Microcarrier (Cytodex-3) | Demonstration of higher osteogenic potency in comparison to static, monolayer cultures | [79] |
| iPS | Human | Fibroblasts | Microcarrier (DE-53) | Microcarrier expansion of hiPSCs and differentiation to neural progenitor cells | [68] |
| ESC | Murine | Embryo | Aggregate culture | Demonstration of controlled, reproducible expansion of mESCs as aggregates in suspension culture | [80] |
| ESC | Murine | Embryo | Microcarrier (Cytodex-3) | Demonstration of microcarrier mESC culture in a stirred-tank bioreactor | [81] |
| ESC | Murine | Embryo | Aggregate culture | Retention of pluripotency following expansion of mESCs as aggregates in suspension culture | [82] |
| ESC | Human | Embryo | Embryoid bodies (EBs) | Comparison of growth and differentiation of hESC-derived EBs in static and stirred suspension cultures | [83] |

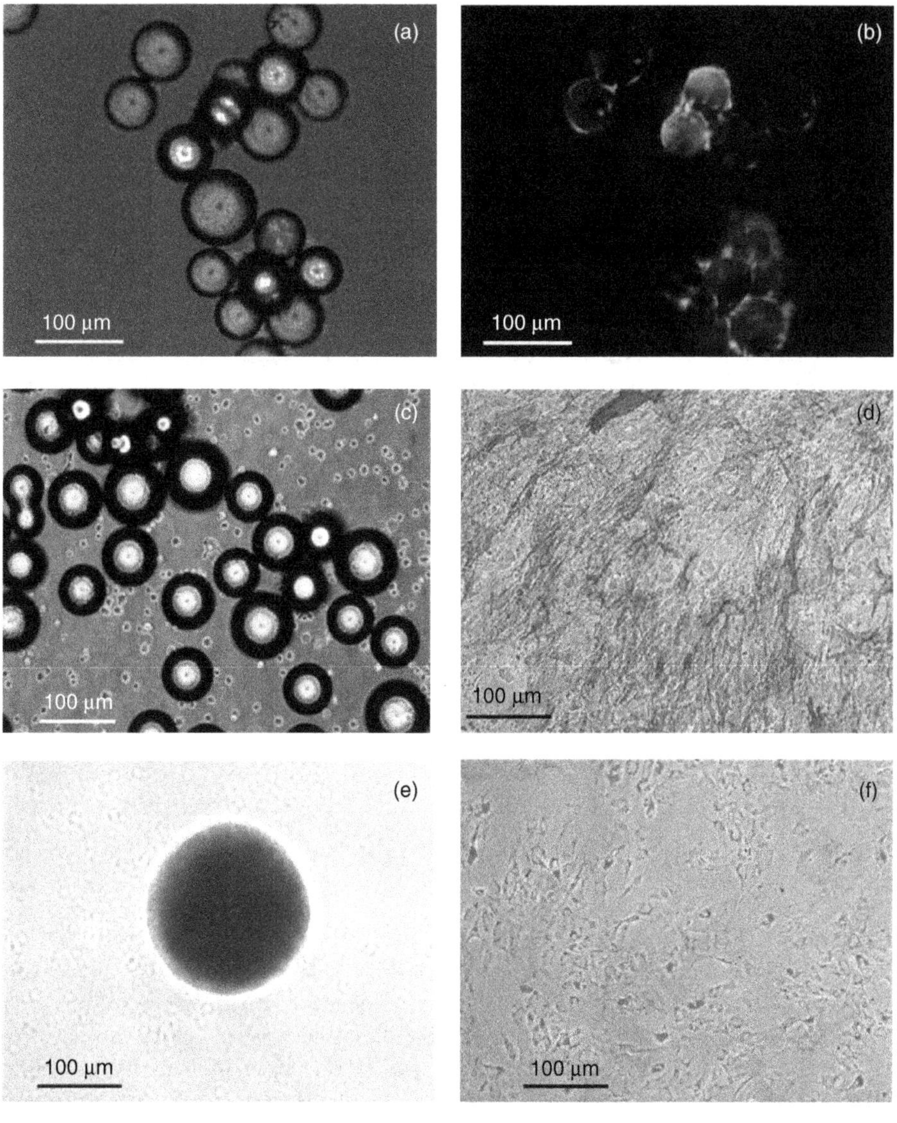


**Figure 4.5** Human mesenchymal stem cells cultured on microcarriers in spinner flasks and subsequently harvested and characterized to determine cell quality with respect to retention of differentiation capacity postharvest and identity. (a) Phase-contrast image of hMSCs attached to Plastic P102-L microcarriers. (b) Fluorescent image of identical microcarriers in the previous image depicting viable cells (green fluorescent calcein-AM) and nonviable cells (red-fluorescent ethidium homodimer-1). (c) Successful detachment of hMSCs from Plastic P102-L microcarriers. (d) Alkaline phosphatase and Von Kossa staining demonstrating osteogenic differentiation. (e) Alcian Blue staining demonstrating chondrogenic differentiation. (f) Oil Red O staining demonstrating adipogenic differentiation. (g) Multiparameter flow cytometry demonstrating cell identity with the dual gating of CD73 (+), CD90 (+), CD105(+), CD 34(−), and HLA-DR (−) for hMSCs postharvest from microcarrier culture.

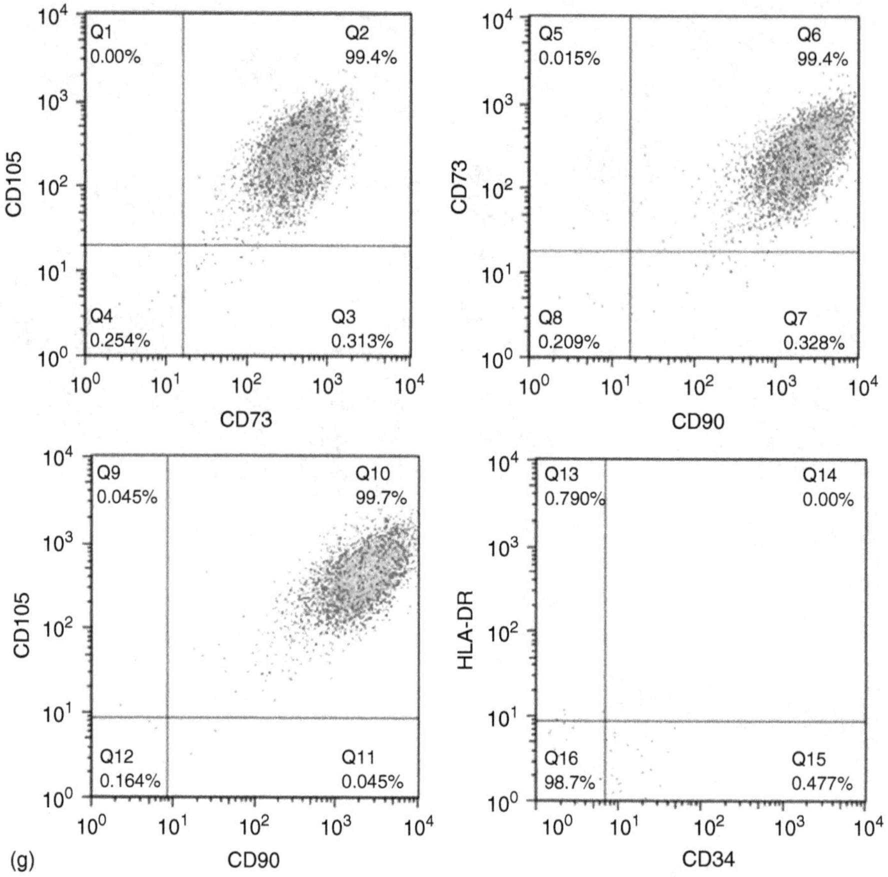


**Figure 4.5** (*Continued*)

In an attempt to increase the cell densities, research has focused on working toward the optimal bioprocess conditions for hMSC microcarrier culture, such as microcarrier type and concentration, dissolved oxygen, cell seeding density, medium composition (including serum and xeno-free), medium exchange regime, bioreactor configuration and agitation strategy amongst others. A summary of these studies is outlined in Table 4.3. For further details on the findings of these studies, the reader is directed to two reviews that summarize their findings [90, 91]. It is worth noting that although these studies have resulted in significant progress toward developing a scalable-hMSC microcarrier expansion system, there is still room for improvement, particularly given that many of these studies have only considered one factor at a time and have not considered a multifactorial approach and the potential effect of each parameter on others.

Similarly, numerous studies have been conducted to investigate the ability to culture human pluripotent stem cells (either hESCs or hiPSCs) on microcarriers in stirred-tank bioreactor systems, with some studies focusing on the cells retaining

**Table 4.3** Summary of cell therapy scale-up bioreactor technology.

| Bioreactor type | Current scale (l) | Manufacturers/companies | References |
|---|---|---|---|
| Rocking motion | 0.3–500 | GE Healthcare, Sartorius, Applikon, Pall | [86] |
| Pneumatically driven | 3–500 | PBS Biotech | [87] |
| Packed and fluidized bed | ≤5 | Eppendorf, Pall, Pluristem | [88, 89] |
| Stirred tank | 0.015–1000 | Sartorius, Applikon, Eppendorf, Pall | [63, 66, 77] |

their ability pluripotency [92, 93], while others seeking to generate a population of specialized cells [68, 94]. With pluripotent stem cells generally recognized to be a more challenging cell type to culture even in monolayer systems in comparison to other cell therapy candidates (e.g., hMSCs and HSCs), it is encouraging that the cells have been able to grow on microcarriers in stirred-tank conditions. Leung and colleagues did note, however, that hESCs can induce differentiation in microcarrier cultures, which may be the result of the hydrodynamic forces the cells experience in an agitated culture environment [95]. Yet, other studies involving human pluripotent stem cells [92, 93] have not reported spontaneous differentiation; as such further investigation will be necessary to identify the true effect of hydrodynamic forces on the cells.

As with the biopharmaceutical industry, there have been concerns raised about the effect of these hydrodynamic forces, in particular, shear stresses experienced by the cells during the expansion process. There is a perception that mammalian cells in general are "shear sensitive" due to their lack of a cell wall [55], and Nienow [55] suggests that this perception adversely impacted the development of free suspension mammalian cell culture. Nienow [55] further adds that the concern for shear sensitivity was excessive, and the majority of industrial processes using mammalian cells now employ stirred-tank bioreactors. This issue of shear sensitivity is also a concern within the cell therapy industry and is oft-cited as a disadvantage when using a stirred-tank bioreactor system [96, 97]. Obviously, as mentioned previously, given the need to retain cell's CQAs and not just viability, extra caution must be taken when considering the effect of hydrodynamic forces on stem cells grown on microcarriers.

The current approach to the understanding of damage to both freely suspended cells and cells on microcarriers is based on a comparison of the size of the suspended entity (cell or microcarrier) to the Kolmogorov microscale of turbulence [55] (where the Kolmogorov microscale of turbulence is an estimation of the size of the smallest eddies to order-of-magnitude accuracy and is a relationship between the local specific energy dissipation rate and the kinematic viscosity). If the size of the entity is small enough compared with the Kolmogorov scale, then damage to the cell should not occur. For growth on microcarriers, the Kolmogorov scale should not be less than ~1/2 to 2/3 of the size of the microcarrier [64, 75]. Since the microscale is reduced by increasing agitation intensity, to ensure that the microscale is not too small, it has been found best to operate the bioreactor at the minimum speed required to just suspend the microcarriers. Under these conditions, it has been shown that the cells maintain their CQA during culture [63, 64].

Much of the research on stem cell microcarrier culture has focused primarily on the expansion aspect of the process. Yet equally as important is the ability to effectively harvest the cells, which essentially requires a two-step approach involving first the detachment of the cells from the microcarrier surface and then the separation of the microcarriers from the cell suspension. Recently, we reported the first instance of a scalable harvesting technique for hMSCs cultured on microcarriers [64]. Much of the work before this study had only harvested milliliter samples for cell characterization and analysis studies, and as such it was still to be proven that a scalable technique could be employed to successfully harvest the cells. In brief, based on underlying theoretical concepts developed for secondary nucleation due to agitation, it can be shown that increases in agitator speed greatly increase the stresses detaching cells from the microcarriers. This leads to their rapid detachment (in about 7 min) but once detached, the cells are significantly smaller than the Kolmogorov scale. Thus, the harvesting involved increasing the agitation speed fivefold compared with that used during culture in the presence of a dissociation reagent to remove the cells from the surface of the microcarrier and subsequently a filtration step to separate the single-cell suspension from the microcarrier suspension [64]. A harvesting efficiency of >95% was obtained using this method and despite the fivefold increase in agitation speed, the harvested cells were viable and retained the key quality attributes outlining the potential to harvest cells of sufficient quantity and quality from microcarriers. This technique has now been used in our laboratory for harvesting three different donor cell lines using two different microcarriers in each case with the cells maintaining their CQAs. Given the underlying theories involved in this approach, the protocol is also essentially scalable.

## 4.7 Future Trends for Microcarrier Culture

With the need to develop regulatory compliant processes, it is becoming increasingly clear within the cell therapy industry that traditional stainless steel stirred-tank bioreactors are being overlooked in favor of single-use stirred-tank bioreactors (SUBs) [98, 99], a trend which is aligned with traditional biopharmaceutical production. This switch eases the regulatory burden of having to validate cleaning and sterilizing procedures and avoids the large capital and infrastructure costs associated with stainless steel stirred-tank bioreactors and their maintenance. It also simplifies the process by reducing the time and cost associated with setup and waste disposal procedures. However, this approach does have its drawbacks with respect to both the ongoing costs of SUBs and, perhaps of greater concern, the reliance upon a single supplier for a key aspect of the process. This limitation has implications with respect to both the reliability of supply and the quality of the SUBs. Any problem encountered by the supplier with respect to the supply of SUBs, by default, becomes an issue for the cell therapy company. Similarly, there will need to be stringent oversight and validation to ensure that the quality of SUBs supplied is consistent, as any variation in the build quality of the SUBs may have a direct impact on the cell therapy process. Another issue might be the ability to use the harvesting protocol described earlier in some SUBs because many of the configurations currently available would be unable to give the short burst of intense agitation required to detach the cells. It will also be important to show that they are able to effectively suspend the microcarriers at suitable low agitation desired to increase cell density.

At present, the number of cells produced on microcarriers in a stirred bioreactor compared with those obtained at confluence in a T-flask suggests that confluence has been achieved on the microcarriers. The current seeding density and microcarrier concentration is itself based on conditions in T-flasks and some studies aimed at optimizing them [48]. Clearly to get higher cell numbers, a higher concentration of microcarriers would help, and studies are being undertaken where more fresh ones are added during cultivation and then relying on bead-to-bead transfer to populate them. It is proving a nontrivial problem to date. In addition, at the current cell densities and scales, sufficient oxygen transfer can be achieved through the upper surface of the medium without sparging. At a certain cell density and scale, increased agitation and/or bubbling aeration will probably be required to satisfy the oxygen demand, again with all the issues of cell culture sensitivity (cell density and CDQs in this case) to fluid dynamic stresses, especially since bubbling is potentially more damaging [55] and the use of the protective surfactant, Pluronic F68, may not be acceptable since the cell is the product.

In addition to the shift toward SUBs, there has also been a focus on developing serum-, xeno-free culture processes. Again, the primary reason oft-cited for this is that of regulatory compliance and the need to remove all animal containing components from the process for safety reasons. In addition to this, however, there are also concerns about the batch-to-batch variability associated with serum, which can have significant implications on the process as well as serum availability and supply issues [100]. As such, there has been a trend toward reducing or eliminating serum from the process [101] and even toward developing completely serum- and xeno-free processes [77].

The ability to perform small-scale development work allows for process optimization to take place without the need to utilize large volumes of costly reagents or medium. Until recently, most small-scale development work for both traditional biopharmaceutical production and cell therapy manufacture was performed in smaller bioreactors (1–3 l) or spinner flasks (usually 100 ml) [55]. More recently, however, there has been an influx of microbioreactor technologies that allow for high-throughput studies using milliliter volumes. A key example of this is the advanced microscale bioreactor (ambr®) developed by TAP Biosystems (now a Sartorius Stedim company, Germany), which is a high-throughput automated bioreactor system comprising up to 48 individually controllable 15 ml stirred-tank bioreactor vessels with the ability to monitor and control pH, temperature, and dissolved oxygen. Its physical characteristics have recently been defined [48], and it has proved successful for biopharmaceutical research and development, and work conducted by various companies shows that it gives a process performance similar to large-scale bioreactor systems [102–104]. Research is currently being conducted by us to identify the amenability of the ambr® for hMSC microcarrier culture expansion (Figure 4.4b), where initial results appear promising and have demonstrated the ability to grow the cells on microcarriers in this system.

## 4.8 Preservation of Cell Therapy Products

The preservation of therapeutically active cells as part of the manufacturing process is unique to the cell therapy industry. This process step is necessary to decouple the production from the storage, shipment, and delivery of cell product to the clinic. It also allows for time to validate the product and complete quality assurance tests, before delivery. Preservation can either be long-term cryopreservation (order of years) or short term (order of days), which will both be required for the development of viable cell therapy products.

Cryopreservation of the cell therapy product will be required for an "off-the-shelf" business model where the product will always be available for delivery. Considering that a number of target clinical indications for cell therapy products will be emergency treatments, the long-term storage of these products is vital to their success. It will also be necessary if the product under development is manufactured from a master cell bank, which must be stored for a long term at a central manufacturing facility. The cryopreservation of cell therapy products has undergone much development, with an overview of 66 FDA submissions for MSC products showing that >80% considered the use of cryopreservation before product storage and transportation [105]. Despite this progress, much work is still required to ensure that the cryopreservation process does not have a detrimental impact on the cell function, with some clinical studies demonstrating reduced product efficacy following cryopreservation [106].

An alternative to this model is the short-term preservation (above 0 °C) of cell therapy products, which can then be delivered "fresh" to the patient without being frozen. This model has been adopted by a number of cell therapy companies, for example, Provenge (Dendreon Corporation, USA) has a shelf-life of 18 h, allowing for time to ship the product back to the patient for delivery. This model does add complexity to the product delivery, however, with a streamlined manufacturing and distribution process required to ensure the cells reach the patient within 18 h. Distributed manufacture of the product provides a viable compromise whereby the master bank is manufactured and cryopreserved at a central facility, which can then be shipped to local manufacturing centers where it is expanded and delivered fresh, reducing the complexity and risk in the distribution network. This also allows time to schedule the delivery procedure at a later date and has been adopted by companies such as Genzyme (USA) for the manufacture and delivery of their autologous chondrocyte therapy, Carticel [107].

## 4.9 Conclusions

The scale-up of stem cells intended for cell therapy applications will be necessary if the cell therapy industry is to come to fruition; however, this is not a trivial task. Although lessons can certainly be learnt from the manufacture of other biopharmaceuticals, there are key differences that provide distinct challenges, namely the living cell forming the basis of the product. This requires a focus on product quality and will necessitate that quality standards be built into the manufacturing facility and process; it is unlikely postexpansion purification techniques will be employed, with cell therapy manufacturers likely to focus on maintaining cell quality, right the way through from isolation of source material from donor to administration to the patient. Herein lies another key difference. Unlike the traditional biopharmaceutical production, there are many considerations, not least the different manufacturing paradigms (autologous, allogeneic, haplobank), which will dictate the manufacturing platform and strategy. As opposed to the default expansion system being a stirred-tank bioreactor, the expansion platform selected will form part of a wider cell production strategy, which must integrate with the isolation, logistic, harvesting, storage and preservation, and delivery processes. Although this poses challenges, it also presents significant opportunity to innovate with respect to technologies, processes, reimbursement strategies, and business models. For example, while autologous therapies place significant burden on transfer of material between donor, manufacturing site, and recipient, there may be opportunities to integrate closed manufacturing platforms within clinical settings. Moreover, given the attractive proposition of personalized medicine, it is likely that healthcare payers will be willing to pay for efficacious autologous therapies, even if the reimbursement values be higher than what healthcare payers are normally accustomed to. With this in mind, although stirred-tank bioreactors are rightly being considered by many and are proving particularly effective for allogeneic applications, it would be incorrect to suggest that only one manufacturing technology will be employed for the production of all cell therapies, but rather it is likely that the cell therapy technology landscape will include an array of manufacturing platforms, each specifically selected for each cell therapy process.

## References


1. Stuckey, D.W., Hingtgen, S.D., Karakas, N., Rich, B.E., and Shah, K. (2015) Engineering toxin-resistant therapeutic stem cells to treat brain tumors. *Stem Cells*, **33**(2):589–600. doi: 10.1002/stem.1874.
2. Mason, C., Brindley, D.A., Culme-Seymour, E.J., and Davie, N.L. (2011) Cell therapy industry: billion dollar global business with unlimited potential. *Regen. Med.*, **6**, 265–272.
3. Alliance for Regenerative Medicine (2013) Regenerative Medicine Annual Report. ARM, Washington, USA.
4. IBIS World (2015) Cell Therapy - Industry Market Research Report. Washington, USA, Report number: IBSS5523960.
5. TriMark Publications (2013) Regenerative Medicine Markets.
6. Health Canada (2012) Summary Basis of Decision (SBD) for PROCHYMAL®.
7. Caplan, A.I. and Correa, D. (2011) The MSC: an injury drugstore. *Cell Stem Cell*, **9**, 11–15.
8. Davie, N.L., Brindley, D.A., Culme-Seymour, E.J., and Mason, C. (2012) Streamlining cell therapy manufacture. *Bioprocess Int.*, **10**, 24–28.
9. Rowley, J., Abraham, E., Campbell, A., Brandwein, H., and Oh, S. (2012) Meeting lot-size challenges of manufacturing adherent cells for therapy. *Bioprocess Int.*, **10**, 16–22.
10. Mason, C. and Dunnill, P. (2007) Lessons for the nascent regenerative medicine industry from the biotech sector. *Regen. Med.*, **2**, 753–756.
11. Carmen, J., Burger, S.R., McCaman, M., and Rowley, J.A. (2012) Developing assays to address identity, potency, purity and safety: cell characterization in cell therapy process development. *Regen. Med.*, **7**, 85–100.
12. Bravery, C.A., Carmen, J., Fong, T., Oprea, W., Hoogendoorn, K.H., Woda, J., Burger, S.R., Rowley, J.A., Bonyhadi, M.L., and Van't Hof, W. (2013) Potency assay development for cellular therapy products: an ISCT review of the requirements and experiences in the industry. *Cytotherapy*, **15**, 9–19.
13. Sharma, S., Raju, R., Sui, S., and Hu, W.S. (2011) Stem cell culture engineering – process scale up and beyond. *Biotechnol. J.*, **6**, 1317–1329.
14. Hourd, P., Chandra, A., Medcalf, N., and Williams, D.J. (2014) Regulatory challenges for the manufacture and scale-out of autologous cell therapies, in *StemBook*, The Stem Cell Research Community.
15. Geron Corporation (2011) Geron Presents Clinical Data Update from GRNOPC1 Spinal Cord Injury Trial.
16. Hare, J.M., Traverse, J.H., Henry, T.D., Dib, N., Strumpf, R.K., Schulman, S.P., Gerstenblith, G., DeMaria, A.N., Denktas, A.E., Gammon, R.S., Hermiller, J.B. Jr., Reisman, M.A., Schaer, G.L., and Sherman, W. (2009) A randomized, double-blind, placebo-controlled, dose-escalation study of intravenous adult human mesenchymal stem cells (prochymal) after acute myocardial infarction. *J. Am. Coll. Cardiol.*, **54**, 2277–2286.
17. Montagnon, B.J., Fanget, B., and Vincent-Falquet, J.C. (1984) Industrial-scale production of inactivated poliovirus vaccine prepared by culture of Vero cells on microcarrier. *Rev. Infect. Dis.*, **6** (Suppl. 2), S341–S344.
18. Rourou, S., van der Ark, A., van der Velden, T., and Kallel, H. (2007) A microcarrier cell culture process for propagating rabies virus in Vero cells grown in a stirred bioreactor under fully animal component free conditions. *Vaccine*, **25**, 3879–3889.
19. Worton, R., Ho, C., and Duff, C. (1977) Chromosome stability in CHO cells. *Somatic Cell Genet.*, **3**, 27–45.
20. Wagner, W., Horn, P., Castoldi, M., Diehlmann, A., Bork, S., Saffrich, R., Benes, V., Blake, J., Pfister, S., Eckstein, V., and Ho, A.D. (2008) Replicative senescence of mesenchymal stem cells: a continuous and organized process. *PLoS One*, **3**, e2213.
21. Draper, J.S., Smith, K., Gokhale, P., Moore, H.D., Maltby, E., Johnson, J., Meisner, L., Zwaka, T.P., Thomson, J.A., and Andrews, P.W. (2004) Recurrent

gain of chromosomes 17q and 12 in cultured human embryonic stem cells. *Nat. Biotechnol.*, **22**, 53–54.

22. Tavassoli, M. and Crosby, W.H. (1968) Transplantation of marrow to extramedullary sites. *Science*, **161**, 54–56.
23. Heathman, T.R.J., Nienow, A.W., McCall, M.J., Coopman, K., Kara, B., and Hewitt, C.J. (2014) The translation of cell-based therapies: clinical landscape and manufacturing challenges. *Regen. Med.*, **10** (1), 49–64. DOI 10.2217/rme.14.73.
24. Glen, K.E., Workman, V.L., Ahmed, F., Ratcliffe, E., Stacey, A.J., and Thomas, R.J. (2013) Production of erythrocytes from directly isolated or Delta1 Notch ligand expanded CD34+ hematopoietic progenitor cells: process characterization, monitoring and implications for manufacture. *Cytotherapy*, **15**, 1106–1117.
25. Caplan, A.I. (1991) Mesenchymal stem cells. *J. Orthop. Res.*, **9**, 641–650.
26. Pittenger, M.F., Mackay, A.M., Beck, S.C., Jaiswal, R.K., Douglas, R., Mosca, J.D., Moorman, M.A., Simonetti, D.W., Craig, S., and Marshak, D.R. (1999) Multilineage potential of adult human mesenchymal stem cells. *Science*, **284**, 143–147.
27. Friedenstein, A.J., Chailakhjan, R.K., and Lalykina, K.S. (1970) The development of fibroblast colonies in monolayer cultures of guinea-pig bone marrow and spleen cells. *Cell Tissue Kinet.*, **3**, 393–403.
28. Friedenstein, A.J., Piatetzky, S. II, and Petrakova, K.V. (1966) Osteogenesis in transplants of bone marrow cells. *J. Embryol. Exp. Morphol.*, **16**, 381–390.
29. Zuk, P.A., Zhu, M., Mizuno, H., Huang, J., Futrell, J.W., Katz, A.J., Benhaim, P., Lorenz, H.P., and Hedrick, M.H. (2001) Multilineage cells from human adipose tissue: implications for cell-based therapies. *Tissue Eng.*, **7**, 211–228.
30. Zuk, P.A., Zhu, M., Ashjian, P., De Ugarte, D.A., Huang, J.I., Mizuno, H., Alfonso, Z.C., Fraser, J.K., Benhaim, P., and Hedrick, M.H. (2002) Human adipose tissue is a source of multipotent stem cells. *Mol. Biol. Cell*, **13**, 4279–4295.
31. Wang, H.S., Hung, S.C., Peng, S.T., Huang, C.C., Wei, H.M., Guo, Y.J., Fu, Y.S., Lai, M.C., and Chen, C.C. (2004) Mesenchymal stem cells in the Wharton's jelly of the human umbilical cord. *Stem Cells*, **22**, 1330–1337.
32. Erices, A., Conget, P., and Minguell, J.J. (2000) Mesenchymal progenitor cells in human umbilical cord blood. *Br. J. Haematol.*, **109**, 235–242.
33. Kogler, G., Sensken, S., Airey, J.A., Trapp, T., Muschen, M., Feldhahn, N., Liedtke, S., Sorg, R.V., Fischer, J., Rosenbaum, C., Greschat, S., Knipper, A., Bender, J., Degistirici, O., Gao, J.Z., Caplan, A.I., Colletti, E., Almeida-Porada, G., Muller, H.W., Zanjani, E., and Wernet, P. (2004) A new human somatic stem cell from placental cord blood with intrinsic pluripotent differentiation potential. *J. Exp. Med.*, **200**, 123–135.
34. von Bahr, L., Batsis, I., Moll, G., Hägg, M., Szakos, A., Sundberg, B., Uzunel, M., Ringden, O., and Le Blanc, K. (2012) Analysis of tissues following mesenchymal stromal cell therapy in humans indicates limited long-term engraftment and no ectopic tissue formation. *Stem Cells*, **30**, 1575–1578.
35. Thomson, J.A., Itskovitz-Eldor, J., Shapiro, S.S., Waknitz, M.A., Swiergiel, J.J., Marshall, V.S., and Jones, J.M. (1998) Embryonic stem cell lines derived from human blastocysts. *Science*, **282**, 1145–1147.
36. Atala, A. (2012) Human embryonic stem cells: early hints on safety and efficacy. *Lancet*, **379**, 689–690.
37. Takahashi, K., Tanabe, K., Ohnuki, M., Narita, M., Ichisaka, T., Tomoda, K., and Yamanaka, S. (2007) Induction of pluripotent stem cells from adult human fibroblasts by defined factors. *Cell*, **131**, 861–872.
38. Yu, J., Vodyanik, M.A., Smuga-Otto, K., Antosiewicz-Bourget, J., Frane, J.L., Tian, S., Nie, J., Jonsdottir, G.A., Ruotti, V., Stewart, R., Slukvin, I.I.,

and Thomson, J.A. (2007) Induced pluripotent stem cell lines derived from human somatic cells. *Science*, **318**, 1917–1920.
39. Taylor, C.J., Bolton, E.M., Pocock, S., Sharples, L.D., Pedersen, R.A., and Bradley, J.A. (2005) Banking on human embryonic stem cells: estimating the number of donor cell lines needed for HLA matching. *Lancet*, **366**, 2019–2025.
40. Nakajima, F., Tokunaga, K., and Nakatsuji, N. (2007) Human leukocyte antigen matching estimations in a hypothetical bank of human embryonic stem cell lines in the Japanese population for use in cell transplantation therapy. *Stem Cells*, **25**, 983–985.
41. Anonymous (2009) Regenerative medicine glossary. *Regen. Med.*, **4**, S1–S88.
42. Rayment, E.A. and Williams, D.J. (2010) Concise review: mind the gap: challenges in characterizing and quantifying cell- and tissue-based therapies for clinical translation. *Stem Cells*, **28**, 996–1004.
43. Bravery, C.A. (2012) Regulation: What are the Real Uncertainties? VALUE Project Final Report: Regenerative Medicine Value Systems: Navigating the Uncertainties, pp. 35–46, http://www.biolatris.com/Biolatris/News&uscore;&&uscore;events&uscore;files/VALUE%20Final%20Report.pdf (accessed 16 October 2015).
44. Binato, R., de Souza Fernandez, T., Lazzarotto-Silva, C., Rocher, B.D., Mencalha, A., Pizzatti, L., Bouzas, L.F., and Abdelhay, E. (2013) Stability of human mesenchymal stem cells during in vitro culture: considerations for cell therapy. *Cell Prolif.*, **46**, 10–22.
45. Dahl, J.A., Duggal, S., Coulston, N., Millar, D., Melki, J., Shahdadfar, A., Brinchmann, J.E., and Collas, P. (2008) Genetic and epigenetic instability of human bone marrow mesenchymal stem cells expanded in autologous serum or fetal bovine serum. *Int. J. Dev. Biol.*, **52**, 1033–1042.
46. Mason, C. and Dunnill, P. (2009) Assessing the value of autologous and allogeneic cells for regenerative medicine. *Regen. Med.*, **4**, 835–853.
47. McCall, M. and Williams, D.J. (2012) What Are the Alternative Manufacturing and Supply Models Available to Regenerative Medicine Companies and How Do the Finance Stack Up? VALUE Project Final Report: Regenerative Medicine Value Systems: Navigating the Uncertainties, pp. 55–65, http://www.biolatris.com/Biolatris/News&uscore;&&uscore;events&uscore;files/VALUE%20Final%20Report.pdf (accessed 16 October 2015).
48. Nienow, A.W., Rielly, C.D., Brosnan, K., Bargh, N., Lee, K., Coopman, K., and Hewitt, C.J. (2013) The physical characterisation of a microscale parallel bioreactor platform with an industrial CHO cell line expressing an IgG4. *Biochem. Eng. J.*, **76**, 25–36.
49. Uccelli, A., Moretta, L., and Pistoia, V. (2006) Immunoregulatory function of mesenchymal stem cells. *Eur. J. Immunol.*, **36**, 2566–2573.
50. Wuchter, P., Bieback, K., Schrezenmeier, H., Bornhauser, M., Muller, L.P., Bonig, H., Wagner, W., Meisel, R., Pavel, P., Tonn, T., Lang, P., Muller, I., Renner, M., Malcherek, G., Saffrich, R., Buss, E.C., Horn, P., Rojewski, M., Schmitt, A., Ho, A.D., Sanzenbacher, R., and Schmitt, M. (2015) Standardization of Good Manufacturing Practice-compliant production of bone marrow-derived human mesenchymal stromal cells for immunotherapeutic applications. *Cytotherapy*. **17**(2):128–39. doi:10.1016/j.jcyt.2014.04.002.
51. Culme-Seymour, E.J. and Mason, C. (2012) 'The Little Purple Book', 2nd edition: cell therapy and regenerative medicine glossary. *Regen. Med.*, **7**, 263–264.
52. Thomas, R.J., Chandra, A., Liu, Y., Hourd, P.C., Conway, P.P., and Williams, D.J. (2007) Manufacture of a human mesenchymal stem cell population using an automated cell culture platform. *Cytotechnology*, **55**, 31–39.

53. Pal, R., Hanwate, M., and Totey, S.M. (2008) Effect of holding time, temperature and different parenteral solutions on viability and functionality of adult bone marrow-derived mesenchymal stem cells before transplantation. *J. Tissue Eng. Regen. Med.*, **2**, 436–444.
54. Fekete, N., Rojewski, M.T., Fürst, D., Kreja, L., Ignatius, A., Dausend, J., and Schrezenmeier, H. (2012) GMP-compliant isolation and large-scale expansion of bone marrow-derived MSC. *PLoS One*, **7**, e43255.
55. Nienow, A.W. (2006) Reactor engineering in large scale animal cell culture. *Cytotechnology*, **50**, 9–33.
56. Meuwly, F., Ruffieux, P.A., Kadouri, A., and von Stockar, U. (2007) Packed-bed bioreactors for mammalian cell culture: bioprocess and biomedical applications. *Biotechnol. Adv.*, **25**, 45–56.
57. Oh, S.K.W. and Choo, A.B.H. (2008) Advances and perspectives in human and mouse embryonic stem cell bioprocessing. *Drug Discovery Today Technol.*, **5**, e125–e130.
58. Want, A.J., Nienow, A.W., Hewitt, C.J., and Coopman, K. (2012) Large-scale expansion and exploitation of pluripotent stem cells for regenerative medicine purposes: beyond the T flask. *Regen. Med.*, **7**, 71–84.
59. Amit, M., Chebath, J., Margulets, V., Laevsky, I., Miropolsky, Y., Shariki, K., Peri, M., Blais, I., Slutsky, G., Revel, M., and Itskovitz-Eldor, J. (2010) Suspension culture of undifferentiated human embryonic and induced pluripotent stem cells. *Stem Cell Rev.*, **6**, 248–259.
60. Fluri, D.A., Tonge, P.D., Song, H., Baptista, R.P., Shakiba, N., Shukla, S., Clarke, G., Nagy, A., and Zandstra, P.W. (2012) Derivation, expansion and differentiation of induced pluripotent stem cells in continuous suspension cultures. *Nat. Methods*, **9**, 509–516.
61. Baraniak, P.R. and McDevitt, T.C. (2012) Scaffold-free culture of mesenchymal stem cell spheroids in suspension preserves multilineage potential. *Cell Tissue Res.*, **347**, 701–711.
62. Ratcliffe, E., Glen, K.E., Workman, V.L., Stacey, A.J., and Thomas, R.J. (2012) A novel automated bioreactor for scalable process optimisation of haematopoietic stem cell culture. *J. Biotechnol.*, **161**, 387–390.
63. Rafiq, Q.A., Brosnan, K.M., Coopman, K., Nienow, A.W., and Hewitt, C.J. (2013) Culture of human mesenchymal stem cells on microcarriers in a 5 l stirred-tank bioreactor. *Biotechnol. Lett.*, **35**, 1233–1245.
64. Nienow, A.W., Rafiq, Q.A., Coopman, K., and Hewitt, C.J. (2014) A potentially scalable method for the harvesting of hMSCs from microcarriers. *Biochem. Eng. J.*, **85**, 79–88.
65. Melero-Martin, J.M. and Al-Rubeai, M. (2007) *In vitro* expansion of chondrocytes, in *Topics in Tissue Engineering*, vol. **3** (eds N. Ashammakhi, R. Reis, and E. Chiellini), University of Oulu, Oulu Finland, pp. 1–37.
66. Oh, S.K., Chen, A.K., Mok, Y., Chen, X., Lim, U.M., Chin, A., Choo, A.B., and Reuveny, S. (2009) Long-term microcarrier suspension cultures of human embryonic stem cells. *Stem Cell Res.*, **2**, 219–230.
67. Fok, E.Y. and Zandstra, P.W. (2005) Shear-controlled single-step mouse embryonic stem cell expansion and embryoid body-based differentiation. *Stem Cells*, **23**, 1333–1342.
68. Bardy, J., Chen, A.K., Lim, Y.M., Wu, S., Wei, S., Weiping, H., Chan, K., Reuveny, S., and Oh, S.K. (2013) Microcarrier suspension cultures for high-density expansion and differentiation of human pluripotent stem cells to neural progenitor cells. *Tissue Eng. Part C Methods*, **19**, 166–180.
69. Schop, D., van Dijkhuizen-Radersma, R., Borgart, E., Janssen, F.W., Rozemuller, H., Prins, H.J., and de Bruijn, J.D. (2010) Expansion of human mesenchymal stromal cells on microcarriers: growth and metabolism. *J. Tissue Eng. Regen. Med.*, **4**, 131–140.
70. dos Santos, F., Andrade, P.Z., Eibes, G., da Silva, C.L., and Cabral, J.M. (2011)

Ex vivo expansion of human mesenchymal stem cells on microcarriers. *Methods Mol. Biol.*, **698**, 189–198.

71. Liu, Y., Liu, T., Fan, X., Ma, X., and Cui, Z. (2006) Ex vivo expansion of hematopoietic stem cells derived from umbilical cord blood in rotating wall vessel. *J. Biotechnol.*, **124**, 592–601.
72. Li, Q., Liu, Q., Cai, H., and Tan, W.S. (2006) A comparative gene-expression analysis of CD34+ hematopoietic stem and progenitor cells grown in static and stirred culture systems. *Cell. Mol. Biol. Lett.*, **11**, 475–487.
73. Frauenschuh, S., Reichmann, E., Ibold, Y., Goetz, P.M., Sittinger, M., and Ringe, J. (2007) A microcarrier-based cultivation system for expansion of primary mesenchymal stem cells. *Biotechnol. Prog.*, **23**, 187–193.
74. Ferrari, C., Balandras, F., Guedon, E., Olmos, E., Chevalot, I., and Marc, A. (2012) Limiting cell aggregation during mesenchymal stem cell expansion on microcarriers. *Biotechnol. Prog.*, **28**, 780–787.
75. Hewitt, C.J., Lee, K., Nienow, A.W., Thomas, R.J., Smith, M., and Thomas, C.R. (2011) Expansion of human mesenchymal stem cells on microcarriers. *Biotechnol. Lett.*, **33**, 2325–2335.
76. Yuan, Y., Kallos, M.S., Hunter, C., and Sen, A. (2010) Improved expansion of human bone marrow-derived mesenchymal stem cells in microcarrier-based suspension culture. *J. Tissue Eng. Regen. Med.* 15;**146**(4):194–7. doi: 10.1016/j.jbiotec.2010.02.015.
77. dos Santos, F., Andrade, P.Z., Abecasis, M.M., Gimble, J.M., Chase, L.G., Campbell, A.M., Boucher, S., Vemuri, M.C., Silva, C.L., and Cabral, J.M. (2011) Toward a clinical-grade expansion of mesenchymal stem cells from human sources: a microcarrier-based culture system under xeno-free conditions. *Tissue Eng. Part C Methods*, **17**, 1201–1210.
78. Eibes, G., dos Santos, F., Andrade, P.Z., Boura, J.S., Abecasis, M.M., da Silva, C.L., and Cabral, J.M. (2010) Maximizing the ex vivo expansion of human mesenchymal stem cells using a microcarrier-based stirred culture system. *J. Biotechnol.*, **146**, 194–197.
79. Goh, T.K., Zhang, Z.Y., Chen, A.K., Reuveny, S., Choolani, M., Chan, J.K., and Oh, S.K. (2013) Microcarrier culture for efficient expansion and osteogenic differentiation of human fetal mesenchymal stem cells. *BioRes. Open Access*, **2**, 84–97.
80. Cormier, J.T., zur Nieden, N.I., Rancourt, D.E., and Kallos, M.S. (2006) Expansion of undifferentiated murine embryonic stem cells as aggregates in suspension culture bioreactors. *Tissue Eng.*, **12**, 3233–3245.
81. Abranches, E., Bekman, E., Henrique, D., and Cabral, J.M. (2007) Expansion of mouse embryonic stem cells on microcarriers. *Biotechnol. Bioeng.*, **96**, 1211–1221.
82. zur Nieden, N.I., Cormier, J.T., Rancourt, D.E., and Kallos, M.S. (2007) Embryonic stem cells remain highly pluripotent following long term expansion as aggregates in suspension bioreactors. *J. Biotechnol.*, **129**, 421–432.
83. Cameron, C.M., Hu, W.S., and Kaufman, D.S. (2006) Improved development of human embryonic stem cell-derived embryoid bodies by stirred vessel cultivation. *Biotechnol. Bioeng.*, **94**, 938–948.
84. Sun, L.-Y., Lin, S.-Z., Li, Y.-S., Harn, H.-J., and Chiou, T.-W. (2011) Functional cells cultured on microcarriers for use in regenerative medicine research. *Cell Transplant.*, **20**, 49–62.
85. Rodrigues, C.A., Fernandes, T.G., Diogo, M.M., da Silva, C.L., and Cabral, J.M. (2011) Stem cell cultivation in bioreactors. *Biotechnol. Adv.*, **29**, 815–829.
86. Correia, C., Serra, M., Espinha, N., Sousa, M., Brito, C., Burkert, K., Zheng, Y., Hescheler, J., Carrondo, M.J., Saric, T., and Alves, P.M. (2014) Combining hypoxia and bioreactor hydrodynamics boosts induced pluripotent stem cell differentiation towards cardiomyocytes. *Stem Cell Rev.*, **10** (6), 786–801.

87. Obom, K.M., Cummings, P.J., Ciafardoni, J.A., Hashimura, Y., and Giroux, D. (2014) Cultivation of mammalian cells using a single-use pneumatic bioreactor system. *J. Vis. Exp.*, **92**, e52008.
88. Prather, W.R., Toren, A., Meiron, M., Ofir, R., Tschope, C., and Horwitz, E.M. (2009) The role of placental-derived adherent stromal cell (PLX-PAD) in the treatment of critical limb ischemia. *Cytotherapy*, **11**, 427–434.
89. Weber, C., Freimark, D., Portner, R., Pino-Grace, P., Pohl, S., Wallrapp, C., Geigle, P., and Czermak, P. (2010) Expansion of human mesenchymal stem cells in a fixed-bed bioreactor system based on non-porous glass carrier--part A: inoculation, cultivation, and cell harvest procedures. *Int. J. Artif. Organs*, **33**, 512–525.
90. Jung, S., Panchalingam, K.M., Wuerth, R.D., Rosenberg, L., and Behie, L.A. (2012) Large-scale production of human mesenchymal stem cells for clinical applications. *Biotechnol. Appl. Biochem.*, **59**, 106–120.
91. Rafiq, Q.A., Coopman, K., and Hewitt, C.J. (2013) Scale-up of human mesenchymal stem cell culture: current technologies and future challenges. *Curr. Opin. Chem. Eng.*, **2**, 8–16.
92. Storm, M.P., Orchard, C.B., Bone, H.K., Chaudhuri, J.B., and Welham, M.J. (2010) Three-dimensional culture systems for the expansion of pluripotent embryonic stem cells. *Biotechnol. Bioeng.*, **107**, 683–695.
93. Phillips, B.W., Horne, R., Lay, T.S., Rust, W.L., Teck, T.T., and Crook, J.M. (2008) Attachment and growth of human embryonic stem cells on microcarriers. *J. Biotechnol.*, **138**, 24–32.
94. Lu, S.J., Kelley, T., Feng, Q., Chen, A., Reuveny, S., Lanza, R., and Oh, S.K. (2013) 3D microcarrier system for efficient differentiation of human pluripotent stem cells into hematopoietic cells without feeders and serum [corrected]. *Regen. Med.*, **8**, 413–424.
95. Leung, H.W., Chen, A., Choo, A.B., Reuveny, S., and Oh, S.K. (2011) Agitation can induce differentiation of human pluripotent stem cells in microcarrier cultures. *Tissue Eng. Part C Methods*, **17**, 165–172.
96. Hambor, J.E. (2012) Bioreactor design and bioprocess controls for industrialized cell processing. *Bioprocess Int.*, **10**, 22–33.
97. Carlos, A.V.R., Tiago, G.F., MariaMargarida, D., da CláudiaLobato, S., and Joaquim, M.S.C. (2012) *Stem Cell Engineering*, CRC Press, pp. 1–28.
98. Szczypka, M., Splan, D., Woolls, H., and Brandwein, H. (2014) Single-use bioreactors and microcarriers. *Bioprocess Int.*, **12**, 54–61.
99. Simaria, A.S., Hassan, S., Varadaraju, H., Rowley, J., Warren, K., Vanek, P., and Farid, S.S. (2014) Allogeneic cell therapy bioprocess economics and optimization: single-use cell expansion technologies. *Biotechnol. Bioeng.*, **111**, 69–83.
100. Brindley, D.A., Davie, N.L., Culme-Seymour, E.J., Mason, C., Smith, D.W., and Rowley, J.A. (2012) Peak serum: implications of serum supply for cell therapy manufacturing. *Regen. Med.*, **7**, 7–13.
101. Jung, S., Sen, A., Rosenberg, L., and Behie, L.A. (2012) Human mesenchymal stem cell culture: rapid and efficient isolation and expansion in a defined serum-free medium. *J. Tissue Eng. Regen. Med.*, **6**, 391–403.
102. Lewis, G., Lugg, R., Lee, K., and Wales, R. (2010) A qualitative and quantitative mimic for early process development. *Bioprocess. J.*, **9**, 22–25.
103. Hsu, W.T., Aulakh, R.P., Traul, D.L., and Yuk, I.H. (2012) Advanced microscale bioreactor system: a representative scale-down model for bench-top bioreactors. *Cytotechnology*, **64**, 667–678.
104. Moses, S., Manahan, M., Ambrogelly, A., and Ling, W.L.W. (2012) Assessment of AMBRTM as a model for high-throughput cell culture process development strategy. *Adv. Biosci. Biotechnol.*, **3**, 918.

105. Mendicino, M., Bailey, A.M., Wonnacott, K., Puri, R.K., and Bauer, S.R. (2014) MSC-based product characterization for clinical trials: an FDA perspective. *Cell Stem Cell*, **14**, 141–145.
106. Moll, G., Alm, J.J., Davies, L.C., von Bahr, L., Heldring, N., Stenbeck-Funke, L., Hamad, O.A., Hinsch, R., Ignatowicz, L., Locke, M., Lönnies, H., Lambris, J.D., Teramura, Y., Nilsson-Ekdahl, K., Nilsson, B., and Le Blanc, K. (2014) Do cryopreserved mesenchymal stromal cells display impaired immunomodulatory and therapeutic properties? *Stem Cells*, **32**, 2430–2442.
107. Coopman, K. and Medcalf, N. (2014) From Production to Patient: Challenges and Approaches for Delivering Cell Therapies, http://www.stembook.org/node/6149 (accessed 31 October 2015).

# 5
# Artificial Liver Bioreactor Design

*Katrin Zeilinger and Jörg C. Gerlach*

## 5.1
## Need for Innovative Liver Therapies

Liver failure as a consequence of chronic progressive or acute liver disease represents a life-threatening situation that may lead to hepatic coma and finally to death if not adequately treated. The available conventional therapies, although often beneficial for alleviation of clinical symptoms, are not effective in causal treatment of acute-on-chronic or acute liver disease. Thus, to date the only efficient therapy of end-stage liver failure is liver transplantation. The introduction of liver transplantation as therapeutic option has significantly reduced the mortality of acute liver failure patients [1]. Progresses made within the last decades in surgical techniques, intensive care, immunosuppressive regimen, and organ preservation methods have made liver transplantation a form of well-established and successful therapy. The 5-year survival rate in the United States was 70.5% for deceased donor transplants performed in 2007 [2]. However, the existing shortage in available donor organs allows no significant expansion of transplantation programs, and the number of patients on the waiting list for transplantation largely exceeds the number of donor organs available for transplantation.

Thus, there is an urgent need for the development of alternative or supplementary therapies to liver transplantation. Two principal therapeutic approaches are conceivable: temporary liver support to bridge the organ function until transplantation or until regeneration of the own liver or permanent organ cure by cell transplantation to repair the damaged tissue. Figure 5.1 illustrates the potential applications of bioreactor technologies in liver therapies.

## 5.2
## Requirements to Liver Support Systems

To provide successful temporary support of the failing organ in end-stage liver disease, the complex metabolic performances of liver cells have to be taken into account. Major functions of the liver are (i) transformation of endogenous

*Bioreactors: Design, Operation and Novel Applications*, First Edition. Edited by Carl-Fredrik Mandenius.

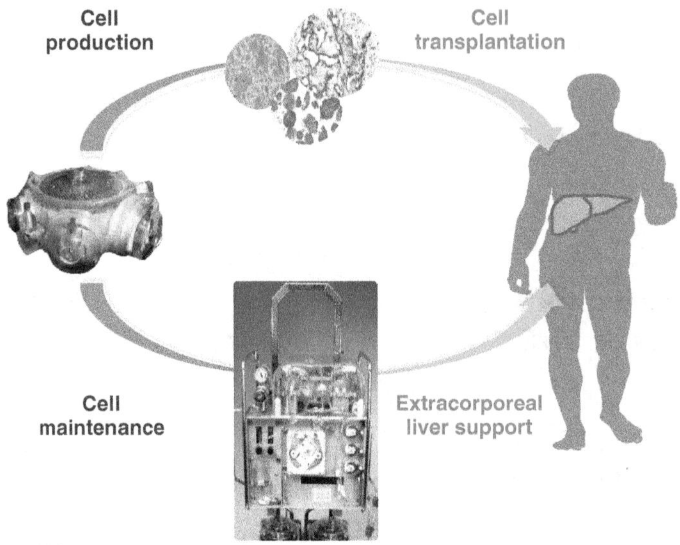


**Figure 5.1** Potential clinical applications of bioreactors for cell maintenance to support the liver function in extracorporeal devices or for production of cells to be transplanted into the injured organ.

metabolites (e.g., ammonia) or exogenous toxic components (e.g., drugs, chemicals) into nontoxic metabolites, (ii) regulation of small molecules, in particular amino acids, carbohydrates, and fatty acids, and (iii) the synthesis of proteins such as albumin or coagulation factors that are released by the liver into the blood. Disturbances in one of several of these functions may lead to toxin accumulation, imbalances in plasma compounds, for example, free amino acids, and/or hypoproteinemia, with detrimental consequences for the patient. Thus, an ideal liver support system would fulfill all main tasks of the human liver in terms of hepatic detoxification, regulation, and synthesis.

Apart from the functionality and efficacy of liver support systems, safety issues have to be considered. In particular, systems used for supporting the liver function in diseased liver need to be stable over the treatment period and show reproducible functions to ensure a standardized quality. In addition, investigation of the biocompatibility of materials used and preventing infection of patients are demanded to ensure the clinical safety of treatments.

## 5.3 Bioreactor Technologies Used in Clinical Trials

### 5.3.1 Artificial Liver Support Systems

Artificial liver support systems are aimed to remove toxins from the blood of patients with liver failure using mechanical and/or chemical adsorption techniques [3, 4]. In order to provide efficient detoxification, most of these systems rely upon dialysis-type detoxification of the patient plasma with adsorbents, using either activated carbon suspended in the dialysate for toxin adsorption, as realized in the Biologic-DT [5, 6], or albumin provided in the dialysis liquid as a physiological scavenger of lipophilic compounds [7–9]. The molecular adsorbent recirculating system (MARS) [10] combines albumin dialysis with a charcoal and anion exchange column, which enables recycling of the albumin molecules after passing the adsorbent column. The system was shown to provide efficient removal of plasma toxins from the plasma in preclinical and clinical studies [10, 11]. A different approach is pursued in the fractionated plasma separation and adsorption (FPSA) detoxification device, also known as Prometheus system [12, 13], that directly conducts plasma through solid adsorbent columns to facilitate the removal of toxins.

Results of several randomized clinical trials of artificial support systems have been published recently. The effect on acute liver failure of albumin dialysis using the MARS system was investigated in a trial including 49 patients treated conventionally and 39 with albumin dialysis therapy [14]. Although a tendency toward higher survival rates was observed in the MARS group, definitive conclusions on the efficacy or safety of the system were not possible. Within the RELIEF trial, 189 patients with acute-on-chronic liver failure were subjected either to MARS treatment ($n = 95$) or to standard therapy ($n = 94$) [15]. The results showed that MARS treatment had no significant effect on the survival rate, while hepatic encephalopathy was more frequently improved in the MARS group as compared with the control. In another randomized clinical study, the FPSA was applied in 77 patients with acute-on-chronic liver failure, and compared with 68 patients treated with standard therapy [16]. Again, the probability of survival was not increased by liver support with FPSA. These results suggest that physical–chemical detoxification of plasma alone is not sufficient to address the liver function in a way that the survival probability is increased. The question of whether the quality of life of patients with acute or acute-on-chronic liver disease can be improved by artificial liver support systems is still open.

### 5.3.2 Bioartificial Liver Support Systems

Bioartificial liver support systems are intended to reflect the functions of vital liver cells to allow for liver-specific metabolism and synthesis performances in addition to plasma detoxification. Clinical therapies with bioartificial liver support systems address a temporary support of the failing organ until liver transplantation is available or until organ recovery in case of acute or acute-on-chronic hepatic disease [17, 18].

In order to adequately compensate the metabolic and regulatory performances of the failing organ in clinical application, bioartificial liver support systems have to

- provide differentiated, human-specific hepatic functions

- procure sufficient cell quantities for efficient liver support in patients
- ensure the stable maintenance of metabolic activities
- enable a reproducible cell quality for standardized clinical applications
- ensure the clinical safety of the system, in particular, with respect to the cell source used
- allow the flow of blood and plasma with a mass exchange that can quantitatively address the required metabolism for the patient and the possible metabolism of the cells
- ensure that problems of blood cell damage and coagulation during blood perfusion can be avoided
- avoid negative interactions with the patients' coagulation system, while anticoagulation may be required

Different systems and cell types have been used for clinical liver support, as reviewed in Refs [19–21]. Table 5.1 gives an overview of bioartificial liver support systems that have been clinically used for extracorporeal liver support. Most of these systems represent two-compartment devices where the cells reside in the space between perfused hollow-fiber capillaries serving for plasma or blood perfusion. For example, the extracorporeal liver assist device (ELAD® [22, 23, 31]) and the bioartificial liver support system (BLSS [24, 25]) are both based on modified dialysis-based cartridges loaded with the human hepatoblastoma cell line C3A (ELAD) or with porcine hepatocytes (BLSS). The HepatAssist system [26, 32] utilizes hollow-fiber cartridges charged with cryopreserved porcine hepatocytes in combination with a charcoal column for plasma detoxification, and the Amsterdam Medical Centre Bioartificial Liver Device (AMC-BAL [27]) utilizes porcine liver cells cultivated on a spirally wound polyester fabric, while oxygen is supplied via hollow fibers. In the radial flow bioreactor [28], primary porcine hepatocytes are cultivated in a nonwoven-mesh scaffold, which is perfused in a radial flow configuration. Our own development, the modular extracorporeal liver support (MELS) consists of a four-compartment hollow fiber bioreactor technology [33, 34], which can be combined with single-pass albumin dialysis for detoxification. The system has been clinically applied with primary porcine [29] or human [30] liver cells.

The results from clinical evaluation of those devices have been described in various reviews [35–37]. Several clinical trials showed a beneficial effect on clinical parameters of bioartificial liver support treatment, including stabilization of hemodynamics and coagulopathy, decreases in ammonia and bilirubin levels and hepatoencephalopathy. However, a validated evaluation of the clinical efficacy is not yet available, due to the lack of randomized clinical studies.

A major precondition to adequately support the failing liver is the provision of sufficient cell numbers. Based on the experience from clinical partial liver resection, a cell mass of 150–300 g has been suggested to be necessary for efficient liver support in patients with liver failure [38]. Most bioartificial liver devices have been using a cell mass in the range of 100–200 g cells, which is at the lower limit of the critical cell mass needed. However, since the functional performance of cells cultured *in vitro* is probably lower than that of cells in their natural environment *in*

**Table 5.1** Bioartificial liver support systems used in clinical trials.

| Bioreactor technology | Cell type used | Indications in clinical trials | Clinical outcome |
|---|---|---|---|
| Hollow fiber–based bioartificial liver device perfused with plasma (ELAD) [22, 23] | Human hepatoblastoma cell line (C3A) | Acute/fulminant liver failure | No significant difference in survival, improvement in galactose elimination and encephalopathy |
| Hollow fiber–based bioartificial liver device perfused with whole blood (BLSS) [24, 25] | Primary porcine hepatocytes | Acute/fulminant liver failure (safety study) | No serious adverse events; treatment well tolerated by patients |
| Hollow fiber–based bioartificial liver with hepatocytes attached to dextran microcarriers (HepatAssist) [26] | Cryopreserved porcine hepatocytes | Acute liver failure/primary nonfunction after transplantation | Tendency toward improved survival, yet not significant |
| Amsterdam Medical Centre Bioartificial Liver Device (AMC-BAL) [27] | Primary porcine hepatocytes | Acute liver failure/bridging to liver transplantation | No severe adverse events, successful bridging to liver transplantation shown |
| Radial flow bioreactor perfused with plasma (RFB-BAL) [28] | Primary porcine hepatocytes | Acute liver failure/bridging to liver transplantation | Improvement of encephalopathy level, decrease in ammonia and transaminases |
| Hollow fiber–based bioartificial liver with integral oxygenation (MELS) [29, 30] | Primary porcine or human liver cells | Acute liver failure/bridging to transplantation | No severe adverse events; in some patients, clinical and/or biochemical improvement |

*vivo*, the needed cell mass in extracorporeal culture devices might be even higher than the cell mass estimated based on *in vivo* data.

The need of increased cell numbers is addressed in the MELS bioreactor [39]. The technology is based on three independent capillary systems for countercurrent plasma perfusion and integral oxygenation in the bioreactor, enabling decentralized but high-efficient nutrient supply and metabolite removal; and high-performance gas exchange/oxygenation within the interwoven capillary subunits, providing more physiological gradients. The construction allows up- or downscaling of the cell compartment by variation of the length and number of capillaries. By this way, different size variants of the technology were constructed, ranging from laboratory systems for *in vitro* research with a cell compartment volume of 8 ml or 2 ml to large-scale clinical systems with a cell compartment of 800 ml, which comprise $\sim 1.4 \times 10^{10}$ cells corresponding to a cell mass of 500 g [40]. Studies on primary human liver cells cultured in bioreactors showed stable maintenance of hepatic functions, including urea and albumin synthesis, galactose and sorbitol uptake, amino acid metabolism, and lidocaine conversion [41–43]. A miniaturized version of the bioreactor technology was successfully applied for drug metabolism studies on primary human hepatocytes under defined perfusion conditions [44, 45]. The structure of the bioreactor technology and the histology of primary human liver cells in the bioreactor cell compartment are shown in Figure 5.2.

To provide controlled medium supply and oxygenation, the bioreactor is integrated into separate tubing systems for medium and gas perfusion (Figure 5.3). Bioreactor capillaries are connected via inflow and outflow ports to a tubing circuit with an integrated pump for medium recirculation through the bioreactor. The system pressure is continuously monitored via pressure sensors, which allows for automated stopping of the pump in case of drastic pressure increase. A separate pump is used for continuous substitution of fresh medium, while used medium is removed from the circuit as a consequence of increased hydrostatic pressure upon the addition of fresh medium. A gas mixture of air, $CO_2$, and additional oxygen, if required, is supplied via electronically controlled mass flow meters. Adjustment of the pH value in the recirculating medium is possible by adapting the $CO_2$ concentration in the air mixture. In addition, the system provides the option for integration of various sensors to control culture conditions, such as pH, oxygen, or metabolic parameters.

## 5.4
## Optimization of Bioartificial Liver Bioreactor Designs

To enable long-term maintenance of cell functionality, culture systems need to approximate the physiological situation of the cells in the natural liver with regard to nutrient supply, metabolite removal, and oxygenation, and they have to support cell adhesion and communication [46, 47]. Oxygenation via gassing of the perfusate performed in the typical two-compartment hollow-fiber systems can

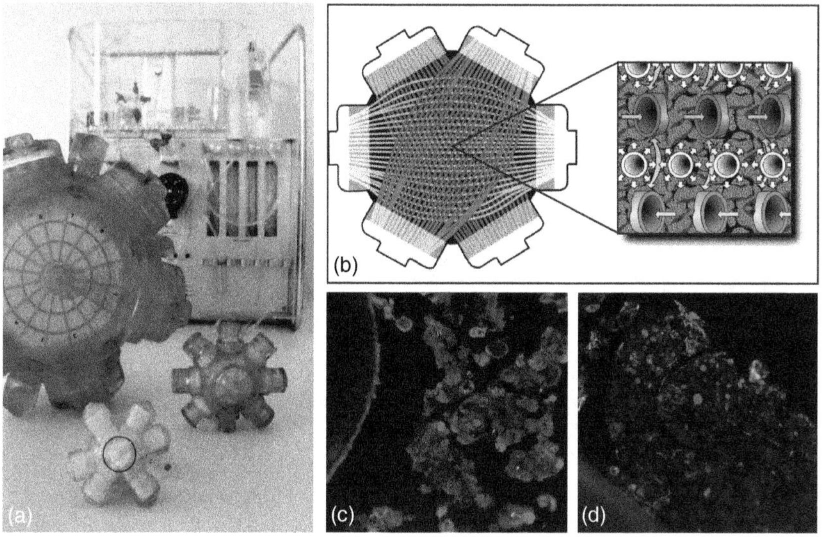


**Figure 5.2** (a) Four-compartment hollow-fiber bioreactors for clinical use in extracorporeal liver support systems (left) or in downscaled versions for laboratory research (right, foreground); (b) structure of the capillary network with independent capillary systems for countercurrent medium/plasma perfusion (blue and red) and oxygenation (yellow); (c) detection of cytochrome 2C9 (red) and the transporter protein multidrug resistance protein 2 (MRP2, green); (d) reorganization of hepatocytes (CK 18, green) and nonparenchymal cells (vimentin, red) between the artificial capillaries in the bioreactor.

generate gas gradients in the cell compartment associated with insufficient oxygen supply of part of the cells. Direct cell compartment perfusion, realized for example, in the AMC-BAL [27] and in the radial flow bioreactor [28] could lead to enhanced exposition of shear stress of the cells, in addition to potential immunological interactions in case of plasma perfusion during clinical application. In contrast, "indirect" perfusion, that is, diffusion through capillary membranes according to solute gradients, as performed in the dialysis-based systems creates mass gradients along the capillary distance, which could impair mass exchange especially in areas near the outflow.

Various bioreactor technologies have been developed to improve oxygen and mass flow characteristics in bioartificial devices. For example, flat membrane bioreactors allowing integral oxygenation were successfully tested *in vitro* and *in vivo* with porcine hepatocytes [48, 49]. Treatment with a bioartificial liver based on the flat membrane technology increased the survival time of pigs in an animal model of fulminant liver failure [50]. Integral oxygenation was also addressed in a coaxial hollow fiber–based bioreactor used for the cultivation of rat hepatocytes [51]. Further approaches include polymer scaffolds with intrinsic channels constructed via 3D printing [52], microarray bioreactors [53, 54], and polyester fiber beds containing autologous biomatrix [55]. While these systems

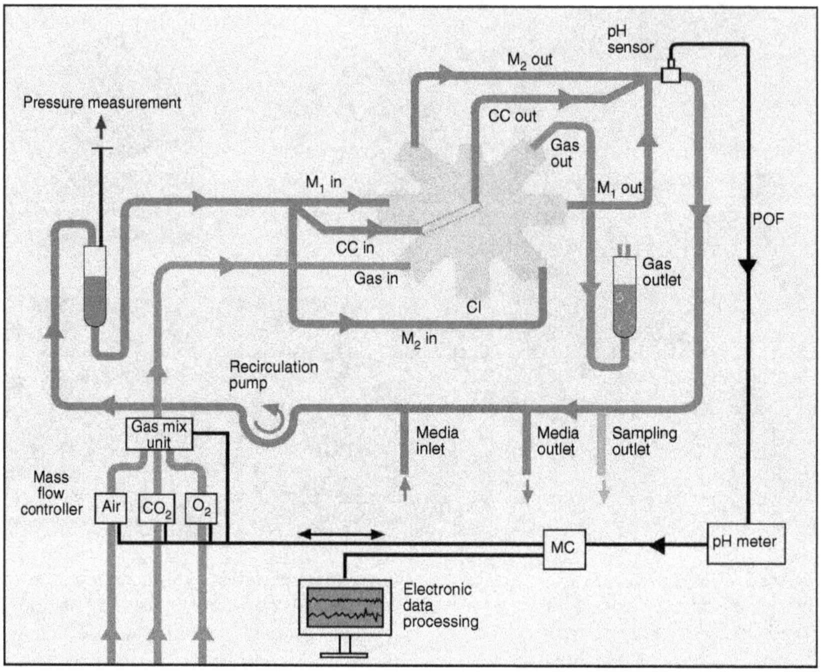


**Figure 5.3** Bioreactor perfusion circuit with tubing for pump-driven medium recirculation through the bioreactor (red: inflow tube sections, blue: outflow tube sections) and for fresh medium inlet (red) as well as used medium outlet (blue). The bioreactor disposes of two medium perfusion capillary systems ($M_1$ in and out, $M_2$ in and out), which are countercurrent perfused to enhance mass exchange. Cells are inoculated via a tube line (CC in) branching from the recirculation circuit. An electronically controlled gas mix unit provides defined flow rates and concentrations of gases in the supplied gas mixture (gray lines, gas in and gas out). Automated pH control is possible by the integration of pH sensors into the perfusion circuit. POF, plastic optical fiber, MC, microcontroller.

show the feasibility of complex culture technologies at a laboratory scale, a major challenge is yet the extension of these designs toward clinical dimensions.

Successful scale-up of liver cell culture for the production of up to $10^{11}$ cells in a volume of 1.1 l was achieved in a fluidized-bed bioreactor with alginate-based spherical beads, which enable the formation of compact aggregates with high cell densities [56]. The device was recently evaluated for the treatment of ischemic acute liver failure in pigs using encapsulated HepG2 cell spheroids [57]. Animals treated with the spheroid-based bioartificial liver showed an improvement in clinical parameters, including significantly lower increases in intracranial pressure, lower ammonia levels, and higher bilirubin conjugation compared with the control group connected to a similar system without cells. Studies on microencapsulation of HepaRG cells in alginate spheroids revealed a polarized organization of hepatocyte-like cells in the spheroids, exhibiting an interconnected bile canalicular network and biliary transporter activity [58]. Dynamic fluidized-bed bioreactors showed a higher viability and better microstructure of encapsulated immortalized hepatocytes as compared with static cell cultivation in a spinner flask system [59].

Combinations of different technologies and materials were also tested for bioartificial liver development. For example, Zhang *et al.* used a polyurethane scaffold integrated with single-layer polyethersulfone hollow fibers for hepatocyte culture [60]. Concentrations of ammonia and unconjugated bilirubin in plasma from patients with liver failure were significantly decreased during 6 h of circulation.

Future research may also address biomechanical aspects of mass exchange by mimicking the liver physiology. The liver architecture and vascularization is more complex than in other organs due to its central role in the metabolism of endogenous and exogenous substances. Blood from the intestines enters the liver via the portal vein to ensure rapid uptake and metabolism (first-pass effect) and leaves the organ via the central vein. In addition to the portal-venous blood, arterial blood passes the sinusoidal capillaries supplying the hepatocytes with oxygen and soluble factors, for example, hormones. The sinusoidal endothelial cells lining the hepatocytes are acting as a selective barrier toward the blood; substance permeation to and from hepatocytes is facilitated by fenestrations arranged in sieve plates, which distinguishes them from the capillaries found in any other organs [61, 62]. The plasma exchange between the sinusoidal space and the hepatocytes is driven by mechanical forces created by the heart pulsation, pressure changes due to the breathing movements, and the regulation of sinusoidal perfusion [63, 64]. In addition to mechanical forces, there are indications that a biological regulation of plasma filtration through the sinusoidal endothelium occurs via contractions of Ito cells (stellate cells), which are located outside the sinusoids in the space of Disse [65]. Changes in the contraction state of Ito cells may influence the interstitial and sinusoidal pressure and thereby regulate transport processes through the vascular lining. These processes may be reflected in the development of future bioreactor generations to enable mass exchange in bioartificial systems close to the *in vivo* situation.

In this respect, metabolic flux analysis approaches might be useful to get a deeper insight into hepatic cell metabolism under physiological and pathophysiological conditions [66–68] and to evaluate the effect of potential influencing factors [69, 70]. In addition, flux balance models may be used for the evaluation and optimization of culture technologies and processing conditions for applications in hepatic tissue engineering and bioartificial liver support [71–73].

## 5.5
## Improvement of Cell Biology in Bioartificial Livers

The clinical use of bioartificial liver devices requires stable maintenance of hepatocytes over several days up to weeks. There is increasing evidence that the functionality of hepatocytes *in vitro* is favored by conditions reflecting the physiological microenvironment of the cells in the organ, including provision of physiological matrix proteins [74, 75], coculture with nonparenchymal cells [76], and supply with soluble factors [77].

For example, it could be shown that rat small hepatocytes have the potential to reconstruct hepatic organoids including bile canalicular networks *in vitro* [78]. Cocultivation of hepatocytes with fibroblasts was successfully tested in micropatterned borosilicate wafers [79]. Enhanced tissue formation by the cells has also been observed in 3D culture systems involving coculture with hepatic nonparenchymal cells in a collagen sponge model [80] or on microporous membranes [81]. The four-compartment hollow-fiber membrane technology employed in the MELS system addresses the cellular needs of 3D tissue density conditions by providing a 3D scaffold made of interwoven fibers for cell adhesion and reorganization in the space between the capillaries. Spontaneous reassembling of primary human hepatocytes and nonparenchymal liver cells between the capillaries was demonstrated, including vascular and bile-duct-like structures characterized by specific marker staining [82, 83]. Moreover, specific transporter proteins involved in the biliary excretion of substances or metabolites were detected and showed a distribution pattern similar to that in native liver tissue [40].

Collagen, Matrigel® or other biological compounds are used frequently to improve cell adhesion and stability. However, their usage in bioartificial liver devices for clinical usage is restricted due to their undefined composition and mostly animal origin. To overcome this limitation, integrated nanostructured self-assembling peptides and a defined combination of growth factors and cytokines were used in a scalable bioartificial liver module [84]. The authors could show that primary rat hepatocytes showed stable albumin and urea production, as well as preservation of phase 1 and phase 2 metabolism, suggesting a stabilizing effect of the used matrix factors.

Studies on fetal liver stem cells have demonstrated the importance of growth factors and cytokines for cell differentiation [85]. Hepatocyte growth factor (HGF), for example, can stimulate hepatocytes to replicate *in vitro*, and oncostatin M, a multifunctional cytokine of the interleukin-6 family, has been shown to induce maturation of fetal mouse hepatocytes *in vitro* [86–88]. In this line of evidence, it was shown that the exposition of cultured hepatocytes with plasma from acute liver failure patients affects hepatocyte metabolism *in vitro*, although data are partly controversial. While some studies indicate a negative impact of acute liver failure plasma exposure on metabolic functions such as ammonia clearance, CYP450 activity, and urea synthesis [89], other groups observed an advantageous effect on hepatic maintenance and regeneration [90]. Although the underlying mechanisms are unclear, it can be assumed from these studies that factors released by the failing liver and/or molecules accumulated in the plasma due to reduced hepatic metabolism may play an important role in tissue formation and cell regeneration *in vitro*. The culture model and the microenvironment offered to the cells seem to be critical factors in those processes. For example, toxic plasma exposure had a more advantageous influence on hepatocytes cultured in a polyurethane foam/spheroid culture system than on those maintained in 2D monolayer cultures [91]. Furthermore, a beneficial effect of plasma incubation was observed in HepG2 cell cultures seeded on polystyrene microcarriers, with a significantly increased cell number and incorporation amino acids into protein compared with culture medium [92]. Increased growth factor expression and cell regeneration was also observed in four-compartment bioreactors after perfusion with acute liver failure plasma during clinical extracorporeal liver support application [93]. These studies emphasize the importance of a physiological environment permissive for exogenous factors to improve cell maintenance and reorganization.

## 5.6 Bioreactors Enabling Cell Production for Transplantation

While extracorporeal systems offer an option for temporary support of the failing liver, methods for intracorporeal application of cells or *in vitro* engineered tissues could be used to achieve permanent recovery of the diseased liver. Transplantation of primary human hepatocytes into the liver or the spleen has been investigated in several clinical studies (reviewed by Strom and Fisher [94], Pietrosi *et al.* [95], Puppi and Dhawan [96]). Therapies were performed to support patients with acute to fulminant liver failure [97] or to cure patients with inherited metabolic disorders, such as genetic hypercholesterolemia, glycogen storage disease [98], urea cycle defects [99], or coagulation factor VII deficiency [100]. These studies showed the technical feasibility of cell transplantation, and in some cases, alleviation of blood parameters and/or clinical improvement was observed. However, the low number of patients included in those studies prevents a definitive evaluation of the efficiency of the therapy.

Major hurdles to a more widespread application of this procedure are (i) the limited availability of primary human liver cells, (ii) the limited transplantable cell volume that can be applied without risking vascular thrombosis, and (iii) the need for immunosuppressive therapy in nonautologous hepatocyte applications. To enable safe clinical applications, the generated cells must be stable during *in vitro* maintenance and after transplantation into the target organ, at least until self-recovery of the tissue, or permanently if self-repair is not possible.

The currently used static 2D culture techniques provide a rather unphysiological environment for the cultured cells. In addition, an upscale of 2D culture models for larger cell amounts is hardly possible. Cell cultivation under 3D conditions is supposed to better approximate the physiological milieu than 2D cultures by enhancing physical cell-to-cell contacts, accumulation of extracellular matrices, and local growth factor delivery [46, 47]. In this context, bioreactor technologies appear important in terms of large-scale cell production under reproducible and controllable *in vitro* conditions. Bioreactor technology development may facilitate cell expansion by providing improved conditions for enhanced growth factor delivery and paracrine stimulation of cell growth.

For example, successful expansion of embryonic stem cells in microcarrier-based culture systems was shown [101]. The importance of cell perfusion was demonstrated both in 2D cultures [102] and in a perfused microcarrier-based bioreactor with oxygen control [103]. However, an upscale of microcarrier cultures is limited by the need of frequent passaging associated with substantial cell losses. The generation of large quantities of hepatocytes from mouse embryonic stem cells was investigated in a rotating simulated microgravity bioreactor supplied with exogenous growth factors and hormones [104]. Embryoid body–derived cells grown in the rotating bioreactor exhibited higher levels of hepatic genes and proteins than cells grown in static culture. Moreover, successful engraftment into the recipient livers of nude mice was shown in an experimental transplantation model [105].

Upscaling of expansion of mouse embryonic stem cells in four-compartment hollow fiber–based bioreactors with a volume of 8 ml or 800 ml resulted in the generation of a cell number of $5 \times 10^9$ cells in one 800 ml bioreactor, as determined by metabolic parameters and DNA/protein quantification [106]. Marker expression analysis confirmed the preservation of pluripotency in the cells during expansion (Figure 5.4).

To make *in vitro* generated cells available for clinical use in cell transplantation, methods for directed cell differentiation, as well as suitable harvesting technologies enabling the generation of vital and functional cells for such applications have to be developed. A major precondition is the availability of a suitable cell source for deriving hepatic cells ensuring clinical safety of applications.

## 5.7 Cell Sources for Bioartificial Liver Bioreactors

A major hurdle to the clinical use of bioartificial liver systems is the lack of efficient and safe cell sources for such therapies. The ideal cell source would be of human origin, would display all typical functions of the liver, and be expandable and free of any clinical risks for the patient. Within the last decade, significant progress has been made in cell sources of hepatocytes for bioartificial liver [107]. Approaches include primary cells, hepatic cell lines, and progenitor or stem cells from different origins. Table 5.2 gives an overview of cell types under investigation for use in bioartificial liver devices, with their respective advantages and limitations.

### 5.7.1 Primary Liver Cells

Primary porcine liver cells have been used in several devices due to their good availability and *in vitro* functionality. A possible risk of porcine endogeneous retrovirus (PERV) transfer during application in patients has been discussed, but to date there is no evidence for PERV infection in patients or for release of infectious particles by porcine liver cells cultured in bioartificial livers [108,

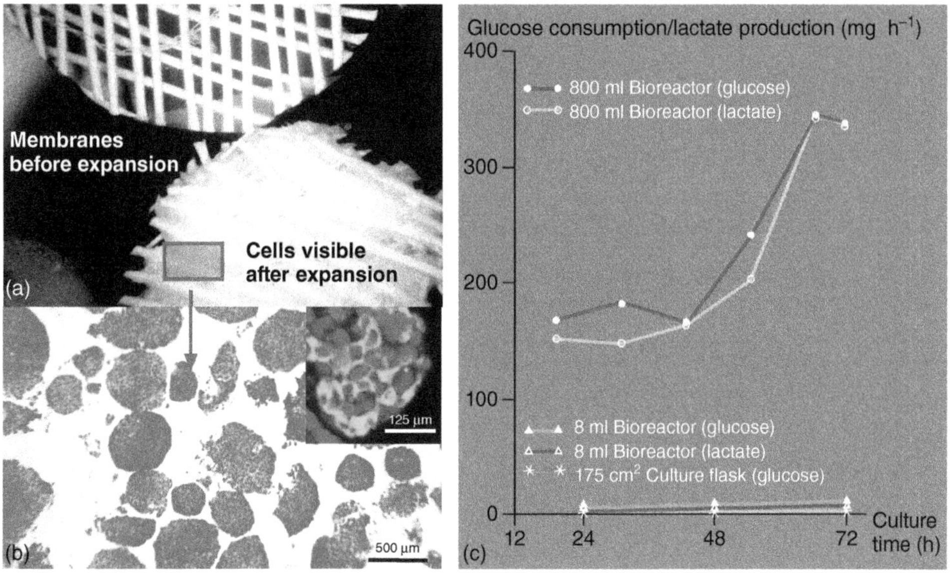


**Figure 5.4** (a) One layer of the capillary membrane network from an 8 ml bioreactor before inoculation of mouse embryonic stem cells (upper left) and after 6 days of culture (lower right); aggregates are visible between the fibers. (b) Toluidine blue staining of cells obtained from an 800 ml bioreactor after day 3. Inset: SSEA-1 immunoreactivity (green) on day 3 in the 800 ml bioreactor culture (blue: DAPI staining of cell nuclei). (c) Glucose consumption and lactate production in the 800 and 8 ml bioreactors versus glucose consumption in 2D control dishes over 3 days. Values are given as consumption or production rates per hour per bioreactor/culture flask. Thus, the extremely different scales of the bioreactors and cell numbers are reflected by these medium parameters (reproduced with minor editing modifications from [106] with permission from Karger Publishers. ©2010 Karger Publishers, Basel, Switzerland).

109]. However, careful surveillance of patients is needed to detect any indications of virus transmission. In addition, the use of xenogeneic cell types is associated with an increased risk of rejection episodes, and their metabolic performance shows some differences to human metabolism due to interspecies differences [110].

Primary human liver cells represent the cell source of choice due to their human-equivalent functionality and clinical safety [111]. In a clinical pilot study performed with the MELS device, primary human liver cells from donor organs discarded due to organ injury (steatosis, fibrosis/cirrhosis, vascular deficiencies) have been used, showing the feasibility of culturing primary human liver cells in specific bioreactor systems [83]. However, the scarce availability of primary human hepatocytes at sufficient quality and quantity for clinical application restricts their use in clinical studies. In addition, the logistics of cell isolation is extremely demanding. In the case of using freshly isolated primary liver cells, at least 48–72 h are needed from cell seeding to therapeutic application to allow for

**Table 5.2** Cell types for potential use in bioartificial liver bioreactors.

| Cell type | | Advantages | Limitations |
|---|---|---|---|
| Primary liver cells | Primary porcine hepatocytes<br>Primary human hepatocytes | Good availability<br>High functionality<br>Human-equivalent functionality | Risk of immune rejection<br>Nonhuman metabolism<br>Scarce availability<br>Complex logistics needed |
| Hepatic cell lines | For example, C3A/HepG2, HepaRG | Rapid cell growth, enabling large-scale production | Partial deviation of metabolic activities |
| Liver-specific stem cells | Adult progenitor cells | Committed to hepatic differentiation<br>High clinical safety | Limited expansion capability<br>Complete functionality not shown yet |
| | Fetal progenitor cells | Higher proliferation than adult stem cells | Premature metabolic activities<br>Ethical concerns |
| Pluripotent stem cells | Embryonic stem cells | Unlimited proliferation capacity | Difficult control of cell differentiation<br>Risk of tumor formation<br>Ethical concerns |
| | Induced pluripotent stem cells | Unlimited proliferation<br>Autologous generation possible<br>Ethically accepted | Difficult control of cell differentiation<br>Risk of tumor formation<br>Possible risks due to genetic manipulation |

cell recovery from isolation and adaptation to the culture environment [41]. In case of acute liver failure, temporary organ support has to be provided immediately, which requires the availability of on-demand systems that can be delivered ready-to-use. Since the lifetime of primary cells in *in vitro* systems is limited, systems based on primary cells have to be constantly available and regularly renewed to be prepared for clinical use should a patient be in need for therapy. Therefore, the possibility of storing culture devices until application would greatly improve logistics of bioartificial liver therapies. Cryopreservation of cells could solve the problem of limited survival time of cells *in vitro*. Various approaches for cryopreservation of hepatocytes have been investigated during the last decades (reviewed by Fuller *et al.* [112]). For example, rat hepatocytes encapsulated in 2% alginate microbeads and subjected to a complex freezing/thawing procedure showed well-preserved functions and were successfully transplanted in an animal experimental model [113]. In a similar approach, Massie *et al.* described a method for storage of alginate-encapsulated liver cell spheroids at −80 °C or −170 °C for up to 1 year [114]. However, upscale of cryopreservation methods to therapeutic cell numbers has not yet been shown.

### 5.7.2
### Hepatic Cell Lines

As an alternative to primary human or porcine liver cells, human hepatic cell lines are under investigation for usage in bioartificial liver systems due to their considerable availability and practicability. The human C3A cell line, a derivative of the HepG2 cell line, has been used in clinical trials with the ELAD system. However, C3A cells show an insufficient level of some important liver functions, including ammonia detoxification and urea synthesis, and thus are not ideally suited to replace the functions of the native organ [115]. Genetic modification of hepatoma cell lines can be one way to increase the functional performance of cells. Recently, it was reported that transfection of HepG2 cells with human augmenter of liver regeneration (hALR) leads to increased $\alpha$-fetoprotein, urea, and albumin production as compared with nontransfected HepG2 cells [116].

The HepaRG cell line derived from human hepatocellular carcinoma represents a promising cell source for bioartificial liver development, since it shows a range of liver-specific functions, including cytochrome P450-dependent metabolism [117, 118]. The cell line has the ability to differentiate into both hepatocytes and biliary cells when treated with dimethyl sulfoxide [119]. HepaRG cells cultured in the AMC bioreactor showed increased functional performances as compared with 2D monolayer cultures in terms of ammonia elimination, urea production, and cytochrome P450 3A4 activity [120]. Treatment of rats with induced liver failure with HepaRG cells cultured in the AMC-BAL increased the survival time of the animals by ~50% as compared with cell-free BAL treatment [121]. In addition, bile acid secretion including hydroxylation, conjugation, and transport of bile salts was demonstrated in the AMC-BAL [122]. Cultivation of HepaRG in four-compartment hollow-fiber bioreactors showed stable metabolic activities of several human-relevant cytochrome P450 enzymes over up to 4 weeks [123]. Histological studies performed in the bioreactor system proved the occurrence of both hepatocyte-like and biliary structures between the capillaries. Thus HepaRG cells could be of interest for bioartificial liver systems for building up a biliary system to drain the bile produced by hepatocytes.

### 5.7.3
### Stem Cells

Stem cells characterized by their potential to grow and to differentiate into specific cell types *in vitro* could solve the problem of scarce cell availability in bioartificial liver development. In addition, the risk of immunological reactions in cell transplantation therapy can be avoided if using autologous cells from the patient [124]. Different types of stem cells are under investigation. In accordance to their differentiation potential, these can been divided in pluripotent stem cells represented by embryonic stem (ES) cells and induced pluripotent stem (iPS) cells, and tissue-specific stem cells ("adult" stem cells) described in the liver as hepatic stem cells [125], small hepatocytes [76], or progenitor cells [126]. The requirements of the specific clinical application should be considered in the choice of the stem cell type to be used.

A major advantage of tissue-specific "adult" over pluripotent stem cells is their lower risk of undesired cell developments, namely tumor formation, and consequently a higher clinical safety, which facilitates the clinical translation of *in vitro* research results.

Various methods for isolation, culture, and differentiation of liver stem cells derived from fetal or adult liver have been reported [127–129]. *In vitro* studies on tissue-derived liver stem cells showed that their culture behavior is largely influenced by the supply of cytokines, growth factors, hormones, and extracellular matrix proteins [88, 130, 131]. Human liver cell cultures exposed to plasma from acute liver failure patients during clinical liver support therapy showed increased evidence of progenitor activation emphasizing the importance of regenerative factors [93]. However, large-scale proliferation of liver progenitor cells has not yet been achieved, which to date restricts clinical applications of the cells.

Due to their better growth ability as compared with adult hepatocytes, fetal liver cells offer the option to be used in liver support systems, but could also be expanded in bioreactors for clinical cell transplantation. Intrasplenic transplantation of fetal liver cells in patients with end-stage chronic liver disease was well tolerated and safe, and showed some positive effects on clinical scores [132]. Studies in perfused four-compartment bioreactors showed the suitability of 3D culture systems to support fetal cell growth and maturation [133]. However, the use of fetal cells is restricted by ethical concerns about the use of fetal tissue.

A further promising approach is based on the use of rat pancreas-derived progenitor cells, which are differentiated into hepatic cells by the addition of dexamethasone [134]. The differentiated cells (B-13/H cells) showed stable expression of hepatic markers, including liver-typical cytochrome P450 enzyme activities, and they showed liver-specific engraftment in mice [135]. Thus, if the method can be successfully transferred to human cells, they may present a easily renewable and cost-efficient source of functional hepatocyte-like cells.

Pluripotent stem cells have the advantage of unlimited proliferation *in vitro* and can thus be used to generate large cell amounts for potential therapeutic application. Protocols for *in vitro* differentiation of human ES cells into hepatic cells are mostly based on a two-step procedure, using various cytokine cocktails in defined time schedules to maturate the cells first into endodermal cells followed by full maturation into hepatocyte-like cells [136–138]. However, the use of human ES cells is under debate due to ethical concerns about the generation of the cells from human blastocysts.

During the last years, human iPS cells generated by genetic reprogramming of cells from adult tissue have gained increasing importance in stem cell research [139, 140]. The use of iPS cells could solve the problem of scarce cell availability for clinical tissue regeneration and organ support. A major benefit of the cells is that they are ethically accepted and can principally be generated from the patient's own cells to be used for autologous therapy. Thus, iPS cells provide the unique option to solve major problems of cell-based medicine: the scarce availability of suitable cells for organ regeneration, and the risk of immune reactions in the case of using xenogeneic cells. Within the scope of developing new cell-based therapies of liver diseases, genetically corrected iPS-derived cells could also be used to repair genetic metabolic diseases, if the safety of the cells can be shown.

The availability of iPS cell-derived cells for clinical use is currently limited by the existing methods used to expand and differentiate the cells *in vitro*. Current protocols for hepatic differentiation of iPS cells are based on the sequential addition of different cocktails of growth factors and hormones [141, 142]. Improved maturation was attempted, for example, by applying a cytokine and small molecule–based protocol for the direct differentiation of ES cells and iPS cells into hepatic cells [143]. However, the functional performance of iPS-derived hepatic cells is still insufficient. Investigations of spontaneous differentiation of mouse or human ES cells in perfused bioreactors suggest that improved cell differentiation and tissue formation can be attained under bioreactor conditions [144, 145]. Studies on hepatic differentiation of human ES cells [146] indicate that 3D culture techniques may also be suitable to promote liver-specific differentiation of iPS cells. In the same line of evidence, it was reported that coculture of iPS cells with endothelial cells and mesenchymal stem cells on a presolidified matrix resulted in the formation of 3D structures that showed high levels of liver-specific functions after transplantation in an experimental mouse model [147].

The expansion of pluripotent stem cells for differentiation is currently limited by the labor-intensive methods to propagate the cells in their undifferentiated state. In addition, control of cell proliferation and direction of cell differentiation for safe clinical application is still a challenge. Thus, improved culture technologies are needed to make the cells available for clinical organ support or transplantation therapies. In this context, the development of specific 3D bioreactor technologies appears to be important also for developing individual culture models for therapeutic applications.

## 5.8 Outlook

Bioartificial liver support systems have been successfully used in initial clinical trials to support the liver function in patients with acute liver failure. However, approval of therapeutic application of such devices as a standard therapy for end-stage liver disease has not yet been realized.

Major challenges in the development of bioartificial liver bioreactors for extracorporeal liver support or cell transplantation can been seen in technical improvements, but also in the optimization of cell biology in bioreactors. The question of suitable cell sources with respect to availability, efficacy, and safety has to be addressed individually for specific clinical therapies. In addition, improvement of cell functionality is required if hepatic cell lines or stem cell-derived cells are used. This demands the further development of culture technologies that meet the needs of physiological tissue formation; promising approaches can be seen in 3D culture models based on different scaffolds and matrix components.

To address these issues, efforts by cell biologists and bioengineers in close cooperation with physicians and medical scientists are needed to enable successful clinical translation of results from *in vitro* research.

## References


1. Caraceni, P. and Van Thiel, D.H. (1995) Acute liver failure. *Lancet*, **345**, 163–169.
2. Scientific Registry of Transplant Recipients (2012) OPTN & SRTR Annual Data Report 2012: Liver, http://www.srtr.org (accessed 16 October 2015).
3. Patzer, J.F. II, (2006) Principles of bound solute dialysis. *Ther. Apher. Dial.*, **10**, 118–124.
4. Patzer, J.F. II, (2008) Thermodynamic considerations in solid adsorption of bound solutes for patient support in liver failure. *Artif. Organs*, accepted **32**(7), 499–508.
5. Hughes, R.D., Pucknell, A., Routley, D. *et al.* (1994) Evaluation of the BioLogic-DT sorbent-suspension dialyser in patients with fulminant hepatic failure. *Int. J. Artif. Organs*, **17**, 657–662.
6. Aladag, M., Gurakar, A., Jalil, S., Wright, H., Alamian, S., Rashwan, S., Sebastian, A. *et al.* (2004) A liver transplant center experience with liver dialysis (Biologic DT) in the management of patients with fulminant hepatic failure: a preliminary report. *Transplant. Proc.*, **36**, 203–205.
7. Kreymann, B., Seige, M., Schweigart, U., Kopp, K.F., and Classen, M. (1999) Albumin dialysis: effective removal of copper in a patient with fulminant Wilson disease and successful bridging to liver transplantation: a new possibility for the elimination of protein-bound toxins. *J. Hepatol.*, **31**, 1080–1085.
8. Patzer, J.F. II,, Safta, S.A., and Miller, R.H. (2006) Slow continuous ultrafiltration with bound solute dialysis. *ASAIO J.*, **52** (1), 47–58.
9. Mitzner, S., Klammt, S., Stange, J., and Schmidt, R. (2006) Albumin regeneration in liver support-comparison of different methods. *Ther. Apher. Dial.*, **10** (2), 108–117. Review. Erratum in: *Ther. Apher. Dial.* (2006) **10** (6) 518.
10. Stange, J., Hassanein, T.I., Mehta, R., Mitzner, S.R., and Bartlett, R.H. (2002) The molecular adsorbents recycling system as a liver support system based on albumin dialysis: a summary of preclinical investigations, prospective, randomized, controlled clinical trial, and clinical experience from 19 centers. *Artif. Organs*, **26**, 103–110.
11. Heemann, U., Treichel, U., Loock, J., Philipp, T., Gerken, G., Malago, M., Klammt, S., Loehr, M., Liebe, S., Mitzner, S., Schmidt, R., and Stange, J. (2002) Albumin dialysis in cirrhosis with superimposed acute liver injury: a prospective, controlled study. *Hepatology*, **36** (4, Pt. 1), 949–958.
12. Kramer, L., Bauer, E., Schenk, P., Steininger, R., Vigl, M., and Mallek, R. (2003) Successful treatment of refractory cerebral oedema in ecstasy/cocaine-induced fulminant hepatic failure using a new high-efficacy liver detoxification device (FPSA-Prometheus). *Wien. Klin. Wochenschr.*, **115** (15-16), 599–603.
13. Rifai, K., Ernst, T., Kretschmer, U., Bahr, M.J., Schneider, A., Hafer, C., Haller, H., Manns, M.P., and Fliser, D. (2003) Prometheus-a new extracorporeal system for the treatment of liver failure. *J. Hepatol.*, **39**, 984–990.
14. Saliba, F., Camus, C., Durand, F., Mathurin, P., Letierce, A., Delafosse, B., Barange, K., Perrigault, P.F., Belnard, M., Ichaï, P., and Samuel, D. (2013) Albumin dialysis with a noncell artificial liver support device in patients with acute liver failure: a randomized, controlled trial. *Ann. Intern. Med.*, **159** (8), 522–531.

15. Bañares, R., Nevens, F., Larsen, F.S., Jalan, R., Albillos, A., Dollinger, M., Saliba, F., Sauerbruch, T., Klammt, S., Ockenga, J., Pares, A., Wendon, J., Brünnler, T., Kramer, L., Mathurin, P., de la Mata, M., Gasbarrini, A., Müllhaupt, B., Wilmer, A., Laleman, W., Eefsen, M., Sen, S., Zipprich, A., Tenorio, T., Pavesi, M., Schmidt, H.H., Mitzner, S., Williams, R., Arroyo, V., and RELIEF Study Group (2013) Extracorporeal albumin dialysis with the molecular adsorbent recirculating system in acute-on-chronic liver failure: the RELIEF trial. *Hepatology*, **57** (3), 1153–1162.
16. Kribben, A., Gerken, G., Haag, S., Herget-Rosenthal, S., Treichel, U., Betz, C., Sarrazin, C., Hoste, E., Van Vlierberghe, H., Escorsell, A., Hafer, C., Schreiner, O., Galle, P.R., Mancini, E., Caraceni, P., Karvellas, C.J., Salmhofer, H., Knotek, M., Ginès, P., Kozik-Jaromin, J., Rifai, K., and HELIOS Study Group (2012) Effects of fractionated plasma separation and adsorption on survival in patients with acute-on-chronic liver failure. *Gastroenterology*, **142** (4), 782–789.
17. Ting, P.P. and Demetriou, A.A. (2000) Clinical experience with artificial liver support systems. *Can. J. Gastroenterol.*, **14** (Suppl. D), 79D–84D.
18. Chan, C., Berthiaume, F., Nath, B.D., Tilles, A.W., Toner, M., and Yarmush, M.L. (2004) Hepatic tissue engineering for adjunct and temporary liver support: critical technologies. *Liver Transpl.*, **10** (11), 1331–1342.
19. Fiegel, H.C., Kaufmann, P.M., Bruns, H., Kluth, D., Horch, R.E., Vacanti, J.P., and Kneser, U. (2008) Hepatic tissue engineering: from transplantation to customized cell-based liver directed therapies from the laboratory. *J. Cell. Mol. Med.*, **12** (1), 56–66. Epub 2007 Nov 16. Review.
20. Gerlach, J.C., Zeilinger, K., and Patzer, J.F. (2008) Bioartificial liver systems: why, what, whither? *Regen. Med.*, **3** (4), 575–595.
21. Bañares, R., Catalina, M.V., and Vaquero, J. (2013) Liver support systems: will they ever reach prime time? *Curr. Gastroenterol. Rep.*, **15** (3), 312.
22. Ellis, A.J., Hughes, R.D., Wendon, J.A., Dunne, J., Langley, P.G., Kelly, J.H., Gislason, G.T., Sussman, N.L., and Williams, R. (1996) Pilot-controlled trial of the extracorporeal liver assist device in acute liver failure. *Hepatology*, **24**, 1446–1451.
23. Millis, J.M., Cronin, D.C., Johnson, R., Conjeevaram, H., Conlin, C., Trevino, S., and Maguire, P. (2002) Initial experience with the modified extracorporeal liver-assist device for patients with fulminant hepatic failure: system modifications and clinical impact. *Transplantation*, **74**, 1735–1746.
24. Mazariegos, G.V., Kramer, D.J., Lopez, R.C., Shakil, A.O., Rosenbloom, A.J., DeVera, M.D., Giraldo, M., Grogan, T.A., Zhu, Y., Fulmer, M.L., Amiot, B.P., and Patzer, J.F. II, (2001) Safety observations in the phase I clinical evaluation of the Excorp Medical BLSS after the first four patients. *ASAIO J.*, **47**, 471–475.
25. Mazariegos, G.V., Patzer, J.F. II,, Lopez, R.C., Giraldo, M., DeVera, M.D., Grogan, T.A., Zhu, Y., Fulmer, M.L., Amiot, B.P., and Kramer, D.J. (2002) First clinical use of a novel bioartificial liver support system (BLSS). *Am. J. Transplant.*, **2**, 260–266.
26. Demetriou, A.A., Brown, R.S. Jr.,, Busuttil, R.W., Fair, J., McGuire, B.M., Rosenthal, P., Am Esch, J.S. II, *et al.* (2004) Prospective, randomized, multicenter, controlled trial of a bioartificial liver in treating acute liver failure. *Ann. Surg.*, **239**, 660–667; discussion 667–670.
27. Van De Kerkhove, M.P., Di Florio, E., Scuderi, V., Mancini, A., Belli, A., Bracco, A., Dauri, M., Tisone, G., Di Nicuolo, G., Amoroso, P., Spadari, A., Lombardi, G., Hoekstra, R., Calise, F., and Chamuleau, R.A. (2002) Phase I clinical trial with the AMC-bioartificial liver. *Int. J. Artif. Organs*, **25**, 950–959.
28. Morsiani, E., Pazzi, P., Puviani, A.C., Brogli, M., Valieri, L., Gorini, P., Scoletta, P., Marangoni, E., Ragazzi, R.,

Azzena, G., Frazzoli, E., Di Luca, D., Cassai, E., Lombardi, G., Cavallari, A., Faenza, S., Pasetto, A., Girardis, M., Jovine, E., and Pinnaet, A.D. (2002) Early experiences with a porcine hepatocyte-based bioartificial liver in acute hepatic failure patients. *Int. J. Artif. Organs*, **25**, 192–202.

29. Gerlach, J.C., Botsch, M., Kardassis, D., Lemmens, P., Schön, M., Janke, J., Puhl, G., Unger, J., Kraemer, M., Busse, B., Böhmer, C., Belal, R., Ingenlath, M., Kosan, M., Kosan, B., Sültmann, J., Patzold, A., Tietze, S., Roissant, R., Müller, C., Mönch, E., Sauer, I.M., and Neuhaus, P. (2001) Experimental evaluation of a cell module for hybrid liver support. *Int. J. Artif. Organs*, **24** (11), 793–798.
30. Sauer, I.M., Zeilinger, K., Pless, G., Kardassis, D., Theruvath, T., Pascher, A., Goetz, M., Neuhaus, P., and Gerlach, J.C. (2003) Extracorporeal liver support based on primary human liver cells and albumin dialysis – treatment of a patient with primary graft nonfunction. *J. Hepatol.*, **39**, 649–653.
31. Sussman, N.L., Chong, M.G., Koussayer, T., He, D.E., Shang, T.A., Whisennand, H.H., and Kelly, J.H. (1992) Reversal of fulminant hepatic failure using an extracorporeal liver assist device. *Hepatology*, **16**, 60–65.
32. Watanabe, F.D., Mullon, C.J., Hewitt, W.R., Arkadopoulos, N., Kahaku, E., Eguchi, S., Khalili, T., Arnaout, W., Shackleton, C.R., Rozga, J., Solomon, B., and Demetriou, A.A. (1997) Clinical experience with a bioartificial liver in the treatment of severe liver failure. A phase I clinical trial. *Ann. Surg.*, **225**, 484–491.
33. Gerlach, J., Klöppel, K., Müller, C., Schnoy, N., Smith, M., and Neuhaus, P. (1993) Hepatocyte aggregate culture technique for bioreactors in hybrid liver support systems. *Int. J. Artif. Organs*, **16** (12), 843–846.
34. Gerlach, J., Schnoy, N., Encke, J., Müller, C., Smith, M., and Neuhaus, P. (1995) Improved hepatocyte in vitro maintenance in a culture model with woven multicompartment capillary systems: electron microscopy studies. *Hepatology*, **22**, 546–552.
35. Patzer, J.F. II, (2001) Advances in bioartificial liver assist devices. *Ann. N.Y. Acad. Sci.*, **944**, 320–333.
36. Kjaergard, L.L., Liu, J., Als-Nielsen, B., and Gluud, C. (2003) Artificial and bioartificial support systems for acute and acute-on-chronic liver failure: a systematic review. *J. Am. Med. Assoc.*, **289**, 217–222.
37. Van De Kerkhove, M.P., Hoekstra, R., Chamuleau, R.A., and Van Gulik, T.M. (2004) Clinical application of bioartificial liver support systems. *Ann. Surg.*, **240**, 216–230.
38. Morsiani, E., Brogli, M., Galavotti, D., Pazzi, P., Puviani, A.C., and Azzena, G.F. (2002) Biologic liver support: optimal cell source and mass. *Int. J. Artif. Organs*, **25** (10), 985–993.
39. Gerlach, J. (1996) Development of a hybrid liver support system – a review. *Int. J. Artif. Organs*, **19**, 645–655.
40. Zeilinger, K., Schreiter, T., Darnell, M., Söderdahl, T., Lübberstedt, M., Dillner, B., Knobeloch, D., Nüssler, A.K., Gerlach, J.C., and Andersson, T. (2011) Scaling down of a clinical 3D perfusion multi-compartment hollow fiber liver bioreactor developed for extracorporeal liver support to an analytical scale device useful for hepatic pharmacological in vitro studies. *Tissue Eng. Part C Methods*, **17** (5), 549–556.
41. Pless, G., Steffen, I., Zeilinger, K., Sauer, I.M., Katenz, E., Kehr, D.C., Roth, S., Mieder, T., Schwartlander, R., Muller, C., Wegner, B., Hout, M.S., and Gerlach, J.C. (2006) Evaluation of primary human liver cells in bioreactor cultures for extracorporeal liver support on the basis of urea production. *Artif. Organs*, **30** (9), 686–694.
42. Guthke, R., Zeilinger, K., Sickinger, S., Schmidt-Heck, W., Buentemeyer, H., Iding, K., Lehmann, J., Pless, G., and Gerlach, J.C. (2006) Dynamics of amino acid metabolism of primary human liver cells in 3D-bioreactors. *Bioprocess Biosyst. Eng.*, **28** (5), 331–340. Epub 2006 Mar 21.
43. Gerlach, J.C., Brayfield, C., Puhl, G., Borneman, R., Müller, C.,

Schmelzer, E., and Zeilinger, K. (2010) Lidocaine/monoethylglycinexylidide test, galactose elimination test, and sorbitol elimination test for metabolic assessment of liver cell bioreactors. *Artif. Organs*, **34** (6), 462–472.
44. Hoffmann, S.A., Müller-Vieira, U., Biemel, K., Knobeloch, D., Heydel, S., Lübberstedt, M., Nüssler, A.K., Andersson, T.B., Gerlach, J.C., and Zeilinger, K. (2012) Analysis of drug metabolism activities in a miniaturized liver cell bioreactor for use in pharmacological studies. *Biotechnol. Bioeng.*, **109** (12), 3172–3181.
45. Lübberstedt, M., Müller-Vieira, U., Biemel, K.M., Darnell, M., Hoffmann, S.A., Knöspel, F., Wönne, E.C., Knobeloch, D., Nüssler, A.K., Gerlach, J.C., Andersson, T.B., and Zeilinger, K. (2015) Serum-free culture of primary human hepatocytes in a miniaturized hollow-fiber membrane bioreactor for pharmacological in vitro studies. *J. Tissue Eng. Regen. Med.*. **9**(9), 1017–1026.
46. Cukierman, E., Pankov, R., and Yamada, K.M. (2002) Cell interactions with three-dimensional matrices. *Curr. Opin. Cell Biol.*, **14** (5), 633–639.
47. Abbott, A. (2003) Cell culture: biology's new dimension. *Nature*, **424** (6951), 870–872.
48. Shito, M., Kim, N.H., Baskaran, H., Tilles, A.W., Tompkins, R.G., Yarmush, M.L., and Toner, M. (2001) In vitro and in vivo evaluation of albumin synthesis rate of porcine hepatocytes in a flat-plate bioreactor. *Artif. Organs*, **25**, 571–578.
49. De Bartolo, L., Jarosch-Von Schweder, G., Haverich, A., and Bader, A. (2000) A novel full-scale flat membrane bioreactor utilizing porcine hepatocytes: cell viability and tissue-specific functions. *Biotechnol. Progr.*, **16**, 102–108.
50. Shito, M., Tilles, A.W., Tompkins, R.G., Yarmush, M.L., and Toner, M. (2003) Efficacy of an extracorporeal flat-plate bioartificial liver in treating fulminant hepatic failure. *J. Surg. Res.*, **111**, 53–62.
51. MacDonald, J.M., Wolfe, S.P., Roy-Chowdhury, I., Kubota, H., and Reid, L.M. (2001) Effect of flow configuration and membrane characteristics on membrane fouling in a novel multicoaxial hollow-fiber bioartificial liver. *Ann. N.Y. Acad. Sci.*, **944**, 334–343.
52. Kim, S.S., Utsunomiya, H., Koski, J.A., Wu, B.M., Cima, M.J., Sohn, J., Mukai, K., Griffith, L.G., and Vacanti, J.P. (1998) Survival and function of hepatocytes on a novel three-dimensional synthetic biodegradable polymer scaffold with an intrinsic network of channels. *Ann. Surg.*, **228**, 8–13.
53. Powers, M.J., Janigian, D.M., Wack, K.E., Baker, C.S., Beer Stolz, D., and Griffith, L.G. (2002) Functional behavior of primary rat liver cells in a three-dimensional perfused microarray bioreactor. *Tissue Eng.*, **8**, 499–513.
54. Powers, M.J., Domansky, K., Kaazempur-Mofrad, M.R., Kalezi, A., Capitano, A., Upadhyaya, A., Kurzawski, P., Wack, K.E., Stolz, D.B., Kamm, R., and Griffith, L.G. (2002) A microfabricated array bioreactor for perfused 3D liver culture. *Biotechnol. Bioeng.*, **78**, 257–269.
55. Ambrosino, G., Varotto, S., Basso, S., Galavotti, D., Cecchetto, A., Carraro, P., Naso, A., De Silvestro, G., Plebani, M., Giron, G., Abatangelo, G., Donato, D., Braga, G.P., Cestrone, A., Marrelli, L., Trombetta, M., Lorenzelli, V., Picardi, A., Valente, M.L., Palu, G., Colantoni, A., Van Thiel, D., Ricordi, C., and D'Amico, D.F. (2002) ALEX (artificial liver for extracorporeal xenoassistance): a new bioreactor containing a porcine autologous biomatrix as hepatocyte support. Preliminary results in an ex vivo experimental model. *Int. J. Artif. Organs*, **25**, 960–965.
56. Erro, E., Bundy, J., Massie, I., Chalmers, S.A., Gautier, A., Gerontas, S., Hoare, M., Sharratt, P., Choudhury, S., Lubowiecki, M., Llewellyn, I., Legallais, C., Fuller, B., Hodgson, H., and Selden, C. (2013) Bioengineering the liver: scale-up and cool chain delivery of the liver cell biomass for clinical targeting in a bioartificial liver support system. *Biores. Open Access*, **2** (1), 1–11.
57. Selden, C., Spearman, C.W., Kahn, D., Miller, M., Figaji, A., Erro, E., Bundy, J.,

Massie, I., Chalmers, S.A., Arendse, H., Gautier, A., Sharratt, P., Fuller, B., and Hodgson, H. (2013) Evaluation of encapsulated liver cell spheroids in a fluidised-bed bioartificial liver for treatment of ischaemic acute liver failure in pigs in a translational setting. *PLoS One*, **8** (12), e82312.

58. Rebelo, S.P., Costa, R., Estrada, M., Shevchenko, V., Brito, C., and Alves, P.M. (2015) HepaRG microencapsulated spheroids in DMSO-free culture: novel culturing approaches for enhanced xenobiotic and biosynthetic metabolism. *Arch. Toxicol.* **89**(8), 1347–1358.
59. Yu, C.B., Pan, X.P., Yu, L., Yu, X.P., Du, W.B., Cao, H.C., Li, J., Chen, P., and Li, L.J. (2014) Evaluation of a novel choanoid fluidized bed bioreactor for future bioartificial livers. *World J. Gastroenterol.*, **20** (22), 6869–6877.
60. Zhang, S., Chen, L., Liu, T., Wang, Z., and Wang, Y. (2014) Integration of single-layer skin hollow fibers and scaffolds develops a three-dimensional hybrid bioreactor for bioartificial livers. *J. Mater. Sci. Mater. Med.*, **25** (1), 207–216.
61. McCuskey, R.S. (1994) The hepatic microvascular system, in Arias IM, Boyer JL, Fausto N, Jakoby WB, Schachter DA, Shafritz DA (eds), Raven Press, New York. *The Liver Biology and Pathobiology*, 3rd edn, 1089–1106.
62. Saxena, R., Theise, N.D., and Crawford, J.M. (1999) Microanatomy of the human liver – exploring the hidden interfaces. *Hepatology*, **30** (6), 1339–1346.
63. Van Der Smissen, P., Breat, F., Crabbe, E., De Zanger, R., and Wisse, E. (1995) The cytoskeleton of the liver sieve in situ: a TEM study, in *Cells of the Hepatic Sinusoid*, vol. **5** (eds E. Wisse, D.L. Knook, and K. Wale), Kupffer Cell Foundation, Leiden, pp. 275–277.
64. Tiniakos, D.G., Lee, J.A., and Burt, A.D. (1996) Innervation of the liver: morphology and function. *Liver*, **16**, 151–160.
65. Kamegaya, Y., Oda, M., Kazemoto, S., Yokomori, H., Kaneko, H., Honda, K., Ishii, H., and Tsuchiya, M. (1995) Evidence for the spontaneous contractility of ITO cells by time-lapse cinematographic and computerized image analysis, in *Cells of the Hepatic Sinusoid*, vol. **5** (eds E. Wisse, D.L. Knook, and K. Wale), Kupffer Cell Foundation.
66. Banta, S., Vemula, M., Yokoyama, T., Jayaraman, A., Berthiaume, F., and Yarmush, M.L. (2007) Contribution of gene expression to metabolic fluxes in hypermetabolic livers induced through burn injury and cecal ligation and puncture in rats. *Biotechnol. Bioeng.*, **97** (1), 118–137.
67. Lee, K., Berthiaume, F., Stephanopoulos, G.N., Yarmush, D.M., and Yarmush, M.L. (2000) Metabolic flux analysis of postburn hepatic hypermetabolism. *Metab. Eng.*, **2**, 312–327.
68. Yokoyama, T., Banta, S., Berthiaume, F., Nagrath, D., Tompkins, R.G., and Yarmush, M.L. (2005) Evolution of intrahepatic carbon, nitrogen, and energy metabolism in a D-galactosamine-induced rat liver failure model. *Metab. Eng.*, **7**, 88–103.
69. Chan, C., Berthiaume, F., Lee, K., and Yarmush, M.L. (2003) Metabolic flux analysis of cultured hepatocytes exposed to plasma. *Biotechnol. Bioeng.*, **81**, 33–50.
70. Chan, C., Berthiaume, F., Lee, K., and Yarmush, M.L. (2003) Metabolic flux analysis of hepatocyte function in hormone- and amino acid-supplemented plasma. *Metab. Eng.*, **5**, 1–15.
71. Sharma, N.S., Ierapetritou, M.G., and Yarmush, M.L. (2005) Novel quantitative tools for engineering analysis of hepatocyte cultures in bioartificial liver systems. *Biotechnol. Bioeng.*, **92** (3), 321–335.
72. Nolan, R.P., Fenley, A.P., and Lee, K. (2006) Identification of distributed metabolic objectives in the hypermetabolic liver by flux and energy balance analysis. *Metab. Eng.*, **8**, 30–45.
73. Uygun, K., Matthew, H.W.T., and Huang, Y. (2007) Investigation of metabolic objectives in cultured hepatocytes. *Biotechnol. Bioeng.*, **97** (3), 622–637.

74. Reid, L.M., Gatmaitan, Z., Arias, I., Ponce, P., and Rojkind, M. (1980) Long-term cultures of normal rat hepatocytes on liver biomatrix. *Ann. N.Y. Acad. Sci.*, **439**, 70–76.
75. Kono, Y., Yang, S., and Roberts, E.A. (1997) Extended primary culture of human hepatocytes in a collagen gel sandwich system. *In Vitro Cell. Dev. Biol. Anim.*, **33**, 467–472.
76. Mitaka, T., Sato, F., Mizuguchi, T., Yokono, T., and Mochizuki, Y. (1999) Reconstruction of hepatic organoid by rat small hepatocytes and hepatic nonparenchymal cells. *Hepatology*, **29** (1), 111–125.
77. Reid, L.M. and Luntz, T.L. (1997) Ex vivo maintenance of differentiated mammalian cells. *Methods Mol. Biol.*, **75**, 31–57.
78. Sudo, R., Kohara, H., Mitaka, T., Ikeda, M., and Tanishita, K. (2005) Coordinated movement of bile canalicular networks reconstructed by rat small hepatocytes. *Ann. Biomed. Eng.*, **33** (5), 696–708.
79. Bhatia, S.N., Yarmush, M.L., and Toner, M. (1997) Controlling cell interactions by micropatterning in co-cultures: hepatocytes and 3T3 fibroblasts. *J. Biomed. Mater. Res.*, **34**, 189–199.
80. Harada, K., Mitaka, T., Miyamoto, S., Sugimoto, S., Ikeda, S., Takeda, H., Mochizuki, Y., and Hirata, K. (2003) Rapid formation of hepatic organoid in collagen sponge by rat small hepatocytes and hepatic nonparenchymal cells. *J. Hepatol.*, **39** (5), 716–723.
81. Sudo, R., Mitaka, T., Ikeda, M., and Tanishita, K. (2005) Reconstruction of 3D stacked-up structures by rat small hepatocytes on microporous membranes. *FASEB J.*, **19** (12), 1695–1697. Epub 2005 Aug 17.
82. Zeilinger, K., Holland, G., Sauer, I.M., Efimova, E., Kardassis, D., Obermayer, N., Liu, M., Neuhaus, P., and Gerlach, J.C. (2004) Time course of primary liver cell reorganization in three-dimensional high-density bioreactors for extracorporeal liver support: an immunohistochemical and ultrastructural study. *Tissue Eng.*, **10** (7), 1113–1124.
83. Gerlach, J.C., Mutig, K., Sauer, I.M., Schrade, P., Efimova, E., Mieder, T., Naumann, G., Grunwald, A., Pless, G., Mas, A., Bachmann, S., Neuhaus, P., and Zeilinger, K. (2003) Use of primary human liver cells originating from discarded grafts in a bioreactor for liver support therapy and the prospects of culturing adult liver stem cells in bioreactors: a morphologic study. *Transplantation*, **76** (5), 781–786.
84. Giri, S., Braumann, U.D., Giri, P., Acikgöz, A., Scheibe, P., Nieber, K., and Bader, A. (2013) Nanostructured self-assembling peptides as a defined extracellular matrix for long-term functional maintenance of primary hepatocytes in a bioartificial liver modular device. *Int. J. Nanomed.*, **8**, 1525–1539.
85. Yoon, J.-H., Lee, H.V.-S., Lee, J.S., Park, J.B., and Kim, C.Y. (1999) Development of non-transformed liver cell line with differentiated-hepatocyte and urea-synthetic functions: applicable for bioartificial liver. *Int. J. Artif. Organs*, **18**, 2127–2136.
86. Kamiya, A., Kinoshita, T., Ito, Y., Matsui, T., Korikawa, Y., Senba, E., Nakashima, K., Taga, T., Yoshida, K., Kishsimoto, T., and Miyajima, A. (1999) Fetal liver development requires a paracrine action of oncostatin M through the gp130 signal transducer. *EMBO J.*, **18**, 2127–2136.
87. Kojima, N., Kinoshita, T., Kamiya, A., Nakaruma, K., Nakashima, K., Taga, T., and Miyajima, A. (2000) Cell density-dependent regulation of hepatic development by a gp130-independent pathway. *Biochem. Biophys. Res. Commun.*, **277**, 152–158.
88. Sakai, Y., Jiang, J., Kojima, N., Kinoshita, T., and Miyajima, A. (2002) Enhanced in vitro maturation of fetal mouse liver cells with oncostatin M, nicotinamide and dimethylsulfoxide. *Cell Transplant.*, **11**, 435–441.
89. Abrahamse, S.L., van de Kerkhove, M.P., Sosef, M.N., Hartman, R., Chamuleau, R.A., and van Gulik, T.M. (2002) Treatment of acute liver failure in pigs reduces hepatocyte function in a bioartificial liver support system. *Int. J. Artif. Organs*, **25** (10), 966–974.

90. Chen, M.F., Hwang, T.L., and Yu, H.C. (1996) Effect of serum from partially hepatectomized cirrhotic and noncirrhotic rats on liver regeneration with primary hepatocyte cultures. *Eur. Surg. Res.*, **28** (6), 413–418.
91. Yamashita, Y., Shimada, M., Tsujita, E., Shirabe, K., Ijima, H., Nakazawa, K., Sakiyama, R., Fukuda, J., Funatsu, K., and Sugimachi, K. (2002) High metabolic function of primary human and porcine hepatocytes in a polyurethane foam/spheroid culture system in plasma from patients with fulminant hepatic failure. *Cell Transplant.*, **11** (4), 379–384.
92. Cunningham, J.M. and Hodgson, H.J. (1992) Microcarrier culture of hepatocytes in whole plasma for use in liver support bioreactors. *Int. J. Artif. Organs*, **15** (3), 162–167.
93. Schmelzer, E., Mutig, K., Schrade, P., Bachmann, S., Gerlach, J.C., and Zeilinger, K. (2009) Effect of human patient plasma ex vivo treatment on gene expression and progenitor cell activation of primary human liver cells in multi-compartment 3D perfusion bioreactors for extra-corporeal liver support. *Biotechnol. Bioeng.*, **103** (4), 817–827.
94. Strom, S. and Fisher, R. (2003) Hepatocyte transplantation: new possibilities for therapy. *Gastroenterology*, **124** (2), 568–571.
95. Pietrosi, G., Vizzini, G.B. *et al.* (2009) Clinical applications of hepatocyte transplantation. *World J. Gastroenterol.*, **15**, 2074–2077.
96. Puppi, J. and Dhawan, A. (2009) Human hepatocyte transplantation overview. *Methods Mol. Biol.*, **481**, 1–16.
97. Strom, S.C., Fisher, R.A. *et al.* (1997) Hepatocyte transplantation as a bridge to orthotopic liver transplantation in terminal liver failure. *Transplantation*, **63**, 559–569.
98. Muraca, M., Gerunda, G. *et al.* (2002) Hepatocyte transplantation as a treatment for glycogen storage disease type 1a. *Lancet*, **359**, 317–318.
99. Meyburg, J., Das, A.M. *et al.* (2009) One liver for four children: first clinical series of liver cell transplantation for severe neonatal urea cycle defects. *Transplantation*, **87**, 636–641.
100. Dhawan, A., Mitry, R.R. *et al.* (2004) Hepatocyte transplantation for inherited factor VII deficiency. *Transplantation*, **78**, 1812–1814.
101. Phillips, B.W., Horne, R., Lay, T.S., Rust, W.L., Teck, T.T., and Crook, J.M. (2008) Attachment and growth of human embryonic stem cells on microcarriers. *J. Biotechnol.*, **138** (1–2), 24–32.
102. Fong, W.J., Tan, H.L., Choo, A., and Oh, S.K. (2005) Perfusion cultures of human embryonic stem cells. *Bioprocess Biosyst. Eng.*, **27** (6), 381–387.
103. Serra, M., Brito, C., Sousa, M.F., Jensen, J., Tostões, R., Clemente, J., Strehl, R., Hyllner, J., Carrondo, M.J., and Alves, P.M. (2010) Improving expansion of pluripotent human embryonic stem cells in perfused bioreactors through oxygen control. *J. Biotechnol.*, **148** (4), 208–215.
104. Wang, Y., Zhang, Y., Zhang, S., Peng, G., Liu, T., Li, Y., Xiang, D., Wassler, M.J., Shelat, H.S., and Geng, Y. (2012) Rotating microgravity-bioreactor cultivation enhances the hepatic differentiation of mouse embryonic stem cells on biodegradable polymer scaffolds. *Tissue Eng. Part A*, **18** (21–22), 2376–2385. doi: 10.1089/ten.TEA.2012.0097. Epub 2012 Sep 24.
105. Zhang, S., Zhang, Y., Chen, L., Liu, T., Li, Y., Wang, Y., and Geng, Y. (2013) Efficient large-scale generation of functional hepatocytes from mouse embryonic stem cells grown in a rotating bioreactor with exogenous growth factors and hormones. *Stem Cell Res. Ther.*, **4** (6), 145.
106. Gerlach, J.C., Lübberstedt, M., Edsbagge, J., Ring, A., Hout, M., Baun, M., Rossberg, I., Knöspel, F., Peters, G., Eckert, K., Wulf-Goldenberg, A., Björquist, P., Stachelscheid, H., Urbaniak, T., Schatten, G., Miki, T., Schmelzer, E., and Zeilinger, K. (2010) Interwoven four-compartment capillary membrane technology for

three-dimensional perfusion with decentralized mass exchange to scale up embryonic stem cell culture. *Cells Tissues Organs*, **192** (1), 39–49.
107. Pan, X.P. and Li, L.J. (2012) Advances in cell sources of hepatocytes for bioartificial liver. *Hepatobiliary Pancreat. Dis. Int.*, **11** (6), 594–605. Review.
108. Irgang, M., Sauer, I.M., Karlas, A., Zeilinger, K., Gerlach, J.C., Kurth, R., Neuhaus, P., and Denner, J. (2003) Porcine endogenous retroviruses: no infection in patients treated with a bioreactor based on porcine liver cells. *J. Clin. Virol.*, **28** (2), 141–154.
109. Di Nicuolo, G., van de Kerkhove, M.P., Hoekstra, R., Beld, M.G., Amoroso, P., Battisti, S., Starace, M., di Florio, E., Scuderi, V., Scala, S., Bracco, A., Mancini, A., Chamuleau, R.A., and Calise, F. (2005) No evidence of in vitro and in vivo porcine endogenous retrovirus infection after plasmapheresis through the AMC-bioartificial liver. *Xenotransplantation*, **12** (4), 286–292.
110. Gerlach, J., Zeilinger, K., Sauer, I.M., Mieder, T., Nauman, G., Grünwald, A., Pless, G., Holland, A., Mas, A., Vienken, J., and Neuhaus, P. (2002) Extracorporeal liver support: porcine or human cell based systems? *Int. J. Artif. Organs*, **25** (10), 1013–1019.
111. Tsiaoussis, J., Newsome, P.N., Nelson, L.J., Hayes, P.C., and Plevris, J.N. (2001) Which hepatocyte will it be? Hepatocyte choice for bioartificial liver support systems. *Liver Transpl.*, **7** (1), 2–10.
112. Fuller, B.J., Petrenko, A.Y., Rodriguez, J.V., Somov, A.Y., Balaban, C.L., and Guibert, E.E. (2013) Biopreservation of hepatocytes: current concepts on hypothermic preservation, cryopreservation, and vitrification. *Cryo Lett.*, **34** (4), 432–452. Review.
113. Aoki, T., Koizumi, T., Kobayashi, Y., Yasuda, D., Izumida, Y., Jin, Z., Nishino, N., Shimizu, Y., Kato, H., Murai, N., Niiya, T., Enami, Y., Mitamura, K., Yamamoto, T., and Kusano, M. (2007) A novel method of cryopreservation of rat and human hepatocytes by using encapsulation technique and possible use for cell transplantation. *Cell Transplant.*, **16** (1), 67–73.
114. Massie, I., Selden, C., Hodgson, H., and Fuller, B. (2013) Storage temperatures for cold-chain delivery in cell therapy: a study of alginate-encapsulated liver cell spheroids stored at −80°c or −170°c for up to 1 year. *Tissue Eng. Part C Methods*, **19** (3), 189–195.
115. Hoekstra, R. and Chamuleau, R.A. (2002) Recent developments on human cell lines for the bioartificial liver. *Int. J. Artif. Organs*, **25**, 985–993.
116. Liu, H., You, S., Rong, Y., Wu, Y., Zhu, B., Wan, Z., Liu, W., Mao, P., and Xin, S. (2013) Newly established human liver cell line: a potential cell source for the bioartificial liver in the future. *Hum. Cell*, **26** (4), 155–161.
117. Lübberstedt, M., Müller-Vieira, U., Mayer, M., Biemel, K.M., Knöspel, F., Knobeloch, D., Nüssler, A.K., Gerlach, J.C., and Zeilinger, K. (2011) HepaRG human hepatic cell line utility as a surrogate for primary human hepatocytes in drug metabolism assessment in vitro. *J. Pharmacol. Toxicol. Methods*, **63** (1), 59–68.
118. Andersson, T.B., Kanebratt, K.P., and Kenna, J.G. (2012) The HepaRG cell line: a unique in vitro tool for understanding drug metabolism and toxicology in human. *Expert Opin. Drug Metab. Toxicol.*, **8** (7), 909–920. doi: 10.1517/17425255.2012.685159. Epub 2012 May 8. Review.
119. Gripon, P., Rumin, S., Urban, S., Le Seyec, J., Glaise, D., Cannie, I., Guyomard, C., Lucas, J., Trepo, C., and Guguen-Guillouzo, C. (2002) Infection of a human hepatoma cell line by hepatitis B virus. *Proc. Natl. Acad. Sci. U.S.A.*, **99** (24), 15655–15660.
120. Nibourg, G.A., Hoekstra, R., van der Hoeven, T.V., Ackermans, M.T., Hakvoort, T.B., van Gulik, T.M., and Chamuleau, R.A. (2013) Increased hepatic functionality of the human hepatoma cell line HepaRG cultured in the AMC bioreactor. *Int. J. Biochem. Cell Biol.*, **45** (8), 1860–1868.
121. Nibourg, G.A., Chamuleau, R.A., van der Hoeven, T.V., Maas, M.A., Ruiter,

A.F., Lamers, W.H., Oude Elferink, R.P., van Gulik, T.M., and Hoekstra, R. (2012) Liver progenitor cell line HepaRG differentiated in a bioartificial liver effectively supplies liver support to rats with acute liver failure. *PLoS One*, 7 (6), e38778.

122. Hoekstra, R., Nibourg, G.A., van der Hoeven, T.V., Plomer, G., Seppen, J., Ackermans, M.T., Camus, S., Kulik, W., van Gulik, T.M., Elferink, R.P., and Chamuleau, R.A. (2013) Phase 1 and phase 2 drug metabolism and bile acid production of HepaRG cells in a bioartificial liver in absence of dimethyl sulfoxide. *Liver Int.*, **33** (4), 516–524. doi: 10.1111/liv.12090. Epub 2013 Feb 7.

123. Darnell, M., Schreiter, T., Zeilinger, K., Urbaniak, T., Söderdahl, T., Rossberg, I., Dillnér, B., Berg, A.L., Gerlach, J.C., and Andersson, T.B. (2011) Cytochrome P450-dependent metabolism in HepaRG cells cultured in a dynamic three-dimensional bioreactor. *Drug Metab. Dispos.*, **39** (7), 1131–1138.

124. Souza, B.S., Nogueira, R.C., Oliveira, S.A., de Freitas, L.A., Lyra, L.G., Ribeiro Dos Santos, R., Lyra, A.C., and Soares, M.B. (2009) Current status of stem cell therapy for liver diseases. *Cell Transplant.* **18**(2), 1261–1279.

125. Theise, N.D., Saxena, R., Portmann, B.C., Thung, S.N., Yee, H., Chiriboga, L., Kumar, A., and Crawford, J.M. (1999) The canals of Hering and hepatic stem cells in humans. *Hepatology*, **30** (6), 1425–1433.

126. Tan, J., Hytiroglou, P., Wieczorek, R., Park, Y.N., Thung, S.N., Arias, B., and Theise, N.D. (2002) Immunohistochemical evidence for hepatic progenitor cells in liver diseases. *Liver*, **22** (5), 365–373.

127. Quante, M. and Wang, T.C. (2009) Stem cells in gastroenterology and hepatology. *Nat. Rev. Gastroenterol. Hepatol.*. **6**(12), 724–737.

128. Schmelzer, E., Wauthier, E., and Reid, L.M. (2006) The phenotypes of pluripotent human hepatic progenitors. *Stem Cells*, **24** (8), 1852–1858. Epub 2006 Apr 20.

129. Schmelzer, E., Zhang, L., Bruce, A., Wauthier, E., Ludlow, J., Yao, H., Moss, N., Melhem, A., McClelland, R., Turner, W.S. *et al.* (2007) Human hepatic stem cells from fetal and postnatal donors. *J. Exp. Med.*, **204** (8), 1973–1987.

130. Stachelscheid, H., Urbaniak, T., Ring, A., Spengler, B., Gerlach, J.C., and Zeilinger, K. (2009) Isolation and characterization of adult human liver progenitors from ischemic liver tissue derived from therapeutic hepatectomies. *Tissue Eng. Part A*, **15** (7), 1633–1643.

131. Kinoshita, T., Sekiguchi, T., Xu, M.J., Ito, Y., Kamiya, A., Tsuji, K., and Nakamura, T. (1999) Hepatic differentiation induced by oncostatin M attenuates fetal liver hematopoiesis. *Proc. Natl. Acad. Sci. U.S.A.*, **96**, 7265–7270.

132. Pietrosi, G., Vizzini, G., Gerlach, J., Chinnici, C., Luca, A., Amico, G., D'Amato, M., Conaldi, P.G., Petri, S.L., Spada, M., Tuzzolino, F., Alio, L., Schmelzer, E., and Gridelli, B. (2015) Phase I-II matched case-control study of human fetal liver cell transplantation for treatment of chronic liver disease. *Cell Transplant.*. **24**(8), 1627–1638.

133. Ring, A., Gerlach, J.C., Peters, G., Pazin, B., Minervini, C., Turner, M.E., Thompson, R.L., Triolo, F., Gridelli, B., and Miki, T. (2010) Hepatic maturation of human fetal hepatocytes in four-compartment 3D perfusion culture. *Tissue Eng. Part C Methods*. **16**(5), 835–845.

134. Wallace, K., Fairhall, E.A., Charlton, K.A., and Wright, M.C. (2010) AR42J-B-13 cell: an expandable progenitor to generate an unlimited supply of functional hepatocytes. *Toxicology*, **278** (3), 277–287.

135. Fairhall, E.A., Charles, M.A., Wallace, K., Schwab, C.J., Harrison, C.J., Richter, M., Hoffmann, S.A., Charlton, K.A., Zeilinger, K., and Wright, M.C. (2013) The B-13 hepatocyte progenitor cell resists pluripotency induction and differentiation to non-hepatocyte cells. *Toxicol. Res.*. **2**, 308–320.

136. D'Amour, K.A. *et al.* (2005) Efficient differentiation of human embryonic

stem cells to definitive endoderm. *Nat. Biotechnol.*, **23**, 1534–1541.

137. Hay, D.C. *et al.* (2008) Efficient differentiation of hepatocytes from human embryonic stem cells exhibiting markers recapitulating liver development in vivo. *Stem Cells*. **26**, 894–902.

138. Agarwal, S. *et al.* (2008) Efficient differentiation of functional hepatocytes from human embryonic stem cells. *Stem Cells*, **26**(5), 1117–1127.

139. Takahashi, K., Tanabe, K., Ohnuki, M., Narita, M., Ichisaka, T., Tomoda, K., and Yamanaka, S. (2007) Induction of pluripotent stem cells from adult human fibroblasts by defined factors. *Cell*, **131** (5), 861–872.

140. Yu, J., Hu, K., Smuga-Otto, K., Tian, S., Stewart, R., Slukvin, I.I., and Thomson, J.A. (2009) Human induced pluripotent stem cells free of vector and transgene sequences. *Science*, **324** (5928), 797–801.

141. Sullivan, G.J., Hay, D.C., Park, I.H., Fletcher, J., Hannoun, Z., Payne, C.M., Dalgetty, D., Black, J.R., Ross, J.A., Samuel, K., Wang, G., Daley, G.Q., Lee, J.H., Church, G.M., Forbes, S.J., Iredale, J.P., and Wilmut, I. (2010) Generation of liver disease-specific induced pluripotent stem cells along with efficient differentiation to functional hepatocyte-like cells. *Hepatology*, **51** (1), 329–335.

142. Chen, Y.F., Tseng, C.Y., Wang, H.W., Kuo, H.C., Yang, V.W., and Lee, O.K. (2012) Rapid generation of mature hepatocyte-like cells from human induced pluripotent stem cells by an efficient three-step protocol. *Hepatology*, **55** (4), 1193–1203.

143. Sgodda, M., Mobus, S., Hoepfner, J., Sharma, A.D., Schambach, A., Greber, B., Ott, M., and Cantz, T. (2013) Improved hepatic differentiation strategies for human induced pluripotent stem cells. *Curr. Mol. Med.*, **13** (5), 842–855.

144. Gerlach, J.C., Hout, M., Edsbagge, J., Björquist, P., Lübberstedt, M., Miki, T., Stachelscheid, H., Schmelzer, E., Schatten, G., and Zeilinger, K. (2010) Dynamic 3D culture promotes spontaneous embryonic stem cell differentiation in vitro. *Tissue Eng. Part C Methods*, **16** (1), 115–121.

145. Stachelscheid, H., Wulf-Goldenberg, A., Eckert, K., Jensen, J., Edsbagge, J., Björquist, P., Rivero, M., Strehl, R., Jozefczuk, J., Prigione, A., Adjaye, J., Urbaniak, T., Bussmann, P., Zeilinger, K., and Gerlach, J.C. (2013) Teratoma formation of human embryonic stem cells in 3D perfusion culture bioreactors. *J. Tissue Eng. Regen. Med.*, **7** (9), 729–741.

146. Miki, T., Ring, A., and Gerlach, J. (2011) Hepatic differentiation of human embryonic stem cells is promoted by three-dimensional dynamic perfusion culture conditions. *Tissue Eng. Part C Methods*, **17** (5), 557–568.

147. Takebe, T., Sekine, K., Enomura, M., Koike, H., Kimura, M., Ogaeri, T., Zhang, R.R., Ueno, Y., Zheng, Y.W., Koike, N., Aoyama, S., Adachi, Y., and Taniguchi, H. (2013) Vascularized and functional human liver from an iPSC-derived organ bud transplant. *Nature*, **499** (7459), 481–484.

# 6
# Bioreactors for Expansion of Pluripotent Stem Cells and Their Differentiation to Cardiac Cells

*Robert Zweigerdt, Birgit Andree, Christina Kropp, and Henning Kempf*

## 6.1
## Introduction

### 6.1.1
### Requirement for Advanced Cell Therapies for Heart Repair

Cardiovascular disorders (CVDs) represent the most prominent causes of premature death in developed countries. Recent statistics demonstrate that 28% of premature deaths in men and 19% in women aged less than 75 years result from CVDs [1]. Atherosclerotic plaques restricting the lumen and flexibility of coronary vessels in the heart often precede myocardial infarction (MI). An ultimate vessel occlusion can trigger MI by interrupting the oxygen and nutrition supply to the affected area. Tissue ischemia and subsequent reperfusion may induce the irreversible loss of billions of cardiomyocytes (CMs) [2]. Since CMs in the mature human heart are cell cycle arrested they do not exhibit prominent proliferation, if any [3]. Recent evidence has also challenged the presence of a stem cell population of relevant regenerative potential in the heart [4]. Consequently, rather than tissue regeneration, an akinetic fibrotic scar is formed in the affected area. This condition typically leads to reduced heart function and might ultimately result in organ failure. The only curative option, heart transplantation, is limited by the lack of donor organs. Despite substantial advancements in the engineering of left ventricular assist devices (LVADs) [5], a machine bridging acute heart failure patients to organ transplantation, the technique is still hampered by severe limitations. These include frequent infections as well as the high risk of thrombosis, which might occur despite the necessity to take blood diluting drugs which, in itself, can provoke undesirable side effects such as uncontrolled bleeding.

Cell therapy has thus been envisioned as a promising strategy for heart repair. The aim is to achieve tissue repair by the ectopic delivery of cells to the organ, either to stimulate the hearts' endogenous regeneration or to directly replace lost CMs by the transplanted donor cells. Despite numerous clinical trials focused on the transplantation of patients own cells derived, for example, from skeletal

*Bioreactors: Design, Operation and Novel Applications*, First Edition. Edited by Carl-Fredrik Mandenius.

muscle, bone marrow, or peripheral blood to the heart limited, only recovery of heart function was achieved [6, 7]. Furthermore, the hypothesized generation of CMs by "transdifferentiation" from patients' adult (stem or progenitor) cells after cardiac delivery has been denied [3]. It is now widely accepted that some specific subtypes of adult stem or progenitor cells, particularly those from the bone marrow, can stimulate heart recovery by the release of paracrine factors triggering cell- and tissue-modulating events such as diminishing MI-induced CMs loss and/or improved tissue revascularization post-MI [8]. However, it remains to be shown that these effects are substantial enough to significantly improve patients' condition in the long run.

### 6.1.2 Pluripotent Stem Cell–Based Strategies for Heart Repair

An alternative concept aims at replacing damaged heart tissue by the transplantation of *in vitro* generated bona fide CMs, alone or in combination with other lineages (Figure 6.1). Cell types that might foster heart repair when co-transplanted with CMs include endothelial cells and pericytes to stimulate graft vascularization [9], connective tissue–forming and extracellular matrix–forming fibroblast [10] as well as mesenchymal stem cells (MSCs), which have been shown to act immunomodulatory [11] and release pro-survival factors [12], thus stimulating donor cell engraftment upon co-transplantation to the heart [13].

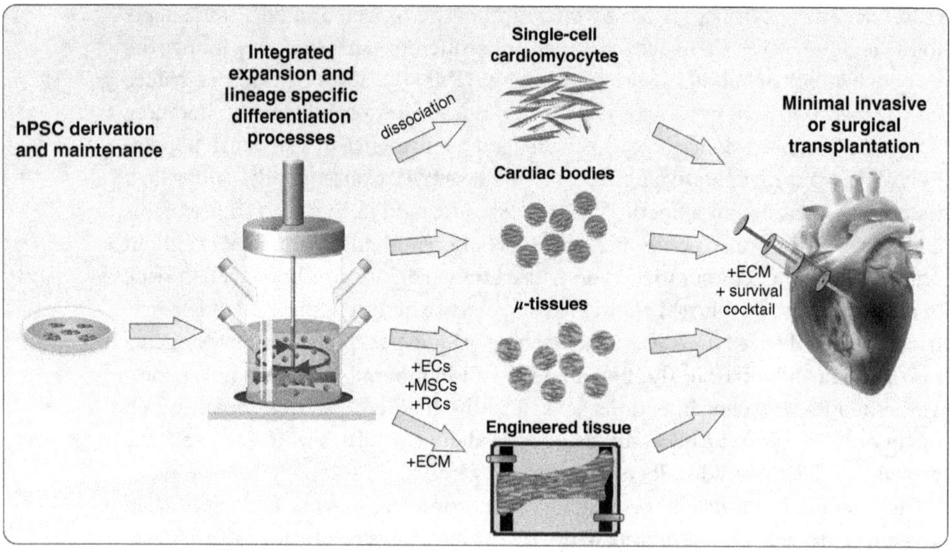


**Figure 6.1** Process strategies for pluripotent stem cell-based heart repair. hPSCs – human pluripotent stem cells; ECs – endothelial cells; MSCs – mesenchymal stem cells; PCs – pericytes; ECM – extracellular matrix.

All these cell types have recently been derived with increasing efficiencies from human pluripotent stem cells (hPSCs), including human embryonic stem cells (hESCs), and human induced pluripotent stem cells (hiPSCs), as demonstrated for endothelial cells [9], pericytes [14], and MSCs [15] as well as bona fide CMs [16]. In addition to their potential to differentiate into essentially any somatic cell type, pluripotent stem cells (PSCs) have the property of unlimited proliferation at the pluripotent state, at appropriate culture conditions. While embryonic stem cells (ESCs) have raised ethical concerns due to their origin from blastocyst-stage human embryos, hiPSCs are derived via somatic cell reprogramming (Figure 6.5). The iPSC technology is not only overcoming ethical issues, it also enables the derivation of patient-specific PSCs, which, in principle, should preclude rejection of iPSC-derived progenies when transplanted back to the patient of origin. Moreover, the technology opens the exciting possibility to establishing patient- and tissue-specific disease models *in vitro.*

With respect to the heart, a plethora of strategies for hPSC-based tissue repair have been suggested, including the direct transmural injection of hPSC-derived CMs alone or in combination with other cells and matrices (Figure 6.1), which was initially tested in rodent models [17, 18]. Alternatively, the transplantation of multilayered cell sheets or more sophisticated tissue-engineered constructs or patches is considered [19–22]. Known roadblocks include the poor retention of injected cells in heart tissue [23], low survival rates *in situ* due to hypoxic, pro-apoptotic, inflammatory, and immunological conditions [24], and the lack of straightforward strategies to control and orchestrate the functional, electromechanical integration of donor CMs with endogenous heart muscle cells [25].

Recently, however, the successful formation of substantial grafts of hESC-derived CMs in infarcted hearts of non-human primates was demonstrated [25]. Furthermore, using tissue engineering approaches or multicellular aggregates, recent work also enabled the implantation and monitoring of hPSC derivatives in porcine models of heart damage [13, 26, 27]. In addition to non-human primates, swine are among the most appropriate models with respect to the physiology and pathophysiology of the human heart including an equivalent beating rate and organ size. While rodents have been very useful to demonstrate the proof of concept for the functional coupling of mouse embryonic stem cell (mESC)-derived CMs to endogenous heart muscle cells [28], mice and rat hearts are considered of limited use to test the electromechanical coupling of hPSC-derived CMs since they have a four- to sixfold higher beating rate compared with human; guinea pigs having a heart rate ~250 beats per minute (bpm) have been suggested as a potential alternative [29] thus ranging at the interphase between rats (~400 bpm) and mice (~600 bpm) on one end of the spectrum and pigs, non-human primates and man (~60–200 bpm) on the other end.

Notably, a first case study of a human patient receiving hESC-derived cardiac progenitor cells was recently published demonstrating that the technology is at the edge to clinical translation [30]. This highlights that, in addition to large cell number production, the standardized processing of hPSCs and their progenies at

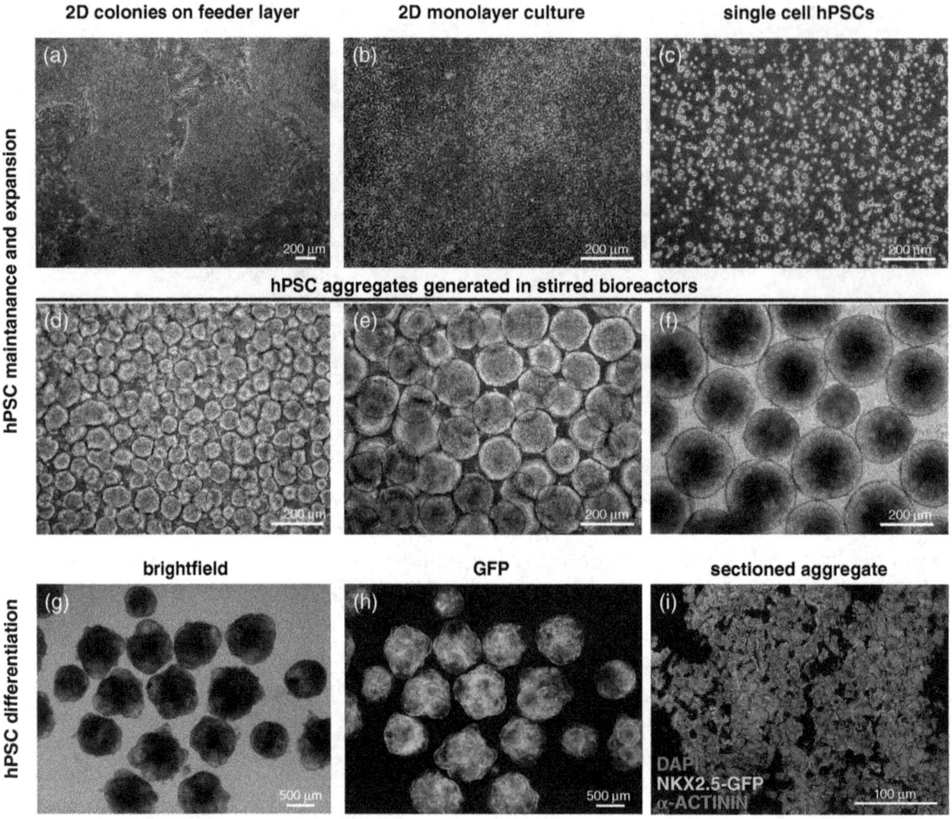


**Figure 6.2** Culture method-dependent morphology of hPSCs. Representative hPSC images of conventional 2D feeder culture (a) and feeder-free culture using chemically defined medium (b). For aggregate generation under fully defined conditions in stirred bioreactors, single-cell hPSCs (c) are inoculated. Formation of spherical aggregates is exemplarily shown over time (d–f) at 1, 3, and 7 days after inoculation, respectively. The 3D aggregates generated in a stirred bioreactor maintain high differentiation potential into beating cardiomyocytes (g, h). Differentiation was performed under fully defined conditions applying small-molecule Wnt pathway modulators. Brightfield image (g) and corresponding NKX2.5-GFP expression (h) on day 7 of differentiation using an HES3-NKX2.5$^{w/GFP}$ reporter cell line [31]. Cryosectioned aggregates were stained against the structural cardiomyocyte marker $\alpha$-actinin (i).

good manufacturing practice (GMP)-compliant conditions are urgently required. Respective cell production processes will be mandatory to facilitate the initiation of meaningful clinical trials aiming at testing feasibility and, most importantly, safety of envisioned cell therapies. However, to perform functional, preclinical testing of hPSC-derived CMs in large animal models readily requires the robust production of billions of cells equivalent to the therapeutic cell doses estimated for man, long ahead of the routine clinical application.

In the course of this chapter, we discuss conventional as well as more advanced strategies for the production of hPSCs at the pluripotent state, particularly focusing on bioreactor-based techniques and process upscaling. These processes aim at providing the "raw material" for the subsequent differentiation into, generally, any required somatic cell type. Here, we highlight the status of cardiac differentiation protocols aiming at the efficient and standardized mass production of functional CMs.

## 6.2 Culture Technologies for Pluripotent Stem Cell Expansion

### 6.2.1 Matrix-Dependent Cultivation in 2D

The conventional maintenance and expansion of undifferentiated hPSCs in laboratory scale depends on the adherence to substrates, which may be referred to as planar or two-dimensional (2D) culture. The cultivation of hPSC colonies is typically performed on conventional dishes or T-flasks, either in co-culture with mitotically inactivated fibroblast cells (referred to as "feeder cells") or on extracellular matrices (Figure 6.2, Table 6.1). Originally, this culture type relied on passaging of semi-dissociated cell colonies as small cell clumps resulting in semi-quantitative, only splitting ratios [32, 59] (Table 6.1).

Such co-culture approaches are cumbersome for routine use and represent excessive hurdles for upscaling [60]. Thus, feeder cell replacement by semi-defined natural matrices (such as matrigel [59]), recombinant proteins (such a laminin [61]), or fully synthetic agents (reviewed by Villa-Diaz *et al.* [62]) enabling hPSC "monolayer"-type culture (Figure 6.2) in combination with defined, non-feeder cell-conditioned culture media [63] represent important breakthroughs in the field. Advances in hPSC culture have also supported automated cultivation by robotic systems [64]. Automation allows for investigator-independent standardizations and higher sample throughput [65]. In particular, the automation of somatic cell reprogramming protocols will facilitate the standardized derivation of large numbers of individual hiPSC clones, thereby supporting large, ongoing projects aiming at the generation and banking of high numbers of patient and disease-specific hiPSC lines such as the European consortium StemBANCC [66].

### 6.2.2 Outscaling hPSC Production in 2D

By multiplying the manual or automated handling of conventional culture dishes or by using multilayered flasks, scale-out, rather than scale-up, of 2D culture for hPSC mass cell production deems generally feasible (Table 6.1). This will, however, still require relatively large space. Since space and time for therapeutic

**Table 6.1** Methods of pluripotent stem cell culturing.

| | Process strategy | Advantages | Disadvantages | References |
|---|---|---|---|---|
| 2D | Feeder based | Easy to handle; easy monitoring; affordable | Space- and labor-intensive; variability in components; poor scalability (only scale-out is possible); heterogeneous culture conditions; poor process control; co-culture is difficult considering reproducibility and product-safety | [32] |
| | Feeder free | Easy to handle; affordable; compatible with xeno-free and chemically defined culture conditions (compared with feeder-based methods) | Space- and labor-intensive; poor scalability (only scale-out is possible); heterogeneous culture conditions; poor process control | [33–35] |
| 3D | Rotary cell culture system | Low shear stress due to the absence of moving parts inside the culture; good mass transfer; homogeneous culture conditions | Poor scalability; cell harvesting and monitoring is difficult | [36–39] |
| | Hydrogel | High cell expansion demonstrated; possibly superior mimic of the *in vivo* stem cell niche | Cell harvesting and monitoring is difficult; additional culture component (additional material costs); increased complexity compared with aggregate-only suspension culture; might have a disposition to local culture heterogeneity compared with homogeneously stirred suspension cultures (microcarrier based or aggregates only); to our best knowledge, no translation to bioreactor suspension culture is shown so far | [40–42] |

| | | | | |
|---|---|---|---|---|
| Encapsulation | | Protection from physical forces; high scalability; possibly superior mimicking of the *in vivo* stem cell niche | Cell harvesting and monitoring is difficult (decapsulation necessary); limited mass transfer and gas diffusion; additional culture component (additional material costs); might have a disposition to local culture heterogeneity compared with homogeneously stirred suspension cultures (microcarrier based or aggregates only) | [43, 44] |
| Microcarrier | | Homogeneous culture conditions; good monitoring and control of process parameters; high scalability; high mass transfer; high cell expansion; cell retention easy | High shear stress; cell harvesting and monitoring is difficult (detaching of cells from the carriers is necessary); control/prevention of microcarrier agglomeration; additional culture components and costs (carrier + carrier coating); potential adsorbtion of growth factors and cytokines to the microcarriers diminishing and obscuring their concentration in the process | [45–47] |
| Aggregates only | Cell clump inoculation, stirred bioreactors | Homogeneous culture conditions; feeder- and matrix-free; good monitoring and control of process parameters; high scalability; high mass transfer | High shear stress; poor reproducibility regarding inoculation procedure; heterogeneity in aggregates; cell retention challenging | [38, 48, 49] |
| | Single-cell inoculation, static conditions | Easy to handle; affordable; feeder- and matrix-free | Space- and labor-intensive; poor scalability (only scale-out is possible); heterogeneous culture conditions; poor process control; long-term use of ROCK inhibitor is necessary; poor mass transfer | [50–52] |
| | Single-cell inoculation, stirred bioreactors | Homogeneous culture conditions; feeder- and matrix-free; good monitoring and control of process parameters; highly reproducible inoculation procedure; high homogeneity in aggregates; high scalability; high mass transfer | High shear stress; initial cell death due to single-cell inoculation; tight control of aggregate-size is necessary; cell retention is challenging; long-term use of ROCK inhibitor is necessary | [53–58] |

cell production in clean room facilities are essential cost factors, such scale-out approaches are obviously restricted by economic aspects of bioprocessing [67]. Another issue of the scale-out approach in conventional plastic-attached culture is the limited ability to monitor and control key process parameters such as the accurate continuous assessment of cell proliferation, online monitoring of the pH and dissolved oxygen (DO), glucose consumption, lactate production, and other valuable process-related factors. This fact makes proper production process development, control, and extension rather challenging.

### 6.2.3
### Hydrogel-Supported Transition to 3D

The use of synthetic, potentially GMP-compatible hydrogels has recently been suggested as means of transition toward three-dimensional (3D) hPSC culture [40], which might reasonably mimic the *in vivo* niche (i.e., the inner cell mass of the blastocyst; Figure 6.5) that transiently supports stem cell proliferation [41]. The hydrogel format supports controlled single-cell seeding and allows for long-term serial expansion of hPSC lines with a high yield of up to $\sim 2.0 \times 10^7$ cells $ml^{-1}$ of hydrogel [40]. Cell seeding in hydrogels might potentially be compatible with a packed-bed-bioreactor format, which might be used to scale-up this technique from the current tissue culture scale to larger process dimension including the use of typical process monitoring tools such as pH and DO measurement and control (see details on packed-bed-bioreactors elsewhere in this book). If compatibility with hPSC mass expansion and differentiation can be demonstrated, the technique might be valuable for both autologous and allogeneic strategies for therapeutic cell production schematically outlined in Figure 6.5.

## 6.3
## 3D Suspension Culture

### 6.3.1
### Advantages of Using Instrumented Stirred Tank Bioreactors

Recently, the field is reaching consensus that 3D suspension culture has great potential to achieve the extensive cell number requirements for hPSC production and differentiation via systematic process optimization and upscaling [60, 68]. In this respect, fully equipped stirred-tank bioreactors provide numerous advantages for the development of efficient suspension culture processes, which are as follows:

1) Impeller-controlled stirring ensures homogeneous mixing and distribution of cells, nutrients, and gases across the whole culture.
2) Impeller design and the stirring speed in addition to the cell inoculation density allow for cell-aggregation control and aggregate development.

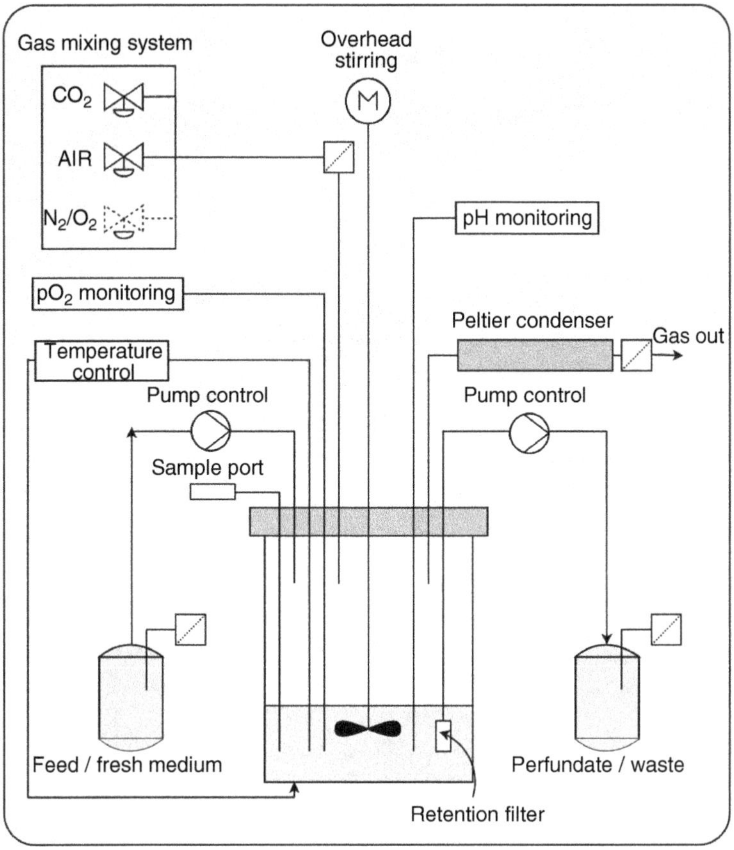


**Figure 6.3** Schematic of stirred instrumented bioreactor technology. The bioreactor is equipped with an eight-blade pitched impeller for magnet-coupled overhead drive stirring, a temperature sensor, pH- and DO-electrodes, as well as a sampling port and a liquid-free exhaust gas condenser. Overlay gassing is performed utilizing a flow-controlled gas mixing system. Perfusion is technically established via peristaltic pumps, a retention system placed at the waste medium stream, which is connected to a waste bottle, as well as of a feed line connected to a fresh medium reservoir.

3) As schematically shown in Figure 6.3, fully instrumented tank reactors (in contrast to simple spinner flasks) allow for monitoring and, importantly, controlling and modulating key process parameters including pH, DO, and potentially online monitoring of the viable biomass. Additional process tools, for example, for the online monitoring of metabolic parameters are readily established for larger bioreactors. Advanced tools to better assess and control stem cell-specific features such as label-free noninvasive monitoring of cells' pluripotency versus differentiation [69] and automated 3D microscopy of suspended cell aggregates [70] are currently under developed and will likely be implemented into stirred bioreactors in the near future (Figure 6.4).

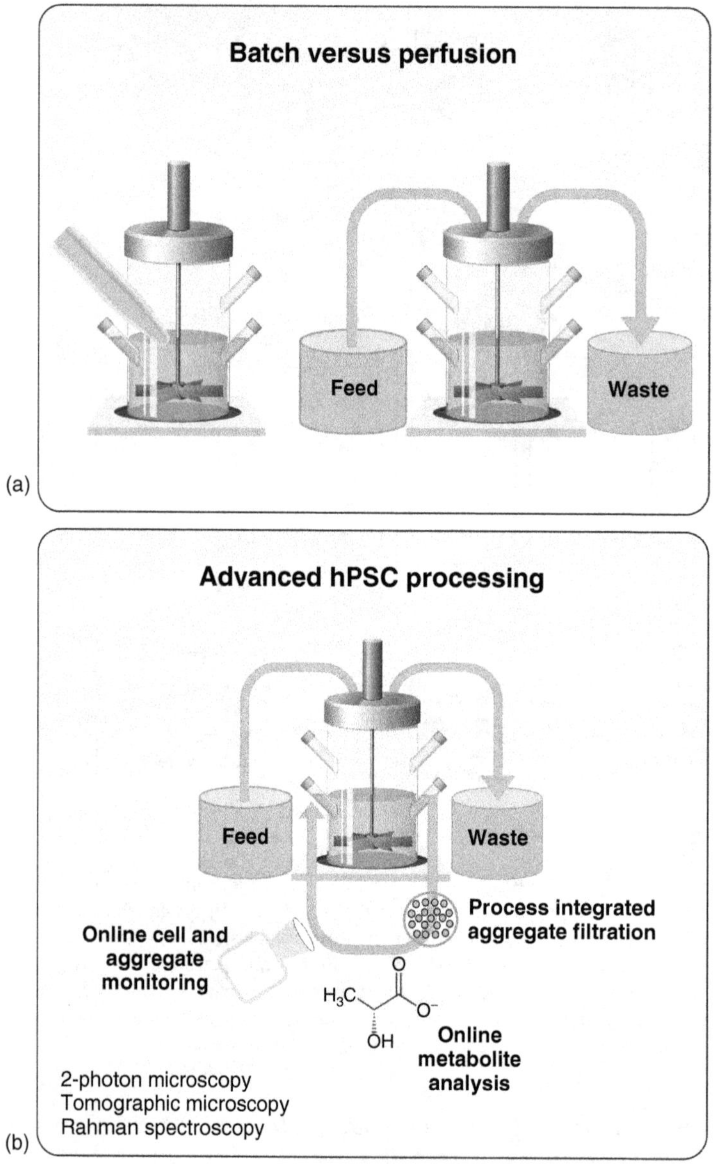


**Figure 6.4** Culture strategies for hPSC processing in stirred-tank bioreactors. (a) In batch-fed processes the entire culture volume is manually exchanged, whereas in perfusion processes the culture medium is continuously replaced by application of an automatic pump device combined with a cell retention system to maintain cells inside the bioreactor. The culture volume is thereby kept constant via applying equal feed and outflow rates. (b) In advanced hPSC bioprocesses, perfused bioreactors would be equipped with an additional external loop for a process-integrated aggregate filtration, online metabolite analysis, as well as an subsequent online cell and aggregate monitoring to enable sophisticated process control.

4) Respective ports allow the simple, regular cell- and medium-sampling for additional offline analysis (Figure 6.3).
5) In- and out-ports in combination with respect to cell retention systems allow for constant medium throughput, thus enabling more flexible and controlled feeding strategies (Figure 6.3). This allows, for example, applying perfusion feeding, which has substantial advantages compared with typical batch feeding by a single medium bolus typically applied in 2D tissue culture and for conventional spinner-flask reactors (see Figure 6.4, and discussion in the later sections).
6) Parallel, small-scale tank bioreactor systems consisting of numerous independently monitored and controlled vessels support the multiparametric process development; the availability of physically equivalent larger reactor systems subsequently supports relative linear process upscaling.
7) Disposable, GMP compliant culture vessels that are meanwhile available at most process dimensions allow for clinical translation.

Due to these advantages, stirred-tank reactor technology is extensively used in biotechnical industries for the optimization and subsequent mass production of recombinant proteins and vaccines from established mammalian cell lines such as CHO, HeLa, and HEK293 cells, whereby established processes in >1000 l scale reactors have been established [71]. However, respective bioprocesses aim at the production of cell-derived products such as proteins or vaccines, only, paying limited attention to the cells' biomedical features. This focus is in stark contrast to the envisioned production of stem cells and their progenies for cell therapies. Due to the relative complex culture conditions for PSCs (as well as other stem cell types such as MSCs and hematopoietic stem cells (HSCs)) at the pluripotent state and the often even more laborious differentiation regimen required for the derivation of most specific cell types, PSC processing in stirred-tank reactors is still at its infancy.

Due to this complexity, current process dimensions are focused on 100 ml scale optimization trials, at most. High costs for stem cell-specific culture and differentiation media, which often include components subject to batch-to-batch variation such as bovine (BSA) or human serum albumin (HSA), further slowdown this field of research.

Finally, transition of established 2D hPSC culture toward stirred suspension has been challenging due to cells' inherent features as further outlined in the following sections.

The mESCs, which have been established in 1981 [72] and thus long ahead of hESCs in 1998 [32] as well as mouse/human iPSCs in 2006/2007 [73, 74], were initially used as a model for PSC process development and upscaling.

It has been well established that mESCs can be cultured as cell aggregates at least for the induction of differentiation, which were then termed embryoid bodies (EBs) due to their multilineage differentiation potential. Early work showed that mESC inoculation as single cells in suspension culture followed by dynamic condition (i.e., rotation or stirring) supports the formation of more homogenous

EBs compared with static controls, which induced detrimental cell- and aggregate-fusion [75, 76]. These observations led first to processes for the combined mass expansion and differentiation of mESCs in stirred, fully instrumented bioreactors in 2000 ml scale [77]. Importantly, this work demonstrated that mESC aggregation can be controlled by stirring conditions, that is, the stirring speed. Utilizing engineered mESC lines these studies also demonstrated the induction of a relatively high cell yield of $\sim 1 \times 10^7$ mESC $ml^{-1}$ during a combined cell expansion and differentiation phase. Moreover, the successful production of up to $\sim 10^9$ murine CMs in a single 2000 ml scale bioreactor run was demonstrated whereby genetic enrichment enabled >99% purity of CMs [77]. Further process optimization was achieve via transition from batch to perfusion feeding and resulted in a ~fivefold increase in CM yield, allowing to harvest up to $4.6 \times 10^9$ CMs in a single 2000 ml scale bioreactor run [78]. However, these early experiments with mESCs relied on the use of mitogen-rich, poorly defined, fetal calf serum (FCS)-supplemented culture media. It should also be noted that mESCs have a significantly higher proliferation rate, corresponding to a threefold shorter population doubling time (PDT) of ~12–16 h compared with ~36 h for hESCs [79]. Moreover, in contrast to their human counterparts, mouse PSCs are also less sensitive regarding cells' viability after dissociation, which supported culture process inoculation by single cell suspensions for subsequent cell aggregation control in stirred bioreactors.

### 6.3.2 Process Inoculation and Passaging Strategies: Cell Clumps Versus Single Cells

Due to their sensitivity to single cell–based passaging – resulting in essentially quantitative cell loss – hPSCs were traditionally passaged as semi-dissociated colonies or clumps in 2D culture (as noted earlier) and inoculation of 3D cultures initially also relied on this strategy. Since these cell clumps tended to stick to each other resulting in extensive fusion, particularly in static suspension culture, encapsulation into matrices has been suggested to control this process [43, 44]. Limitations of this approach include the addition of extra matrices to process, requirement of special equipment, and the limited control on how many cells or clumps will be entrapped in a single capsule. Another issue is the uncontrolled burst of such capsules, for example, in response to cell proliferation or shear stress in bioreactors, further limiting control and reproducibility of the method. More recently a functional polymer termed "gelan gum" was reported to avoid fusion of aggregates in suspension [42]. Although the underlying concept is sound, the method did not support single-cell seeding but required mechanical disruption of cell aggregates by processing through a strainer for process inoculation and passaging.

In contrast, passaging of single cells allows, in principle, for fully defined process inoculation at a preoptimized cell density. The strategy thus supports experimental control in basic science and substantially facilitates the development of standard operation procedures (SOPs) required for GMP-compliant processes as well. The other advantage of single cell–based passaging is the entire dissociation of colonies (2D) and aggregates (from 3D) at every expansion step, which disrupts the prolonged intercellular interaction in the core of respective colonies or aggregates. Full dissociation disrupts established cell–cell contacts, counteracting the formation of local gradients and the development of undesired culture heterogeneity, which potentially triggers uncontrolled differentiation [52].

However, prerequisite for single cell–based passaging of hPSCs was the development of advanced culture media and particularly the supplementation of the chemical compound Y-27632, a potent Rho-associated kinase (ROCK) inhibitor (ROCKi), which seems to transiently support single hPSC vitality, creating a window for hPSC re-aggregation after seeding [80].

### 6.3.3
### Microcarriers or Matrix-Free Suspension Culture: Pro and Contra

One successful strategy for hPSC transition to suspension culture is the use of microcarriers (MCs). These are particles manufactured from different materials available in divergent shapes and sizes, which might be used with our without further matrix coating. MCs were introduced to bioprocessing to provide anchorage-dependent cells, which are difficult to adapt to suspension culture, with a "floating surface" for attachment and, in parallel, to extensively elevate the surface area upon transition to 3D. The expansion of hPSCs on MCs has been successfully demonstrated in static culture dishes [45, 81] followed by stirred conditions including spinner flasks [82] and instrumented stirred tank bioreactors [55]. Despite this success, investigators have noted the tendency of hPSC to stick to each other (rather than to the prescreened types of microcarriers), thereby inducing an additional level of culture heterogeneity (Table 6.1). Stiff MC types might also increase shear stress, particularly at stirred conditions, and an extensive lag-phase was noted upon transition of hPSC to MCs in bioreactors [55, 82]. The approach also requires the potentially cumbersome removal of microcarriers from clinical-grade cell preparations before clinical applications.

Utilizing the chemical ROCKi in combination with enzymatic dissociation into single cells for more standardized inoculation [83], we and others have demonstrated the expansion of undifferentiated hPSCs as matrix-free cell-only aggregates in static suspension and in stirred spinner flasks [50–52, 84]. In these proof-of-concept studies, maintenance of pluripotency and karyotype stability over numerous passages (i.e., repeated cycles of aggregate dissociation and re-aggregation) was demonstrated. Subsequently, transfer to instrumented stirred-tank bioreactors in 100 ml scale (parallel DASGIP reactor system, Julich, Germany) was conducted including the initial optimization of the cell inoculation densities and stirring conditions (such as the impeller design and the stirring speed) [54]. Key process parameters including growth kinetics, pH, DO, and lactate concentration were monitored over the 7-day process duration [54]. Compared with suspension culture on plastic dishes, where cells are more concentrated on the bottom of the dish due to gravity (rather than being homogeneously distributed in full 3D), inoculated at $\sim 0.5-1 \times 10^5$ cell ml$^{-1}$ [50–52] about five- to tenfold higher cell seeding at $\sim 5 \times 10^5$ cells ml$^{-1}$ was required for successful aggregate formation in impeller-stirred reactors [54]. Cell harvest densities of about $2 \times 10^6$ cells $ml^{-1}$ corresponding to a three- to sixfold expansion per passage were subsequently achieved at stirred conditions [50–52, 84]. Thus, despite the general success of the method, these data request for substantial process optimization with respect to the cell yield per milliliter medium and to the overall process yield as well.

### 6.3.4
### Optimization and Current Limitations of hPSC Processing in Stirred Bioreactors

Applying the matrix-free aggregate approach, several optimization trials were recently performed in the batch-feeding mode (usually performing entire medium change once per day; Figure 6.4) equivalent to the typical feeding strategy in conventional 2D culture. Thorough investigations on the inoculation density and the agitation rate in a two-parameter, three-level factorial experimental design were recently published [85], whereas other investigators have separately tested more parameters including the DO tension in instrumented stirred-tank reactors [86]. Although a close interdependency of individual parameters was demonstrated, the overall progress with respect to the maximally achieved cell harvest density was limited, suggesting the presence of suboptimal or detrimental process conditions.

Applying 2D monolayer and MC-based 3D culture, Chen *et al.* [87] performed investigations into metabolic aspects of hPSC cultivation. Utilizing three conventional hPSC culture media, the study suggests that neither glucose nor amino acid limitations seems to limit hPSC proliferation when aiming at higher cell densities in suspension culture, but rather the accumulation of metabolic by-products such as ammonium and/or lactate. Given that hPSCs rely mainly on glycolysis (i.e., in contrast to somatic cells in which metabolism is dominated by oxidative phosphorylation) to cover their demand for energy and metabolic building blocks [88], substantial lactate accumulation and culture acidification are expected consequences of increasing cell density. This might represent a significant roadblock for achieving higher specific process yields particularly in the batch-feeding mode, which has been shown to result is stark oscillating process patterns upon the daily medium exchange in instrumented bioreactors [54, 85–87].

To generate a more homogenous culture environment and achieve high cell density, perfusion feeding is well established in industrial-scale biotechnology [89, 90]. Initial studies on hPSCs have indeed shown that perfusion feeding can have substantial impact on PSC expansion and differentiation in 2D [91] as well as in 3D instrumented reactors [55]. Perfusion is characterized by a continuous exchange of medium in the culture vessel by fresh medium, which typically relies on cell retention devices to avoid washout of cells from the bioreactor alongside with the continuous medium exchange (Figure 6.3). Perfusion further allows better process automation and improved control of the culture environment including pH, DO tension, and the concentration of nutrients and waste products. Furthermore, the self-conditioning ability of cells by the secretion of stimulating growth factors, which might be controlled by the flow rate in perfusion, could permit reducing expensive medium components [92].

Recently, it was shown for mouse PSCs that perfusion can indeed improve process performance and cell yields in instrumented tank reactors achieving cell densities of $>2 \times 10^7$ $ml^{-1}$ via increasing the perfusion rate to up to ~fivefold process volumes/day [53] reflecting earlier results in 2000 ml in instrumented reactors by our group comparing batch- and perfusion-feeding [78]. However, as noted earlier, very limited studies on hPSC perfusion cultures are yet available and the exact mechanisms by which perfusion improves PSC yields are not well understood.

## 6.4 Autologous Versus Allogeneic Cell Therapies: Practical and Economic Considerations for hPSC Processing

The first hPSC-based cell treatments have recently been applied to patients. These include the use of hESC-derived-retinal pigment epithelial cells (and more recently of hiPSC-derived-retinal pigment epithelial cells [93]) to treat age-related macular degeneration in the eye [94], the announced transplantation of hESC-derived insulin producing cells to overcome insulin-dependence of type1 diabetes by the company ViaCyte (http://viacyte.com/clinical/clinical-trials/), and the first heart failure patient receiving hESC-derived cardiac progenitor cells as noted in the introduction [30].

In contrast to hESCs, hiPSCs are compatible with an allogeneic cell therapy concept. For this to occur individual, patient-specific hiPSC lines have to be derived, characterized, banked, expanded, and differentiated such that the resulting functional progenies can be transplanted back to the donor patient for organ repair (Figure 6.5). The strategy, as per definition, should allow successful, rejection-free cell transplantation without the use of immune suppressive drugs, but this is an area of ongoing debate [95–97]. Such hiPSC-dependent autologous approach would require the development of relative small-scale processes, allowing for the generation of a disease-dependent treatment dose for a single patient, only, but which will include the production of extra cells for quality controls, *in vitro* potency assays, and to eventually repeat a cell treatment, if required. Ideally, such small-scale processing would be highly automated, require minimal (clean room) space, and support the parallel cell production for multiple patients, most likely in disposable bioreactors. Apparently, the autologous approach will be relatively time and cost intensive and will not be the rational strategy for all diseases. For example, it creates no advantage to treat autoimmune-induced disorders such as pancreatic $\beta$-cell loss in diabetes type1 patients by their own autologous cells.

An allogeneic cell therapy refers to the production of non-patient-specific cells for transplantation into recipients that are not-related to the hPSC donor and will likely require immune-suppressive treatments equivalent to conventional whole-organ transplantation. However, the level of immune suppression will likely depend on HLA matching, which can be established by the use of several

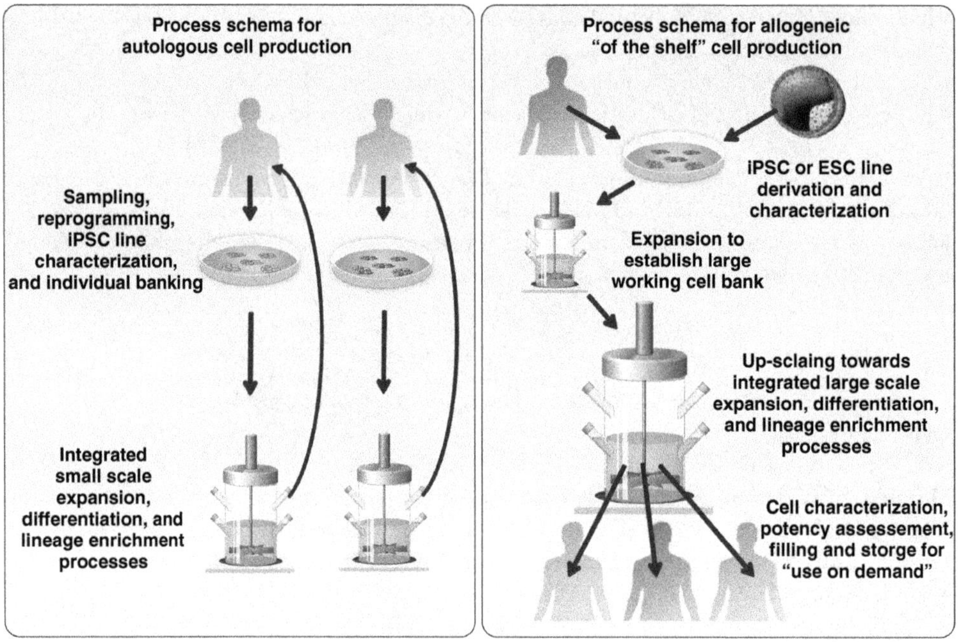


**Figure 6.5** Process schemas for autologous and allogeneic cell production.

hPSC lines (rather than using a single, "universal" cell line across all patients) to cover patient population of a specific country or region as discussed in more detail elsewhere [98].

However, the bulk production of larger allogenic cell batches to produce "of the shelf" cell therapy product will require the establishment of large-scale cell production (Figure 6.5), which, if available, should substantially reduce cost for bulk cell production and characterization. Despite these general statements, limited experience on cost estimations for hPSC–progeny production exists, given the low maturation grade of the field [67].

Process optimization is also the key to reduce the future "cost of goods" (COG), a factor that can be defined as the sum of all costs (including all materials and maintenance) to manufacture a defined stem cell product [67]. In addition to the demonstration of the therapeutic success of advanced stem cell therapies, COG is an important factor regarding the ultimate socioeconomic acceptance on novel therapies [67].

## 6.5 Upscaling hPSC Cardiomyogenic Differentiation in Bioreactors

Expanding this discussion of hPSC production at the pluripotent state, we provide a brief status of upscaling cardiomyogenic differentiation of PSCs. This not only reflects our own experience and knowhow but also represents one of the most progressed fields in hPSC mass production, reflecting the enormous therapeutic and socioeconomic potential of this research.

In general, one can subdivide sequential versus integrated production processes. Sequential production refers to the independent processing of cells at pluripotent state in the first step, which are then differentiated in a second, rather different, approach. This might include, for example, the transition from 2D culture for cell expansion to 3D differentiation. Traditionally, cell expansion strategies (outlined earlier) have been developed and optimized independently of differentiation protocols. In contrast, integrated processing aims at the development of closely connected "single-step" approaches based on direct transition from hPSC production into lineage differentiation. This might be established by switching specific media, only, which control the different steps of the process but without gross adaptation of the bioreactor environment.

Aiming at mimicking signals known from developmental biology, cardiac differentiation of mouse and human PSCs has been developed by combining recombinant factors such as Activin, BMPs, WNTs, and VEGF, reviewed elsewhere [99, 100].

To establish chemically defined culture conditions and significantly reduced costs, small molecules were tested such as the compound SB203580, a potent p38 MAPK inhibitor. Supplementation of SB203580 to EBs differentiated in tissue culture scale suspension culture resulted in up to 10% CM induction from hESCs in a concentration-dependent manner [101, 102], which allowed enrichment to >99% purity by genetic selection [103, 104].

More recently, >80% cardiomyocyte induction was achieved from hESC and hiPSC lines by the use of the GSK3$\beta$ inhibitor CHIR99021 triggering WNT pathway activation. Subsequently, the protocol employs supplementation of chemical WNT pathway inhibitors such as IWP or IWR at later stages [105], reflecting a pattern of biphasic WNT activity known from heart development. Initially, the protocol was developed in a 2D monolayer format for hPSC expansion and differentiation [106] limiting its application to upscaling in stirred reactors.

To overcome these limitations, we have recently established a multiwell platform for hPSC cultivation and differentiation in low attachment 12-well dishes to facilitate process optimization in higher throughput. By showing that the cardiac induction closely depended on the transient CHIR99021 treatment step, the work enabled direct combination of hPSC expansion from single cell–inoculated aggregates with subsequent cardiac differentiation into an integrated "one-step" process [16]. Upscaling to rotated Erlenmeyer flasks in 20 ml scale confirmed the robustness of the protocol and applicability to several established hESC [31] and hiPSC lines [107] and supported ultimate transition into stirred, instrumented tank reactors. The integrated process in 100 ml scale allowed to directly combine single cell–inoculated, aggregate-based hPSC expansion followed by the induction of up to 85% pure CMs in defined media enabling the production of 40–50 million CMs in a single process run [16]. A detailed description of the experimental procedure including all steps for bioreactor setup, calibration, and troubleshooting was recently accepted for publication as well [108].

At its current efficiency, the process could potentially be scaled up to 2000 ml for the production of ∼$1 \times 10^9$ human CMs in a single batch, which has been discussed as an appropriate therapeutic cell dose to replace the loss of MI-induced CMs [25, 60]. Interestingly, this calculated yield closely reflects our previous results achieved for mESC expansion and cardiac differentiation demonstrating production of $1-5 \times 10^9$ CMs, in 2000 ml scale bioreactors. However, recent work on hPSC cardiac differentiation in a fully equipped tank reactor showed that process modifications, for example, by DO control, can substantially boost CM yield [109], suggesting an extensive future optimization potential.

## 6.6
## Conclusion

Taken together, the current status in the field of PSC processing suggests that suspension culture in instrumented bioreactors has a great potential for clinical and industrial-scale process development but significant progress and upscaling steps are still required.

Beyond modifying the inoculation density, stirring speed, and oxygen concentration, to our best knowledge, no significant process parameters such as pH control by automated feedback loops have yet been published for hPSC processing.

Moreover, Figure 6.4 presents a summary of potential process monitoring and control tools that might be highly valuable to facilitate hPSC processing and control. Some of these tools such as advanced online monitoring of hPSC aggregate development in 4D (i.e., monitoring of 3D aggregates at subcellular resolution in real time) are currently under development. Other technologies, including the online monitoring of metabolites, such as glucose, lactate, and ammonium concentration via automated process sampling, are generally established for process analysis in industrial-scale bioreactors. But they have not yet been applied to the relatively small-scale processing of hPSCs, which were mainly performed in 100 ml scale fully equipped reactors and, to our best knowledge, have not yet exceeded a 300 ml culture approach in simple spinner flasks, with the exemption of mESC culture and differentiation in 2000 ml instrumented reactors as mentioned earlier. Closely integrating metabolic analysis, ideally online, will be of great value to leverage hPSC processing. Moreover, as also shown in Figure 6.4, novel, process-integrated tools and techniques for better control of cell aggregation and more homogeneous aggregate development will be most valuable for increasing process efficiency and reproducibility, since controlling aggregate size, development, and homogeneity is known to play a role in lineage differentiation and process efficiency [16, 109, 110].

## List of Abbreviations

| | |
|---|---|
| bpm | beats per minute |
| BSA | bovine- serum albumin |
| CVD | cardiovascular disorders |
| CMs | cardiomyocytes |
| COG | cost of goods |
| DO | dissolved oxygen |
| EBs | embryoid bodies |
| FCS | fetal calf serum |
| GMP | good manufacturing practice |
| HSC | hematopoietic stem cells |
| hESC | human embryonic stem cells |
| hiPSC | human induced pluripotent stem cells |
| hPSC | human pluripotent stem cells |
| HAS | human serum albumin |
| LVAD | left ventricular assist devices |
| MSC | mesenchymal stem cells |
| MC | microcarierr |
| mESC | mouse embryonic stem cells |
| MI | myocardial infarction |
| ROCKi | Rho-associated kinase inhibitor |
| SOP | standard operation procedures |
| 3D | three-dimensional |
| 2D | two dimensional |

## References

1. Bhatnagar, P., Wickramasinghe, K., Williams, J., Rayner, M., and Townsend, N. (2015) The epidemiology of cardiovascular disease in the UK 2014. *Heart*, **101**, 1182–1189.
2. Zweigerdt, R. (2007) The art of cobbling a running pump--will human embryonic stem cells mend broken hearts? *Semin. Cell Dev. Biol.*, **18**, 794–804.
3. Soonpaa, M.H., Rubart, M., and Field, L.J. (2013) Challenges measuring cardiomyocyte renewal. *Biochim. Biophys. Acta*, **1833**, 799–803.
4. van Berlo, J.H., Kanisicak, O., Maillet, M., Vagnozzi, R.J., Karch, J., Lin, S.C., Middleton, R.C., Marban, E., and Molkentin, J.D. (2014) c-kit+ cells minimally contribute cardiomyocytes to the heart. *Nature*, **509**, 337–341.
5. Rojas, S.V., Avsar, M., Uribarri, A., Hanke, J.S., Haverich, A., and Schmitto, J.D. (2015) A new era of ventricular assist device surgery: less invasive procedures. *Minerva Chir.*, **70**, 63–68.
6. Donndorf, P., Strauer, B.E., Haverich, A., and Steinhoff, G. (2013) Stem cell therapy for the treatment of acute myocardial infarction and chronic ischemic heart disease. *Curr. Pharm. Biotechnol.*, **14**, 12–19.
7. Murry, C.E., Field, L.J., and Menasche, P. (2005) Cell-based cardiac repair: reflections at the 10-year point. *Circulation*, **112**, 3174–3183.
8. Korf-Klingebiel, M., Reboll, M.R., Klede, S., Brod, T., Pich, A., Polten, F., Napp, L.C., Bauersachs, J., Ganser, A., Brinkmann, E. *et al.* (2015) Myeloid-derived growth factor (C19orf10)

mediates cardiac repair following myocardial infarction. *Nat. Med.*, **21**, 140–149.

9. Orlova, V.V., van den Hil, F.E., Petrus-Reurer, S., Drabsch, Y., Ten Dijke, P., and Mummery, C.L. (2014) Generation, expansion and functional analysis of endothelial cells and pericytes derived from human pluripotent stem cells. *Nat. Protoc.*, **9**, 1514–1531.
10. Kensah, G., Roa Lara, A., Dahlmann, J., Zweigerdt, R., Schwanke, K., Hegermann, J., Skvorc, D., Gawol, A., Azizian, A., Wagner, S. *et al.* (2013) Murine and human pluripotent stem cell-derived cardiac bodies form contractile myocardial tissue in vitro. *Eur. Heart J.*, **34**, 1134–1146.
11. Flemming, A., Schallmoser, K., Strunk, D., Stolk, M., Volk, H.D., and Seifert, M. (2011) Immunomodulative efficacy of bone marrow-derived mesenchymal stem cells cultured in human platelet lysate. *J. Clin. Immunol.*, **31**, 1143–1156.
12. Arslan, F., Lai, R.C., Smeets, M.B., Akeroyd, L., Choo, A., Aguor, E.N., Timmers, L., van Rijen, H.V., Doevendans, P.A., Pasterkamp, G. *et al.* (2013) Mesenchymal stem cell-derived exosomes increase ATP levels, decrease oxidative stress and activate PI3K/Akt pathway to enhance myocardial viability and prevent adverse remodeling after myocardial ischemia/reperfusion injury. *Stem Cell Res.*, **10**, 301–312.
13. Kawamura, M., Miyagawa, S., Fukushima, S., Saito, A., Toda, K., Daimon, T., Shimizu, T.M.P., Okano, T., and Sawa, Y. (2015) Xenotransplantation of bone marrow-derived human mesenchymal stem cell sheets attenuates left ventricular remodeling in a porcine ischemic cardiomyopathy model. *Tissue Eng. Part A*, **21**, 2272–2280.
14. Dar, A., Domev, H., Ben-Yosef, O., Tzukerman, M., Zeevi-Levin, N., Novak, A., Germanguz, I., Amit, M., and Itskovitz-Eldor, J. (2012) Multipotent vasculogenic pericytes from human pluripotent stem cells promote recovery of murine ischemic limb. *Circulation*, **125**, 87–99.
15. Chen, T.S., Arslan, F., Yin, Y., Tan, S.S., Lai, R.C., Choo, A.B., Padmanabhan, J., Lee, C.N., de Kleijn, D.P., and Lim, S.K. (2011) Enabling a robust scalable manufacturing process for therapeutic exosomes through oncogenic immortalization of human ESC-derived MSCs. *J. Transl. Med.*, **9**, 47.
16. Kempf, H., Olmer, R., Kropp, C., Ruckert, M., Jara-Avaca, M., Robles-Diaz, D., Franke, A., Elliott, D.A., Wojciechowski, D., Fischer, M. *et al.* (2014) Controlling expansion and cardiomyogenic differentiation of human pluripotent stem cells in scalable suspension culture. *Stem Cell Rep.*, **3**, 1132–1146.
17. Dai, W., Field, L.J., Rubart, M., Reuter, S., Hale, S.L., Zweigerdt, R., Graichen, R.E., Kay, G.L., Jyrala, A.J., Colman, A. *et al.* (2007) Survival and maturation of human embryonic stem cell-derived cardiomyocytes in rat hearts. *J. Mol. Cell. Cardiol.*, **43**, 504–516.
18. van Laake, L.W., Passier, R., Monshouwer-Kloots, J., Verkleij, A.J., Lips, D.J., Freund, C., den Ouden, K., Ward-Van Oostwaard, D., Korving, J., Tertoolen, L.G. *et al.* (2007) Human embryonic stem cell-derived cardiomyocytes survive and mature in the mouse heart and transiently improve function after myocardial infarction. *Stem Cell Res.*, **1**, 9–24.
19. Andree, B., Bela, K., Horvath, T., Lux, M., Ramm, R., Venturini, L., Ciubotaru, A., Zweigerdt, R., Haverich, A., and Hilfiker, A. (2014) Successful re-endothelialization of a perfusable biological vascularized matrix (BioVaM) for the generation of 3D artificial cardiac tissue. *Basic Res. Cardiol.*, **109**, 441.
20. Hirt, M.N., Hansen, A., and Eschenhagen, T. (2014) Cardiac tissue engineering: state of the art. *Circ. Res.*, **114**, 354–367.
21. Kensah, G., Gruh, I., Viering, J., Schumann, H., Dahlmann, J., Meyer, H., Skvorc, D., Bar, A., Akhyari, P., Heisterkamp, A. *et al.* (2011) A novel miniaturized multimodal bioreactor for continuous in situ assessment of

bioartificial cardiac tissue during stimulation and maturation. *Tissue Eng. Part C Methods*, **17**, 463–473.

22. Matsuura, K., Masuda, S., Haraguchi, Y., Yasuda, N., Shimizu, T., Hagiwara, N., Zandstra, P.W., and Okano, T. (2011) Creation of mouse embryonic stem cell-derived cardiac cell sheets. *Biomaterials*, **32**, 7355–7362.
23. Martens, A., Rojas, S.V., Baraki, H., Rathert, C., Schecker, N., Zweigerdt, R., Schwanke, K., Rojas-Hernandez, S., Martin, U., Saito, S. *et al.* (2014) Substantial early loss of induced pluripotent stem cells following transplantation in myocardial infarction. *Artif. Organs*, **38**, 978–984.
24. Laflamme, M.A., Chen, K.Y., Naumova, A.V., Muskheli, V., Fugate, J.A., Dupras, S.K., Reinecke, H., Xu, C., Hassanipour, M., Police, S. *et al.* (2007) Cardiomyocytes derived from human embryonic stem cells in pro-survival factors enhance function of infarcted rat hearts. *Nat. Biotechnol.*, **25**, 1015–1024.
25. Chong, J.J., Yang, X., Don, C.W., Minami, E., Liu, Y.W., Weyers, J.J., Mahoney, W.M., Van Biber, B., Cook, S.M., Palpant, N.J. *et al.* (2014) Human embryonic-stem-cell-derived cardiomyocytes regenerate non-human primate hearts. *Nature*, **510**, 273–277.
26. Templin, C., Zweigerdt, R., Schwanke, K., Olmer, R., Ghadri, J.R., Emmert, M.Y., Muller, E., Kuest, S.M., Cohrs, S., Schibli, R. *et al.* (2012) Transplantation and tracking of human-induced pluripotent stem cells in a pig model of myocardial infarction: assessment of cell survival, engraftment, and distribution by hybrid single photon emission computed tomography/computed tomography of sodium iodide symporter transgene expression. *Circulation*, **126**, 430–439.
27. Ye, L., Chang, Y.H., Xiong, Q., Zhang, P., Zhang, L., Somasundaram, P., Lepley, M., Swingen, C., Su, L., Wendel, J.S. *et al.* (2014) Cardiac repair in a porcine model of acute myocardial infarction with human induced pluripotent stem cell-derived cardiovascular cells. *Cell Stem Cell*, **15**, 750–761.
28. Rubart, M., Wang, E., Dunn, K.W., and Field, L.J. (2003) Two-photon molecular excitation imaging of Ca2+ transients in Langendorff-perfused mouse hearts. *Am. J. Physiol. Cell Physiol.*, **284**, C1654–C1668.
29. Shiba, Y., Filice, D., Fernandes, S., Minami, E., Dupras, S.K., Biber, B.V., Trinh, P., Hirota, Y., Gold, J.D., Viswanathan, M. *et al.* (2014) Electrical integration of human embryonic stem cell-derived cardiomyocytes in a guinea pig chronic infarct model. *J. Cardiovasc. Pharmacol. Ther.*, **19**, 368–381.
30. Menasche, P., Vanneaux, V., Hagege, A., Bel, A., Cholley, B., Cacciapuoti, I., Parouchev, A., Benhamouda, N., Tachdjian, G., Tosca, L. *et al.* (2015) Human embryonic stem cell-derived cardiac progenitors for severe heart failure treatment: first clinical case report. *Eur. Heart J.*, **36**, 2011–2017.
31. Elliott, D.A., Braam, S.R., Koutsis, K., Ng, E.S., Jenny, R., Lagerqvist, E.L., Biben, C., Hatzistavrou, T., Hirst, C.E., Yu, Q.C. *et al.* (2011) NKX2-5(eGFP/w) hESCs for isolation of human cardiac progenitors and cardiomyocytes. *Nat. Methods*, **8**, 1037–1040.
32. Thomson, J.A., Itskovitz-Eldor, J., Shapiro, S.S., Waknitz, M.A., Swiergiel, J.J., Marshall, V.S., and Jones, J.M. (1998) Embryonic stem cell lines derived from human blastocysts. *Science*, **282**, 1145–1147.
33. Chen, K.G., Mallon, B.S., Hamilton, R.S., Kozhich, O.A., Park, K., Hoeppner, D.J., Robey, P.G., and McKay, R.D. (2012) Non-colony type monolayer culture of human embryonic stem cells. *Stem Cell Res.*, **9**, 237–248.
34. Braam, S.R., Denning, C., Matsa, E., Young, L.E., Passier, R., and Mummery, C.L. (2008) Feeder-free culture of human embryonic stem cells in conditioned medium for efficient genetic modification. *Nat. Protoc.*, **3**, 1435–1443.
35. Ludwig, T.E., Bergendahl, V., Levenstein, M.E., Yu, J., Probasco, M.D., and Thomson, J.A. (2006) Feeder-independent culture of human embryonic stem cells. *Nat. Methods*, **3** (8), 637–646.

36. Côme, J., Nissan, X., Aubry, L., Tournois, J., Girard, M., Perrier, A.L., Peschanski, M., and Cailleret, M. (2008) Improvement of culture conditions of human embryoid bodies using a controlled perfused and dialyzed bioreactor system. *Tissue Eng. Part C Methods*, **14** (4), 289–298.
37. Wang, X., Wei, G., Yu, W., Zhao, Y., Yu, X., and Ma, X. (2006) Scalable producing embryoid bodies by rotary cell culture system and constructing engineered cardiac tissue with ES-derived cardiomyocytes in vitro. *Biotechnol. Progr.*, **22**, 811–818.
38. Yirme, G., Amit, M., Laevsky, I., Osenberg, S., and Itskovitz-Eldor, J. (2008) Establishing a dynamic process for the formation, propagation, and differentiation of human embryoid bodies. *Stem Cells Dev.*, **17**, 1227–1242.
39. Gerecht-Nir, S., Cohen, S., and Itskovitz-Eldor, J. (2004) Bioreactor cultivation enhances the efficiency of human embryoid body (hEB) formation and differentiation. *Biotechnol. Bioeng.*, **86**, 493–502.
40. Lei, Y. and Schaffer, D.V. (2013) A fully defined and scalable 3D culture system for human pluripotent stem cell expansion and differentiation. *Proc. Natl. Acad. Sci. U.S.A.*, **110**, E5039–E5048.
41. Lou, Y.R., Kanninen, L., Kuisma, T., Niklander, J., Noon, L.A., Burks, D., Urtti, A., and Yliperttula, M. (2014) The use of nanofibrillar cellulose hydrogel as a flexible three-dimensional model to culture human pluripotent stem cells. *Stem Cells Dev.*, **23** (4), 380–392.
42. Otsuji, T.G., Bin, J., Yoshimura, A., Tomura, M., Tateyama, D., Minami, I., Yoshikawa, Y., Aiba, K., Heuser, J.E., Nishino, T. *et al.* (2014) A 3D sphere culture system containing functional polymers for large-scale human pluripotent stem cell production. *Stem Cell Rep.*, **2**, 734–745.
43. Randle, W.L., Cha, J.M., Hwang, Y.S., Chan, K.L., Kazarian, S.G., Polak, J.M., and Mantalaris, A. (2007) Integrated 3-dimensional expansion and osteogenic differentiation of murine embryonic stem cells. *Tissue Eng.*, **13** (12), 2957–2970.
44. Serra, M., Correia, C., Malpique, R., Brito, C., Jensen, J., Bjorquist, P., Carrondo, M.J., and Alves, P.M. (2011) Microencapsulation technology: a powerful tool for integrating expansion and cryopreservation of human embryonic stem cells. *PLoS One*, **6**, e23212.
45. Oh, S.K., Chen, A.K., Mok, Y., Chen, X., Lim, U.M., Chin, A., Choo, A.B., and Reuveny, S. (2009) Long-term microcarrier suspension cultures of human embryonic stem cells. *Stem Cell Res.*, **2**, 219–230.
46. Bardy, J., Chen, A.K., Lim, Y.M., Wu, S., Wei, S., Weiping, H., Chan, K., Reuveny, S., and Oh, S.K. (2013) Microcarrier suspension cultures for high-density expansion and differentiation of human pluripotent stem cells to neural progenitor cells. *Tissue Eng. Part C Methods*, **19** (2), 166–180.
47. Fok, E.Y. and Zandstra, P.W. (2005) Shear-controlled single-step mouse embryonic stem cell expansion and embryoid body-based differentiation. *Stem Cells*, **23**, 1333–1342.
48. Hunt, M.M.,, Meng, G., Rancourt, D.E., Gates, I.D., and Kallos, M.S. (2014) Factorial experimental design for the culture of human embryonic stem cells as aggregates in stirred suspension bioreactors reveals the potential for interaction effects between bioprocess parameters. *Tissue Eng.* Part C Methods **20**, 76–89.
49. Amit, M., Chebath, J., Margulets, V., Laevsky, I., Miropolsky, Y., Shariki, K., Peri, M., Blais, I., Slutsky, G., Revel, M., and Itskovitz-Eldor, J. (2010) Suspension culture of undifferentiated human embryonic and induced pluripotent stem cells. *Stem Cell Rev.*, **6**, 248–259.
50. Olmer, R., Haase, A., Merkert, S., Cui, W., Palecek, J., Ran, C., Kirschning, A., Scheper, T., Glage, S., Miller, K., Curnow, E.C., Hayes, E.S., and Martin, U. (2010) Long term expansion of undifferentiated human iPS and ES cells in suspension culture using a defined medium. *Stem Cell Res.*, **5**, 51–64.
51. Singh, H., Mok, P., Balakrishnan, T., Rahmat, S.N., and Zweigerdt, R. (2010)

Up-scaling single cell-inoculated suspension culture of human embryonic stem cells. *Stem Cell Res.*, **4**, 165–179.
52. Zweigerdt, R., Olmer, R., Singh, H., Haverich, A., and Martin, U. (2011) Scalable expansion of human pluripotent stem cells in suspension culture. *Nat. Protoc.*, **6** (5), 689–700.
53. Baptista, R.P., Fluri, D.A., and Zandstra, P.W. (2013) High density continuous production of murine pluripotent cells in an acoustic perfused bioreactor at different oxygen concentrations. *Biotechnol. Bioeng.*, **110**, 648–655.
54. Olmer, R., Lange, A., Selzer, S., Kasper, C., Haverich, A., Martin, U., and Zweigerdt, R. (2012) Suspension culture of human pluripotent stem cells in controlled, stirred bioreactors. *Tissue Eng. Part C Methods*, **18** (10), 772–784.
55. Serra, M., Brito, C., Sousa, M.F., Jensen, J., Tostoes, R., Clemente, J., Strehl, R., Hyllner, J., Carrondo, M.J., and Alves, P.M. (2010) Improving expansion of pluripotent human embryonic stem cells in perfused bioreactors through oxygen control. *J. Biotechnol.*, **148**, 208–215.
56. Fernandes-Platzgummer, A., Diogo, M.M., Lobato da Silva, C., and Cabral, J.M.S. (2014) Maximizing mouse embryonic stem cell production in a stirred tank reactor by controlling dissolved oxygen concentration and continuous perfusion operation. *Biochem. Eng. J.*, **82**, 81–90.
57. Chen, V.C., Couture, S.M., Ye, J., Lin, Z., Hua, G., Huang, H.I., Wu, J., Hsu, D., Carpenter, M.K., and Couture, L.A. (2012) Scalable GMP compliant suspension culture system for human ES cells. *Stem Cell Res.*, **8**, 388–402.
58. Wang, Y., Chou, B.K., Dowey, S., He, C., Gerecht, S., and Cheng, L. (2013) Scalable expansion of human induced pluripotent stem cells in the defined xeno-free E8 medium under adherent and suspension culture conditions. *Stem Cell Res.*, **11**, 1103–1116.
59. Crook, J.M., Peura, T.T., Kravets, L., Bosman, A.G., Buzzard, J.J., Horne, R., Hentze, H., Dunn, N.R., Zweigerdt, R., Chua, F. *et al.* (2007) The generation of six clinical-grade human embryonic stem cell lines. *Cell Stem Cell*, **1**, 490–494.
60. Zweigerdt, R. (2009) Large scale production of stem cells and their derivatives. *Adv. Biochem. Eng. Biotechnol.*, **114**, 201–235.
61. Rodin, S., Antonsson, L., Niaudet, C., Simonson, O.E., Salmela, E., Hansson, E.M., Domogatskaya, A., Xiao, Z., Damdimopoulou, P., Sheikhi, M. *et al.* (2014) Clonal culturing of human embryonic stem cells on laminin-521/E-cadherin matrix in defined and xeno-free environment. *Nat. Commun.*, **5**, 3195.
62. Villa-Diaz, L.G., Ross, A.M., Lahann, J., and Krebsbach, P.H. (2013) Concise review: the evolution of human pluripotent stem cell culture: from feeder cells to synthetic coatings. *Stem Cells*, **31**, 1–7.
63. Chen, G., Gulbranson, D.R., Hou, Z., Bolin, J.M., Ruotti, V., Probasco, M.D., Smuga-Otto, K., Howden, S.E., Diol, N.R., Propson, N.E. *et al.* (2011) Chemically defined conditions for human iPSC derivation and culture. *Nat. Methods*, **8**, 424–429.
64. Terstegge, S., Laufenberg, I., Pochert, J., Schenk, S., Itskovitz-Eldor, J., Endl, E., and Brustle, O. (2007) Automated maintenance of embryonic stem cell cultures. *Biotechnol. Bioeng.*, **96**, 195–201.
65. Thomas, R.J., Anderson, D., Chandra, A., Smith, N.M., Young, L.E., Williams, D., and Denning, C. (2009) Automated, scalable culture of human embryonic stem cells in feeder-free conditions. *Biotechnol. Bioeng.*, **102**, 1636–1644.
66. Morrison, M., Klein, C., Clemann, N., Collier, D.A., Hardy, J., Heibetaerer, B., Cader, M.Z., Graf, M., and Kaye, J. (2015) StemBANCC: governing access to material and data in a large stem cell research consortium. *Stem Cell Rev.*, **11**, 681–687.
67. Jenkins, M.J. and Farid, S.S. (2015) Human pluripotent stem cell-derived products: advances towards robust, scalable and cost-effective manufacturing strategies. *Biotechnol. J.*, **10**, 83–95.

**68.** Chen, K.G., Mallon, B.S., McKay, R.D., and Robey, P.G. (2014) Human pluripotent stem cell culture: considerations for maintenance, expansion, and therapeutics. *Cell Stem Cell*, **14**, 13–26.

**69.** Ilin, Y. and Kraft, M.L. (2015) Secondary ion mass spectrometry and Raman spectroscopy for tissue engineering applications. *Curr. Opin. Biotechnol.*, **31**, 108–116.

**70.** Lorbeer, R.A., Heidrich, M., Lorbeer, C., Ramirez Ojeda, D.F., Bicker, G., Meyer, H., and Heisterkamp, A. (2011) Highly efficient 3D fluorescence microscopy with a scanning laser optical tomograph. *Opt. Express*, **19**, 5419–5430.

**71.** Bleckwenn, N.A. and Shiloach, J. (2004) Large-scale cell culture, in *Current Protocols in Immunology* (eds J.E. Coligan *et al.*)Appendix 1, Appendix 1U. Wiley Online Library

**72.** Evans, M.J. and Kaufman, M.H. (1981) Establishment in culture of pluripotential cells from mouse embryos. *Nature*, Evans MJ, Kaufman MH. (5819) **292**, 154–156.

**73.** Takahashi, K., Tanabe, K., Ohnuki, M., Narita, M., Ichisaka, T., Tomoda, K., and Yamanaka, S. (2007) Induction of pluripotent stem cells from adult human fibroblasts by defined factors. *Cell*, **131**, 861–872.

**74.** Takahashi, K. and Yamanaka, S. (2006) Induction of pluripotent stem cells from mouse embryonic and adult fibroblast cultures by defined factors. *Cell*, **126**, 663–676.

**75.** Zandstra, P.W., Bauwens, C., Yin, T., Liu, Q., Schiller, H., Zweigerdt, R., Pasumarthi, K.B., and Field, L.J. (2003) Scalable production of embryonic stem cell-derived cardiomyocytes. *Tissue Eng.*, **9**, 767–778.

**76.** Zweigerdt, R., Burg, M., Willbold, E., Abts, H., and Ruediger, M. (2003) Generation of confluent cardiomyocyte monolayers derived from embryonic stem cells in suspension: a cell source for new therapies and screening strategies. *Cytotherapy*, **5**, 399–413.

**77.** Schroeder, M., Niebruegge, S., Werner, A., Willbold, E., Burg, M., Ruediger, M., Field, L.J., Lehmann, J., and Zweigerdt, R. (2005) Differentiation and lineage selection of mouse embryonic stem cells in a stirred bench scale bioreactor with automated process control. *Biotechnol. Bioeng.*, **92**, 920–933.

**78.** Niebruegge, S., Nehring, A., Bar, H., Schroeder, M., Zweigerdt, R., and Lehmann, J. (2008) Cardiomyocyte production in mass suspension culture: embryonic stem cells as a source for great amounts of functional cardiomyocytes. *Tissue Eng. Part A*, **14**, 1591–1601.

**79.** Ginis, I., Luo, Y., Miura, T., Thies, S., Brandenberger, R., Gerecht-Nir, S., Amit, M., Hoke, A., Carpenter, M.K., Itskovitz-Eldor, J. *et al.* (2004) Differences between human and mouse embryonic stem cells. *Dev. Biol.*, **269**, 360–380.

**80.** Watanabe, K., Ueno, M., Kamiya, D., Nishiyama, A., Matsumura, M., Wataya, T., Takahashi, J.B., Nishikawa, S., Muguruma, K., and Sasai, Y. (2007) A ROCK inhibitor permits survival of dissociated human embryonic stem cells. *Nat. Biotechnol.*, **25**, 681–686.

**81.** Phillips, B.W., Lim, R.Y., Tan, T.T., Rust, W.L., and Crook, J.M. (2008) Efficient expansion of clinical-grade human fibroblasts on microcarriers: cells suitable for ex vivo expansion of clinical-grade hESCs. *J. Biotechnol.*, **134**, 79–87.

**82.** Chen, A.K., Chen, X., Choo, A.B., Reuveny, S., and Oh, S.K. (2010) Expansion of human embryonic stem cells on cellulose microcarriers. *Curr. Protoc. Stem Cell Biol.*, Chapter 1, doi: 10.1002/9780470151808.sc01c11s14, Unit 1C 11.

**83.** Ellerstrom, C., Strehl, R., Noaksson, K., Hyllner, J., and Semb, H. (2007) Facilitated expansion of human embryonic stem cells by single-cell enzymatic dissociation. *Stem Cells*, **25**, 1690–1696.

**84.** Amit, M., Laevsky, I., Miropolsky, Y., Shariki, K., Peri, M., and Itskovitz-Eldor, J. (2011) Dynamic suspension culture for scalable expansion of undifferentiated human pluripotent stem cells. *Nat. Protoc.*, **6**, 572–579.

**85.** Hunt, M.M., Meng, G., Rancourt, D.E., Gates, I.D., and Kallos, M.S. (2014) Factorial experimental design for the culture of human embryonic stem cells as aggregates in stirred suspension bioreactors reveals the potential for interaction effects between bioprocess parameters. *Tissue Eng. Part C Methods,* **20,** 76–89.

86. Abbasalizadeh, S., Larijani, M.R., Samadian, A., and Baharvand, H. (2012) Bioprocess development for mass production of size-controlled human pluripotent stem cell aggregates in stirred suspension bioreactor. *Tissue Eng. Part C Methods,* **18,** 831–851.
87. Chen, X., Chen, A., Woo, T.L., Choo, A.B., Reuveny, S., and Oh, S.K. (2010) Investigations into the metabolism of two-dimensional colony and suspended microcarrier cultures of human embryonic stem cells in serum-free media. *Stem Cells Dev.,* **19,** 1781–1792.
88. Moussaieff, A., Rouleau, M., Kitsberg, D., Cohen, M., Levy, G., Barasch, D., Nemirovski, A., Shen-Orr, S., Laevsky, I., Amit, M. *et al.* (2015) Glycolysis-mediated changes in acetyl-CoA and histone acetylation control the early differentiation of embryonic stem cells. *Cell Metab.,* **21,** 392–402.
89. Chu, L. and Robinson, D.K. (2001) Industrial choices for protein production by large-scale cell culture. *Curr. Opin. Biotechnol.,* **12,** 180–187.
90. Tao, Y., Shih, J., Sinacore, M., Ryll, T., and Yusuf-Makagiansar, H. (2011) Development and implementation of a perfusion-based high cell density cell banking process. *Biotechnol. Progr.,* **27,** 824–829.
91. Fong, W.J., Tan, H.L., Choo, A., and Oh, S.K. (2005) Perfusion cultures of human embryonic stem cells. *Bioprocess Biosyst. Eng.,* **27,** 381–387.
92. Castilho, L.R. and Medronho, R.A. (2002) Cell retention devices for suspended-cell perfusion cultures. *Adv. Biochem. Eng. Biotechnol.,* **74,** 129–169.
93. Kanemura, H., Go, M.J., Shikamura, M., Nishishita, N., Sakai, N., Kamao, H., Mandai, M., Morinaga, C., Takahashi, M., and Kawamata, S. (2014) Tumorigenicity studies of induced pluripotent stem cell (iPSC)-derived retinal pigment epithelium (RPE) for the treatment of age-related macular degeneration. *PLoS One,* **9,** e85336.
94. Schwartz, S.D., Hubschman, J.P., Heilwell, G., Franco-Cardenas, V., Pan, C.K., Ostrick, R.M., Mickunas, E., Gay, R., Klimanskaya, I., and Lanza, R. (2012) Embryonic stem cell trials for macular degeneration: a preliminary report. *Lancet,* **379,** 713–720.
95. Okita, K., Nagata, N., and Yamanaka, S. (2011) Immunogenicity of induced pluripotent stem cells. *Circ. Res.,* **109,** 720–721.
96. Zhao, T., Zhang, Z.N., Rong, Z., and Xu, Y. (2011) Immunogenicity of induced pluripotent stem cells. *Nature,* **474,** 212–215.
97. Araki, R., Uda, M., Hoki, Y., Sunayama, M., Nakamura, M. *et al.* (2013), *Nature* **494**:100–104
98. Laura, Riolobos, Roli, K Hirata, Cameron, J., Turtle1,, Pei-Rong, Wang1, German, G., Gornalussel, Maja Zavajlevski1, Stanley R, Riddell and David W, Russell1 (2013) HLA Engineering of Human Pluripotent Stem Cells, *Molecular Therapy,* **21** 6, 1232–1241. doi: 10.1038/mt.2013.59.
99. Burridge, P.W., Keller, G., Gold, J.D., and Wu, J.C. (2012) Production of de novo cardiomyocytes: human pluripotent stem cell differentiation and direct reprogramming. *Cell Stem Cell,* **10,** 16–28.
100. Mummery, C.L., Zhang, J., Ng, E.S., Elliott, D.A., Elefanty, A.G., and Kamp, T.J. (2012) Differentiation of human embryonic stem cells and induced pluripotent stem cells to cardiomyocytes: a methods overview. *Circ. Res.,* **111,** 344–358.
101. Graichen, R., Xu, X., Braam, S.R., Balakrishnan, T., Norfiza, S., Sieh, S., Soo, S.Y., Tham, S.C., Mummery, C., Colman, A. *et al.* (2008) Enhanced cardiomyogenesis of human embryonic stem cells by a small molecular inhibitor of p38 MAPK. *Differentiation,* **76,** 357–370.
102. Kempf, H., Lecina, M., Ting, S., Zweigerdt, R., and Oh, S. (2011) Distinct regulation of mitogen-activated protein kinase activities is coupled with enhanced cardiac differentiation of

human embryonic stem cells. *Stem Cell Res.*, 7, 198–209.

**103.** Schwanke, K., Merkert, S., Kempf, H., Hartung, S., Jara-Avaca, M., Templin, C., Gohring, G., Haverich, A., Martin, U., and Zweigerdt, R. (2014) Fast and efficient multitransgenic modification of human pluripotent stem cells. *Hum. Gene Ther. Methods*, **25**, 136–153.

**104.** Xu, X.Q., Soo, S.Y., Sun, W., and Zweigerdt, R. (2009) Global expression profile of highly enriched cardiomyocytes derived from human embryonic stem cells. *Stem Cells*, **27**, 2163–2174.

**105.** Lian, X., Zhang, J., Azarin, S.M., Zhu, K., Hazeltine, L.B., Bao, X., Hsiao, C., Kamp, T.J., and Palecek, S.P. (2013) Directed cardiomyocyte differentiation from human pluripotent stem cells by modulating Wnt/beta-catenin signaling under fully defined conditions. *Nat. Protoc.*, **8**, 162–175.

**106.** Lian, X., Hsiao, C., Wilson, G., Zhu, K., Hazeltine, L.B., Azarin, S.M., Raval, K.K., Zhang, J., Kamp, T.J., and Palecek, S.P. (2012) Robust cardiomyocyte differentiation from human pluripotent stem cells via temporal modulation of canonical Wnt signaling. *Proc. Natl. Acad. Sci. U.S.A.*, **109**, E1848–E1857.

**107.** Haase, A., Olmer, R., Schwanke, K., Wunderlich, S., Merkert, S., Hess, C., Zweigerdt, R., Gruh, I., Meyer, J., Wagner, S. *et al.* (2009) Generation of induced pluripotent stem cells from human cord blood. *Cell Stem Cell*, **5**, 434–441.

**108.** Kempf, H. *et al.* (2015) *Nat. Protoc.*, Cardiac differentiation of human pluripotent stem cells in scalable suspension culture. *Nat Protoc.* 2015 Sep;10(9):1345–61. doi: 10.1038/nprot.2015.089. Epub 2015 Aug 13.

**109.** Niebruegge, S., Bauwens, C.L., Peerani, R., Thavandiran, N., Masse, S., Sevaptisidis, E., Nanthakumar, K., Woodhouse, K., Husain, M., Kumacheva, E. *et al.* (2009) Generation of human embryonic stem cell-derived mesoderm and cardiac cells using size-specified aggregates in an oxygen-controlled bioreactor. *Biotechnol. Bioeng.*, **102**, 493–507.

**110.** Dahlmann, J., Kensah, G., Kempf, H., Skvorc, D., Gawol, A., Elliott, D.A., Drager, G., Zweigerdt, R., Martin, U., and Gruh, I. (2013) The use of agarose microwells for scalable embryoid body formation and cardiac differentiation of human and murine pluripotent stem cells. *Biomaterials*, **34**, 2463–2471.

# 7 Culturing Entrapped Stem Cells in Continuous Bioreactors

*Rui Tostoes and Paula M. Alves*

## 7.1 Introduction

Stem cells are cells with the ability to self-renew and differentiate into mature, functional cell types. The applications of stem cells can be broadly divided into two areas: as cell sources for *in vitro* models (both for disease models and toxicology studies) and for cell therapies. As *in vitro* models, stem cells can be used for unraveling the fundamental mechanisms of disease onset, progression, and mitigation or cure and as a tool for toxicity testing. In both cases, it is desirable that the stem cells are differentiated into cells of interest, which may be terminally differentiated cells (typically for *in vitro* tests and in some cases for therapies) or expandable progenitors (more likely to be used in therapies). As therapies, stem cells are a part of regenerative medicine and cellular therapy (reviewed elsewhere in this book; see Chapter 4), with ongoing clinical trials in areas such as diabetes, graft versus host disease (GvHD), and acute lymphoblastic leukemia, among others.

The culture of stem cells is mostly performed using flat surfaces such as T-flasks in temperature- and atmosphere-controlled incubators, which do not control the physicochemical composition of the culture medium. The use of bioreactors to culture stem cells has two main advantages over such systems: it enables monitoring and control of culture parameters such as dissolved oxygen (DO) and pH, and it allows for an easier and more straightforward scalability, which ranges from hundreds of milliliters to $10^4$ l. While scalability is critical for stem cell therapies, which require high cell numbers and are desirable for the process economics of stem cell production in general, the environmental control that bioreactors enable is of paramount importance for any stem cell culture since these cultures are already affected by cell biology-related variables. It is worthwhile to note that there is a decreasing degree of control when comparing a stirred-tank bioreactor with DO and pH measurement and control in the culture medium bulk with a plug-flow bioreactor, which may have the same controls in the medium inlet and outlet but with limited knowledge of the values of these parameter near the cellular microenvironment.

*Bioreactors: Design, Operation and Novel Applications*, First Edition. Edited by Carl-Fredrik Mandenius.

In cell cultures, the cells are constantly consuming nutrients (and signaling specific growth factor and/or cytokines) and producing metabolites; in stem cell cultures, the half-life of the soluble factors used in the growth or differentiation media ranges from hours to days [1, 2]. This is the reason for non-scalable stem cell cultures (i.e., performed in T-flasks) generally requiring a full or partial medium exchange at intervals of 1–4 days, whether it is for expansion or differentiation purposes. The impact of these changes on culture medium composition is proportional to the time of culture; that is, it affects long-term cultures more severely than the short-term ones. The use of a continuous bioreactor adds culture medium composition to the controlled parameters within the vessel and such control enables to change the culture media as the stem cells progress into a desired cell type.

The entrapment of stem cells in a material provides a three-dimensional (3D) culture environment, which mimics the *in vivo* cellular environment, where the cells are in contact with each other and the extracellular matrix, unlike the regular T-flask culture. In addition, the requirement for mixing and thus fluid flow in the bioreactors generates shear; the shear rate is defined as the non-normal pressure component arising from the fluid flow. This is defined as $\tau = \partial v/\partial y$, where $\tau$ is the shear rate and the right side of the equation represents the gradient of change of the velocity vector parallel to the cell surface with the distance $y$ from this same surface. The shear stress is given by the product of the shear rate by the kinematic viscosity of the solution. The impact of shear on stem cell viability, proliferation, and differentiation has been extensively documented, as is discussed later. The entrapment of stem cells is a widely used technology to mitigate these shear-related effects of bioreactor cultures and when coupled to a continuous medium feed it provides a more controlled stem cell culture environment. From a clinical perspective, the entrapment material can provide a physical protection that stops allogeneic cells from being targeted by the host immune system and also prevents the implanted cells to proliferate in the host's body.

## 7.2
## Materials Used in Stem Cell Entrapment

The most critical physical parameters of the materials used for stem cell entrapment materials are obtained in stress–strain tests. These tests have established standards that can be found in the International Organization for Standardization (ISO) and American Society for Testing and Materials (ASTM) websites. The stress–strain curves obtained from these tests relate the normalized force per unit area (stress) applied to the material with the resulting relative displacement of the material (strain). One of the most important parameters obtained in such tests is the Young's modulus ($E$, in force per unit area), which is the slope in the elastic region of the stress–strain curve and is a measure of the material's stiffness, known to affect the intracellular signaling [3] mainly via the integrin receptors and the associated focal adhesion kinases. Recently, Engler *et al.* [4] have shown that the matrix stiffness can direct the differentiation of human mesenchymal stem cells (hMSCs), to neuronal, myogenic, and osteogenic cellular fates, depending on whether the polyacrylamide substrate has a lower or higher stiffness. Thus, the Young's modulus is a critical mechanical property of the materials used to culture stem cells.

These materials can be classified into synthetic or natural materials, and both these groups can be further subdivided into polymer-, peptide-, and ceramic-based synthetic materials, and protein, polysaccharide, and complex natural materials (Table 7.1). The use of synthetic materials possesses one advantage over natural materials: a higher degree of control during the manufacturing process that can simplify the quality control (QC) and the regulatory approval processes in the case of therapeutic applications, since the composition of the synthetic materials is well defined. However, if the natural materials are abundant and the extraction/preparation/sterilization/endotoxin removal process is simple, the process economics may be more favorable compared with their synthetic counterparts. In addition, complex natural materials such as Matrigel have many extracellular matrix (ECM) proteins and embedded growth factors that provide a very favorable microenvironment for stem cell culture.

Regardless of their origins, the use of materials in stem cell culture has the main goal of increasing the similarity between the culture and the *in vivo* environments.

## 7.3 Synthetic Materials

### 7.3.1 Polymers

Synthetic polymers offer the possibility to control the composition, chemical and physical properties as well as the rate of degradation of the material. Polyethylene glycol (PEG) is approved by the US Food and Drug Administration (FDA) to be used as an excipient and drug conjugate, the most notable example of it being the PEG–interferon-$\alpha$ conjugate (PegIntron®), which increases the solubility and half-life of the protein in the human body [5]. The unmodified PEG-diacrylate monomers are typically polymerized by ultraviolet (UV) irradiation, forming a polymer that has low cell adhesion and low protein binding [6], and is fully hydrated in aqueous solution. The introduction of covalent chemical modifications before the polymerization reaction can add further functionalities such as cell adhesion. For instance, the crosslinking of the integrin-binding Arg-Gly-Asp (RGD) peptide motif into PEG hydrogels increases the cell spreading (adhesion) onto the polymer [7] and the addition of functional chemical groups can also be used to direct the differentiation of hMSCs [8]. The degradation of PEG hydrogels can also be tuned by crosslinking the polymer with cleavable peptide linkers as shown by Anderson *et al.* who demonstrated that PEG hydrogels with

**Table 7.1** Characteristics of the most common materials used in the cultures of entrapped stem cells.

| Material | | Origin | Manufacture | Regulatory | Stem cell applications | References |
|---|---|---|---|---|---|---|
| Polymers | PEG | Synthetic | Scalable chemical synthesis from ethylene oxide; used in several industries including medical devices and pharmaceutical. | Used in the pharmaceutical industry as an excipient (US FDA approved). | hMSC differentiation in PEG capsules by tethering chemical functional groups into the gel. Encapsulation of hMSCs with enzymatically cleavable PEG hydrogel. | [8, 9] |
| | PLGA | | Scalable chemical synthesis from the monomers glycolic and lactic acid; used in medical device industry. | Used in the medical device industry due to biocompatibility and biodegradability; US FDA-approved medical device material. | As a substrate for 2D hMSC culture in combinations with other polymers. As macroporous beads for human amniotic fluid stem cells; beads and cells injected in animal model and shown to be effective for cardiac therapy. | [10, 11, 87] |
| Peptide | | | Solid-phase synthesis from amino acids (AA); this process yields a maximum peptide length of less than 100 AA. | Enfuvirtide® is a synthetic peptide for HIV treatment; despite synthetic these peptides require a Biologics License Application (BLA) for the US FDA approval. | 3D-scaffold culture of hepatic progenitor cell line increased hepatic functionality of the cells; hESCs cultured on a Synthemax® surface successfully differentiated into cardiomyocytes. | [14–16] |
| Ceramics | | | Fusion of organic and inorganic components with heat; scalable and well-known process. | Used as a medical device (e.g., femoral head replacement); US FDA-approved medical device material. | Mainly used as an implantable scaffold with seeded hMSCs differentiated toward bone tissue. | [19, 20, 88] |

| | | | | | | |
|---|---|---|---|---|---|---|
| Protein | Collagen | Natural | Collagen I is typically isolated from bovine or murine sources and the chemical extraction process is good. | Easily sourced but difficult to sterilize, leading to a lack of GMP grade material, even though collagen elicits a low immune response *in vivo*. Apligraft® is a collagen-containing product approved by the US FDA for wound healing. | Neural precursor primary cell culture in 3D gel; encapsulation of hMSCs in collagen–agarose beads; mESC culture and differentiation. | [22, 24, 25, 89] |
| | Fibrin | | Fibrin is generated when fibrinogen is cleaved by thrombin; this is ideal for an autologous process where scale is not an issue. | Given its autologous nature the regulatory process should not be a limiting factor. Fibrin sealants for bleeding during surgery are US FDA approved as BLAs. | hMSC culture for bone defect corrections and as an aerosol for wound healing; used for neural differentiation of mESCs. | [28–31] |
| | Silk | | Raw material for the textile industry makes sourcing simple; the manufacture process allows for fine tuning of properties such as degradation rate. | Already used as sutures; cell-free scaffolds for connective tissue repair have been approved by the US FDA. Unlike collagen it can be steam and radiation sterilized. | hMSC culture for undifferentiated expansion, chondrogenic and osteogenic differentiation; typical culture involves 3D scaffolds. | [32–35] |
| Carbohydrates | Agarose | | Derived from the agar extracted from seaweed and red algae; the process is simple and involves boiling the washed raw material, filtering the resulting gel, and drying it. Used in the food industry. | The nonanimal origin is a regulatory advantage; it is currently used in the oral formulation of Theophylline® for respiratory disease. | hMSC culture for chondrogenic differentiation; mESC and hESC culture for expansion and hematopoietic differentiation. | [36–38] |

(continued overleaf)

**Table 7.1** (*Continued*)

| Material | | Origin | Manufacture | Regulatory | Stem cell applications | References |
|---|---|---|---|---|---|---|
| | Alginate | | Extracted from brown seaweeds similarly to agarose; it is also used in the food, cosmetic, and pharmaceutical industry. | Used in pharmaceuticals for stomach pain (Gaviscon®); the Acticoat® alginate dressing is a medical device (510 K granted by the US FDA) for wound healing indications. | hMSC culture for chondrogenic differentiation; hESC expansion and EB differentiation. | [39, 90, 91] |
| | Hyaluran | | Hyaluran can be extracted from tissues such as the umbilical cord or produced by recombinant microorganisms; the latter delivers higher yields with lower contaminants than the extraction process [92]. | Used as synovial fluid replacement in osteoarthritis treatment. The effectiveness of injecting HA and hMSCs for intervertebral disc repair has been demonstrated in a recent \|mesoblast Phase II trial. | hMSC chondrogenic and osteogenic differentiation in HA hydrogels. Expansion and differentiation (three germ layers) of hESCs. | [93–95] |
| Complex | Matrigel | | Matrigel is derived from tumors that are propagated in mice colonies. This method is hardly scalable and very difficult to be made GMP compliant. | Matrigel has an animal origin, is complex, and its composition is not defined. It is unlikely that the regulatory authorities will allow it to be included in any cell therapy process. | Used for hPSC expansion and differentiation into hepatocytes and retinal progenitors. | [46–48, 96, 97] |

increased degradation rate also increase the efficiency of hMSC differentiation into osteogenic, chondrogenic, and adipogenic lineages [9].

The synthetic polymer poly(lactic-co-glycolic acid) (PLGA) is a copolymer of the glycolic and lactic acid monomers and similar to PEG it can be functionalized and its degradation rate can be modified chemically. It is widely used in the clinical applications as a US FDA-approved biodegradable material; examples of such applications are surgical sutures (Vicryl®, Johnson and Johnson), grafts, and prosthetic devices (Degradable Solution AG, Switzerland). The Langer Lab at the MIT has taken full advantage of the combinatorial chemistry that the synthetic polymers offer and has combined 4 modified forms of PLGA with 24 different polymers [10]. The resulting new copolymers were seeded with hMSCs, and their cell–polymer interaction was studied. Recently, Huang *et al.* [11] have synthesized PLGA porous beads, which support human amniotic fluid stem cell (hAFSC) growth; these beads allowed the cells to grow three-dimensionally without nutrient limitations and to deposit ECM components such as collagen III and fibronectin. These beads containing hAFSC were injected into rats that had suffered myocardial infarction, and these animals showed significant improvements in cardiac function compared with those administered dissociated cells without the beads.

There has been a significant development in entrapment materials made from polytetrafluoroethylene (PTFE), which is a commonly used material in the medical device industry. The most notable application of PTFE to entrapped stem cell culture are the Encaptra® macrocapsules commercialized by ViaCyte in combination with human embryonic stem cell (hESC)-derived insulin-producing cells. This combination product, VC-01™, is set to begin a Phase 1/2a clinical trial in mid-2014.

### 7.3.2 Peptides

Peptides are composed of short sequences of amino acids and typically synthesized in standardized solid-phase chemical reaction between the C-terminus and N-terminus of the amino acids [12]. Zhang *et al.* [13] synthesized a peptide consisting of an alternating sequence of positively and negatively charged amino acids with alternating hydrophobic and hydrophilic side chains; this peptide, called RADA16 after its amino acid sequence, has the ability to self-assemble into stable $\beta$-sheets with a fibrilar structure very close to the ECM and with 99% water content. The geometry of the material can range from a membranous layer to hydrogel scaffolds, which support the growth and differentiation of hepatocyte progenitors [14] and neural stem cells [15]. In the latter work, Zhang *et al.* have further functionalized the RADA16 scaffold through the covalent attachment of the Ile-Lys-Val-Ala-Val (IKVAV) peptide sequence, which is found in laminin and also binds to integrin receptors. The latest commercial application of this material is Corning's peptide acrylate surface (Synthemax®)-coated cell culture flasks and beads [16]. Human embryonic stem cells were cultured in the Synthemax® surface and successfully differentiated into cardiomyocytes. Unlike RADA16, Synthemax® is structurally composed of acrylate surfaces and the synthetic peptides, derived from ECM motifs, are mostly used to functionalize the substrate for a given cell fate, such as the differentiation of hESC to oligodendrocytes [17].

### 7.3.3 Ceramic

Ceramic materials are composed of inorganic materials that are blended with water and an organic binder to make the final ceramic product. These materials have been historically used in orthopedic surgery due to their high Young's modulus (in comparison with other biomaterials), where they are classified as inert or bioactive [18]. The bioactive ceramics have osteoconductive properties that accelerate bone healing; these properties have been used to produce ceramic scaffolds with hMSC-derived osteoblasts via osteogenic differentiation. Toquet *et al.* were among the first to demonstrate that it was possible to expand bone marrow–derived hMSCs on macroporous calcium phosphate pellets and subsequently differentiate these cells into an osteogenic fate [19]. The osteogenic activity of calcium phosphate when used in hMSC differentiation is due to the fact that calcium is a major component of the bone, in the form of hydroxyapatite. When cubic scaffolds with an 80%–20% mixture of hydroxyapatite and tricalcium phosphate (TCP) were seeded with hMSCs and ectopically implanted into mice, the bone formation rate was significantly higher than using any of the individual materials [20]. These examples demonstrate the importance of mimicking the *in vivo* environment and how the materials that contact the cells are an important factor in providing such cues.

## 7.4 Natural Materials

### 7.4.1 Proteins

Naturally occurring proteins offer the opportunity to mimic the *in vivo* microenvironment that stem cells experience with a limited (or absent) chemical modification. The most abundant protein in the mammalian body is collagen, and its main function *in vivo* is to provide stiffness to the extracellular matrix; the linkage between cells and collagen can be direct or it can be achieved via the fibronectin molecule, which binds collagen and the cell surface molecule integrin via the RGD motif. The collagen isoforms I, II, III, and V are fibrous proteins located in the interstitial area of the tissues, whereas collagen IV is a part of the basal layer of most epithelial tissues [21]. As a biomaterial, collagen is biodegradable, biocompatible, has low immunogenicity, and can be extracted from virtually any mammal. Its main disadvantage is the difficulty in sterilization, similar to any other protein-based materials, since steam and radiation can easily change the protein conformation. In the case of collagen, acid treatment is the most common sterilization method while ethanol immersion is also an alternative.

Type I collagen is the isoform used in the vast majority of cell culture applications; these include neural stem cell [22, 23], hMSC [24], and mouse ESC [25] cultures. In the latter work, Battista *et al.* have shown that mouse embryonic stem cell (mESC) embryoid bodies (EBs) differentiated is inhibited with increasing collagen concentrations, probably due to an increasing stiffness of the gels. The effect of uniaxial tension on the osteogenic differentiation of hMSCs was demonstrated by Sumanasinghe *et al.* [26] to increase the gene expression of the bone morphogenic protein 2 (BMP-2) by 10% compared with the nonstrained control. The best example of a commercial application involving collagen is Organogenisis' Apligraft®, which is a regenerative medicine product for wound healing whose formulation includes collagen.

Fibrin is a protein involved in blood clotting; it is formed as thrombin interacts with fibrinogen and forms the fibrin polymer [27]. This protein is ideal for therapeutic applications since its raw materials can be extracted from a patient's blood to make the *ex vivo* fibrin synthesis, which constitutes a degradable autologous product that should elicit no immune response when it is reimplanted into the patient. The literature reports several therapeutic applications of fibrin gels for bone defects using hMSCs [28, 29] and more recently Falanga *et al.* have reported the application of a mixture of hMSCs with fibrin in a spray for wound healing [30]. This protein has also been used for the neural differentiation of mESCs [31] and for building artificial heart valves containing primary human myofibroblasts [27].

The use of silk as a material originated in the textile industry (which produces about $4 \times 10^5$ tons of silk annually) and has migrated to the medical field decades ago as silk sutures started being used. The silk protein is one of the most resilient in nature and its Young's modulus ranges from 5 to 17 GPa (compared with collagen that has 2–50 MPa) [32]. The main advantage of silk over the remaining natural biomaterials is that steam sterilization and gamma irradiation do not change its structure. Given its stiffness, this material, functionalized with the RGD peptide, has already been used for the osteogenic differentiation of hMSC and proven to upregulate the expression of bone markers compared with collagen [33]. The culture and chondrogenic differentiation of hMSCs was also achieved in 3D porous silk scaffolds [34], and also the undifferentiated expansion of this cell type in electrospun silk nanofibers has been demonstrated [35].

### 7.4.2
### Polysaccharides

Polysaccharides are, similar to proteins, a natural component of the ECM. The main advantage of polysaccharides over proteins is that the starting monomers are smaller and thus easier to manipulate and sterilize. In their unmodified form, these carbohydrates do not bind the cells. Instead, they force the cellular populations into forming cell–cell contacts once these are contained inside the material [36].

Agarose is a polymer extracted from algae or seaweeds and is structurally composed by the monomers D-galactose and 3,6-anhydro-L-galactopyranose. Historically, it has been used for the electrophoresis of nucleic acids. When agarose is dissolved in aqueous solution it can be solubilized by heating the mixture; once this mixture cools to the gelling temperature, it becomes solid. To resolubilize the gel, it is necessary to heat it to the melting temperature, which is significantly higher than the gelling temperature. Agarose gels with low gelling/melting temperature are used to make this process compatible with cell entrapment, and by changing the concentration of agarose in the final mixture it is possible to fine tune the stiffness of the material.

The culture of hMSC in 2% (w/v) agarose gels was shown to be as efficient as the gold standard pellet culture for the chondrogenic differentiation of these stem cells [37].The chondrogenic differentiation of MSC in 2% agarose gels was also studied by Mauck *et al.*; in this case, the cells were entrapped in $4 \times 1.5$ mm (diameter × thickness) disks and subject to cyclic compressive loading, which resulted in an increase in the amount of glycosaminoglycans (GAGs) produced after 4 weeks, when compared with a nonloaded control [38]. The latter work exemplifies the importance of the forces applied to stem cell cultures and how biomaterials can be used to enable such processes.

Alginates are polymers composed of $\beta$-D-mannuronate (M) and $\alpha$-L-guluronate (G) residues, which are extracted from brown algae. Their advantages and extraction process are similar to agarose, but alginate forms an anionic viscous solution that gels only in the presence of divalent cations such as calcium or barium (the crosslinking agents). The effect of alginate in stem cell processing has been thoroughly studied by Serra *et al.* where hESCs were cultured either as 3D multicellular spheroids or adherent to microcarriers in spinner vessels. These two culture configurations were compared for their ability to expand undifferentiated hESCs with or without encapsulation in 1.1% alginate [39]. The results showed that without alginate encapsulation hESC aggregates did not survive throughout the 2-week culture period. For hESCs immobilized in microcarriers, alginate encapsulation yielded a 40% increase in the final cell number. The results of this experimental matrix suggest that while shear stress is known to be detrimental for ESC cultures, the main effect of alginate encapsulation is the individualization of the culture "units" (whether aggregates or microcarriers). In the absence of alginate encapsulation these units clump together, leading to cell death in aggregates and limited proliferation in microcarriers due to nutrient limitations in the core of the aggregated units. In another work, agarose and alginate scaffolds have been compared with a commercial gelatin (a form of collagen) scaffold, Surgifoam®, for their ability to differentiate human adipose – derived stem cells (hADSCs) into cartilage [40]. The cells tended to adhere and spread in the gelatin scaffold, adopting an elongated configuration, whereas in agarose and alginate scaffold the cells remained rounded and only interacted among them. This led to significant differences in the type of cartilage obtained when using gelatin compared with the algae-derived materials; moreover, the final cell number was significantly higher in Surgifoam®. These results can be generalized to highlight the difference between bioactive materials that physically interact with the stem cells and materials that are biologically inert, as is the case for (unmodified) agarose and alginate.

Hyaluran (or hyaluronic acid, HA) is a nonsulfated glycosaminoglycan, which is ubiquitous throughout the ECM and similar to collagen it has a low immune reactivity and is biodegradable. It is constituted by monomers of *N*-acetylglucosamine and glucuronic acid, stimulates cell proliferation and migration [41], and it has an enriched distribution in the mammalian embryo [42], which was the main reason for its application to stem cell culture. Hyaluran has been used in the medical industry for the treatment of osteoarthritis (OA) as a replacement for the low elasticity synovial fluid of the OA patients [43]. This established medical practice is being modified for nonvertebral disk regeneration using cellular therapy [44], where human mesenchymal precursor cells (commercialized by Mesoblast, AU) are injected with hyaluronan to repair the damaged disks.

### 7.4.3
### Complex

The natural materials mentioned earlier have a defined composition and as such are amenable to regulatory compliance. However, one of the most used materials in human pluripotent stem cell (hPSC) culture (which includes embryonic and induced pluripotent stem cells) is Matrigel. This material is natural, complex, and poorly defined since it is derived from Engelbreth–Holm–Swarm (EHS) tumors grown in mice, that is, it also has an animal origin, making it an unlikely candidate to be accepted by the regulatory authorities for any therapeutic indication. The good performance of this material is likely to be related with the presence of several ECM components (collagen IV, laminin, and entactin) with a basement membrane character and the presence of several soluble factors, either free or associated with the ECM, such as basic fibroblast growth factor (bFGF), epidermal growth factor (EGF), or insulin-like growth factor-1 (IGF-1) [45]. Matrigel coatings have provided a substrate for hESC expansion in perfused stirred-tank reactors [46] and for the hepatic differentiation of hESC [47], for example. The Sasai lab [48] has differentiated hESC EBs into optic cup-like structures in one of the most successful examples of *in vitro* recapitulation of morphogenesis. In this work, the authors have directed the differentiation using a time-variant growth factor cocktail, while the EBs were entrapped in growth factor-reduced Matrigel, and recently published work [49] has demonstrated that Matrigel is the critical component of this process (i.e., some of the soluble factors such as the Wnt inhibitor Noggin could be removed from the protocol).

Recently, the use of decellularized tissue from cadaveric organs has shown promising results for the culture of stem cells in tubular organs such as the trachea [50] or larynx [51] and in more complex organs such as the liver [52] or heart [53]. This technique has the main advantage of providing a built-in vascularized system and tissue-specific ECM. As with other complex natural materials, these scaffolds may lose their properties during the sterilization process; still, the benefit of using decellularized tissue is high enough to justify further work to overcome any regulatory hurdles.

## 7.5 Manufacturing and Regulatory Constraints

The culture of stem cells entrapped in a material can have two general types of application:

1) implantation of the combined product or the stem cells alone into patients (therapeutic);
2) *in vitro* models of healthy or diseased tissue for drug development.

In the case of the therapeutic application of the cultured stem cells, it is necessary to account for the regulatory and manufacturing constraints as summarized in Table 7.2. In general, using a cellular product alone is less complex from the regulatory and manufacturing point of view than a cellular product within a material, especially given the fact that it is difficult to classify the combined product as a biologic drug or a medical device alone. For this reason, use of these combined products must have a strong clinical motivation such as an increase in potency,

**Table 7.2** Regulatory and manufacturing constraints of therapeutic applications of entrapped stem cells.

| | | | |
|---|---|---|---|
| Regulatory | Regulator | US FDA – regulated by the CBER. according to 21 CFR | EMA (EU) – according to the ATMP regulation EC 1394/2007 |
| | Clinical trials | Investigational New Drug (IND) or Investigational Device Exemption (IDE) | EudraLex Vol. 10 |
| | Manufacturing | Biologics License Application (BLA) or Premarket Approval/510(k) | EudraLex Vol. 4, Annex II. |
| Manufacturing | Cells + material<br>The combination product typically requires a minimum 4-day culture before implantation [50, 98] and the cells may be of autologous or allogeneic origin. | | Cells<br>If the stem cells are cultured entrapped in a material, it is necessary to remove the material before implantation. To adopt this strategy, the entrapped stem cell culture must provide a significant fold increase in product potency to offset the additional cost and manufacturing complexity. |

which offsets the increase in the production cost due to the additional regulatory burden and manufacturing complexity. Highlighted in Table 7.2 are the main regulatory and manufacturing hurdles for the therapeutic application of entrapped stem cells. From the regulatory side, the main regulatory bodies in the United States and European Union must be notified, and their guidelines on cell or combined products must be complied to have an authorization for clinical trials and manufacturing/commercialization. The regulatory strategy must be planned years before any IND submission, and a good starting point before achieving actual regulatory compliance is to implement a quality management system based on the ISO guidelines, ISO 9001. In fact, the guidelines of the US FDA, defined in the Code of Federal Regulations 21 (21 CFR) and of the European Medicine Agency (EMA), defined in the EudraLex legislation, have been extensively harmonized with the ISO guidelines for Quality Systems. There are also fast-track approval routes either for indications where no alternative treatment is available such as the Orphan Drug designation in the United States or for small-scale hospital interventions, the Hospital Exemption in the EU.

For *in vitro* models, the regulatory and manufacturing constraints are less stringent. On the regulatory side, there is a general requirement for these tests to be performed within Good Laboratory Practice. In Europe, the EMA's Committee for Medicinal Products for Human Use (CHMP) defines the guidelines for preclinical tests and the US FDA has an analogous guidance document (reference M3(R2)). *In vitro* tests using stem cell–derived hepatocytes, for instance, can be used for administration, distribution, metabolism, excretion (ADME) and toxicity test of potential drug candidates [54]. These regulatory guidelines do not apply if the *in vitro* test is a screen for candidate molecules that can enhance the hepatic differentiation of stem cells [55].

From the manufacturing point of view, it may also be necessary to remove the encapsulated stem cells from the material in instances where an analysis of the intracellular contents is required (such as RNA) or if the material interferes with the substance being tested. The main difference between therapeutic and *in vitro* applications, with regard to manufacturing, is that the cells produced for *in vitro* tests tend to be terminally differentiated since they are supposed to mimic the cell (either healthy or diseased) behavior in the human body. The implication of this is that the culture of entrapped stem cells is more likely to be long-term for *in vitro* tests, which require terminally differentiated cells.

The manufacture of scaffolds for entrapped stem cell culture is likely to be impacted by the development of additive manufacturing, namely in the form of 3D printing. This technology has been used to print tissues using multicellular aggregates as the building blocks of tubular structures; these spheroids give rise to compact structures through tissue fusion thus forming vessel-like structures [56]. While this technology is novel, the "bio-ink" used is still based on a mixture of cells with the previously described materials, such as collagen [56], Matrigel, alginate, and agarose [57, 58]. One of the most relevant developments in this area has been the work by Christopher Chen's lab [59], where a carbohydrate glass network was 3D printed and encapsulated in a fibrin gel containing the cells to be cultured. The glass–carbohydrate network can be easily dissolved by culture medium leaving empty channels that can be coated with endothelial cells. The most relevant achievement in this work is the controlled design of a vascular system, which will enable an efficient mass transfer when the combination product is implanted.

## 7.6 Mass Transfer in the Entrapment Material

A detailed analysis of transport phenomena in bioreactors can be found in the chapter "Flow pattern analysis and mass transfer in bioreactors" of this book. For this chapter, it is useful to start by analyzing the diffusional mass transfer of soluble factors inside the material containing the entrapped stem cells. The main goal of a preliminary analysis of the diffusion of soluble factors (i.e., nutrients, metabolites, growth factors, cytokines) is to determine whether any given factor will be depleted or accumulated within the cellular microenvironment in the absence of convective flow inside the material. This diffusion phenomenon can be described by the one-dimensional form of Fick's second law plus a consumption (or production) term:

$$\frac{dC}{dt} = D\frac{d^2C}{dx^2} + Q \tag{7.1}$$

where $C$ is the concentration of the soluble factor in amount per unit volume, $t$ is the time, $x$ is the dimension where diffusion proceeds, $D$ is the diffusion coefficient in unit area per unit time, and $Q$ is the volumetric reaction rate of the soluble factor. From this equation, it can be concluded that the concentration gradient from the outside to the interior of the material is the driving force for mass transport, while the diffusion coefficient is the proportionality constant that affects the magnitude of the rate of change of concentration with time, and the volumetric reaction rate can add or subtract to the soluble factor depending on the spatial location of this reaction (usually colocalized with the cells, which may be homogeneously or heterogeneously distributed throughout the material). The diffusion coefficient $D$ is usually measured in an aqueous solution but when the cells are entrapped the effective diffusion coefficient of the soluble factor in the material ($D_{eff}$) is lower than $D$. The parameter $D_{eff}$ is a function of the physicochemical characteristics of the entrapping material, which can change over time (depending on its degradation rate). The pore structure (porosity and tortuosity) and charge of the material act as a barrier for the diffusion of any soluble molecule, making this a complex system to analyze without experimental data; nevertheless, there are examples of such calculations in the literature. The work performed by Curcio *et al.* aimed at elucidating the reason for a limited cellular penetration inside decellularized tracheas after implantation [60], the hypothesis being that oxygen was the limiting soluble factor. These authors have developed a computational model to simulate the oxygen distribution inside the trachea construct, and the effective oxygen diffusion was calculated to include the porosity and tortuosity of the construct as a function of the volumetric cell density and the void fraction, respectively. Both these parameters were based on assumptions, given the difficulty of measuring the cell concentration and empty space inside a construct. One of the experimental methods used to calculate an effective diffusion coefficient is to actually monitor the release or absorption of a model soluble factor from or into the material by measuring the concentration of the model molecule over time. This concentration time profile, together with the geometry of the material, can be used to calculate $D_{eff}$ by using textbook numerical recipes [61]. This method was used by Tostoes *et al.* to measure the effective diffusion coefficient of bovine serum albumin in ultra-high viscosity alginate capsules [62]; the measured $D_{eff}$ was two orders of magnitude less than the reported value for the diffusion of albumin in water. In this work, the authors have also estimated whether this soluble factor would accumulate in the capsules when it was being produced by encapsulated rat hepatocytes by using a modified form of the Damköhler number:

$$Da = \frac{\text{Reaction rate}}{\text{Mass transfer rate}} = \frac{X_V \times q_{alb}}{D_{eff} \times (C_{in} - C_{out})} \times R^2 \tag{7.2}$$

where $q_{alb}$ is the specific albumin synthesis rate, in unit mass per cells per unit time, $X_V$ is the viable cell concentration in cells per unit volume, $D$ is the diffusion coefficient of albumin in alginate, in unit area per unit time, $R$ is the average radius of the alginate capsule, and $C_{out}$ and $C_{in}$ are the albumin concentrations in the culture medium bulk and the albumin concentration inside the alginate capsule, respectively, in unit mass per unit volume. This formula can be modified according to the geometry of the cell scaffold such that it informs the design of the same scaffold; similarly, the reaction rate may depend only on the concentration of the soluble factor or it can be a higher-order reaction if, for instance, the soluble factor is a ligand that binds to a receptor at the cell surface to activate an intracellular signaling pathway, as is the case for the fibroblast growth factor system. The interactions between the ECM, soluble factors, and the stem cells within the entrapped material have been reviewed elsewhere [63] but such complex adsorption, desorption, and chemical reactions have a critical effect on the effective diffusion coefficient of the soluble factors [64] and are one of the most important areas to be investigated for stem cell culture and differentiation strategies to evolve into a more defined processes.

This analysis can be performed according to a process flow diagram (PFD) which will inform the process design, as depicted in Figure 7.1.

In the presence of diffusional limitations, it is possible to change the process and/or the scaffold. In the latter case, the change may involve increasing the porosity such that $D_{eff}$ also increases [65], decreasing the diffusional distances by changing the construct geometry [66] or tethering the limiting factor to the material [33, 67]. These changes are possible and offer a wide range of solutions for diffusional problems; however, as discussed earlier, changes to the materials are constrained by regulatory and manufacturing limitations. Another approach to the problem is to change the process to enable a higher mass transfer to

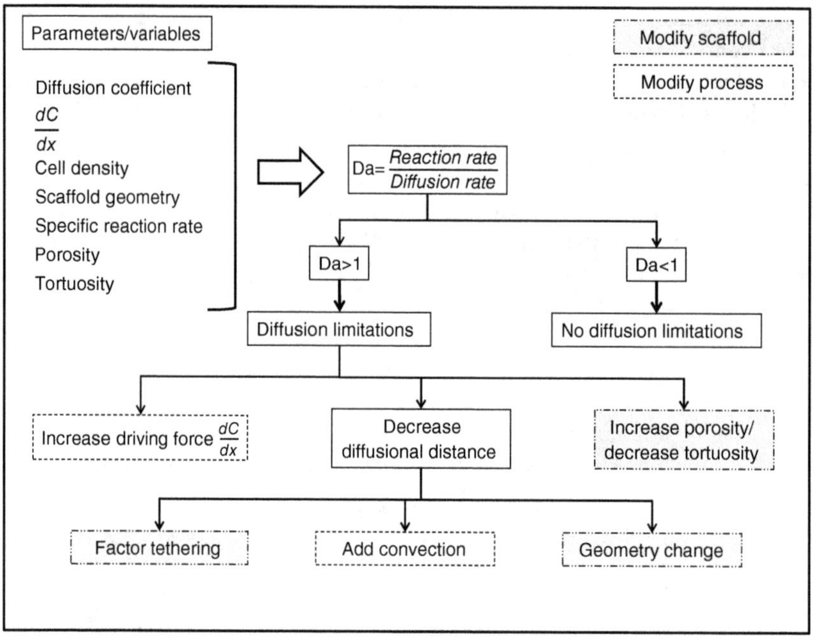


**Figure 7.1** Process flow diagram for a mass transfer-based process design.

the interior of the construct, which can be achieved by the use of continuous bioreactors. Depending on how much the mass transfer must be enhanced the continuous bioreactor can provide a maximum driving force $dC/dx$, while the mass transfer inside the construct is kept diffusional or the construct itself can be perfused with culture medium flown through it.

## 7.7 Continuous Bioreactors for Entrapped Stem Cell Culture

When the entrapped stem cells are cultured in batch systems, the driving force for nutrient entry and metabolite exit in the material construct containing the stem cells ($dC/dx$ in Eq. 7.2) is continuously dropping. In continuous systems with cell retention, this driving force can be kept nearly constant. There are two general types of bioreactors that can be used in a continuous operation: "perfectly" mixed and plug-flow bioreactors (as detailed in Chapter 1). An example of a perfectly mixed bioreactor is a continuous stirred-tank reactor (CSTR) with a cell retention mechanism [68], whereas a plug-flow reactor (PFR) configuration would correspond to a construct containing cells, which is part of a fluidic circuit that flows through the porous material that contains the cells [69]. The PFR configuration enables fresh medium to be forced into the scaffold porous structure, thus creating a convective flow, which minimizes the diffusional distances while being kept below critical shear rate values [69]. In addition, this type of flow is very efficient to seed the cells into the scaffolds in the beginning of the culture. The work by Wendt *et al.* established a comparison between well plates (static) spinner flasks (stirred) and a PFR (perfused) chondrocyte cell seeding. These authors demonstrated that the PFR system allowed for a significantly more uniform and efficient cell seeding. Bone marrow stromal cells were also shown to be more uniformly distributed in ceramic scaffolds when using perfused seeding, in comparison with the static control [70]. It is important to remember that the constructs that can be perfused by PFRs are porous, that is, this culture system is not appropriate for hydrogel scaffolds, which have a smaller hydraulic permeability.

The main drawback of PFR reactors is the difficulty to measure and control the pH and DO inside the scaffold; moreover, it is very difficult to measure cell growth, which makes any optimization of these processes mostly empirical. These problems can be minimized by having low residence times (1 – 10 s) of the culture medium in the scaffold [71, 72] and monitoring the parameters in the outlet while applying the control in the inlet. The requirement to have low residence times in the PFR implies that the length of the construct perfused by the culture medium is kept within the millimeter scale, which makes the system difficult to scale-up. Most PFR systems used in entrapped stem cell cultures have medium recirculation onto a reservoir, which increases the overall residence time of the whole system to several days, that is, close to the CSTR levels. Thus, while the mass transfer in the construct is convective the total culture medium used is kept at the same cost as a CSTR. Overall, the PFR systems are not amenable to scale-up but provide adequate conditions for seeding and growing cells for an autologous therapeutic product. In fact, the lack of sensors in any closed processing system simplifies the regulatory problem since that for each sensor there is an associated failure mode, or risk, that needs to be mitigated. For these reasons, the main therapeutic application of this continuous reactor has been in autologous bone tissue engineering [69, 71, 72].

Continuous stirred-tank reactors, unlike PFRs, have higher residence times (1 – 4 days) and the mass transfer inside the material is diffusional, which limits the size of the scaffolds to hundreds of micrometers to minimize the $dC/dx$ (Eq. 7.2) gradients. On the contrary, it is possible to take samples from the culture broth and thus measure the cell growth and culture medium composition. The pH and DO measurement and control are performed on the culture broth that is homogeneously mixed, that is, the DO and pH in the vicinity of the scaffold is known. Another advantage of the CSTR compared with PFRs is its scalability. When considering allogeneic cellular therapies, where scalability enables the establishment of economies of scale, a CSTR tends to be the most appropriate choice for obtaining large cell numbers under a controlled system. The Zandstra group from the University of Toronto has demonstrated that a CSTR prevents the fluctuation of the culture medium composition between discrete medium exchanges in alginate-encapsulated mESC expansion; one often overlooked advantage of continuous systems over discrete medium exchanges is that it avoids process stresses such as interrupting agitation to allow the constructs to settle before replacing the medium.

One of the most difficult aspects when comparing the performance of different reactor configurations is the coupling between mass transfer and residence time. In a PFR, the mass transfer and the residence time are both coupled because the culture medium flow rate (or the Superficial velocity = Flow rate/Scaffold area) provides both media exchange and convective mass transport. In a CSTR, the mass transfer is uncoupled from the residence time because the impeller rotation provides the mass transfer, whereas the perfusion flow rate provides the residence time (or medium replacement rate); these concepts are explored in more detail in Chapter 10. Considering this issue, Zhang *et al.* have compared the effect of the different mass transfer mechanisms in a PFR-type reactor, a stirred-tank vessel, a rotating-wall vessel, and a biaxial reactor (BXR) with the same residence time between the four reactors [73] for human fetal MSC culture in a polycaprolactone–tricalcium phosphate (PCL–TCP) scaffold. The results showed that the BXR produced more cells at a faster grow rate with more ectopic bone formation in immunodeficient mice compared with the other three types of reactor. Interestingly, the authors have also noticed the presence of "perfusion holes" in the PFR reactor, raising the hypothesis that the scaffold could be mechanically degraded by the fluid flow in this type of reactor. The BXR had been previously shown to enhance the fluid flow inside the scaffolds 20- to 40-fold, due to its bidirectional motion, while keeping the shear stress levels below 2 Pa [74] and is one of the most recent innovations in the continuous bioreactor field.

Many of the recombinant protein bioprocessing principles have been taken up by the nascent stem cell bioprocessing field; in recombinant protein production, when cells are cultured in a continuous bioreactor with cell retention, the higher the flow the more the growth of the cells [75], since the main goal is to remove metabolites and add nutrients to the cell culture; typically, the perfusion flow rate is only limited by the added process stresses originating from the cell retention mechanism or dilution of the secreted product. In stem cell bioprocessing, although it is still necessary to add nutrients and remove metabolites, it may also be important to retain soluble factors or ECM proteins that are produced by the cells that have a positive effect on the final cell type wanted. This idea has been demonstrated by Junho and Ma by building a bioreactor system where a porous polyethylene terephthalate (PET) scaffold seeded with hMSCs was cultured with medium perfused through the scaffold or tangentially to it, thus providing a convective or diffusional mass transfer inside the scaffold, respectively [76]. On comparing these systems with each other, it was found that the tangential perfusion system favored the expansion of undifferentiated hMSCs, while the flow-through perfusion culture was enriched in osteogenic markers such as BMP-2, ALP, and RUNX2 (Figure 7.2). The authors have further correlated this differential cell fate with an increase in the ECM proteins fibronectin, vitronectin, and collagen type I in the tangential flow. Thus, for different cellular fates, the diffusional limitations to mass transfer may be beneficial or detrimental.

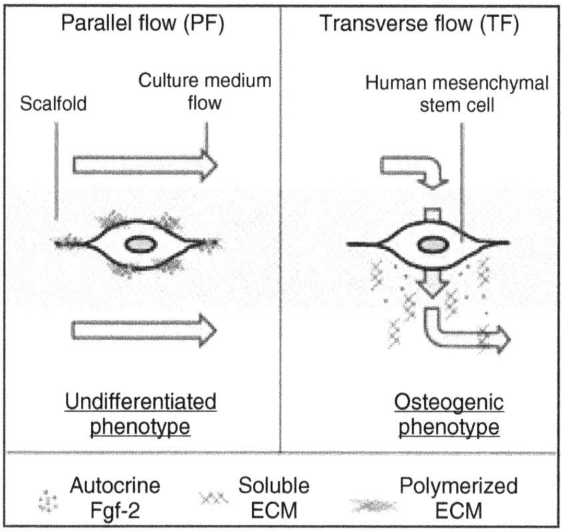


**Figure 7.2** The effect of autocrine factor washout in the differentiation of hMSCs. This figure depicts the findings in [76], where the authors demonstrate that the convective flow of culture medium through the hMSC-containing scaffold removes ECM components (Collagen I, Fibronectin and Vitronectin) and the bFGF soluble factor that it is bound to the ECM. When these cell constructs were cultivated in parallel flow the diffusional mass transfer allowed for an accumulation of the ECM and the bound Fgf-2 molecules which in turn maintained the undifferentiated phenotype of the cultured hMSCs.

We have begun this chapter by highlighting how entrapped stem cells are protected from the shear forces that mass transfer mechanisms generate; however, there are instances when these shear forces are an important component of the differentiation process of the stem cells into an intended cell type. These processes include cardiac [77, 78] and endothelial differentiation of ESCs [79] and chondrogenic differentiation of hMSCs [38]. The Langer group developed a pulsatile flow bioreactor to develop functional arteries *in vitro*, which imparted periodic radial stress to a PLGA scaffold with seeded smooth muscle cells (SMCs) [80]; this radial stress mimics the perpendicular force applied to arteries as the heart pumps blood through the vessels; as the culture progressed the PLGA scaffold deteriorated and was replaced by an SMC blood vessel with a Young's module close to the *in vivo* values. This bioreactor was also used to expand hMSCs and differentiate these cells into SMCs [81]; to achieve this, these authors have optimized the expansion and differentiation process by manipulating the soluble factors, ECMs, and the radial stress applied to the culture for each of these culture stages. For the expansion phase, it was beneficial to remove the radial stress, add the mitogenic growth factor platelet-derived growth factor (PDGF) and to coat the scaffold with fibronectin. For the differentiation phase, the PDGF was removed while TGF-$\beta$ was added to the culture medium, and the cyclic radial strain mechanism was turned on. This global optimization of a stem cell culture process is an example of how physical forces, soluble factors, ECM proteins, and time are critical process variables for producing a desired cell type.

## 7.8 Future Perspectives

The cultures of entrapped stem cells using continuous bioreactors allow for a control of the soluble factor composition, which is a part of the cellular microenvironment. It is known from the literature that the composition of the ECMs, such as the soluble factors, also changes throughout the development of the human embryo and that these changes have a functional role in the cellular phenotype [82, 83]. Starting from this point, it is important to distinguish between entrapped systems that intend to functionally replace the ECMs, as is the case with the RGD motif or the commercial Synthemax® substrate (Corning), and entrapped systems consisting of nonfunctionalized hydrogels. If the former system is being used, the associated intracellular signaling pathways (such as integrin signaling for RGD) will remain constant, while the differentiation process may require different ECM materials throughout the entire period for different signaling pathways. The work published by Young and Engler reports the development of a PEG–HA hydrogel, which stiffens throughout the cardiomyocyte differentiation of precardiac cells; these hydrogels allowed for a threefold improvement in the differentiation process compared with constant stiffness polyacrylamide hydrogels [84]. Thus, these defined material approaches provide a very significant control of single expansion or differentiation steps, and it is expected that temporal control of the functionality of the material will be a strong research area in future. The symmetrical approach is to rely on the cells to produce the appropriate ECM; a remarkable example of this is the evolution of protocols for the differentiation of hPSCs into retinal progenitor cells (RPCs). One of the first efficient protocols in this differentiation process consisted of separated undifferentiated expansion of hESCs, formation of EBs, and plating of the EBs on Matrigel [85]. In 2012, Nakano *et al.* reported a similar method, where the soluble factor addition and withdrawal was matched to the eye development and the EB formation step was performed embedded in Matrigel, achieving optic cup-like structures via morphogenetic movements typical of embryonic development [48]; finally, Robin Ali's group at UCL demonstrated that it was possible to differentiate hiPS into $Crx^+$ RPC if the EBs were cultured inside Matrigel throughout the whole differentiation process. Ultimately, the materials used for stem cell entrapment must be tailored to the specific cell manufacturing process and may need to be changed throughout culture time; it is critical to consider that the cells have the endogenous ability to produce (and degrade) ECM.

The endogenous ability of the stem cells to produce paracrine factors and ECM creates a requirement to have appropriate sensor technology that enables the measurement of soluble factors and posterior actuation to control the cell manufacturing process. Such sensors would be as essential to cell therapy manufacturing as the glucose and lactate sensors still are for the recombinant protein production industry. Recently, Csaszar *et al.* have demonstrated that online monitoring and control of TGF-$\beta$ in hematopoietic stem cell (HSC) cultures led to higher final concentrations of $CD34^+$ cells by maintaining the TGF-$\beta$ levels below a critical threshold [86]. The ability to monitor and control the critical soluble factors in solution is likely to yield order of magnitude increases in the productivity of cell manufacturing processes.

## References

1. Williams, J.B. and Napoli, J.L. (1985) Metabolism of retinoic acid and retinol during differentiation of F9 embryonal carcinoma cells. *Proc. Natl. Acad. Sci. U.S.A.*, **82**, 4658–4662.
2. Benoit, D.S.W. and Anseth, K.S. (2005) Heparin functionalized PEG gels that modulate protein adsorption for hMSC adhesion and differentiation. *Acta Biomater.*, **1**, 461–470.
3. Wang, N., Butler, J.P., and Ingber, D.E. (1993) Mechanotransduction across the cell-surface and through the cytoskeleton. *Science*, **260**, 1124–1127.
4. Engler, A.J., Sen, S., Sweeney, H.L., and Discher, D.E. (2006) Matrix elasticity directs stem cell lineage specification. *Cell*, **126**, 677–689.
5. Caliceti, P. and Veronese, F.M. (2003) Pharmacokinetic and biodistribution properties of poly(ethylene glycol)-protein conjugates. *Adv. Drug Delivery Rev.*, **55**, 1261–1277.
6. Du, H., Chandaroy, P., and Hui, S.W. (1997) Grafted poly-(ethylene glycol) on lipid surfaces inhibits protein adsorption and cell adhesion. *Biochim. Biophys. Acta*, **1326**, 236–248.
7. Hern, D.L. and Hubbell, J.A. (1998) Incorporation of adhesion peptides into nonadhesive hydrogels useful for tissue resurfacing. *J. Biomed. Mater. Res.*, **39**, 266–276.
8. Benoit, D.S.W., Schwartz, M.P., Durney, A.R., and Anseth, K.S. (2008) Small functional groups for controlled differentiation of hydrogel-encapsulated human mesenchymal stem cells. *Nat. Mater.*, **7**, 816–823.
9. Anderson, S.B., Lin, C.-C., Kuntzler, D.V., and Anseth, K.S. (2011) The performance of human mesenchymal stem cells encapsulated in cell-degradable polymer-peptide hydrogels. *Biomaterials*, **32**, 3564–3574.
10. Anderson, D.G., Putnam, D., Lavik, E.B., Mahmood, T.A., and Langer, R. (2005) Biomaterial microarrays: rapid, microscale screening of polymer-cell interaction. *Biomaterials*, **26**, 4892–4897.
11. Huang, C.C., Wei, H.J., Yeh, Y.C., Wang, J.J., Lin, W.W., Lee, T.Y., Hwang, S.M., Choi, S.W., Xia, Y., Chang, Y., and Sung, H.W. (2012) Injectable PLGA porous beads cellularized by hAFSCs for cellular cardiomyoplasty. *Biomaterials*, **33**, 4069–4077.
12. Merrifield, R.B. (1963) Solid phase peptide synthesis. 1. Synthesis of a tetrapeptide. *J. Am. Chem. Soc.*, **85**, 2149–2154.
13. Zhang, S.G., Holmes, T.C., Dipersio, C.M., Hynes, R.O., Su, X., and Rich, A. (1995) Self-complementary oligopeptide matrices support mammalian-cell attachment. *Biomaterials*, **16**, 1385–1393.
14. Semino, C.E., Merok, J.R., Crane, G.G., Panagiotakos, G., and Zhang, S.G. (2003) Functional differentiation of hepatocyte-like spheroid structures from putative liver progenitor cells in three-dimensional peptide scaffolds. *Differentiation*, **71**, 262–270.
15. Zhang, Z.X., Zheng, Q.X., Wu, Y.C., and Hao, D.J. (2010) Compatibility of neural stem cells with functionalized self-assembling peptide scaffold in vitro. *Biotechnol. Bioprocess Eng.*, **15**, 545–551.
16. Melkoumian, Z., Weber, J.L., Weber, D.M., Fadeev, A.G., Zhou, Y., Dolley-Sonneville, P., Yang, J., Qiu, L., Priest, C.A., Shogbon, C., Martin, A.W., Nelson, J., West, P., Beltzer, J.P., Pal, S.,

and Brandenberger, R. (2010) Synthetic peptide-acrylate surfaces for long-term self-renewal and cardiomyocyte differentiation of human embryonic stem cells. *Nat. Biotechnol.*, **28**, 606–610.

17. Li, Y., Gautam, A., Yang, J., Qiu, L., Melkoumian, Z., Weber, J., Telukuntla, L., Srivastava, R., Whiteley, E.M., and Brandenberger, R. (2013) Differentiation of oligodendrocyte progenitor cells from human embryonic stem cells on vitronectin-derived synthetic peptide acrylate surface. *Stem Cells Dev.*, **22**, 1497–1505.
18. Hamadouche, M. and Sedel, L. (2000) Ceramics in orthopaedics. *J. Bone Joint Surg. Br.*, **82-B**, 1095–1099.
19. Toquet, J., Rohanizadeh, R., Guicheux, J., Couillard, S., Passuti, N., Daculsi, G., and Heymann, D. (1999) Osteogenic potential in vitro of human bone marrow cells cultured on macroporous biphasic calcium phosphate ceramic. *J. Biomed. Mater. Res.*, **44**, 98–108.
20. Arinzeh, T.L., Tran, T., McAlary, J., and Daculsi, G. (2005) A comparative study of biphasic calcium phosphate ceramics for human mesenchymal stem-cell-induced bone formation. *Biomaterials*, **26**, 3631–3638.
21. Parenteau-Bareil, R., Gauvin, R., and Berthod, F. (2010) Collagen-based biomaterials for tissue engineering applications. *Materials*, **3**, 1863–1887.
22. O'Connor, S.M., Stenger, D.A., Shaffer, K.M., Maric, D., Barker, J.L., and Ma, W. (2000) Primary neural precursor cell expansion, differentiation and cytosolic Ca2+ response in three-dimensional collagen gel. *J. Neurosci. Methods*, **102**, 187–195.
23. Dai, W., Kawazoe, N., Lin, X., Dong, J., and Chen, G. (2010) The influence of structural design of PLGA/collagen hybrid scaffolds in cartilage tissue engineering. *Biomaterials*, **31**, 2141–2152.
24. Batorsky, A., Liao, J.H., Lund, A.W., Plopper, G.E., and Stegemann, J.P. (2005) Encapsulation of adult human mesenchymal stem cells within collagen-agarose microenvironments. *Biotechnol. Bioeng.*, **92**, 492–500.
25. Battista, S., Guarnieri, D., Borselli, C., Zeppetelli, S., Borzacchiello, A., Mayol, L., Gerbasio, D., Keene, D.R., Ambrosio, L., and Netti, P.A. (2005) The effect of matrix composition of 3D constructs on embryonic stem cell differentiation. *Biomaterials*, **26**, 6194–6207.
26. Sumanasinghe, R.D., Bernacki, S.H., and Loboa, E.G. (2006) Osteogenic differentiation of human mesenchymal stem cells in collagen matrices: Effect of uniaxial cyclic tensile strain on bone morphogenetic protein (BMP-2) mRNA expression. *Tissue Eng.*, **12**, 3459–3465.
27. Jockenhoevel, S., Zund, G., Hoerstrup, S.P., Chalabi, K., Sachweh, J.S., Demircan, L., Messmer, B.J., and Turina, M. (2001) Fibrin gel -- advantages of a new scaffold in cardiovascular tissue engineering. *Eur. J. Cardiothorac. Surg.*, **19**, 424–430.
28. Bensaid, W., Triffitt, J.T., Blanchat, C., Oudina, K., Sedel, L., and Petite, H. (2003) A biodegradable fibrin scaffold for mesenchymal stem cell transplantation. *Biomaterials*, **24**, 2497–2502.
29. Catelas, I., Sese, N., Wu, B.M., Dunn, J.C.Y., Helgerson, S., and Tawil, B. (2006) Human mesenchymal stem cell proliferation and osteogenic differentiation in fibrin gels in vitro. *Tissue Eng.*, **12**, 2385–2396.
30. Falanga, V., Iwamoto, S., Chartier, M., Yufit, T., Butmarc, J., Kouttab, N., Shrayer, D., and Carson, P. (2007) Autologous bone marrow-derived cultured mesenchymal stem cells delivered in a fibrin spray accelerate healing in murine and human cutaneous wounds. *Tissue Eng.*, **13**, 1299–1312.
31. Willerth, S.M., Arendas, K.J., Gottlieb, D.I., and Sakiyama-Elbert, S.E. (2006) Optimization of fibrin scaffolds for differentiation of murine embryonic stem cells into neural lineage cells. *Biomaterials*, **27**, 5990–6003.
32. Vepari, C. and Kaplan, D.L. (2007) Silk as a biomaterial. *Prog. Polym. Sci.*, **32**, 991–1007.
33. Meinel, L., Karageorgiou, V., Hofmann, S., Fajardo, R., Snyder, B., Li, C.M., Zichner, L., Langer, R., Vunjak-Novakovic, G., and Kaplan, D.L. (2004) Engineering bone-like tissue in vitro using human bone marrow stem

cells and silk scaffolds. *J. Biomed. Mater. Res. A*, **71A**, 25–34.
34. Wang, Y.Z., Kim, U.J., Blasioli, D.J., Kim, H.J., and Kaplan, D.L. (2005) In vitro cartilage tissue engineering with 3D porous aqueous-derived silk scaffolds and mesenchymal stem cells. *Biomaterials*, **26**, 7082–7094.
35. Jin, H.J., Chen, J.S., Karageorgiou, V., Altman, G.H., and Kaplan, D.L. (2004) Human bone marrow stromal cell responses on electrospun silk fibroin mats. *Biomaterials*, **25**, 1039–1047.
36. Dang, S.M., Gerecht-Nir, S., Chen, J., Itskovitz-Eldor, J., and Zandstra, P.W. (2004) Controlled, scalable embryonic stem cell differentiation culture. *Stem Cells*, **22**, 275–282.
37. Huang, C.Y.C., Reuben, P.M., D'Ippolito, G., Schiller, P.C., and Cheung, H.S. (2004) Chondrogenesis of human bone marrow-derived mesenchymal stem cells in agarose culture. *Anat. Rec. A Discov. Mol. Cell. Evol. Biol.*, **278A**, 428–436.
38. Mauck, R.L., Byers, B.A., Yuan, X., and Tuan, R.S. (2007) Regulation of cartilaginous ECM gene transcription by chondrocytes and MSCs in 3D culture in response to dynamic loading. *Biomech. Model. Mechanobiol.*, **6**, 113–125.
39. Serra, M., Correia, C., Malpique, R., Brito, C., Jensen, J., Bjorquist, P., Carrondo, M.J.T., and Alves, P.M. (2011) Microencapsulation technology: a powerful tool for integrating expansion and cryopreservation of human embryonic stem cells. *PLoS One*, **6**, e23212.
40. Awad, H.A., Wickham, M.Q., Leddy, H.A., Gimble, J.M., and Guilak, F. (2004) Chondrogenic differentiation of adipose-derived adult stem cells in agarose, alginate, and gelatin scaffolds. *Biomaterials*, **25**, 3211–3222.
41. Frantz, C., Stewart, K.M., and Weaver, V.M. (2010) The extracellular matrix at a glance. *J. Cell Sci.*, **123**, 4195–4200.
42. Toole, B.P. (2001) Hyaluronan in morphogenesis. *Semin. Cell Dev. Biol.*, **12**, 79–87.
43. Moreland, L.W. (2003) Intra-articular hyaluronan (hyaluronic acid) and hylans for the treatment of osteoarthritis: mechanisms of action. *Arthritis Res. Ther.*, **5**, 54–67.
44. Ghosh, P., Moore, R., Vernon-Roberts, B., Goldschlager, T., Pascoe, D., Zannettino, A., Gronthos, S., and Itescu, S. (2012) Immunoselected STRO-3(+) mesenchymal precursor cells and restoration of the extracellular matrix of degenerate intervertebral discs Laboratory investigation. *J. Neurosurg. Spine*, **16**, 479–488.
45. Hughes, C.S., Postovit, L.M., and Lajoie, G.A. (2010) Matrigel: a complex protein mixture required for optimal growth of cell culture. *Proteomics*, **10**, 1886–1890.
46. Serra, M., Brito, C., Sousa, M.F., Jensen, J., Tostoes, R., Clemente, J., Strehl, R., Hyllner, J., Carrondo, M.J., and Alves, P.M. (2010) Improving expansion of pluripotent human embryonic stem cells in perfused bioreactors through oxygen control. *J. Biotechnol.*, **148**, 208–215.
47. Cai, J., Zhao, Y., Liu, Y., Ye, F., Song, Z., Qin, H., Meng, S., Chen, Y., Zhou, R., Song, X., Guo, Y., Ding, M., and Deng, H. (2007) Directed differentiation of human embryonic stem cells into functional hepatic cells. *Hepatology*, **45**, 1229–1239.
48. Nakano, T., Ando, S., Takata, N., Kawada, M., Muguruma, K., Sekiguchi, K., Saito, K., Yonemura, S., Eiraku, M., and Sasai, Y. (2012) Self-formation of optic cups and storable stratified neural retina from human ESCs. *Cell Stem Cell*, **10**, 771–785.
49. Gonzalez-Cordero, A., West, E.L., Pearson, R.A., Duran, Y., Carvalho, L.S., Chu, C.J., Naeem, A., Blackford, S.J.I., Georgiadis, A., Lakowski, J., Hubank, M., Smith, A.J., Bainbridge, J.W.B., Sowden, J.C., and Ali, R.R. (2013) Photoreceptor precursors derived from three-dimensional embryonic stem cell cultures integrate and mature within adult degenerate retina. *Nat. Biotechnol.*, **31**, 741–747.
50. Macchiarini, P., Jungebluth, P., Go, T., Asnaghi, M.A., Rees, L.E., Cogan, T.A., Dodson, A., Martorell, J., Bellini, S., Parnigotto, P.P., Dickinson, S.C., Hollander, A.P., Mantero, S., Conconi, M.T., and Birchall, M.A. (2008) Clinical transplantation of a tissue-engineered airway. *Lancet*, **372**, 2023–2030.

51. Hou, N., Cui, P., Luo, J., Ma, R., and Zhu, L. (2011) Tissue-engineered larynx using perfusion-decellularized technique and mesenchymal stem cells in a rabbit model. *Acta Otolaryngol.*, **131**, 645–652.
52. Baptista, P.M., Siddiqui, M.M., Lozier, G., Rodriguez, S.R., Atala, A., and Soker, S. (2011) The use of whole organ decellularization for the generation of a vascularized liver organoid. *Hepatology*, **53**, 604–617.
53. Akhyari, P., Aubin, H., Gwanmesia, P., Barth, M., Hoffmann, S., Huelsmann, J., Preuss, K., and Lichtenberg, A. (2011) The quest for an optimized protocol for whole-heart decellularization: a comparison of three popular and a novel decellularization technique and their diverse effects on crucial extracellular matrix qualities. *Tissue Eng. Part C Methods*, **17**, 915–926.
54. Guguen-Guillouzo, C., Corlu, A., and Guillouzo, A. (2010) Stem cell-derived hepatocytes and their use in toxicology. *Toxicology*, **270**, 3–9.
55. Zhu, S., Wurdak, H., Wang, J., Lyssiotis, C.A., Peters, E.C., Cho, C.Y., Wu, X., and Schultz, P.G. (2009) A small molecule primes embryonic stem cells for differentiation. *Cell Stem Cell*, **4**, 416–426.
56. Norotte, C., Marga, F.S., Niklason, L.E., and Forgacs, G. (2009) Scaffold-free vascular tissue engineering using bioprinting. *Biomaterials*, **30**, 5910–5917.
57. Fedorovich, N.E., Dewijn, J.R., Verbout, A.J., Alblas, J., and Dhert, W.J.A. (2008) Three-dimensional fiber deposition of cell-laden, viable, patterned constructs for bone tissue printing. *Tissue Eng. Part A*, **14**, 127–133.
58. Gaetani, R., Doevendans, P.A., Metz, C.H.G., Alblas, J., Messina, E., Giacomello, A., and Sluijtera, J.P.G. (2012) Cardiac tissue engineering using tissue printing technology and human cardiac progenitor cells. *Biomaterials*, **33**, 1782–1790.
59. Miller, J.S., Stevens, K.R., Yang, M.T., Baker, B.M., Nguyen, D.H., Cohen, D.M., Toro, E., Chen, A.A., Galie, P.A., Yu, X., Chaturvedi, R., Bhatia, S.N., and Chen, C.S. (2012) Rapid casting of patterned vascular networks for perfusable engineered three-dimensional tissues. *Nat. Mater.*, **11**, 768–774.
60. Curcio, E., Macchiarini, P., and De Bartolo, L. (2010) Oxygen mass transfer in a human tissue-engineered trachea. *Biomaterials*, **31**, 5131–5136.
61. Crank, J. (1975) *The Mathematics of Diffusion*, 2nd edn, Clarendon Press, Oxford.
62. Tostoes, R.M., Leite, S.B., Miranda, J.P., Sousa, M., Wang, D.I., Carrondo, M.J., and Alves, P.M. (2011) Perfusion of 3D encapsulated hepatocytes--a synergistic effect enhancing long-term functionality in bioreactors. *Biotechnol. Bioeng.*, **108**, 41–49.
63. Discher, D.E., Mooney, D.J., and Zandstra, P.W. (2009) Growth factors, matrices, and forces combine and control stem cells. *Science*, **324**, 1673–1677.
64. Williams, C.M., Mehta, G., Peyton, S.R., Zeiger, A.S., Van Vliet, K.J., and Griffith, L.G. (2011) Autocrine-controlled formation and function of tissue-like aggregates by primary hepatocytes in micropatterned hydrogel arrays. *Tissue Eng. Part A*, **17**, 1055–1068.
65. Li, H., Wijekoon, A., and Leipzig, N.D. (2012) 3D differentiation of neural stem cells in macroporous photopolymerizable hydrogel scaffolds. *PLoS One*, **7**, e48824.
66. Radisic, M., Deen, W., Langer, R., and Vunjak-Novakovic, G. (2005) Mathematical model of oxygen distribution in engineered cardiac tissue with parallel channel array perfused with culture medium containing oxygen carriers. *Am. J. Physiol. Heart Circ. Physiol.*, **288**, H1278–H1289.
67. Fan, V.H., Au, A., Tamama, K., Littrell, R., Richardson, L.B., Wright, J.W., Wells, A., and Griffith, L.G. (2007) Tethered epidermal growth factor provides a survival advantage to mesenchymal stem cells. *Stem Cells*, **25**, 1241–1251.
68. Bauwens, C., Yin, T., Dang, S., Peerani, R., and Zandstra, P.W. (2005) Development of a perfusion fed bioreactor for embryonic stem cell-derived cardiomyocyte generation: oxygen-mediated enhancement of cardiomyocyte output. *Biotechnol. Bioeng.*, **90**, 452–461.

69. Goldstein, A.S., Juarez, T.M., Helmke, C.D., Gustin, M.C., and Mikos, A.G. (2001) Effect of convection on osteoblastic cell growth and function in biodegradable polymer foam scaffolds. *Biomaterials*, **22**, 1279–1288.
70. Wendt, D., Marsano, A., Jakob, M., Heberer, M., and Martin, I. (2003) Oscillating perfusion of cell suspensions through three-dimensional scaffolds enhances cell seeding efficiency and uniformity. *Biotechnol. Bioeng.*, **84**, 205–214.
71. Braccini, A., Wendt, D., Jaquiery, C., Jakob, M., Heberer, M., Kenins, L., Wodnar-Filipowicz, A., Quarto, R., and Martin, I. (2005) Three-dimensional perfusion culture of human bone marrow cells and generation of osteoinductive grafts. *Stem Cells*, **23**, 1066–1072.
72. Datta, N., Pham, Q.P., Sharma, U., Sikavitsas, V.I., Jansen, J.A., and Mikos, A.G. (2006) In vitro generated extracellular matrix and fluid shear stress synergistically enhance 3D osteoblastic differentiation. *Proc. Natl. Acad. Sci. U.S.A.*, **103**, 2488–2493.
73. Zhang, Z.Y., Teoh, S.H., Teo, E.Y., Khoon Chong, M.S., Shin, C.W., Tien, F.T., Choolani, M.A., and Chan, J.K. (2010) A comparison of bioreactors for culture of fetal mesenchymal stem cells for bone tissue engineering. *Biomaterials*, **31**, 8684–8695.
74. Singh, H., Teoh, S.H., Low, H.T., and Hutmacher, D.W. (2005) Flow modelling within a scaffold under the influence of uni-axial and bi-axial bioreactor rotation. *J. Biotechnol.*, **119**, 181–196.
75. Hiller, G.W., Clark, D.S., and Blanch, H.W. (1993) Cell retention-chemostat studies of hybridoma cells - analysis of hybridoma growth and metabolism in continuous suspension-culture on serum-free medium. *Biotechnol. Bioeng.*, **42**, 185–195.
76. Kim, J. and Ma, T. (2012) Perfusion regulation of hMSC microenvironment and osteogenic differentiation in 3D scaffold. *Biotechnol. Bioeng.*, **109**, 252–261.
77. Schmelter, M., Ateghang, B., Helmig, S., Wartenberg, M., and Sauer, H. (2006) Embryonic stem cells utilize reactive oxygen species as transducers of mechanical strain-induced cardiovascular differentiation. *FASEB J.*, **20**, 1182–1184.
78. Correia, C., Serra, M., Espinha, N., Sousa, M., Brito, C., Burkert, K., Zheng, Y., Hescheler, J., Carrondo, M.J., Saric, T., and Alves, P.M. (2014) Combining hypoxia and bioreactor hydrodynamics boosts induced pluripotent stem cell differentiation towards cardiomyocytes. *Stem Cell Rev.*, **10**, 786–801.
79. Yamamoto, K., Sokabe, T., Watabe, T., Miyazono, K., Yamashita, J.K., Obi, S., Ohura, N., Matsushita, A., Kamiya, A., and Ando, J. (2005) Fluid shear stress induces differentiation of Flk-1-positive embryonic stem cells into vascular endothelial cells in vitro. *Am. J. Physiol. Heart Circ. Physiol.*, **288**, H1915–H1924.
80. Niklason, L.E., Gao, J., Abbott, W.M., Hirschi, K.K., Houser, S., Marini, R., and Langer, R. (1999) Functional arteries grown in vitro. *Science*, **284**, 489–493.
81. Gong, Z. and Niklason, L.E. (2008) Small-diameter human vessel wall engineered from bone marrow-derived mesenchymal stem cells (hMSCs). *FASEB J.*, **22**, 1635–1648.
82. Sheppard, A.M., Hamilton, S.K., and Pearlman, A.L. (1991) Changes in the distribution of extracellular-matrix components accompany early morphogenetic events of mammalian cortical development. *J. Neurosci.*, **11**, 3928–3942.
83. Brown, N.H. (2011) Extracellular matrix in development: insights from mechanisms conserved between invertebrates and vertebrates. *Cold Spring Harbor Perspect. Biol.*, **3**, a005082.
84. Young, J.L. and Engler, A.J. (2011) Hydrogels with time-dependent material properties enhance cardiomyocyte differentiation in vitro. *Biomaterials*, **32**, 1002–1009.
85. Lamba, D.A., Karl, M.O., Ware, C.B., and Reh, T.A. (2006) Efficient generation of retinal progenitor cells from human embryonic stem cells. *Proc. Natl. Acad. Sci. U.S.A.*, **103**, 12769–12774.
86. Csaszar, E., Chen, K., Caldwell, J., Chan, W., and Zandstra, P.W. (2014) Real-time monitoring and control of soluble signaling factors enables enhanced progenitor cell outputs from human cord blood

stem cell cultures. *Biotechnol. Bioeng.*, **111**, 1258–1264.

87. Cowan, C.M., Shi, Y.Y., Aalami, O.O., Chou, Y.F., Mari, C., Thomas, R., Quarto, N., Contag, C.H., Wu, B., and Longaker, M.T. (2004) Adipose-derived adult stromal cells heal critical-size mouse calvarial defects. *Nat. Biotechnol.*, **22**, 560–567.
88. Liu, Q., Cen, L., Yin, S., Chen, L., Liu, G., Chang, J., and Cui, L. (2008) A comparative study of proliferation and osteogenic differentiation of adipose-derived stem cells on akermanite and beta-TCP ceramics. *Biomaterials*, **29**, 4792–4799.
89. Brannvall, K., Bergman, K., Wallenquist, U., Svahn, S., Bowden, T., Hilborn, J., and Forsberg-Nilsson, K. (2007) Enhanced neuronal differentiation in a three-dimensional collagen-hyaluronan matrix. *J. Neurosci. Res.*, **85**, 2138–2146.
90. Gerecht-Nir, S., Cohen, S., Ziskind, A., and Itskovitz-Eldor, J. (2004) Three-dimensional porous alginate scaffolds provide a conducive environment for generation of well-vascularized embryoid bodies from human embryonic stem cells. *Biotechnol. Bioeng.*, **88**, 313–320.
91. Ma, H.L., Hung, S.C., Lin, S.Y., Chen, Y.L., and Lo, W.H. (2003) Chondrogenesis of human mesenchymal stem cells encapsulated in alginate beads. *J. Biomed. Mater. Res. A*, **64**, 273–281.
92. Boeriu, C.G., Springer, J., Kooy, F.K., van den Broek, L.A.M., and Eggink, G. (2013) Production methods for hyaluronan. *Int. J. Carbohydr. Chem.*, **2013**, 14.
93. Chung, C. and Burdick, J.A. (2009) Influence of three-dimensional hyaluronic acid microenvironments on mesenchymal stem cell chondrogenesis. *Tissue Eng. Part A*, **15**, 243–254.
94. Gerecht, S., Burdick, J.A., Ferreira, L.S., Townsend, S.A., Langer, R., and Vunjak-Novakovic, G. (2007) Hyaluronic acid hydrogen for controlled self-renewal and differentiation of human embryonic stem cells. *Proc. Natl. Acad. Sci. U.S.A.*, **104**, 11298–11303.
95. Kim, J., Kim, I.S., Cho, T.H., Lee, K.B., Hwang, S.J., Tae, G., Noh, I., Lee, S.H., Park, Y., and Sun, K. (2007) Bone regeneration using hyaluronic acid-based hydrogel with bone morphogenic protein-2 and human mesenchymal stem cells. *Biomaterials*, **28**, 1830–1837.
96. Ludwig, T.E., Bergendahl, V., Levenstein, M.E., Yu, J., Probasco, M.D., and Thomson, J.A. (2006) Feeder-independent culture of human embryonic stem cells. *Nat. Methods*, **3**, 637–646.
97. Oh, S.K.W., Chen, A.K., Mok, Y., Chen, X., Lim, U.M., Chin, A., Choo, A.B.H., and Reuveny, S. (2009) Long-term microcarrier suspension cultures of human embryonic stem cells. *Stem Cell Res.*, **2**, 219–230.
98. Atala, A., Bauer, S.B., Soker, S., Yoo, J.J., and Retik, A.B. (2006) Tissue-engineered autologous bladders for patients needing cystoplasty. *Lancet*, **367**, 1241–1246.

# 8
# Coping with Physiological Stress During Recombinant Protein Production by Bioreactor Design and Operation

*Pau Ferrer and Francisco Valero*

## 8.1
## Major Physiological Stress Factors in Recombinant Protein Production Processes

### 8.1.1
### Physiological Constraints Imposed by High-Cell-Density Cultivation Conditions

*A priori*, three operational modes can be considered for the production of recombinant proteins, namely batch, fed-batch, and continuous cultivation. For economic feasibility, any cultivation method has to meet several criteria, including high volumetric productivity, high final product concentration, stability and reproducibility of the process, low-cost substrates (in the case of bulk proteins), as well as legal constraints, for example, those related to the application of recombinant DNA technology.

Batch cultures are generally not suited for large-scale production. High substrate (sugar) concentration at the start of the cultivation causes catabolite repression phenomena. Furthermore, osmotic sensitivity of cells (or substrate toxicity, for example, when methanol is used as a carbon source in methylotrophic yeast systems [1]) imposes a limit to the initial concentration of nutrients. In addition, since growth rate cannot be controlled, batch cultivation becomes limited by oxygen. High growth rates may also cause overflow metabolism, leading to by-product formation, even in the presence of oxygen; for instance, alcoholic fermentation is triggered in some yeasts such as *Saccharomyces cerevisiae* [2, 3], acetate formation in *Escherichia coli* [4, 5], or lactate and ammonium excretion in mammalian cell cultures [6–8]. Moreover, nutrient exhaustion and by-product accumulation lead to apoptosis (programmed cell death) in mammalian cell systems. Conversely, growth rate in fed-batch and continuous cultivations can be modulated by controlled addition of substrate, thereby avoiding by-product formation and catabolite repression phenomena, as well as

*Bioreactors: Design, Operation and Novel Applications*, First Edition. Edited by Carl-Fredrik Mandenius.

allowing for the adaptation of oxygen uptake and heat generation to the bioreactor performance. However, prolonged continuous cultivations may result in accumulation of less- or nonproducing cell populations and, therefore, decreased productivity [9].

As a result, fed-batch cultivation is often the preferred operational mode for large-scale production of heterologous proteins. Notably, fed-batch processes are typically performed at low growth rate when operated under substrate-limiting conditions (as discussed in Section 8.4). Therefore, effects of a low growth rate on the cell physiology and product formation rates of cells are of key importance and have a big impact on the specific productivity [10, 11]. For example, cellular responses such as nutrient starvation may be elicited, thereby contributing to the metabolic stress of the host cells. Recent transcriptomic studies in yeasts have revealed that growth rate regulates core processes such as protein synthesis and secretion, as well as stress response [12]. Notably, a substantial part of the substrate carbon is expended to meet maintenance-energy requirements under growth-limiting conditions such as those found in fed-batch processes. High maintenance requirements go at the expense of biomass and product formation and, therefore, are not desired in heterologous protein production [10]. Maintenance requirements may be significantly influenced by environmental parameters such as temperature, pH, and medium composition [10].

Moreover, recombinant protein production processes are typically carried out in high-cell-density cultures (HCDC) to maximize space–time yields and volumetric productivities. In HCDC, cells often suffer several physiological challenges, including limited availability of oxygen, accumulation of carbon dioxide to levels that may diminish growth rate and stimulate by-product formation (e.g., acetate in *E. coli*), heat generation, and reduced mixing efficiency [13, 14]. Such environmental stresses in fed-batch HCDC do not occur merely as transient responses to sudden changes (shock), but rather as long-term adaptation of producing cells to suboptimal conditions [15]. In fact, cells are exposed to progressively increasing energy limitation, undergoing increased cell death and lysis, segregation into viable but nonculturable states, and altered transcriptional profiles at high cell densities [16–18]. Therefore, induction of recombinant protein expression under these "basal" stress conditions imposes an extra, overlying, intrinsic stress to the system.

A further complication in prokaryotic hosts such as *E. coli* are quorum-sensing mechanisms, that is, growth-regulatory intercellular signaling systems where a soluble extracellular signal molecule that upon reaching a critical concentration in the medium enables synchronous triggering of population-level response, for instance, in response to an environmental stress or nutrient limitation [19]. DeLisa *et al.* [20] demonstrated that *E. coli* communicates the stress or burden imposed by the accumulation of heterologous protein. Several quorum sensing–like mechanisms have been described in yeasts [21]. Nevertheless, it is still unknown how these mechanisms may interfere with protein production in these microorganisms (Figure 8.1).

### 8.1.2
### Metabolic and Physiologic Constraints Imposed by High-Level Expression of Recombinant Proteins

It has been long recognized and documented that high-level expression of heterologous proteins have a direct impact on host metabolism, often negatively affecting growth parameters such as growth rate, biomass yield, and specific substrate consumption rate, for example, in bacteria [22, 23], yeast [24–27], and mammalian cells. These hindrances are responsible for limiting the amount of foreign protein that can be produced from the organism. On one hand, foreign gene expression leads to an increase in specific transcription and translation, which may become limiting at very high levels due to the depletion of precursors and energy, as described for *E. coli* [28–30]. Also, a metabolic burden related to plasmid maintenance has been described [31, 32]. Furthermore, *E. coli* elicits a stress response as a result of overexpression of recombinant proteins that triggers upregulation of stress genes, including heat shock, chaperone, and protease-encoding genes [18, 33–36]. Here, the formation of product aggregates (inclusion bodies) and misfolded products, resulting from high synthesis rates, has been shown to be an important factor [37, 38]. Nonetheless, the transcriptional profile of stress responses to high-level expression at high-cell-density cultures is different from low-cell-density cultures, as discussed in Section 8.2.

Burden associated with metabolic demand for recombinant protein synthesis has also been reported for high-level expression of intracellular products in yeasts [29, 39–41]. However, considering the moderate specific productivities of secreted proteins often achieved in yeast systems, limitations in amino acid synthesis, transcription, and translation do not seem to be a major bottleneck. Still, the energy demand is significantly higher for secreted proteins, as folding, glycosylation, and secretion are energy-intensive pathways. In fact, a metabolic burden related to protein secretion has been observed in yeast, even at low-to-medium expression levels [40–43]. There is extensive evidence for limitations in membrane translocation, signal sequence processing, and folding within the endoplasmatic reticulum (ER). Cellular stress responses to unfolded proteins (unfolded protein response, UPR) have been described for yeasts, filamentous fungi, and higher eukaryotes [38, 44], and have been shown to be triggered upon overexpression of a secreted recombinant protein, thereby increasing the metabolic demand of the folding and secretory processes. UPR is also connected to protein degradation via the ER-associated protein degradation (ERAD) pathway, a process that detects misfolded proteins and redirects them to proteasomal degradation in the cytosol, acting as a stress relief pathway. In this context, recent measurements of intracellular fluxes of secreted proteins in yeast using isotopic labeling strategies have revealed that a significant fraction of the synthesized recombinant protein (more than 50%) is actually degraded [45]. Moreover, UPR regulates several metabolic processes such as lipid and amino acid synthesis, as well as downregulating the expression of secreted proteins in fungi [46, 47].

Overall, folding, posttranslational modification, and transport processes are expensive in terms of energy and carbon-building blocks.

Importantly, the interaction of external stresses (e.g., temperature, pH, osmolarity, oxygen availability) and intrinsic stress triggered by protein overexpression – particularly by folding and secretion – plays a major role in the physiological constraints of a production system, which share similar patterns among different classes of cell factories [15, 21, 38] (Figure 8.1), as discussed in Section 8.2.

### 8.1.3 Physiological Constraints in Large-Scale Cultures

In addition, a major problem in large-scale recombinant microbial and cell cultures is the existence of heterogeneous environmental conditions generated by industrial large-scale bioreactors, for instance, temperature, pH, dissolved oxygen tension, or substrate concentration fluctuations [48–51]. Such nonideal hydrodynamic conditions in large-scale bioreactors have been studied in scale-down bioreactors, which consist of a mixed part connected to a nonmixed part by a recirculation pump, thereby mimicking hydrodynamic conditions encountered at the large scale [48, 52]. Heterogeneous conditions such as substrate gradients in large-scale bioreactor cultivations of recombinant *E. coli* have been shown to lower cell yield and increase by-product formation [53], as well as cell lysis, which resulted in product proteolysis [54]. Oscillating substrate concentration and dissolved oxygen levels led to diminished glucose uptake, ethanol formation, and an altered amino acid synthesis in *Bacillus subtilis* [55], whereas oscillating oxygen-limiting conditions decreased specific growth rate and increased by-product formation in *Pichia pastoris* [56]. Besides, inhomogeneities in production-scale bioreactors may also affect product quality in mammalian cell cultures, where heterogeneous conditions in dissolved oxygen have been reported to affect N-glycosylation of several recombinant proteins in mammalian cell cultures (Figure 8.1) [57, 58].

## 8.2 Monitoring Physiological Stress and Metabolic Load as a Tool for Bioprocess Design and Optimization

Monitoring of recombinant protein production processes are the key for process and systems understanding and, therefore, key to rational bioprocess design and optimization [59]. Characterization of the metabolic impact during recombinant protein production processes can be often done through the estimation of classic physiological parameters such as growth yields, maximum specific growth rates, as well as substrate specific consumption or by-product-specific production rates. These indicators are based on the measurement of variables such as biomass concentration, oxygen consumption, and $CO_2$ production rates or substrate and by-product concentration. Nonetheless, these parameters do not reveal the extent and mechanistic aspects of the underlying physiologic changes.

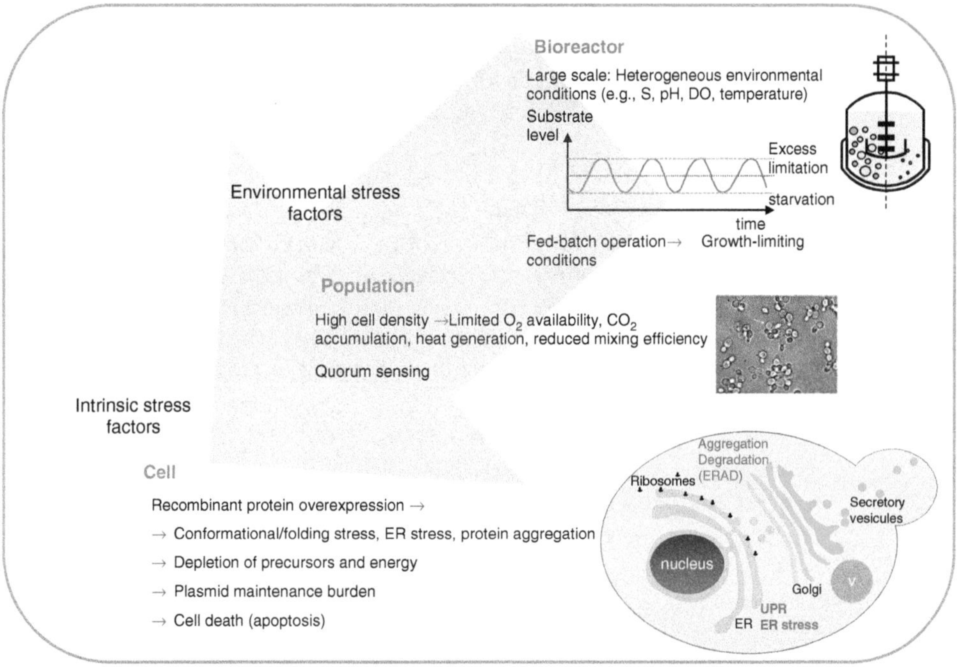


**Figure 8.1** Environmental and intrinsic stress factors affecting recombinant protein production.

In addition to standard monitoring of physical and chemical variables, a series of offline techniques and online process-monitoring devices allow for estimating and measuring physiological parameters such as cell viability, recombinant product, as well as reporter metabolites or other cell components that are responsive to the physiological adaptation of the cells to recombinant protein expression [59–65]. Moreover, transcriptomics, proteomics, metabolomics, and fluxomics are also being used as a tool for system-level understanding of cell response and adaptation to recombinant protein production, both as knowledge base for rational strain improvement (metabolic engineering) and for the identification of stress reporter genes suitable for bioprocess optimization.

### 8.2.1 Monitoring of Physiological Responses to Recombinant Gene Expression Using Flow Cytometry

Recombinant protein production brings cells to stress states and may lead to cell death during the production process. Flow cytometry has been extensively used to determine the fraction of dead cells by differential staining of cells with propidium iodine, from bacteria and yeast to mammalian cells. By triple fluorochrome staining using propidium iodide, ethidium bromide, and bis-oxonol, it is possible to discriminate between undamaged, damaged (membrane depolarized), and dead cells [66, 67]. Flow cytometry with a range of fluorescent markers has also been used to study cell cycle progression and programmed cell death (apoptosis) events during bioreactor mammalian cell cultures, shown to be a valuable tool for optimization of culture strategies and process analytical technologies [68].

This technique can also be applied to measure a wide variety of complex metabolic traits [67], for instance, the detection of the accumulation of reactive oxygen species (ROS) and *in vivo* redox conditions. Oxidative protein folding and secretion can overburden the redox capacity of the endoplasmatic reticulum (ER), resulting in the accumulation of ROS and misfolded/unfolded proteins in this compartment. Staining of ROS can be typically achieved using different fluorescent probes such as DHE (dihydroethidium), DHR (dihydrorhodamine 123), and DCF (2′,7′-dichlorodidydrofluorescein), whereas a recent study has demonstrated that *in vivo* redox conditions in the ER and cytosol of yeast cells can be monitored by targeting redox-sensitive GFP (roGFPs) variants to the respective organelles [69, 70]. The latter procedure is an example of a genetically encoded biosensor, which does not require the addition of exogenous chemicals. Such biosensors can also be designed in such a way that the expression of the fluorescent protein is under transcriptional control of a promoter whose activity is regulated by the intracellular signal of interest, for example, stress-responsive promoters (reviewed in [71] and [72]). For instance, the fluorescence signal of GFP expressed in *E. coli* under a stress-responsive promoter can be measured by 2D multiwavelength fluorescence spectroscopy [73], or flow cytometry [62], thereby allowing *in vivo* online or at-line stress monitoring, as an alternative to direct measurement of mRNA levels of selected stress genes [48, 60].

Surface and intracellular protein components (including inclusion bodies) can also be stained through immunofluorescent techniques for subsequent flow cytometry analysis. In order to access intracellular targets, cells are gently permeabilized. Intracellular immunostaining procedures have been used to monitor and quantify inclusion body formation in *E. coli* [74] or recombinant product retained in the cells and UPR markers such as the BiP protein in yeast [75, 76]. BiP is a chaperone of the HSP70 class that plays an important role in the unfolded protein stress response. Otherwise, GFP and GFP fusion proteins can be used to correlate single-cell and population physiology to recombinant protein expression levels, providing a valuable tool to optimize cultivation/induction strategies, as discussed in Section 8.4 [77, 78].

Notably, flow cytometry is compatible with the implementation of multiparameter (multiplexed) analyses [65, 79–81], improving the discrimination of physiological states, as well as providing valuable information on the phenotypic heterogeneity of a microbial population. Furthermore, these procedures may be potentially automated by connecting a flow cytometer directly to a bioreactor, offering a very powerful tool for monitoring physiological properties of both microbial and mammalian cells under process-related conditions [65, 68, 82, 83].

### 8.2.2
### Monitoring of Reporter Metabolites

The concentration of medium components and metabolites secreted by the cells such as sugars, amino acids, and by-products (acetate, ethanol, lactate, etc.) can be typically quantified by HPLC methods, providing useful information about the metabolism of the cell. In addition, *in situ* probing technologies based on, for example, spectroscopy-based methods and optical analyses can deliver information on changes of the physiological/state resulting from protein production. However, the interpretation or assignment of signal changes to physiology-related parameters is not straightforward [30, 59, 71].

In addition, methods for monitoring of reporter metabolites such as guanosine tetraphosphate (ppGpp), the central signal molecule of the stringent response system in bacteria, and cyclic adenosine monophosphate (cAMP), a signal molecule involved in metabolite repression, have been established. This has allowed to study the correlation between recombinant gene expression levels or the plasmid content of the cell and concentration levels of these metabolites, yielding valuable information on the interrelation between recombinant protein production and physiological state [28, 59, 61, 84, 85]. Quantification of ATP, ADP, and AMP intracellular pools are also another source of information about the metabolic impart of recombinant protein synthesis and secretion, since adenylate pools directly influence these energy-consuming cellular processes. As an example, recombinant protein production in *E. coli* using strong temperature-inducible promoters caused a transient drop of the adenylate energy charge ($AEC = [ATP + 1/2ADP] / [ATP + ADP + AMP]$) – an indicator of the energetic status of the cells – just after temperature upshift [86, 87]. Conversely, a sustained decrease in AEC has been reported along the methanol induction phase of HCDC of *P. pastoris* when using the methanol-inducible AOX promoter (Table 8.1) [99].

### 8.2.3
### Omics Analytical Tools to Assess the Impact of Recombinant Protein Production on Cell Physiology

Global analysis of the host cell metabolism by means of omics analytical platforms such as transcriptomics, proteomics, metabolomics, and fluxomics can now be used to investigate the physiological effect of both environmental stresses and recombinant biosynthesis, guiding the identification of potential metabolic targets for cell engineering, selection of growth conditions, and cultivation strategies favoring recombinant protein production, or identification of stress-responsive marker genes (Table 8.2). In particular, these analytical tools have been increasingly used to characterize the global physiological changes in bacteria, mainly in *E. coli* (reviewed by Carneiro *et al.* [5]) and *Bacillus* sp. [106, 108, 122–126], as well as eukaryotic host systems, including yeast [15, 42, 43, 109, 127], filamentous fungi [118, 119, 128], and mammalian cells [120, 129–132].

**Table 8.1** Stress markers and corresponding technique used for monitoring of the physiological state of recombinant protein production processes.

| Physiological characteristic/stress factor | Probes/markers (examples) | Organism | Technique | References |
|---|---|---|---|---|
| Membrane integrity, cell viability, cell death (necrosis) | Differential staining of undamaged, damaged (membrane depolarized) and dead cells | Bacteria, yeast, mammalian cells | Fluorochrome staining (e.g., propidium iodide, ethidium bromide, bio-oxonol and other dyes that intercalate with double stranded nucleic acids) and flow cytometry | [68, 83, 88–90] |
| Cell cycle phase, apoptosis | Externalization of phosphatidylserine (plasma membrane integrity assay); mitochondrial membrane potential; caspase activation; nuclear apoptosis (DNA fragmentation, nuclear chromatin condensation); intracellular pH | Mammalian cells | Fluorochrome staining (e.g., Annexin-V for plasma membrane assay, JC-1 as mitochondrial membrane indicator, etc.) and flow cytometry | [68, 83, 91, 92] |
| Oxidative stress | Reactive oxygen species (ROS) | Bacteria, yeast | Fluorochrome staining (e.g., dihydroethidium, DHE, dihydrorhodamine123, DHR, 2′,7′-dichlorodidydrofluorescein, DCF) and flow cytometry | [67, 70, 93] |

| | | | | |
|---|---|---|---|---|
| Redox state. Endoplasmatic reticulum redox status (ER oxidative stress) | Redox-sensitive GFP (roGFP) expressed in the cytosol or ER; Reduced to oxidized glutathione ratio (GSH/GSSH); Transcriptional levels of ER stress marker genes (e.g., antioxidative glutathione reductase *GSH1*, and thioredoxin reductase, *TRR1* genes) | Yeast | Flow cytometry<br>LC/MS–MS<br>Alloxan method (for reduced glutathione, GSH)<br>Measurement of mRNA levels by quantitative real-time PCR | [69, 70]<br>[70, 94]<br>[93]<br>[93] |
| Unfolded protein response (UPR)/ER stress | BiP (HSP70 class chaperone), Pdi (protein disulfide isomerase) Transcriptional levels of BiP-encoding (*KAR2*) and Pdi-encoding (*PDI1*) genes | Yeast (*P. pastoris*) and filamentous fungi (*Trichoderma reesei*) | Intracellular immunofluorescent staining and flow cytometry<br>Measurement of mRNA levels by: Northern analysis; Sandwich hybridization assay; TRAC (transcript analysis with aid of affinity capture) and capillary electrophoresis, quantitative real-time PCR | [75, 76]<br>[93, 95–97] |
| Glucose limitation or glucose-starvation-like status due to high metabolic load of recombinant protein production | Intra/extracellular cAMP levels, a molecule involved in metabolite repression (regulation of glucose starvation response) | *E. coli* | HPLC | [59, 61] |

(*continued overleaf*)

**Table 8.1** *continued*

| Physiological characteristic/stress factor | Probes/markers (examples) | Organism | Technique | References |
|---|---|---|---|---|
| General stress | Intracellular guanosine-3′,5′-tetraphosphate (ppGGpp) levels. Central molecule of the stringent response system (regulation of starvation responses) | *E. coli* | HPLC | [61, 85, 98] |
| Energy state | Adenylate energy charge, AEC | *E. coli*<br>*P. pastoris* | HPLC or LC–MS measurement of ATP, ADP and AMP intracellular concentrations | [61, 86, 87]<br>[40, 99] |
| General stress, for example, caused by bioreactor mixing inefficiency (generation of sugar and dissolved oxygen level perturbations) | General stress response transcription factors $\sigma^S$ and $\sigma^{70}$ levels<br>Transcriptional levels of *rpoS* gene (coding for $\sigma^S$)<br>Transcriptional levels of stress marker genes induced by heat stress (*clpB*, *dnaK*), glucose starvation (*uspA csiE*), glucose excess (*ackA*), oxygen limitation (*frd* and *pfl*), and osmotic stress (*proU*) | *E. coli* | Measurement of marker $\sigma$ (prpoS) protein levels by Western analysis<br>Expression of a fusion *rpoS:GFP* gene and flow cytometry monitoring prpoS:gfp fusion protein<br>Measurement of mRNA levels by slot-blot hybridization analysis<br>Expression of GFP gene under the control of a stress-responsive promoter (e.g., *dnaK* promoter or *csiE*) and expression monitoring of gfp levels by flow cytometry or 2D multiwavelength fluorescence spectroscopy | [61]<br>[62]<br>[48, 60, 65]<br>[63, 73] |
| Product aggregation (inclusion bodies formation) | Inclusion bodies | *E. coli* | Immunofluorescent techniques (intracellular immunostaining) and flow cytometry | [74] |

**Table 8.2** Selected examples of omics studies of impact of recombinant protein production on host physiology and its interaction with environmental factors.

| Host organism | Omics analyses | Recombinant protein product | Description | References |
|---|---|---|---|---|
| *E. coli* | DNA microarrays | Chloramphenicol acetyltransferase (CAT); insulin-like growth factor-I fusion protein (IGF-$I_f$)<br>None | Comparative analyses of recombinant and reference (mock plasmid or plasmid-free) strains to detect differential transcriptional profiles. Comparison of noninducing and IPTG-inducing or temperature-inducing conditions.<br>Comparative analyses of a reference strain grown in different carbon sources. | [100–102]<br>[103] |
| *E. coli* | Proteomics (Two-dimensional electrophoresis, 2D-DIGE and protein identification by MALDI-MS or ESI-MS/MS) | Antibody fragment Phosphogluconolactonase (PGL) from *Pseudomonas aeruginosa* | Comparative analyses of recombinant and reference (mock plasmid or plasmid-free) strains to detect differential transcriptional profiles. | [104, 105] |
| *E. coli* | $^{13}$C-based metabolic flux analysis (MFA) | $\beta$-Aminopeptidase DmpA from *Ochrobactrum anthropi* | Effect of IPTG induction of recombinant protein production: Comparison of noninducing versus inducing conditions at different IPTG concentrations | [29] |
| *Bacillus megaterium* | $^{13}$C-MFA | Hydrolase from *Thermobifida fusca* | Comparative analyses of a production strain grown in different carbon sources, under noninducing and inducing conditions (xylose-inducible *xylA* promoter). | [106] |

(*continued overleaf*)

**Table 8.2** *continued*

| Host organism | Omics analyses | Recombinant protein product | Description | References |
|---|---|---|---|---|
| *E. coli* | Combined DNA microarrays and proteomics (2D-DIGE and protein identification by MALDI-MS or ESI-MS/MS) | Green fluorescent protein 3.1 variant (GFP-mut3.1) or human Cu/Zn superoxide dismutase (hSOD)<br>$\alpha$-Glucosidase from *S. cerevisiae*<br>None | Comparative analyses of recombinant and reference (mock plasmid or plasmid-free) strains to detect differential transcriptional profiles.<br>Comparison of noninducing and IPTG-inducing conditions. Impact of recombinant protein species.<br>Effect of plasmid DNA maintenance on host metabolism: Comparative analysis of plasmid free versus plasmid-bearing cells | [33, 35]<br>[107] |
| *B. subtilis* | Combined DNA microarrays and proteomics (2D-DIGE and protein identification by MALDI-MS or ESI-MS/MS) | None | Effect of nitrogen source on cell physiology during fed-batch cultivations | [108] |
| *E. coli* | Combined DNA microarrays and $^{13}C$-based MFA | None | Effects of plasmid DNA on host metabolism: Comparison of a reference (plasmid free) strain, and strains containing low- and high- copy number plasmids | [31] |
| *E. coli* | $^{13}C$-based MFA and metabolomics | Human fibroblast growth factor (hFGF-2) | Response to temperature-induced recombinant protein synthesis | [87] |
| *P. pastoris*<br>*P. pastoris/S. cerevisiae* | DNA microarrays<br>RNA sequencing<br>DNA microarrays | Human serum albumin (HSA)<br>Human lysozyme variants<br>Antibody fragment | Comparison of a recombinant protein producing strain growing at different specific growth rates;<br>Comparison of strains producing different variants of a protein with different degrees of misfolding<br>Cross-species comparison of the effect of oxygen supply in reference (mock plasmid) and recombinant protein producing strains | [12]<br>[109]<br>[110] |

| | | | | |
|---|---|---|---|---|
| *P. pastoris* | Proteomics (2D-DIGE combined with protein identification by ESI-MS/MS or MALDI-ToF MS | Antibody fragment<br>Hepatitis B virus surface antigen (HBsAg); Insulin precursor | Comparison of reference (mock vector) and producing strain under different growth temperatures<br>Cellular response to induction by substrate shift (glycerol to methanol) in a recombinant protein producing strain; comparison of a control and recombinant strain growth after substrate switch to methanol | [111]<br>[112, 113] |
| *P. pastoris* | $^{13}C$-based MFA | Antibody fragment, *R. oryzae* lipase, and so on. | Several studies comparing producer versus non producer strains, high versus low producing cells, and comparison of a producing strain and reference strain grown under different conditions (single, mixed substrates, growth rates, oxygen levels, etc.) | Reviewed in [43] |
| *S. pombe* | $^{13}C$-based MFA | Maltase | Comparison of a set of strains producing different amounts of recombinant protein, and a reference (non producing) strain | [42] |
| *P. pastoris* | Combined DNA microarrays and proteomics (2D-PAGE and protein identification with LC-ESI-QTOF MS/MS) | Antibody fragment | Comparison of reference (mock vector) and producing strain growing in media of different osmolarities | [114] |
| *P. pastoris* | Combined DNA microarrays, proteomics (2D-PAGE and protein identification with LC-ESI-QTOF MS/MS) and $^{13}C$-based MFA | Antibody fragment | Comparison of reference (mock vector) and producing strain growing under different oxygen levels | [115] |

(continued overleaf)

**Table 8.2** *continued*

| Host organism | Omics analyses | Recombinant protein product | Description | References |
|---|---|---|---|---|
| *S. cerevisiae* | DNA microarrays combined with MFA | Superoxide Dismutase (SOD) | Comparison of a producer and non producer reference strain at different growth phases of a batch culture | [116] |
| *Aspergillus nidulans* | DNA microarrays | Bovine chymosin | Comparison of a producer and non producer reference strain | [117] |
| *T. reesei* | Combined transcriptomics (DNA microarrays and TRAC, transcript analysis with aid of affinity capture) and proteomics (2D-DIGE and LC-MS/MS for protein identification) | Native cellulases | Effects of the low growth rate protein production phenotype:<br>Correlation-based analysis of gene and protein expression with specific extracellular protein production rate: Comparison of a producer strain at two different growth rates and low/high cell density. | [118] |
| *Aspergillus niger* | $^{13}C$-based MFA | Fructofuranosidase | Comparison of a wild type and a recombinant protein producer strain | [119] |
| CHO and HEK cells | DNA microarrays | CHO cells: human interferon gamma (IFN-$\gamma$)<br>HEK cells: GFP | Investigation of the transcriptional regulation apoptotic signaling pathways (CHO cells) and the metabolic changes (HEK cells) associated with the batch and fed-batch cultures:<br>observation of the physiological changes along different culture phases (mid-exponential, late exponential and stationary phases) | [120, 121] |

## 8.3 Design and Operation Strategies to Minimize/Overcome Problems Associated with Physiological Stress and Metabolic Load

In general, design of cultivation strategies aimed to attain maximal yields of recombinant protein often follow the principle of selection of cultivation modes that allow prolonging the production phase, at controlled growth rates, as well as minimizing the induction of stress events. Therefore, it is important to fine-tune the production rate of recombinant protein, the capability of the host cell metabolism, and the cultivation conditions, thereby minimizing potential bottlenecks in the metabolism.

### 8.3.1 Overcoming Overflow Metabolism and Substrate Toxicity

As discussed in Section 8.1, acetate accumulation in *E. coli* cultivation, the major product of overflow metabolism of this organism, is primarily connected to specific growth rate and substrate (glucose) uptake rate [4, 133]. High specific glucose consumption rates ($q_s$) promote acetate formation, regardless of the availability of oxygen in the culture. Acetate has an inhibitory effect on both biomass growth and protein synthesis, as well as the general stress response (reviewed in [4, 134]). Under glucose limitation, the value of $q_s$ is directly related to the specific growth rate (provided there is no other limiting nutrient besides glucose). Therefore, fed-batch cultivation strategies controlling nutrient feeding [5, 13, 14, 135] are very efficient to prevent overflow metabolism, or avoiding underfeeding causing transient starvation periods, as well as minimizing metabolic disturbances caused by the induction of protein expression. The simplest feeding strategies can involve the application of constant feeding rates, increased feeding (stepwise), or exponential feeding rates without feedback control (i.e., open loop control). The latter generally allows cells to grow at constant specific growth rate by feeding glucose (or another carbon source) as a growth-limiting nutrient [13, 14]. In this way, acetate formation can be minimized or even avoided by controlling the specific growth rate below a certain threshold value, which may vary depending on the strain and growth medium, but it typically ranges between 0.15 and 0.35 $h^{-1}$ [4, 14]. Notably, exponential feeding strategies allow to maintain glucose concentration in the cultivation broth at near-zero values without fluctuations, thus minimizing perturbations on the cell metabolism [14]. Alternatively, there are nutrient-feeding methods with feedback control: the pH-stat or DO (dissolved oxygen)-stat methods are based on the online monitoring of changes in pH or DO, where a nutrient feed is activated when pH or DO increase as the growth-controlling carbon source becomes depleted. These feedback-controlled feeding methods tend to be conservative, that is, result in a growth rate below the threshold for acetate production. This advantage has made these strategies widely employed for HCDC of *E. coli* (reviewed in [4, 14]).

*S. cerevisiae*, a fermentative (Crabtree positive) yeast, produces ethanol at high glucose concentrations, even in the presence of oxygen. Hence, similar fed-batch strategies (i.e., limiting glucose feeding) have been developed for this cell factory [136–138]. Notably, although protein production in industry is generally based on fed-batch cultivations due to the risk of genetic instability and contamination of continuous cultures, insulin is produced in *S. cerevisiae* by Novo Nordisk in continuous cultivation [139]. Crabtree-negative yeast such as *P. pastoris* offer *a priori* the advantage of a reduced production of by-products (mostly, ethanol and arabitol) [140, 141].

In the case of mammalian cells, batch culture is limited by nutrient exhaustion (mainly glucose and glutamine) and toxic metabolite accumulation (primarily lactate and ammonium), which results in cell death. As for microbial systems, alterative fed-batch and perfusion processes have been developed to overcome this drawback. In particular, controlled feeding of glucose and glutamine has been widely used in fed-batch cultures of hybridoma cells and other cell lines, using either open or closed control loop systems [142–149].

By-product formation is also affected by the composition of the culture medium. For instance, replacement of glucose by other carbon sources such as mannose or fructose, or combining glucose with amino acid in the feed media has been shown to reduce acetate and increase protein yield (reviewed in [4]). Also, glycerol has been demonstrated to be superior to glucose in reducing acetate formation during the recombinant protein induction phase of *E. coli* HCDCs [150]. Replacement of glucose by glycerol in the batch phase of *P. pastoris* fed-batch fermentation cultivation has also been shown to reduce by-product (ethanol and arabitol) formation [141]. The substitution of glucose by galactose and glutamine by glutamate produces a metabolic flow redistribution leading to slower rates of lactate and ammonium formation in mammalian cells. Essentially, both galactose and glutamate are metabolized at a slower rate (i.e., resulting in lower cell growth rates), thereby minimizing by-product formation [151]. The combination of this approach with careful reformulation of the amino acid composition of the growth medium in fed-batch cultures of CHO cells resulted in an increased culture viability, longevity, and protein production [152].

Overall, although fed-batch-controlled methods provide efficient means to prevent overflow metabolism, there is no general protocol or rule for selecting the optimal feeding strategy to achieve maximum productivity of a given protein [1, 14]. In this context, high-throughput microreactors systems (e.g., based on shaken microtiter plates) providing mixing and gas transfer conditions similar to those found in conventional bioreactors, as well as incorporating monitoring and active control devices, offer a powerful tool for fed-batch process development [153]. Moreover, these systems can be combined with slow glucose (or other limiting substrate) release enzymatic systems [154–156], expanding the range of substrate feeding strategies, for example, in *E. coli* and yeast [156].

In some cases, the substrate can be toxic even at relatively low concentrations. For example, high-cell-density cultivation of *P. pastoris* using the classic methanol-inducible expression systems requires careful control of the inducing substrate. High methanol concentration in the culture broth causes an inhibitory effect on growth affecting drastically the productivity of the process [157]. Reactive oxygen species such as hydrogen peroxide and other peroxidated molecules are generated during oxidation of methanol, which need to be removed to minimize cell damage. In fact, catalase, which removes hydrogen peroxide, and a glutathione peroxidase or peroxiredoxin, which removes peroxidated molecules, have been reported to be increased strongly in the methanol fed-batch phase of *P. pastoris* cultivations, concomitant with increased turnover of peroxisomes [113]. Oxidative stress caused by methanol may be relieved by the addition of antioxidants such as ascorbic acid [158].

Moreover, methanol-feeding strategies – particularly for $Mut^+$ (methanol utilization) phenotype cells, that is, cells with the wild-type methanol assimilation capacity – have to take into account the high oxygen consumption requirements and the considerable heat generated by methanol metabolism at high cell densities. Under these conditions, bioreactor oxygen transfer capacity is often unable to sustain the oxygen metabolic demand. The classic and simplest methanol feeding strategies are associated with a dissolved oxygen DO-stat control, maintaining a minimal level around 20% [159]. Although different DO-stat control strategies have been developed [157], methanol is not monitored and accumulation of the substrate could be the cause of growth inhibition. Alternatively, oxygen limitation has been successfully applied to control the methanol uptake during the induction phase (i.e., the limiting nutrient is oxygen instead of methanol), allowing for an improvement in production while avoiding the high demand of the oxygen [160]. Notably, oxygen limited fed-batch (OLFB) strategies have been applied to the production of monoclonal antibodies in glycoengineered *P. pastoris* strains using oxygen uptake rate as a scale-up parameter [161–163]. OLFB has also been reported to reduce cell lysis and increase the quality of the final product [164], as well as prolong the production phase of the target protein [165].

Another common strategy to control methanol levels is the $\mu$-stat control. A preprogrammed exponential methanol-feeding rate profile is obtained from mass balance equations in a classical open-loop control strategy [1]. In this strategy, the culture is always under methanol-limiting conditions [166]. The success of the optimal specific growth rate depends on the nature of the target protein, but it is a simple approach very easy to implement. It has been applied successfully with different proteins, obtaining generally the best production at low specific growth rates [157, 159]. However, this is not a general rule. For instance, this strategy resulted in very low yields and productivities when producing a recombinant *Rhizopus oryzae* lipase, which is known to trigger the UPR [167].

All the strategies for methanol control described so far have the potential problem that methanol concentration is not online measured. Methanol can suffer fluctuations along the induction phase affecting negatively the productivity of the bioprocess. Thus, different methanol closed-loop control strategies have been developed to avoid this problem [159]. In general, the maximal productivity is obtained in the range of $2–4\,g\,l^{-1}$ of methanol [167–169].

### 8.3.2
### Improving the Energy and Building Block Supply

Microbial cell factory platforms such as *E. coli* and yeast can be grown to high cell densities using defined medium with for example, glucose or glycerol. However, high levels of protein production impose an increased amino acid demand. This implies a higher demand for *de novo* amino acid synthesis, that is, for amino acid biosynthetic precursors and required energy [170, 171]. Direct supplementation of amino acids to the growth medium results in higher recombinant protein titers in *E. coli* [172]. Similarly, several studies [39, 173–175] have reported improved protein production in different yeast prototrophic strains growing on defined media by supplementing them with specific amino acids. Coherently, rich complex media have a significant positive effect on recombinant protein production, as shown for several yeasts [176].

As described in Section 8.2, high-level expression of heterologous proteins and their secretion also cause an increased energy demand, which is supplied by the central carbon metabolism, resulting in significant metabolic flux redistributions. The use of auxiliary carbon substrates, that is, the simultaneous utilization of two distinct substrates, has been proven to increase growth yields in several yeasts, most likely because the amount of available energy from the substrates is higher when cells are grown on the substrate mixture than when they are grown on each carbon source individually [42, 177]. The use of mixed substrates has been proved to boost recombinant protein production in yeast. Supplementing acetate to glucose or glycerol minimal media of recombinant *Schizosaccharomyces pombe* secreting a model protein (maltase) in aerobic chemostat cultures improved protein secretion. Acetate cofeeding allowed for an increased carbon flux through the TCA cycle as well as increased mitochondrial NADPH production [42]. Similarly, supplementation of methanol-growing recombinant *P. pastoris* with sorbitol or glycerol has been shown to boost productivities of recombinant proteins [178, 179]. Typically, fed-batch high-cell-density cultivations of recombinant *P. pastoris* based on the methanol-regulated AOX (alcohol oxidase) promoter have three phases: a growth batch phase using glycerol as carbon source, followed by transition batch phase with mixed glycerol/methanol feed, and the induction phase where methanol is used as sole carbon source [1]. The key point to maximize productivity and to minimize metabolic burden of the bioprocess is the induction phase. Some studies have suggested that the adenylate energy charge of *P. pastoris* is affected by recombinant protein expression under methanol-inducing conditions [99, 180]. In particular, Plantz *et al.* [99] reported that energy charge (EC) dropped from 0.75 (just after the transition from the glycerol batch to methanol induction phase) to 0.6 at the end of the methanol-feeding phase, mainly due to a steady decrease of ATP and concomitant increase of ADP along this phase. This observation correlated with a decrease in the recombinant protein-specific production rate. Since adenylate pools influence protein synthesis and secretion, these studies support the hypothesis that methanol as sole carbon source cannot sustain the carbon and energy demands to sustain both growth and recombinant secreted protein production for extended periods. In contrast, mixed glucose/methanol steady-state chemostat cultures were able to maintain stable recombinant protein secretion levels, correlated with high and steady EC values of about 0.9 [40]. This might explain the improved productivity levels observed in *P. pastoris* fed-batch cultures using methanol/multicarbon source feeding strategies, that is, by using mixed substrates high EC levels may be maintained along all production phase.

### 8.3.3 Expression Strategies and Recombinant Gene Transcriptional Tuning for Stress Minimization

The use of inducible (regulated) expression systems has been pointed as a key tool to minimize the metabolic burden. This allows separating the growth phase (repressed state) from the production phase (induced state). Importantly, HCDC fed-batch processes allow for such cultivation schemes, including the optimization of parameters such as the growth rate, induction time, concentration of inducer, and so on. A generically applicable principle to minimize the stress caused by high recombinant protein expression levels is to reduce the transcriptional rate thus allowing recombinant protein translation and folding to proceed more slowly [59, 77]. For instance, this has been accomplished in *E. coli* by limiting the amount of inducer (e.g., IPTG) supplied to the process or by reducing the temperature [14, 78, 181, 182]. A physiologically tolerable induction level allows for a prolonged product formation period and, therefore, a higher specific concentration of product can be obtained at the end of the process [59]. Under such conditions, stress related to protein misfolding and aggregation during induction phase is minimized.

Reduced growth temperatures have been reported to increase recombinant protein production in various host organisms, including bacteria, yeast and filamentous fungi [38, 182]. Proteomic analyses of the effect of temperature on recombinant *P. pastoris* chemostat cultures revealed that folding stress is generally decreased at lower cultivation temperatures, enabling more efficient heterologous protein secretion [111]. Moreover, this study suggested that reduced demand for protein folding and degradation might also lead to lower energy demand. In addition, implementation of temperature-limited fed-batch cultures of *P. pastoris* allowed to drastically reduce cell death values and protein proteolysis [164, 165, 183–185].

Besides temperature, protein synthesis rates are also modulated by growth rate [12, 95, 118], so careful study of the relationship between recombinant protein production and specific growth rate should be performed.

Overall, the selection of cultivation strategies that allow prolonging the production phase and minimizing the induction of stress responses is a key to obtain high yields of recombinant protein. Since protein folding and secretion are affected by multiple environmental factors, including temperature, pH, osmolarity, and oxidative stress, as well as cultivation parameters such as medium composition, induction time, inducer concentration, and growth rate, such selection may be time consuming. The combination of high-throughput microreactor systems mentioned here combined with online monitoring of stress markers may offer a powerful tool for the fast development of fed-batch cultivation strategies minimizing stress and maximizing recombinant protein production. For instance, green fluorescent protein (GFP) fusion proteins have been used as reporters to select stress-minimizing (i.e., increased protein folding resulting in higher fluorescence signal) fed-batch *E. coli* high-cell-density strategies, including the optimization inducer (IPTG) concentration, induction point, temperature, and growth medium composition [77]. Similarly, an approach based on GFP fused to a stress-responsive promoter has been reported for on-line stress monitoring in *E. coli* that should also enable fast selection of optimal production conditions [59].

## 8.4 Bioreactor Design Considerations to Minimize Shear Stress

Stress effects caused by shear forces due to the mechanical agitation in the reactor could require special attention. This is, in particular, the situation with mammalian cells when cultivated in stirred-tank bioreactors at large scale. These cells are very sensitive to shear stress. This property imposes a number of specific constraints on the design and operation of animal cell bioreactors, as it is well established that shear may result in cell death or nonlethal physiological responses [186–190]. Moreover, in case of anchorage-dependent cells, a support material with large surface-ratio needs to be provided, so these cells are usually fixed on microcarriers. Therefore, the choice of agitation conditions in this type of process has to maintain cells (or microcarriers) in complete suspension and homogenized culture medium, while limiting mechanical constraints generated by the hydrodynamics on the cells. Membrane bioreactors and microencapsulation methods have also been developed for simultaneous cell cultivation, product concentration, and toxic by-product removal. Many of these systems behave as perfusion systems, where cells are retained in the reactor, medium is added continuously or semicontinuously, and spent medium (containing toxic metabolites) is removed [191, 192].

Mammalian cells in a stirred-tank bioreactors are subjected to shear stress or turbulence within the liquid phase, as well as shear forces associated with the gas–liquid interface [51]. Liquid-associated shear stress can result in cell death (necrosis), cell apoptosis, and nonlethal physiological responses [51]. The latter can affect cell functions, physiology, surface content of cellular receptors, recombinant protein production levels, as well as inducing apoptosis. These changes can have an impact on growth rate, productivity, and product quality.

To reduce shear damage to mammalian cells in the bioreactor, low shear impellers have been designed, such as down- or up-pumping wide blade hydrofoils and "elephant ear" impellers (Figure 8.2). Moreover, compared with microbial fermentation processes, the power input usually recommended for cell culture is extremely low [51]. The immobilization in gel beads (alginate, collagen,

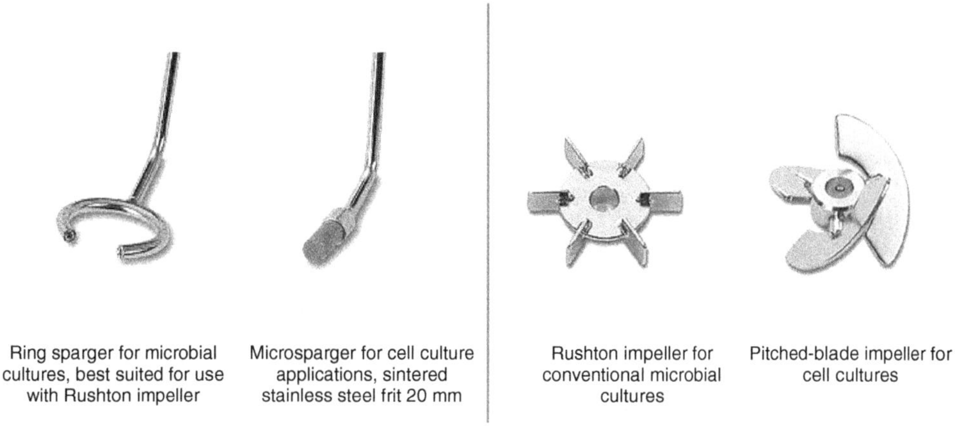


**Figure 8.2** Aeration spargers and impellers commonly used for laboratory scale microbial and cell culture applications.

polyacrylamide), microencapsulation of mammalian cells, and the use of such systems in a packed – or fluidized bed configuration can also minimize shear effects on cells, as well as enhancing productivities [148, 192]. Reactors based on wave-induced agitation with aeration in the headspace, that is, wave bioreactors, have been implemented for different cell systems in reactor volumes up to 200 l. In these reactors, shear stress under aerated conditions has been reported to be insignificant for cell damage.

Cells at the gas–liquid interface are susceptible to damage, as the breakage of air bubbles is particularly destructive to cells that accumulate at the interface of a gas bubble and medium [191]. Many shear protecting agents such as Pluronic F-68 are commonly used to prevent cells from accumulating at the gas–liquid interface [189]. Other additives such as polyethylene glycol (PEG), polyvinyl alcohol (PVA), derivatized celluloses (Methocels), dextrans, and so on can also minimize hydrodynamic damage of cells [51]. When shear protectants are used, sparging of gas is the simplest and acceptable method for gas exchange. Nonetheless, cell lines differ in shear sensitivity. For instance, several industrially relevant cell lines such as CHO cells, widely used for recombinant protein production, are relatively insensitive to cell damage [193].

### Acknowledgments

This work has been supported by the project CTQ2013-42391-R of the Spanish Ministry of Economy and Competitivity (MINECO), and the Catalan Government (Research Group 2014-SGR-452 and Xarxa de Referència en Biotecnologia).

We thank Prof. Carles Solà and Prof. Francesc Gòdia for useful comments and critically reading the manuscript, and we also thank Mr Javier Lobo (Sartorius, Spain) for providing pictures of impellers and spargers for cell culture bioreactors.

## References


1. Cos, O., Ramon, R., Montesinos, J.L., and Valero, F. (2006) Operational strategies, monitoring and control of heterologous protein production in the methylotrophic yeast *Pichia pastoris* under different promoters: a review. *Microb. Cell Fact.*, **5**, 17.
2. Fiechter, A., Fuhrmann, G.F., and Käppeli, O. (1981) Regulation of glucose metabolism in growing yeast cells. *Adv. Microb. Physiol.*, **22**, 123–183.
3. van Hoek, P., van Dijken, J.P., and Pronk, J.T. (1998) Effect of specific growth rate on fermentative capacity of baker's yeast. *Appl. Environ. Microbiol.*, **64**, 4226–4233.
4. Eiteman, M.A. and Altman, E. (2006) Overcoming acetate in *Escherichia coli* recombinant protein fermentations. *Trends Biotechnol.*, **24**, 530–536.
5. Carneiro, S., Ferreira, E.C., and Rocha, I. (2013) Metabolic responses to recombinant bioprocesses in *Escherichia coli*. *J. Biotechnol.*, **164**, 396–408.
6. Hayter, P.M., Curling, E.M., Baines, A.J., Jenkins, N., Salmon, I., Strange, P.G., and Bull, A.T. (1991) Chinese hamster ovary cell growth and interferon production kinetics in stirred batch culture. *Appl. Microbiol. Biotechnol.*, **34**, 559–564.
7. Hassell, T., Gleave, S., and Butler, M. (1991) Growth inhibition in animal cell culture. The effect of lactate and ammonia. *Appl. Biochem. Biotechnol.*, **30**, 29–41.
8. Ozturk, S.S., Riley, M.R., and Palsson, B.O. (1992) Effects of ammonia and lactate on hybridoma growth, metabolism, and antibody production. *Biotechnol. Bioeng.*, **39**, 418–431.
9. Kazemi-Seresht, A., Cruz, A.L., de Hulster, E., Hebly, M., Palmqvist, E.A., van Gulik, W., Daran, J.M., Pronk, J., and Olsson, L. (2013) Long-term adaptation of *Saccharomyces cerevisiae* to the burden of recombinant insulin production. *Biotechnol. Bioeng.*, **110**, 2749–2763.
10. Hensing, M.C., Rouwenhorst, R.J., Heijnen, J.J., van Dijken, J.P., and Pronk, J.T. (1995) Physiological and technological aspects of large-scale heterologous-protein production with yeasts. *Antonie Van Leeuwenhoek*, **67**, 261–279.
11. Maurer, M., Kuhleitner, M., Gasser, B., and Mattanovich, D. (2006) Versatile modeling and optimization of fed batch processes for the production of secreted heterologous proteins with *Pichia pastoris*. *Microb. Cell Fact.*, **5**, 37.
12. Rebnegger, C., Graf, A.B., Valli, M., Steiger, M.G., Gasser, B., Maurer, M., and Mattanovich, D. (2014) In *Pichia pastoris*, growth rate regulates protein synthesis and secretion, mating and stress response. *Biotechnol. J.*, **9**, 511–525.
13. Lee, S.Y. (1996) High cell-density culture of *Escherichia coli*. *Trends Biotechnol.*, **14**, 98–105.
14. Choi, J.H., Keum, K.C., and Lee, S.Y. (2006) Production of recombinant proteins by high cell density culture of *Escherichia coli*. *Chem. Eng. Sci.*, **61**, 876–885.
15. Graf, A., Dragosits, M., Gasser, B., and Mattanovich, D. (2009) Yeast systems biotechnology for the production of heterologous proteins. *FEMS Yeast Res.*, **9**, 335–348.
16. Andersson, L., Strandberg, L., and Enfors, S.O. (1996) Cell segregation and lysis have profound effects on the growth of *Escherichia coli* in high cell density fed batch cultures. *Biotechnol. Progr.*, **12**, 190–195.
17. Hewitt, C.J., Nebe-Von, C.G., Nienow, A.W., and McFarlane, C.M. (1999) Use of multi-staining flow cytometry to characterise the physiological state of *Escherichia coli* W3110 in high cell density fed-batch cultures. *Biotechnol. Bioeng.*, **63**, 705–711.
18. Gill, R.T., DeLisa, M.P., Valdes, J.J., and Bentley, W.E. (2001) Genomic analysis of high-cell-density recombinant *Escherichia coli* fermentation and "cell conditioning" for improved recombinant protein yield. *Biotechnol. Bioeng.*, **72**, 85–95.

19. Huisman, G.W. and Kolter, R. (1994) Sensing starvation: a homoserine lactone-dependent signaling pathway in *Escherichia coli*. *Science*, **265**, 537–539.
20. DeLisa, M.P., Valdes, J.J., and Bentley, W.E. (2001) Quorum signaling via AI-2 communicates the "metabolic burden" associated with heterologous protein production in *Escherichia coli*. *Biotechnol. Bioeng.*, **75**, 439–450.
21. Mattanovich, D., Gasser, B., Hohenblum, H., and Sauer, M. (2004) Stress in recombinant protein producing yeasts. *J. Biotechnol.*, **113**, 121–135.
22. Bentley, W.E., Mirjalili, N., Andersen, D.C., Davis, R.H., and Kompala, D.S. (1990) Plasmid-encoded protein: the principal factor in the "metabolic burden" associated with recombinant bacteria. *Biotechnol. Bioeng.*, **35**, 668–681.
23. Glick, B.R. (1995) Metabolic load and heterologous gene expression. *Biotechnol. Adv.*, **13**, 247–261.
24. Gorgens, J.F., van Zyl, W.H., Knoetze, J.H., and Hahn-Hagerdal, B. (2001) The metabolic burden of the PGK1 and ADH2 promoter systems for heterologous xylanase production by *Saccharomyces cerevisiae* in defined medium. *Biotechnol. Bioeng.*, **73**, 238–245.
25. Vigentini, I., Brambilla, L., Branduardi, P., Merico, A., Porro, D., and Compagno, C. (2005) Heterologous protein production in *Zygosaccharomyces bailii*: physiological effects and fermentative strategies. *FEMS Yeast Res.*, **5**, 647–652.
26. Cos, O., Serrano, A., Montesinos, J.L., Ferrer, P., Cregg, J.M., and Valero, F. (2005) Combined effect of the methanol utilization (Mut) phenotype and gene dosage on recombinant protein production in *Pichia pastoris* fed-batch cultures. *J. Biotechnol.*, **116**, 321–335.
27. Heyland, J., Fu, J., Blank, L.M., and Schmid, A. (2010) Quantitative physiology of *Pichia pastoris* during glucose-limited high-cell density fed-batch cultivation for recombinant protein production. *Biotechnol. Bioeng.*, **107**, 357–368.
28. Sanden, A.M., Prytz, I., Tubulekas, I., Forberg, C., Le, H., Hektor, A., Neubauer, P., Pragai, Z., Harwood, C., Ward, A., Picon, A., De Mattos, J.T., Postma, P., Farewell, A., Nystrom, T., Reeh, S., Pedersen, S., and Larsson, G. (2003) Limiting factors in *Escherichia coli* fed-batch production of recombinant proteins. *Biotechnol. Bioeng.*, **81**, 158–166.
29. Heyland, J., Blank, L.M., and Schmid, A. (2011) Quantification of metabolic limitations during recombinant protein production in *Escherichia coli*. *J. Biotechnol.*, **155**, 178–184.
30. Hoffmann, F. and Rinas, U. (2004) Stress induced by recombinant protein production in *Escherichia coli*. *Adv. Biochem. Eng. Biotechnol.*, **89**, 73–92.
31. Wang, Z., Xiang, L., Shao, J., Wegrzyn, A., and Wegrzyn, G. (2006) Effects of the presence of ColE1 plasmid DNA in *Escherichia coli* on the host cell metabolism. *Microb. Cell Fact.*, **5**, 34.
32. Ow, D.S., Lee, R.M., Nissom, P.M., Philp, R., Oh, S.K., and Yap, M.G. (2007) Inactivating FruR global regulator in plasmid-bearing *Escherichia coli* alters metabolic gene expression and improves growth rate. *J. Biotechnol.*, **131**, 261–269.
33. Jurgen, B., Lin, H.Y., Riemschneider, S., Scharf, C., Neubauer, P., Schmid, R., Hecker, M., and Schweder, T. (2000) Monitoring of genes that respond to overproduction of an insoluble recombinant protein in *Escherichia coli* glucose-limited fed-batch fermentations. *Biotechnol. Bioeng.*, **70**, 217–224.
34. Andersson, L., Yang, S., Neubauer, P., and Enfors, S.O. (1996) Impact of plasmid presence and induction on cellular responses in fed batch cultures of *Escherichia coli*. *J. Biotechnol.*, **46**, 255–263.
35. Durrschmid, K., Reischer, H., Schmidt-Heck, W., Hrebicek, T., Guthke, R., Rizzi, A., and Bayer, K. (2008) Monitoring of transcriptome and proteome profiles to investigate the cellular response of *E. coli* towards

recombinant protein expression under defined chemostat conditions. *J. Biotechnol.*, **135**, 34–44.
36. Aris, A., Corchero, J.L., Benito, A., Carbonell, X., Viaplana, E., and Villaverde, A. (1998) The expression of recombinant genes from bacteriophage lambda strong promoters triggers the SOS response in *Escherichia coli*. *Biotechnol. Bioeng.*, **60**, 551–559.
37. Lesley, S.A., Graziano, J., Cho, C.Y., Knuth, M.W., and Klock, H.E. (2002) Gene expression response to misfolded protein as a screen for soluble recombinant protein. *Protein Eng.*, **15**, 153–160.
38. Gasser, B., Saloheimo, M., Rinas, U., Dragosits, M., Rodriguez-Carmona, E., Baumann, K., Giuliani, M., Parrilli, E., Branduardi, P., Lang, C., Porro, D., Ferrer, P., Tutino, M.L., Mattanovich, D., and Villaverde, A. (2008) Protein folding and conformational stress in microbial cells producing recombinant proteins: a host comparative overview. *Microb. Cell Fact.*, **7**, 11.
39. Heyland, J., Fu, J., Blank, L.M., and Schmid, A. (2011) Carbon metabolism limits recombinant protein production in *Pichia pastoris*. *Biotechnol. Bioeng.*, **108**, 1942–1953.
40. Jordà, J., Rojas, H.C., Carnicer, M., Wahl, A., Ferrer, P., and Albiol, J. (2014) Quantitative metabolomics and instationary 13C-metabolic flux analysis reveals impact of recombinant protein production on trehalose and energy metabolism in *Pichia pastoris*. *Metabolites*, **4**, 281–299.
41. Jordà, J., Jouhten, P., Camara, E., Maaheimo, H., Albiol, J., and Ferrer, P. (2012) Metabolic flux profiling of recombinant protein secreting *Pichia pastoris* growing on glucose:methanol mixtures. *Microb. Cell Fact.*, **11**, 57.
42. Klein, T., Lange, S., Wilhelm, N., Bureik, M., Yang, T.H., Heinzle, E., and Schneider, K. (2014) Overcoming the metabolic burden of protein secretion in *Schizosaccharomyces pombe* – a quantitative approach using 13C-based metabolic flux analysis. *Metab. Eng.*, **21**, 34–45.
43. Ferrer, P. and Albiol, J. (2014) (1)(3)C-based metabolic flux analysis of recombinant *Pichia pastoris*. *Methods Mol. Biol.*, **1191**, 291–313.
44. Schroder, M. and Kaufman, R.J. (2005) The mammalian unfolded protein response. *Annu. Rev. Biochem.*, **74**, 739–789.
45. Pfeffer, M., Maurer, M., Kollensperger, G., Hann, S., Graf, A.B., and Mattanovich, D. (2011) Modeling and measuring intracellular fluxes of secreted recombinant protein in *Pichia pastoris* with a novel 34S labeling procedure. *Microb. Cell Fact.*, **10**, 47.
46. Pakula, T.M., Laxell, M., Huuskonen, A., Uusitalo, J., Saloheimo, M., and Penttila, M. (2003) The effects of drugs inhibiting protein secretion in the filamentous fungus *Trichoderma reesei*. Evidence for down-regulation of genes that encode secreted proteins in the stressed cells. *J. Biol. Chem.*, **278**, 45011–45020.
47. Al-Sheikh, H., Watson, A.J., Lacey, G.A., Punt, P.J., MacKenzie, D.A., Jeenes, D.J., Pakula, T., Penttila, M., Alcocer, M.J., and Archer, D.B. (2004) Endoplasmic reticulum stress leads to the selective transcriptional downregulation of the glucoamylase gene in *Aspergillus niger*. *Mol. Microbiol.*, **53**, 1731–1742.
48. Enfors, S.O., Jahic, M., Rozkov, A., Xu, B., Hecker, M., Jurgen, B., Kruger, E., Schweder, T., Hamer, G., O'Beirne, D., Noisommit-Rizzi, N., Reuss, M., Boone, L., Hewitt, C., McFarlane, C., Nienow, A., Kovacs, T., Tragardh, C., Fuchs, L., Revstedt, J., Friberg, P.C., Hjertager, B., Blomsten, G., Skogman, H., Hjort, S., Hoeks, F., Lin, H.Y., Neubauer, P., van der Lans, R., Luyben, K., Vrabel, P., and Manelius, A. (2001) Physiological responses to mixing in large scale bioreactors. *J. Biotechnol.*, **85**, 175–185.
49. Sandoval-Basurto, E.A., Gosset, G., Bolivar, F., and Ramirez, O.T. (2005) Culture of *Escherichia coli* under dissolved oxygen gradients simulated in a two-compartment scale-down system: metabolic response and production of recombinant protein. *Biotechnol. Bioeng.*, **89**, 453–463.

50. Lara, A.R., Leal, L., Flores, N., Gosset, G., Bolivar, F., and Ramirez, O.T. (2006) Transcriptional and metabolic response of recombinant *Escherichia coli* to spatial dissolved oxygen tension gradients simulated in a scale-down system. *Biotechnol. Bioeng.*, **93**, 372–385.
51. Zhang, H., Wang, W., Quan, C., and Fan, S. (2010) Engineering considerations for process development in mammalian cell cultivation. *Curr. Pharm. Biotechnol.*, **11**, 103–112.
52. Lejeune, A., Delvigne, F., and Thonart, P. (2010) Influence of bioreactor hydraulic characteristics on a *Saccharomyces cerevisiae* fed-batch culture: hydrodynamic modelling and scale-down investigations. *J. Ind. Microbiol. Biotechnol.*, **37**, 225–236.
53. Bylund, F., Collet, E., Enfors, S.O., and Larsson, G. (1998) Substrate gradient formation in the large-scale bioreactor lowers cell yield and increases by-product formation. *Bioprocess. Eng.*, **18**, 171–180.
54. Bylund, F., Castan, A., Mikkola, R., Veide, A., and Larsson, G. (2000) Influence of scale-up on the quality of recombinant human growth hormone. *Biotechnol. Bioeng.*, **69**, 119–128.
55. Junne, S., Klingner, A., Kabisch, J., Schweder, T., and Neubauer, P. (2011) A two-compartment bioreactor system made of commercial parts for bioprocess scale-down studies: impact of oscillations on *Bacillus subtilis* fed-batch cultivations. *Biotechnol. J.*, **6**, 1009–1017.
56. Lorantfy, B., Jazini, M., and Herwig, C. (2013) Investigation of the physiological response to oxygen limited process conditions of *Pichia pastoris* Mut(+) strain using a two-compartment scale-down system. *J. Biosci. Bioeng.*, **116**, 371–379.
57. Restelli, V., Wang, M.D., Huzel, N., Ethier, M., Perreault, H., and Butler, M. (2006) The effect of dissolved oxygen on the production and the glycosylation profile of recombinant human erythropoietin produced from CHO cells. *Biotechnol. Bioeng.*, **94**, 481–494.
58. Serrato, J.A., Hernandez, V., Estrada-Mondaca, S., Palomares, L.A., and Ramirez, O.T. (2007) Differences in the glycosylation profile of a monoclonal antibody produced by hybridomas cultured in serum-supplemented, serum-free or chemically defined media. *Biotechnol. Appl. Biochem.*, **47**, 113–124.
59. Striedner, G. and Bayer, K. (2013) An advanced monitoring platform for rational design of recombinant processes. *Adv. Biochem. Eng. Biotechnol.*, **132**, 65–84.
60. Schweder, T., Kruger, E., Xu, B., Jurgen, B., Blomsten, G., Enfors, S.O., and Hecker, M. (1999) Monitoring of genes that respond to process-related stress in large-scale bioprocesses. *Biotechnol. Bioeng.*, **65**, 151–159.
61. Lin, H., Hoffmann, F., Rozkov, A., Enfors, S.O., Rinas, U., and Neubauer, P. (2004) Change of extracellular cAMP concentration is a sensitive reporter for bacterial fitness in high-cell-density cultures of *Escherichia coli*. *Biotechnol. Bioeng.*, **87**, 602–613.
62. Delvigne, F., Boxus, M., Ingels, S., and Thonart, P. (2009) Bioreactor mixing efficiency modulates the activity of a prpoS::GFP reporter gene in *E. coli*. *Microb. Cell Fact.*, **8**, 15.
63. Delvigne, F., Brognaux, A., Francis, F., Twizere, J.C., Gorret, N., Sorensen, S.J., and Thonart, P. (2011) Green fluorescent protein (GFP) leakage from microbial biosensors provides useful information for the evaluation of the scale-down effect. *Biotechnol. J.*, **6**, 968–978.
64. Reischer, H., Schotola, I., Striedner, G., Potschacher, F., and Bayer, K. (2004) Evaluation of the GFP signal and its aptitude for novel on-line monitoring strategies of recombinant fermentation processes. *J. Biotechnol.*, **108**, 115–125.
65. Brognaux, A., Han, S., Sorensen, S.J., Lebeau, F., Thonart, P., and Delvigne, F. (2013) A low-cost, multiplexable, automated flow cytometry procedure for the characterization of microbial stress dynamics in bioreactors. *Microb. Cell Fact.*, **12**, 100.
66. Mattanovich, D. and Borth, N. (2006) Applications of cell sorting in biotechnology. *Microb. Cell Fact.*, **5**, 12.

67. Tracy, B.P., Gaida, S.M., and Papoutsakis, E.T. (2010) Flow cytometry for bacteria: enabling metabolic engineering, synthetic biology and the elucidation of complex phenotypes. *Curr. Opin. Biotechnol.*, **21**, 85–99.
68. Kuystermans, D., Avesh, M., and Al-Rubeai, M. (2014) Online flow cytometry for monitoring apoptosis in mammalian cell cultures as an application for process analytical technology. *Cytotechnology*.
69. Delic, M., Mattanovich, D., and Gasser, B. (2010) Monitoring intracellular redox conditions in the endoplasmic reticulum of living yeasts. *FEMS Microbiol. Lett.*, **306**, 61–66, DOI: 10.1007/s10616-014-9791-3.
70. Delic, M., Rebnegger, C., Wanka, F., Puxbaum, V., Haberhauer-Troyer, C., Hann, S., Kollensperger, G., Mattanovich, D., and Gasser, B. (2012) Oxidative protein folding and unfolded protein response elicit differing redox regulation in endoplasmic reticulum and cytosol of yeast. *Free Radic. Biol. Med.*, **52**, 2000–2012.
71. Delvigne, F. and Goffin, P. (2014) Microbial heterogeneity affects bioprocess robustness: dynamic single-cell analysis contributes to understanding of microbial populations. *Biotechnol. J.*, **9**, 61–72.
72. Schweder, T. (2011) Bioprocess monitoring by marker gene analysis. *Biotechnol. J.*, **6**, 926–933.
73. Nemecek, S., Marisch, K., Juric, R., and Bayer, K. (2008) Design of transcriptional fusions of stress sensitive promoters and GFP to monitor the overburden of *Escherichia coli* hosts during recombinant protein production. *Bioprocess. Biosyst. Eng.*, **31**, 47–53.
74. Sundstrom, H., Wallberg, F., Ledung, E., Norrman, B., Hewitt, C.J., and Enfors, S.O. (2004) Segregation to non-dividing cells in recombinant *Escherichia coli* fed-batch fermentation processes. *Biotechnol. Lett.*, **26**, 1533–1539.
75. Hohenblum, H., Gasser, B., Maurer, M., Borth, N., and Mattanovich, D. (2004) Effects of gene dosage, promoters, and substrates on unfolded protein stress of recombinant *Pichia pastoris*. *Biotechnol. Bioeng.*, **85**, 367–375.
76. Resina, D., Maurer, M., Cos, O., Arnau, C., Carnicer, M., Marx, H., Gasser, B., Valero, F., Mattanovich, D., and Ferrer, P. (2009) Engineering of bottlenecks in *Rhizopus oryzae* lipase production in *Pichia pastoris* using the nitrogen source-regulated FLD1 promoter. *New Biotechnol.*, **25**, 396–403.
77. Wyre, C. and Overton, T.W. (2014) Use of a stress-minimisation paradigm in high cell density fed-batch *Escherichia coli* fermentations to optimise recombinant protein production. *J. Ind. Microbiol. Biotechnol.*, **41**, 1391–1404.
78. Sevastsyanovich, Y., Alfasi, S., Overton, T., Hall, R., Jones, J., Hewitt, C., and Cole, J. (2009) Exploitation of GFP fusion proteins and stress avoidance as a generic strategy for the production of high-quality recombinant proteins. *FEMS Microbiol. Lett.*, **299**, 86–94.
79. Hewitt, C.J., Nebe-Von, C.G., Axelsson, B., McFarlane, C.M., and Nienow, A.W. (2000) Studies related to the scale-up of high-cell-density *E. coli* fed-batch fermentations using multiparameter flow cytometry: effect of a changing microenvironment with respect to glucose and dissolved oxygen concentration. *Biotechnol. Bioeng.*, **70**, 381–390.
80. Hewitt, C.J. and Nebe-von-Caron, G. (2001) An industrial application of multiparameter flow cytometry: assessment of cell physiological state and its application to the study of microbial fermentations. *Cytometry*, **44**, 179–187.
81. Hewitt, C.J. and Nebe-von-Caron, G. (2004) The application of multiparameter flow cytometry to monitor individual microbial cell physiological state. *Adv. Biochem. Eng. Biotechnol.*, **89**, 197–223.
82. Abu-Absi, N.R., Zamamiri, A., Kacmar, J., Balogh, S.J., and Srienc, F. (2003) Automated flow cytometry for acquisition of time-dependent population data. *Cytometry A*, **51**, 87–96.
83. Sitton, G. and Srienc, F. (2008) Mammalian cell culture scale-up and fed-batch control using automated

flow cytometry. *J. Biotechnol.*, **135**, 174–180.
84. Schweder, T., Lin, H.Y., Jurgen, B., Breitenstein, A., Riemschneider, S., Khalameyzer, V., Gupta, A., Buttner, K., and Neubauer, P. (2002) Role of the general stress response during strong overexpression of a heterologous gene in *Escherichia coli*. *Appl. Microbiol. Biotechnol.*, **58**, 330–337.
85. Neubauer, P., Ahman, M., Tornkvist, M., Larsson, G., and Enfors, S.O. (1995) Response of guanosine tetraphosphate to glucose fluctuations in fed-batch cultivations of *Escherichia coli*. *J. Biotechnol.*, **43**, 195–204.
86. Hoffmann, F., Weber, J., and Rinas, U. (2002) Metabolic adaptation of *Escherichia coli* during temperature-induced recombinant protein production: 1. Readjustment of metabolic enzyme synthesis. *Biotechnol. Bioeng.*, **80**, 313–319.
87. Wittmann, C., Weber, J., Betiku, E., Kromer, J., Bohm, D., and Rinas, U. (2007) Response of fluxome and metabolome to temperature-induced recombinant protein synthesis in *Escherichia coli*. *J. Biotechnol.*, **132**, 375–384.
88. Hohenblum, H., Borth, N., and Mattanovich, D. (2003) Assessing viability and cell-associated product of recombinant protein producing *Pichia pastoris* with flow cytometry. *J. Biotechnol.*, **102**, 281–290.
89. Hewitt, C.J., Nebe-Von, C.G., Nienow, A.W., and McFarlane, C.M. (1999) The use of multi-parameter flow cytometry to compare the physiological response of *Escherichia coli* W3110 to glucose limitation during batch, fed-batch and continuous culture cultivations. *J. Biotechnol.*, **75**, 251–264.
90. David, F., Berger, A., Hansch, R., Rohde, M., and Franco-Lara, E. (2011) Single cell analysis applied to antibody fragment production with *Bacillus megaterium*: development of advanced physiology and bioprocess state estimation tools. *Microb. Cell Fact.*, **10**, 23.
91. Sitton, G. and Srienc, F. (2008) Growth dynamics of mammalian cells monitored with automated cell cycle staining and flow cytometry. *Cytometry A*, **73**, 538–545.
92. Ishaque, A. and Al-Rubeai, M. (1998) Use of intracellular pH and annexin-V flow cytometric assays to monitor apoptosis and its suppression by bcl-2 over-expression in hybridoma cell culture. *J. Immunol. Methods*, **221**, 43–57.
93. Zhu, T., Guo, M., Zhuang, Y., Chu, J., and Zhang, S. (2011) Understanding the effect of foreign gene dosage on the physiology of *Pichia pastoris* by transcriptional analysis of key genes. *Appl. Microbiol. Biotechnol.*, **89**, 1127–1135.
94. Haberhauer-Troyer, C., Delic, M., Gasser, B., Mattanovich, D., Hann, S., and Koellensperger, G. (2013) Accurate quantification of the redox-sensitive GSH/GSSG ratios in the yeast *Pichia pastoris* by HILIC-MS/MS. *Anal. Bioanal. Chem.*, **405**, 2031–2039.
95. Pakula, T.M., Salonen, K., Uusitalo, J., and Penttila, M. (2005) The effect of specific growth rate on protein synthesis and secretion in the filamentous fungus *Trichoderma reesei*. *Microbiology*, **151**, 135–143.
96. Resina, D., Bollok, M., Khatri, N.K., Valero, F., Neubauer, P., and Ferrer, P. (2007) Transcriptional response of *P. pastoris* in fed-batch cultivations to *Rhizopus oryzae* lipase production reveals UPR induction. *Microb. Cell Fact.*, **6**, 21.
97. Gasser, B., Maurer, M., Rautio, J., Sauer, M., Bhattacharyya, A., Saloheimo, M., Penttila, M., and Mattanovich, D. (2007) Monitoring of transcriptional regulation in *Pichia pastoris* under protein production conditions. *BMC Genomics*, **8**, 179.
98. Cserjan-Puschmann, M., Kramer, W., Duerrschmid, E., Striedner, G., and Bayer, K. (1999) Metabolic approaches for the optimisation of recombinant fermentation processes. *Appl. Microbiol. Biotechnol.*, **53**, 43–50.
99. Plantz, B.A., Sinha, J., Villarete, L., Nickerson, K.W., and Schlegel, V.L. (2006) *Pichia pastoris* fermentation

optimization: energy state and testing a growth-associated model. *Appl. Microbiol. Biotechnol.*, **72**, 297–305.

100. Haddadin, F.T. and Harcum, S.W. (2005) Transcriptome profiles for high-cell-density recombinant and wild-type *Escherichia coli*. *Biotechnol. Bioeng.*, **90**, 127–153.
101. Harcum, S.W. and Haddadin, F.T. (2006) Global transcriptome response of recombinant *Escherichia coli* to heat-shock and dual heat-shock recombinant protein induction. *J. Ind. Microbiol. Biotechnol.*, **33**, 801–814.
102. Choi, J.H., Lee, S.J., Lee, S.J., and Lee, S.Y. (2003) Enhanced production of insulin-like growth factor I fusion protein in *Escherichia coli* by coexpression of the down-regulated genes identified by transcriptome profiling. *Appl. Environ. Microbiol.*, **69**, 4737–4742.
103. Oh, M.K. and Liao, J.C. (2000) Gene expression profiling by DNA microarrays and metabolic fluxes in *Escherichia coli*. *Biotechnol. Progr.*, **16**, 278–286.
104. Aldor, I.S., Krawitz, D.C., Forrest, W., Chen, C., Nishihara, J.C., Joly, J.C., and Champion, K.M. (2005) Proteomic profiling of recombinant *Escherichia coli* in high-cell-density fermentations for improved production of an antibody fragment biopharmaceutical. *Appl. Environ. Microbiol.*, **71**, 1717–1728.
105. Wang, Y., Wu, S.L., Hancock, W.S., Trala, R., Kessler, M., Taylor, A.H., Patel, P.S., and Aon, J.C. (2005) Proteomic profiling of *Escherichia coli* proteins under high cell density fed-batch cultivation with overexpression of phosphogluconolactonase. *Biotechnol. Progr.*, **21**, 1401–1411.
106. Furch, T., Wittmann, C., Wang, W., Franco-Lara, E., Jahn, D., and Deckwer, W.D. (2007) Effect of different carbon sources on central metabolic fluxes and the recombinant production of a hydrolase from *Thermobifida fusca* in *Bacillus megaterium*. *J. Biotechnol.*, **132**, 385–394.
107. Ow, D.S., Nissom, P.M., Philp, R., Oh, S.K., and Yap, M.G. (2006) Global transcriptional analysis of metabolic burden due to plasmid maintenance in *Escherichia coli* DH5alpha during batch fermentation. *Enzyme Microb. Technol.*, **39**, 391–398.
108. Jurgen, B., Tobisch, S., Wumpelmann, M., Gordes, D., Koch, A., Thurow, K., Albrecht, D., Hecker, M., and Schweder, T. (2005) Global expression profiling of *Bacillus subtilis* cells during industrial-close fed-batch fermentations with different nitrogen sources. *Biotechnol. Bioeng.*, **92**, 277–298.
109. Hesketh, A.R., Castrillo, J.I., Sawyer, T., Archer, D.B., and Oliver, S.G. (2013) Investigating the physiological response of *Pichia* (Komagataella) *pastoris* GS115 to the heterologous expression of misfolded proteins using chemostat cultures. *Appl. Microbiol. Biotechnol.*, **97**, 9747–9762.
110. Baumann, K., Dato, L., Graf, A.B., Frascotti, G., Dragosits, M., Porro, D., Mattanovich, D., Ferrer, P., and Branduardi, P. (2011) The impact of oxygen on the transcriptome of recombinant *S. cerevisiae* and *P. pastoris* – a comparative analysis. *BMC Genomics*, **12**, 218.
111. Dragosits, M., Stadlmann, J., Albiol, J., Baumann, K., Maurer, M., Gasser, B., Sauer, M., Altmann, F., Ferrer, P., and Mattanovich, D. (2009) The effect of temperature on the proteome of recombinant *Pichia pastoris*. *J. Proteome Res.*, **8**, 1380–1392.
112. Vanz, A.L., Nimtz, M., and Rinas, U. (2014) Decrease of UPR- and ERAD-related proteins in *Pichia pastoris* during methanol-induced secretory insulin precursor production in controlled fed-batch cultures. *Microb. Cell Fact.*, **13**, 23.
113. Vanz, A.L., Lunsdorf, H., Adnan, A., Nimtz, M., Gurramkonda, C., Khanna, N., and Rinas, U. (2012) Physiological response of *Pichia pastoris* GS115 to methanol-induced high level production of the hepatitis B surface antigen: catabolic adaptation, stress responses, and autophagic processes. *Microb. Cell Fact.*, **11**, 103.
114. Dragosits, M., Stadlmann, J., Graf, A., Gasser, B., Maurer, M., Sauer, M., Kreil, D.P., Altmann, F., and Mattanovich, D. (2010) The response to unfolded protein is involved in osmotolerance

of *Pichia pastoris*. *BMC Genomics*, **11**, 207.

115. Baumann, K., Carnicer, M., Dragosits, M., Graf, A.B., Stadlmann, J., Jouhten, P., Maaheimo, H., Gasser, B., Albiol, J., Mattanovich, D., and Ferrer, P. (2010) A multi-level study of recombinant *Pichia pastoris* in different oxygen conditions. *BMC Syst. Biol.*, **4**, 141.
116. Diaz, H., Andrews, B.A., Hayes, A., Castrillo, J., Oliver, S.G., and Asenjo, J.A. (2009) Global gene expression in recombinant and non-recombinant yeast *Saccharomyces cerevisiae* in three different metabolic states. *Biotechnol. Adv.*, **27**, 1092–1117.
117. Sims, A.H., Gent, M.E., Lanthaler, K., Dunn-Coleman, N.S., Oliver, S.G., and Robson, G.D. (2005) Transcriptome analysis of recombinant protein secretion by *Aspergillus nidulans* and the unfolded-protein response in vivo. *Appl. Environ. Microbiol.*, **71**, 2737–2747.
118. Arvas, M., Pakula, T., Smit, B., Rautio, J., Koivistoinen, H., Jouhten, P., Lindfors, E., Wiebe, M., Penttila, M., and Saloheimo, M. (2011) Correlation of gene expression and protein production rate – a system wide study. *BMC Genomics*, **12**, 616.
119. Driouch, H., Melzer, G., and Wittmann, C. (2012) Integration of in vivo and in silico metabolic fluxes for improvement of recombinant protein production. *Metab. Eng.*, **14**, 47–58.
120. Lee, Y.Y., Wong, K.T., Nissom, P.M., Wong, D.C., and Yap, M.G. (2007) Transcriptional profiling of batch and fed-batch protein-free 293-HEK cultures. *Metab. Eng.*, **9**, 52–67.
121. Wong, D.C., Wong, K.T., Lee, Y.Y., Morin, P.N., Heng, C.K., and Yap, M.G. (2006) Transcriptional profiling of apoptotic pathways in batch and fed-batch CHO cell cultures. *Biotechnol. Bioeng.*, **94**, 373–382.
122. Pohl, S., Tu, W.Y., Aldridge, P.D., Gillespie, C., Hahne, H., Mader, U., Read, T.D., and Harwood, C.R. (2011) Combined proteomic and transcriptomic analysis of the response of *Bacillus anthracis* to oxidative stress. *Proteomics*, **11**, 3036–3055.
123. Nicolas, P., Mader, U., Dervyn, E., Rochat, T., Leduc, A., Pigeonneau, N., Bidnenko, E., Marchadier, E., Hoebeke, M., Aymerich, S., Becher, D., Bisicchia, P., Botella, E., Delumeau, O., Doherty, G., Denham, E.L., Fogg, M.J., Fromion, V., Goelzer, A., Hansen, A., Hartig, E., Harwood, C.R., Homuth, G., Jarmer, H., Jules, M., Klipp, E., Le, C.L., Lecointe, F., Lewis, P., Liebermeister, W., March, A., Mars, R.A., Nannapaneni, P., Noone, D., Pohl, S., Rinn, B., Rugheimer, F., Sappa, P.K., Samson, F., Schaffer, M., Schwikowski, B., Steil, L., Stulke, J., Wiegert, T., Devine, K.M., Wilkinson, A.J., van Dijl, J.M., Hecker, M., Volker, U., Bessieres, P., and Noirot, P. (2012) Condition-dependent transcriptome reveals high-level regulatory architecture in *Bacillus subtilis*. *Science*, **335**, 1103–1106.
124. Buescher, J.M., Liebermeister, W., Jules, M., Uhr, M., Muntel, J., Botella, E., Hessling, B., Kleijn, R.J., Le, C.L., Lecointe, F., Mader, U., Nicolas, P., Piersma, S., Rugheimer, F., Becher, D., Bessieres, P., Bidnenko, E., Denham, E.L., Dervyn, E., Devine, K.M., Doherty, G., Drulhe, S., Felicori, L., Fogg, M.J., Goelzer, A., Hansen, A., Harwood, C.R., Hecker, M., Hubner, S., Hultschig, C., Jarmer, H., Klipp, E., Leduc, A., Lewis, P., Molina, F., Noirot, P., Peres, S., Pigeonneau, N., Pohl, S., Rasmussen, S., Rinn, B., Schaffer, M., Schnidder, J., Schwikowski, B., van Dijl, J.M., Veiga, P., Walsh, S., Wilkinson, A.J., Stelling, J., Aymerich, S., and Sauer, U. (2012) Global network reorganization during dynamic adaptations of *Bacillus subtilis* metabolism. *Science*, **335**, 1099–1103.
125. Jurgen, B., Hanschke, R., Sarvas, M., Hecker, M., and Schweder, T. (2001) Proteome and transcriptome based analysis of *Bacillus subtilis* cells overproducing an insoluble heterologous protein. *Appl. Microbiol. Biotechnol.*, **55**, 326–332.
126. Voigt, B., Schroeter, R., Schweder, T., Jurgen, B., Albrecht, D., van Dijl, J.M.,

Maurer, K.H., and Hecker, M. (2014) A proteomic view of cell physiology of the industrial workhorse *Bacillus licheniformis*. *J. Biotechnol.*, **191**, 139–149.

127. Gasser, B., Prielhofer, R., Marx, H., Maurer, M., Nocon, J., Steiger, M., Puxbaum, V., Sauer, M., and Mattanovich, D. (2013) *Pichia pastoris*: protein production host and model organism for biomedical research. *Future Microbiol.*, **8**, 191–208.

128. Jacobs, D.I., Olsthoorn, M.M., Maillet, I., Akeroyd, M., Breestraat, S., Donkers, S., van der Hoeven, R.A., van den Hondel, C.A., Kooistra, R., Lapointe, T., Menke, H., Meulenberg, R., Misset, M., Muller, W.H., van Peij, N.N., Ram, A., Rodriguez, S., Roelofs, M.S., Roubos, J.A., van Tilborg, M.W., Verkleij, A.J., Pel, H.J., Stam, H., and Sagt, C.M. (2009) Effective lead selection for improved protein production in *Aspergillus niger* based on integrated genomics. *Fungal Genet. Biol.*, **46** (Suppl. 1), S141–S152.

129. Griffin, T.J., Seth, G., Xie, H., Bandhakavi, S., and Hu, W.S. (2007) Advancing mammalian cell culture engineering using genome-scale technologies. *Trends Biotechnol.*, **25**, 401–408.

130. Wong, D.C., Wong, K.T., Nissom, P.M., Heng, C.K., and Yap, M.G. (2006) Targeting early apoptotic genes in batch and fed-batch CHO cell cultures. *Biotechnol. Bioeng.*, **95**, 350–361.

131. Kim, J.Y., Kim, Y.G., and Lee, G.M. (2012) CHO cells in biotechnology for production of recombinant proteins: current state and further potential. *Appl. Microbiol. Biotechnol.*, **93**, 917–930.

132. Kildegaard, H.F., Baycin-Hizal, D., Lewis, N.E., and Betenbaugh, M.J. (2013) The emerging CHO systems biology era: harnessing the 'omics revolution for biotechnology. *Curr. Opin. Biotechnol.*, **24**, 1102–1107.

133. van de Walle, M. and Shiloach, J. (1998) Proposed mechanism of acetate accumulation in two recombinant *Escherichia coli* strains during high density fermentation. *Biotechnol. Bioeng.*, **57**, 71–78.

134. Arnold, C.N., McElhanon, J., Lee, A., Leonhart, R., and Siegele, D.A. (2001) Global analysis of *Escherichia coli* gene expression during the acetate-induced acid tolerance response. *J. Bacteriol.*, **183**, 2178–2186.

135. Shiloach, J. and Fass, R. (2005) Growing *E. coli* to high cell density--a historical perspective on method development. *Biotechnol. Adv.*, **23**, 345–357.

136. Mendoza-Vega, O., Sabatie, J., and Brown, S.W. (1994) Industrial production of heterologous proteins by fed-batch cultures of the yeast *Saccharomyces cerevisiae. FEMS Microbiol. Rev.*, **15**, 369–410.

137. Pham, H.T., Larsson, G., and Enfors, S.O. (1998) Growth and energy metabolism in aerobic fed-batch cultures of *Saccharomyces cerevisiae*: simulation and model verification. *Biotechnol. Bioeng.*, **60**, 474–482.

138. Alberghina, L., Porro, D., Martegani, E., and Ranzi, B.M. (1991) Efficient production of recombinant DNA proteins in *Saccharomyces cerevisiae* by controlled high-cell-density fermentation. *Biotechnol. Appl. Biochem.*, **14**, 82–92.

139. Diers, I.V., Rasmussen, E., Larsen, P.H., and Kjaersig, I.L. (1991) Yeast fermentation processes for insulin production. *Bioprocess Technol.*, **13**, 166–176.

140. Baumann, K., Maurer, M., Dragosits, M., Cos, O., Ferrer, P., and Mattanovich, D. (2008) Hypoxic fed-batch cultivation of *Pichia pastoris* increases specific and volumetric productivity of recombinant proteins. *Biotechnol. Bioeng.*, **100**, 177–183.

141. Garcia-Ortega, X., Ferrer, P., Montesinos, J.L., and Valero, F. (2013) Fed-batch operational strategies for recombinant Fab production with *Pichia pastoris* using the constitutive GAP promoter. *Biochem. Eng. J.*, **79**, 172–181.

142. Ljunggren, J. and Haggstrom, L. (1994) Catabolic control of hybridoma cells by glucose and glutamine limited fed

batch cultures. *Biotechnol. Bioeng.*, **44**, 808–818.
143. Zhou, W., Rehm, J., and Hu, W.S. (1995) High viable cell concentration fed-batch cultures of hybridoma cells through on-line nutrient feeding. *Biotechnol. Bioeng.*, **46**, 579–587.
144. Sauer, P.W., Burky, J.E., Wesson, M.C., Sternard, H.D., and Qu, L. (2000) A high-yielding, generic fed-batch cell culture process for production of recombinant antibodies. *Biotechnol. Bioeng.*, **67**, 585–597.
145. Birch, J.R. and Racher, A.J. (2006) Antibody production. *Adv. Drug Deliv. Rev.*, **58**, 671–685.
146. Yang, J.D., Lu, C., Stasny, B., Henley, J., Guinto, W., Gonzalez, C., Gleason, J., Fung, M., Collopy, B., Benjamino, M., Gangi, J., Hanson, M., and Ille, E. (2007) Fed-batch bioreactor process scale-up from 3-L to 2,500-L scale for monoclonal antibody production from cell culture. *Biotechnol. Bioeng.*, **98**, 141–154.
147. Voisard, D., Meuwly, F., Ruffieux, P.A., Baer, G., and Kadouri, A. (2003) Potential of cell retention techniques for large-scale high-density perfusion culture of suspended mammalian cells. *Biotechnol. Bioeng.*, **82**, 751–765.
148. Lecina, M., Tintó, A., Gálvez, J., Gòdia, F., and Cairó, J.J. (2011) Continuous perfusion culture of encapsulated hybridoma cells. *J. Chem. Technol. Biotechnol.*, **86**, 1555–1564.
149. Casablancas, A., Gámez, X., Lecina, M., Solà, C., Cairó, J.J., and Gòdia, F. (2013) Comparison of control strategies for fed-batch culture of hybridoma cells based on on-line monitoring of oxygen uptake rate, optical density and glucose concentration. *J. Chem. Technol. Biotechnol.*, **88**, 1680–1689.
150. Luo, Q., Shen, Y.L., Wei, D.Z., and Cao, W. (2006) Optimization of culture on the overproduction of TRAIL in high-cell-density culture by recombinant *Escherichia coli*. *Appl. Microbiol. Biotechnol.*, **71**, 184–191.
151. Altamirano, C., Paredes, C., Cairo, J.J., and Godia, F. (2000) Improvement of CHO cell culture medium formulation: simultaneous substitution of glucose and glutamine. *Biotechnol. Progr.*, **16**, 69–75.
152. Altamirano, C., Paredes, C., Illanes, A., Cairo, J.J., and Godia, F. (2004) Strategies for fed-batch cultivation of t-PA producing CHO cells: substitution of glucose and glutamine and rational design of culture medium. *J. Biotechnol.*, **110**, 171–179.
153. Funke, M., Buchenauer, A., Mokwa, W., Kluge, S., Hein, L., Muller, C., Kensy, F., and Buchs, J. (2010) Bioprocess control in microscale: scalable fermentations in disposable and user-friendly microfluidic systems. *Microb. Cell Fact.*, **9**, 86.
154. Ukkonen, K., Mayer, S., Vasala, A., and Neubauer, P. (2013) Use of slow glucose feeding as supporting carbon source in lactose autoinduction medium improves the robustness of protein expression at different aeration conditions. *Protein Expr. Purif.*, **91**, 147–154.
155. Siurkus, J., Panula-Perala, J., Horn, U., Kraft, M., Rimseliene, R., and Neubauer, P. (2010) Novel approach of high cell density recombinant bioprocess development: optimisation and scale-up from microliter to pilot scales while maintaining the fed-batch cultivation mode of *E. coli* cultures. *Microb. Cell Fact.*, **9**, 35.
156. Hemmerich, J., Adelantado, N., Barrigon, J.M., Ponte, X., Hormann, A., Ferrer, P., Kensy, F., and Valero, F. (2014) Comprehensive clone screening and evaluation of fed-batch strategies in a microbioreactor and lab scale stirred tank bioreactor system: application on *Pichia pastoris* producing *Rhizopus oryzae* lipase. *Microb. Cell Fact.*, **13**, 36.
157. Valero, F. (2013) Bioprocess engineering of *Pichia pastoris*, an exciting host eukaryotic cell expression system, in *Protein Engineering – Technology and Application*, 1st edn (ed. T. Ogawa), InTech, Croatia Rijeka, pp. 3–32.
158. Xiao, A., Zhou, X., Zhou, L., and Zhang, Y. (2006) Improvement of cell viability and hirudin production by

ascorbic acid in *Pichia pastoris* fermentation. *Appl. Microbiol. Biotechnol.*, **72**, 837–844.

159. Potvin, G., Ahmad, A., and Zhang, Z. (2012) Bioprocess engineering aspects of heterologous protein production in *Pichia pastoris*: a review. *Biochem. Eng. J.*, **64**, 91–105.

160. Khatri, N.K. and Hoffmann, F. (2006) Impact of methanol concentration on secreted protein production in oxygen-limited cultures of recombinant *Pichia pastoris*. *Biotechnol. Bioeng.*, **93**, 871–879.

161. Ye, J., Ly, J., Watts, K., Hsu, A., Walker, A., McLaughlin, K., Berdichevsky, M., Prinz, B., Sean, K.D., d'Anjou, M., Pollard, D., and Potgieter, T. (2011) Optimization of a glycoengineered *Pichia pastoris* cultivation process for commercial antibody production. *Biotechnol. Progr.*, **27**, 1744–1750.

162. Potgieter, T.I., Kersey, S.D., Mallem, M.R., Nylen, A.C., and d'Anjou, M. (2010) Antibody expression kinetics in glycoengineered *Pichia pastoris*. *Biotechnol. Bioeng.*, **106**, 918–927.

163. Berdichevsky, M., d'Anjou, M., Mallem, M.R., Shaikh, S.S., and Potgieter, T.I. (2011) Improved production of monoclonal antibodies through oxygen-limited cultivation of glycoengineered yeast. *J. Biotechnol.*, **155**, 217–224.

164. Charoenrat, T., Ketudat-Cairns, M., Stendahl-Andersen, H., Jahic, M., and Enfors, S.O. (2005) Oxygen-limited fed-batch process: an alternative control for *Pichia pastoris* recombinant protein processes. *Bioprocess. Biosyst. Eng.*, **27**, 399–406.

165. Surribas, A., Stahn, R., Montesinos, J.L., Enfors, S.O., Valero, F., and Jahic, M. (2007) Production of a *Rhizopus oryzae* lipase from *Pichia pastoris* using alternative operational strategies. *J. Biotechnol.*, **130**, 291–299.

166. Dabros, M., Schuler, M.M., and Marison, I.W. (2010) Simple control of specific growth rate in biotechnological fed-batch processes based on enhanced online measurements of biomass. *Bioprocess. Biosyst. Eng.*, **33**, 1109–1118.

167. Barrigón, J.M., Montesinos, J.L., and Valero, F. (2013) Searching the best operational strategies for *Rhizopus oryzae* lipase production in *Pichia pastoris* Mut+ phenotype: methanol limited or methanol non-limited fed-batch cultures? *Biochem. Eng. J.*, **75**, 47–54.

168. Yamawaki, S., Matsumoto, Y., Ohnishi, Y., Kumada, N., Shiomi, T., Katsuda, E.K., Lee, S., and Katoh, S. (2007) Production of single-chain variable fragment antibody (scFv) in fed-batch and continuous cultures of *Pichia pastoris* by two different methanol feeding methods. *J. Biosci. Bioeng.*, **104**, 403–407.

169. Barrigón, J.M., Ramon, R., Rocha, I., Valero, F., Ferreira, E.C., and Montesinos, J.L. (2012) State and specific growth estimation in heterologous protein production by *Pichia pastoris*. *AIChE J.*, **58**, 2967–2979.

170. Kaleta, C., Schauble, S., Rinas, U., and Schuster, S. (2013) Metabolic costs of amino acid and protein production in *Escherichia coli*. *Biotechnol. J.*, **8**, 1105–1114.

171. Raiford, D.W., Heizer, E.M. Jr.,, Miller, R.V., Akashi, H., Raymer, M.L., and Krane, D.E. (2008) Do amino acid biosynthetic costs constrain protein evolution in *Saccharomyces cerevisiae*? *J. Mol. Evol.*, **67**, 621–630.

172. Harcum, S.W., Ramirez, D.M., and Bentley, W.E. (1992) Optimal nutrient feed policies for heterologous protein production. *Appl. Biochem. Biotechnol.*, **34–35**, 161–173.

173. Gorgens, J.F., van Zyl, W.H., Knoetze, J.H., and Hahn-Hagerdal, B. (2005) Amino acid supplementation improves heterologous protein production by *Saccharomyces cerevisiae* in defined medium. *Appl. Microbiol. Biotechnol.*, **67**, 684–691.

174. Gorgens, J.F., Passoth, V., van Zyl, W.H., Knoetze, J.H., and Hahn-Hagerdal, B. (2005) Amino acid supplementation, controlled oxygen limitation and sequential double induction improves heterologous xylanase production by *Pichia stipitis*. *FEMS Yeast Res.*, **5**, 677–683.

175. van , R.E., den HR, Smith, J., van Zyl, W.H., and Gorgens, J.F. (2012) The metabolic burden of cellulase expression by recombinant *Saccharomyces cerevisiae* Y294 in aerobic batch culture. *Appl. Microbiol. Biotechnol.*, **96**, 197–209.
176. Hahn-Hagerdal, B., Karhumaa, K., Larsson, C.U., Gorwa-Grauslund, M., Gorgens, J., and van Zyl, W.H. (2005) Role of cultivation media in the development of yeast strains for large scale industrial use. *Microb. Cell Fact.*, **4**, 31.
177. Babel, W., Brinkmann, U., and Müller, R.H. (1993) The auxiliary substrate concept – an approach for overcoming limits of microbial performances. *Acta Microbiol.*, **13**, 211–242.
178. Ramon, R., Ferrer, P., and Valero, F. (2007) Sorbitol co-feeding reduces metabolic burden caused by the overexpression of a *Rhizopus oryzae* lipase in *Pichia pastoris*. *J. Biotechnol.*, **130**, 39–46.
179. Jungo, C., Schenk, J., Pasquier, M., Marison, I.W., and von Stockar, U. (2007) A quantitative analysis of the benefits of mixed feeds of sorbitol and methanol for the production of recombinant avidin with *Pichia pastoris*. *J. Biotechnol.*, **131**, 57–66.
180. Katakura, Y., Zhang, W., Guoqiang, O., Takeshi, K., Kishimoto, M., Goto, Y., and Suga, K.-I. (1998) Effect of methanol concentration on the production of $\beta$2-glycoprotein I domain V by a recombinant *Pichia pastoris*: a simple system for the control of methanol concentration using a semiconductor gas sensor. *J. Ferment. Bioeng.*, **86**, 482–487.
181. Striedner, G., Cserjan-Puschmann, M., Potschacher, F., and Bayer, K. (2003) Tuning the transcription rate of recombinant protein in strong *Escherichia coli* expression systems through repressor titration. *Biotechnol. Progr.*, **19**, 1427–1432.
182. Dragosits, M., Frascotti, G., Bernard-Granger, L., Vazquez, F., Giuliani, M., Baumann, K., Rodriguez-Carmona, E., Tokkanen, J., Parrilli, E., Wiebe, M.G., Kunert, R., Maurer, M., Gasser, B., Sauer, M., Branduardi, P., Pakula, T., Saloheimo, M., Penttila, M., Ferrer, P., Luisa, T.M., Villaverde, A., Porro, D., and Mattanovich, D. (2011) Influence of growth temperature on the production of antibody Fab fragments in different microbes: a host comparative analysis. *Biotechnol. Progr.*, **27**, 38–46.
183. Svensson, M., Svensson, I., and Enfors, S.O. (2005) Osmotic stability of the cell membrane of *Escherichia coli* from a temperature-limited fed-batch process. *Appl. Microbiol. Biotechnol.*, **67**, 345–350.
184. Svensson, M., Han, L., Silfversparre, G., Haggstrom, L., and Enfors, S.O. (2005) Control of endotoxin release in *Escherichia coli* fed-batch cultures. *Bioprocess. Biosyst. Eng.*, **27**, 91–97.
185. Jahic, M., Wallberg, F., Bollok, M., Garcia, P., and Enfors, S.O. (2003) Temperature limited fed-batch technique for control of proteolysis in *Pichia pastoris* bioreactor cultures. *Microb. Cell Fact.*, **2**, 6.
186. Cherry, R.S. and Papoutsakis, E.T. (1988) Physical mechanisms of cell damage in microcarrier cell culture bioreactors. *Biotechnol. Bioeng.*, **32**, 1001–1014.
187. Chisti, Y. (2000) Animal-cell damage in sparged bioreactors. *Trends Biotechnol.*, **18**, 420–432.
188. van der Pol, L. and Tramper, J. (1998) Shear sensitivity of animal cells from a culture-medium perspective. *Trends Biotechnol.*, **16**, 323–328.
189. Nienow, A.W. (2006) Reactor engineering in large scale animal cell culture. *Cytotechnology*, **50**, 9–33.
190. Papoutsakis, E.T. (1991) Media additives for protecting freely suspended animal cells against agitation and aeration damage. *Trends Biotechnol.*, **9**, 316–324.
191. Shuler, M.L. and Kargi, F. (2002) *Bioprocess Engineering*, 2nd edn, Prentice Hall, Upper Saddle River, NJ.
192. Chu, L. and Robinson, D.K. (2001) Industrial choices for protein production by large-scale cell culture. *Curr. Opin. Biotechnol.*, **12**, 180–187.

193. Nienow, A.W., Langheinrich, C., Stevenson, N.C., Emery, A.N., Clayton, T.M., and Slater, N.K. (1996) Homogenisation and oxygen transfer rates in large agitated and sparged animal cell bioreactors: some implications for growth and production. *Cytotechnology*, **22**, 87–94.

# 9
# Design, Applications, and Development of Single-Use Bioreactors

*Nico M.G. Oosterhuis and Stefan Junne*

## 9.1
## Introduction

During the last decade, single-use bioreactors (SUBs) have successfully been introduced in biopharmaceutical production processes and process development [1]. A SUB is made of a plastic bag, which is mounted on a shaking platform or is surrounded with a housing in order to stabilize its shape when operated under high pressure. The two parts as a whole serve as a bioreactor, which more or less keeps the same features as traditional stirred-tank bioreactors, although limitations in the achievable oxygen mass transfer rates and cultivation sizes exist. In order to achieve a certain stability of the plastic bags, special types of high-density polyethylene films[1)] are usually applied. All connections are made with flexible tubing; disposable filters ensure sterility and sensors are disposable in most cases as well. The bags are sterilized by $\gamma$-radiation. Hence, they are already presterilized at the site where the cultivation is performed. The bags can just be emptied after their utilization before the downstream processing and are thrown away. A SUB is classified based on the mixing technology employed, for example, (i) SUBs fixed in tank-liners (with various constructions for the stirrer device) are called "stirred," (ii) SUBs fixed on a longitudinal shaker are called "wave-mixed," and (iii) SUBs connected to an orbital shaker are called "orbital-shaken."

There are several advantages of SUBs, especially relevant at biopharmaceutical production and the corresponding certification and validation procedures, which reduce the administrative effort at the production site. These are mainly the elimination of cleaning requirements, fast turnaround, flexibility, and reduced risks of (cross-)contamination. In addition, since no *in situ* steam sterilization is needed, lower investment and operational costs are required. This allows a faster

1) Mostly coextrusion films, which consist of a backing layer to strengthen the film and to avoid diffusion of oxygen and carbon dioxide are applied.

*Bioreactors: Design, Operation and Novel Applications,* First Edition. Edited by Carl-Fredrik Mandenius.

increase in capacity without massive investments. Thus, the restriction in size (the maximum volume of SUBs is still limited to a very few cubic meters) can be counteracted by parallelization, although feasibility and economic viability have to be investigated case-by-case. Negative factors, which were observed during the early years of the application of SUBs, such as the risk of breakage of plastic materials and the migration of leachables and extractables, have become less relevant due to the improvement of materials and design of the equipment. However, sometimes these problems still can occur and are objects of further research and development.

Industry leaders have been interviewed to indicate the degree (percentage) to which the processes (upstream part) rely on the application of single-use technology in a study performed by Langer [1]. Although in commercial production still stainless steel equipment is widely used (based on the prior installed and thus already existing capacity before the introduction of single-use technology), SUBs are clearly dominant in processes for the production of clinical material.

Since the introduction of the "wave" technology in the 1990s [2], several types of bioreactors have been introduced [3, 4]. Although, in first instance, only rocking-type bioreactors were used, stirred systems, including a disposable shaft and stirrer blades, have become more and more important. Till date, stirred systems are available from sizes of 20 to 2000 l, and even larger systems have been introduced as beta systems at some selected users. Rocking-type systems are available in sizes with working volumes of <1 l up to a few hundreds of liters.

Alternative bioreactor designs are applied as well, such as orbital-shaken systems [5] and hybrid systems, where air bubbles are used to achieve mixing [6].

Besides, especially for laboratory application, also polycarbonate vessels are available with working volumes of 100 ml up to 15 l. Recently, the application of single-use material in the microliter and low-milliliter range is becoming more important as bioreactors. The highest degree of parallelization is obtained in a microwell plate. However, it was hardly able to interpret this as a bioreactor in the first versions introduced to the market. The purpose was also not the application of microbial cultivation. However, since then, new developments opened possibilities to apply such single-use technologies for the small-scale cultivation [7].

Most SUBs introduced so far are only applied for processes based on mammalian cell culture. As (aerobic) microbial processes usually require at least a tenfold higher power-input for mixing and mass transfer between the liquid and the gas phase than mammalian processes, the present SUBs are not suited to create such conditions. Approximately 40% of all biopharmaceutical products that are in Phase III of clinical development are based on microbial processes. This concerns a wide variety of products such as therapeutic and diagnostic proteins, small-molecular-weight therapeutic proteins, human and veterinary vaccines, and numerous other products. Hence, there is also a strong demand for single-use equipment, which is applicable to microbial processes, as the advantages of using SUBs also account for these kinds of processes.

When designing a SUB, there are some major hurdles to overcome, which are as follows:

1) Mixing and mass transfer should be sufficient to support the growth of cells or microorganisms at comparable level as in traditional bioreactor equipment.
2) Reactors (and processes) have to be scalable.
3) In order to control the process, preferably single-use sensors have to be applied for a variety of parameters. These sensors need to possess characteristics similar to those for traditional bioreactors.
4) All materials need to be resistant to $\gamma$-radiation.
5) Costs of materials should be low enough so that it is financially acceptable to dispose the bags.

Especially, mixing and gas mass transfer are critical for the design of SUBs and their application for microbial application. Table 9.1 provides an overview of different types of currently available SUBs and the volumetric oxygen mass transfer coefficient ($k_L a$-value) achieved in the various systems.

## 9.2 Design Challenges of Single-Use Bioreactors

### 9.2.1 Material Choice and Testing

The main component of an SUB is the disposable bag. Mostly produced from special types of polyethylene films, these bags serve as the real reaction chamber. The material is in contact with the cells, the media, and the product.

Modern bag types are produced with multilayer films, which are mostly coextruded using different extrusion technologies (blown film, cast film, lamination, etc.). The contact layer is in most cases a polyethylene-type material (low-density, LDPE; linear low-density, LLDPE; high-density, HDPE; medium-density, MDPE or ultralow-density, ULDPE). Also, blends of these PE materials or EVA (copolymer of ethylene and vinyl acetate) are applied as contact layer material. The product layer should possess favorable features concerning extractables and leachables (discussed later). As outer layer, materials with a high thermostability (for welding) and good mechanical properties (puncture resistance, dart drop, tensile strength) are applied, including PA (Nylon6) and PET. This material shows good barrier properties against oxygen, water vapor, flavor, odor, and solvents, respectively.

Several mechanical tests are usually performed to ensure proper functionality of the bag film. These tests include the investigation of crack resistance, tensile strength, puncture resistance, dart impact tests, and tear resistance, among others. All these tests are described in uniform and standardized testing protocols (ASTM, ISO). In addition, physical investigations such as $O_2$, $CO_2$, and water

**Table 9.1** Overview of currently available single-use bioreactors and their specific working volumes, mixing mechanisms, and oxygen gas mass transfer characteristics.

| Reactor name | Volume (l) | Bagtype | Mixing | Distributor | $k_L a$ ($h^{-1}$) | References |
|---|---|---|---|---|---|---|
| WAVE bioreactors | 1–500 | Pillow | Rocking | GE Healthcare Life Sciences | <10 | [8] |
| BIOSTAT® CultiBag RM | 1–100 | Pillow | Rocking | Sartorius Stedim Biotech | <22 | [9] |
| XRS20 bioreactor | 2–20 | Pillow | Biaxial rocking | Pall | 73 | www.pall.com |
| Appliflex | 1–25 | Pillow | Rocking | Applikon Biotechnology | <24 | [8, 10] |
| CELL-tainer® – 20 | 0.2–25 | Pillow or square | 2D rocking | Celltainer Biotech | 550 | [11] |
| CELL-tainer® – 200 | 5–200 | Pillow or square | 2D rocking | Celltainer Biotech | 400 | [11] |
| HyPerforma Single-Use Fermentor (S.U.F.) | 30–300 | Tankliner | Stirred | Thermo-Fischer | n.a. | www.thermoscientific.com |
| BIOSTAT® Cultibag STR200 | 50 1000 | Tankliner | Stirred | Sartorius Stedim Biotech | <175 | [12, 13] |
| HyClone™ – Single-use bioreactor | 50 1000 | Tankliner | Stirred | Thermo-Fischer (Hyclone) | <15 | [14] |
| XDR™ Single-use bioreactor | 40–2000 | Tankliner | Stirred | GE Healthcare Life Sciences/Xcellerex™ | <80 | [8], www.gelifesciences.com |

| | | | | | | |
|---|---|---|---|---|---|---|
| Mobius® CellReady | 50–200 | Tankliner | Stirred | Merck Millipore | <60 | [15] |
| Allegro™ STR200 | 20–200 | Square 3D | Stirred | Pall | <35 | www.pall.com |
| BaySHAKE® | 30 | Square | Oscillating | Bayer Healthcare | 20 | [8] |
| Nucleo™ Single-use bioreactor | 50–100 | Square 3D | Paddle | ATMI/Pall | <20 | www.pall.com |
| SBX reactor | 200 | Tankliner | Orbital shaker | Kuhner/ExcellGene | <30 | [16] |
| CellMaker Regular™ | 1–50 | Bubble column | Rotating sparger | Cellexus | n.a. | www.cellexus.com |
| Air-wheel bioreactor | 3–500 | Special shape | Rotating wheel | PBS Biotech | <20 | [8] |

**Table 9.2** Overview of required test procedures of film material for the application as bag in single-use bioreactors.

| | | |
|---|---|---|
| Water for injection (WFI) | pH value Conductivity TOC Weight loss analysis | Volatile extractables Semivolatile extractables Nonvolatile extractables |
| WFI + HCl (pH < 2) | pH value Conductivity TOC Weight loss analysis | Volatile extractables Semivolatile extractables Nonvolatile extractables Metal screening |
| WFI + NaOH (pH < 9) | pH value Conductivity TOC Weight loss analysis | Volatile extractables Semivolatile extractables Nonvolatile extractables |
| Pure ethanol | Weight loss analysis | Volatile extractables Semivolatile extractables Nonvolatile extractables |

vapor resistance tests are performed based on suitable and standardized testing methods [17].

For biopharmaceuticals, the absence of potentially toxic materials is crucial. Regulatory institutions such as the EMA (Europe) and the US FDA (USA) have published guidelines. It is required that production equipment "should not present any hazards to the product" and materials should not be "reactive, additive, or absorptive" [18]. In industries, it is a common practice that the vendor of the bags or equipment performs evaluations and validations to confirm the absence of extractables[2)] and leachables[3)] to ensure product safety. Although the level of leachables, which originate from the bioreactor, detected in the final drug product is rather low, and the product is usually intensively purified, the industry is aware of possible problems and does not take any risk in this matter.

In order to test the materials properly, several test procedures have been developed, including storage of solutions of low, neutral, and high pH-value as well as pure ethanol. Solvent extraction methods and gas chromatography (GC/MS) are applied to identify volatile and semivolatile crucial components. Nonvolatile components are usually identified by certified liquid chromatography procedures [19]. Testing of the materials is within the vendors' responsibility and requires proper analytical skills. Table 9.2 provides an overview of required test procedures of film material [19].

It is important to note that all tests have to be performed after $\gamma$-radiation at levels of >40 kGy (preferably up to 50 kGy) and also after aging of the bags (a minimum of 4 months).

2) Chemical compounds that migrate from any product contact material under exaggerated conditions of time and temperature.
3) Chemical compounds, typically a subset of extractables, that migrate into the drug formulation from any product contact material as a result of direct contact under normal process conditions or accelerated storage conditions and found in the final drug product.

In order to save validation costs and to minimize risks, the industry prefers the application of one single type of film material throughout the whole process, which includes media preparation and storage bags, buffer bags, and (intermediate) drug product bags.

Before the release of single-use bags for regular deliveries by the vendor, bags are subjected to testing as well. Tests such as sealing strength, visual check, particle testing, and integrity test (overpressure test) are performed for each batch of bags according to uniform methods (ASTM, USP, ISO). Bioburden and bacterial endotoxin tests are performed periodically. During $\gamma$-radiation of the bags, dose monitoring is performed in order to ensure that all bags have been exposed to a minimum level of radiation (>25 kGy) and are not exposed to a very high level of radiation (<45 kGy).

### 9.2.2 Sterilization

Although sterilization by ethylene oxide is applied in the medical industry, sterilization by $\gamma$-radiation is the method of choice in the biopharma industry. A major advantage is that the sterilization does not leave any remainders. As long as a minimum dosage is guaranteed (>25 kGy), complete sterility is ensured. To lower the bioburden level as well as particle level before sterilization, bags are manufactured in class 10 000 clean rooms. The combination of the dosage level and the maximum bioburden level is widely accepted and has been validated.

A main disadvantage of ionizing radiation methods ($\gamma$-, $\beta$-radiation) is that all materials have to be $\gamma$-resistant. This constraint limits the choice of materials that can be applied for sterile processing. Polyethylene, which is commonly used as film material and as material to construct connection pillars among others, can oxidize due to $\gamma$-radiation in the presence of oxygen during sterilization. Materials such as Evatane® (copolymerized ethylene–vinyl acetate) or Engage® (a polyolefin elastomer) possess better $\gamma$-resistance features.

### 9.2.3 Sensors and Sampling

$\gamma$-Radiation can strongly affect sensor properties, especially those of material with optical features. Although several suppliers provide pH value and DO sensors based on optical sensing technologies, still limitations and restrictions exist for their application [20, 21]. The validity of pH-value measurements of optical sensors is mostly limited to a range between 6.5 and 8.5, which is sufficient for most cell culture processes. However, for microbial processes, this range can be considered as being very narrow. Apart from the limited applicable range, also the nonlinearity of the calibration curve and its shift after $\gamma$-radiation can create limitations of the applications of these sensors. Developments are ongoing in applying rather traditional, electrochemical pH sensors [22] or modified, dry electrochemical pH sensors, which can resist $\gamma$-radiation and long-term dry storage (Gymetrics SA, Lausanne). Several optical or enzymatical systems are integrated into shake-flasks and other single-use devices, such as the PreSens technology (PreSens Precision Sensing GmbH, Regensburg, Germany) and the C-CIT sensors (C-CIT, Wädenswil, Switzerland) [23].

Besides standard parameters, sensors for impedance measurements were also applied, which determine the capacitance of cells. The cell vitality is a suitable parameter to estimate growth and productivity. The devices can be coupled with flow-through cells. The electrodes can also be attached to the bag surface. Noninvasive radiofrequency measurements were applied to estimate temperature and, more importantly, conductivity [24]. Nacke *et al.* described the application of a microwave spectroscopy sensor suitable for continuous monitoring of the fermentation media in SUBs. Between frequencies of 300 and 10 GHz, the permittivity and conductivity were determined noninvasively by the integration of a dielectric window as mechanical port [25].

A suitable position of the sensors is essential in shaken bioreactors. One solution is to introduce the sensing element via silicone tubing from the top of the bioreactor bag. However, the sensor may fall dry or float at the top of the fluid when mounted in such a way, which will interrupt the measuring signal. Another solution is to locate a sensor at the bottom of the bag. In order to avoid the sensor falling dry when the liquid floats to the side of the bag, the sensor can be located in a small measurement "cup," which is mounted slightly below the bottom plastic film of the bag. Fluid stays in the cup during the movement of the bag, even at very low operating volumes. The liquid in the cup is refreshed during the movement cycle [26].

A so-called "Kleenpak® Sterile Connector" is used for many stirred SUBs. A traditional sensor can be mounted into the bioreactor under sterile conditions. However, the sensors need to be autoclaved in advance. An advantage of using such a connector in combination with traditional bioreactor sensors is that established and validated transmitters and control systems can be used as well. A major disadvantage is that the sensor, in this case, is not made for single-use and validation procedures have to be performed in place.

For sampling, many different systems are used, including sample connectors, where a sterile syringe can be connected to fully prepared, sterile sampling bags [27]. Mechanisms and tools have been developed in order to carry out cell-free sampling by using membrane systems (Figure 9.1), which are applicable in sterilizable bioreactors as well [28]. When applying such methods, an enzymatic detection of glucose and lactate becomes possible, as well. Hence, the two compounds, which are of greatest interest during a cell culture process, can be monitored *in line* at SUBs.

### 9.2.4
### Challenges for Scale-Up and Scale-Down of Single-Use Bioreactors

Bioprocess development, either in biopharmaceutical or other bioprocesses, demands for an acceleration of the scale-up process. In addition, due to numerous growing technologies that are leading to a broad range of feasible processes with

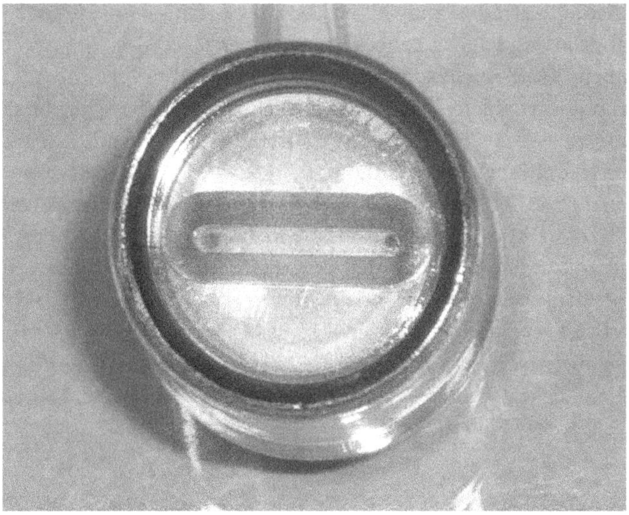

**Figure 9.1** Membrane device for cell-free sampling.

various cell lines and microbial strains, the variety of different applications and cultivation conditions especially in research and initial process development steps is rather increasing than declining. Based on this fact, two necessities arise: (i) first, the scalability of bioreactors is of significant interest in order to ensure similar conditions already in the lab-scale in comparison to the large scale [8], and (ii) the implementation of various disposable systems, which can fulfill the needs of multiple cultivation purposes and an integration into an end-to-end single-use manufacturing [29]. Thereby, an end-to-end manufacturing means that single-use systems are applied throughout all process steps, thus creating an additional surplus, since the rather simple construction platforms of SUBs allow a tailored design much easier than traditional stirred-tank systems. An example of this is the easy addition of light sources that turns a common SUB into a photobioreactor. Another is the lack of several requirements for the installed infrastructure, for example, steam is not needed, which allows the fast installation of SUBs if required. One major drawback that still restricts the application of SUBs is the limited sensor capacity. Many sensors, which have been designed for conventional stirred-tank reactors, cannot be mounted to SUBs. There might be many reasons for nonreusability, including the measurement device itself or the costs of it, the weight, the dimensions or simply the weakness against $\gamma$-radiation. However, a suitable equipment of sensors is necessary in the context of the US FDA initiative to facilitate the integration of process analytical tools (PATs) and in parallel to ensure quality by design at the level of process development and optimization [30]. Besides suitable sensors, suitable ports have to be integrated, including those for *in line* measurement solutions that enable a flexible connection of a variety of sensors similar to those in traditional stirred-tank bioreactors.

With respect to these constraints and the increasing effort to cultivate under most similar conditions in any scale of process development (see also Chapters 2 and 10), the scale-up and scale-down suitability is important. However, there are at least three groups of disposable reactor concepts: (i) stirred SUBs, (ii) orbital-shaken SUBs, and (iii) wave-mixed SUBs. Each group has specific advantages and disadvantages with respect to scalability. Although the stirred SUBs can be scaled similar to conventional reactors, this is a more challenging task for wave-mixed and orbital-shaken reactors due to the lack of knowledge. The shaken reactors, however, can be operated much more flexibly with regard to the filling volume due to the absence of stirrers. Orbital-shaken systems can be scaled-down to the milliliter – or theoretically even to the microliter scale, if the knowledge basis for such a scale-down was increased. This scalability to the microliter scale is certainly more difficult for stirred or wave-mixed systems. An overview of available scales for each reactor system is provided in Figure 9.2.

#### 9.2.4.1
#### Scalability of Stirred Single-Use Bioreactors

Due to the similarity to the well-established traditional stirred-tank reactors, power input and mixing numbers, their influence during scale-up, and their determination are quite established. However, the scalability is restricted, since stirring at a very small scale is not feasible in many systems on the market. The application in the large scale is restricted to a few cubic meters, since mechanical stability and, thus, considerable stirring rates cannot be obtained at a larger scale.

A typical and traditional method for scale-up is the maintenance of the order of magnitude of key parameters such as the height-to-diameter ratio, volumetric power input, and the related mixing time and gas mass transfer coefficient. The derivation of these parameters is well known for traditional stirred-tank reactors. The same approaches are valid for similar single-use stirred bioreactors. The dimensionless power input in stirred systems can be derived as Newton number:

$$Ne = \rho_L n^3 d^5 \tag{9.1}$$

The volumetric power input in stirred bioreactors can be estimated based on the torque measurement at the stirrer shaft:

$$\frac{P}{V_L} = \frac{(M - M_0)u}{V} \tag{9.2}$$

The turbulent flow regime is achieved at Reynolds numbers approximately above 1000. The Reynolds number for stirred SUBs is the same as that valid for conventional stirred-tank reactors:

$$Re = \frac{nd}{\nu} \tag{9.3}$$

The $k_L a$ value is considered to be the most important parameter for microbial aerobic applications. The volumetric oxygen demand of a cultivation (oxygen uptake rate = OUR) has to be satisfied by the cultivation system and the $k_L a$ value

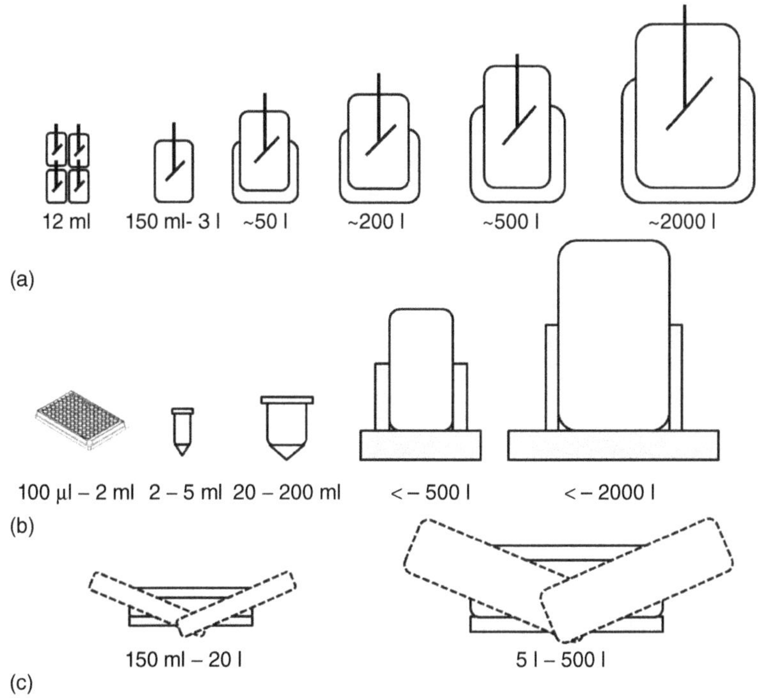


**Figure 9.2** Scales commercially available for different single-use bioreactor concepts. (a) Stirred single-use systems are available from the milliliter scale for parallel cultivation up to the cubic meter scale for production. The lab scale system can be operated with a flexible volume. At larger scale, operation at varying liquid volumes during operation affects the power input and is restricted by the position of the stirrers. (b) Orbital-shaken bioreactors can be scaled-up very easily from the microwell plate scale to the production scale. Film formation along the wall is crucial for the achievement of a certain gas mass transfer; optimal filling levels have to be maintained. (c) Wave-mixed reactors require more space when scaled-up. Therefore, a scale larger than a few hundred liters is not conducted so far. Since the bags can be easily modified in size without changing drastically the power input even during operation, each bioreactor is applicable in a comparably broad volume range.

achievable in it, which is the product of the liquid mass transfer coefficient $k_\mathrm{L}$ and the interfacial area $a$:

$$\mathrm{OUR} = k_\mathrm{L}a(c^*_{\mathrm{O}_2} - c_{\mathrm{O}_2}) \tag{9.4}$$

The OUR relies indirectly on the biomass concentration $c_X$ and the specific oxygen uptake rate $q_{\mathrm{O}_2}$. In the case of directly aerated systems, the oxygen mass transfer coefficient depends mainly on the average bubble size and the gas volume fraction, both of which determine the interfacial gas–liquid area.

In case no direct sparging is applied, but surface aeration, the interfacial area can be estimated simply based on the geometric dimensions, which determine the surface area of the liquid phase:

$$a = \frac{\pi D^2}{V} \tag{9.5}$$

The mixing number $\Phi$ is a parameter, which is in general suitable for the characterization of stirred systems. It is defined as the time, in which a tracer substance is distributed equally in the liquid phase. Usually, the achievement of a value of 95% of equal distribution is considered as being sufficient for the determination of the mixing time. Although any method that follows a colorization or decolorization can be considered to be precise, since dead space and other areas of poor mixing can be identified, the application of pH sensors or other monitoring devices are considered suitable as well. The major drawback of such methods – the degree of mixing is only measured at the position of the sensor – is counteracted by the easiness and the readiness, since bioreactors are usually equipped with suitable sensors anyway.

As the mixing time is an important parameter for the operation of bioprocesses, it is also an important scale-up criterion. The avoidance of gradients, for example, at the addition of acid and base for pH regulation or substrate at fed-batch processes, is of great importance. Usually, mixing times comparable to traditional stirred-tank reactors are achieved in small-scale SUBs with a working volume of up to a few liters [8]. Mixing times of 3 s were measured at the UniVessel SU (Sartorius Stedim Biotech SA, Goettingen, Germany) when the maximum power input was applied. The lowest mixing times, which are achievable at stirred SUBs, were found to be in the range of 20 s up to a scale of 200 l. Maximum mixing times of 200 s are achieved at the lowest power input at 1000 l. Attempts have been undertaken to derive equations for the estimation of mixing time in stirred SUBs [8, 31], resulting in Eq. (9.6):

$$\tau_{\mathrm{m}} = 3.5\left(\frac{P}{V_L}\right)^{-\frac{1}{3}}\left(\frac{d}{D}\right)^{-\frac{1}{3}}\left(\frac{H}{D}\right)^{2.43} D^{\frac{2}{3}} \tag{9.6}$$

Usually, the achievable power input is comparably high due to the low distance between the stirrer and the wall and the low fluid volume in lab scale. As shown in Table 9.1, the largest stirred SUB reach cubic meter scale. Thus, the key question arises whether scalability exists towards this comparably with large scale, but also with the very small scale, in which technical and geometrical limitations might hinder suitable operation conditions.

Among the smallest sizes of stirred SUBs available is a microbioreactor system made of a polymer-based block, which is mixed directly by a magnetic stirrer. The cultivation volume is 150 μl, the DO, pH value as well as the optical density can be measured *online*. Microvalves allow the addition of feed solution or the removal of culture broth [32]. The BioReactor 48 system (2mag AG, Munich, Germany) enables to run 48 cultivations at a scale of 8–12 ml in an automated manor if the system was coupled to a liquid-handling robot. The cultivation chambers are made of plastic and considered disposable. The agitation and the aeration rates are controlled. Monitoring of the pH value and DO is feasible through spots at the bottom of the cultivation chamber (PreSens). An industrial riboflavin production process was carried out in a fed-batch mode. The product yields matched the results obtained in the 3 l scale, which proves the feasibility of scale-down to milliliter range [33].

The scalability of mid-size stirred SUBs should be comparable with traditional stirred bioreactors. Loeffelholz *et al.* performed an engineering characterization of the UniVessel SU and BIOSTAT CultiBag STR (both Sartorius Stedim Biotech) [8]. The height-to-diameter ratio of the BIOSTAT CultiBag STR line was similar for various scales (50, 200, 500, and 1000 l, respectively) [34]. In this case, the power input increases for small volumes compared with large volumes. A maximum power-input-to-volume ($PV^{-1}$) ratio for the BIOSTAT CultiBag STR was determined to be 240 W m$^{-3}$, 133 W m$^{-3}$, and 73 W m$^{-3}$ for 50, 200, and 1000 l bioreactor, respectively [8]. The UniVessel SU exhibited a higher $PV^{-1}$ ratio of 430 W m$^{-3}$ mainly due to the lower cultivation volume of maximal 2 l. A $PV^{-1}$ ratio of up to 150 W m$^{-3}$ is considered suitable for cell culture applications, and the stirred SUBs fulfill this requirement. However, the optimal combination of different stirrer types for mixing and gas mass transfer remains crucial for SUBs, since the stirring rates are restricted due to mechanical reasons and cost efficiency at larger scale. It was shown that the $PV^{-1}$ ratio drops by more than half if two segment blade impellers were used instead of a combination of a segment blade impeller (top) and Rushton turbine (bottom) [8].

#### 9.2.4.2
#### Scalability of Orbital-Shaken Single-Use Bioreactors

The successful application of orbital-shaken SUBs for the cultivation of mammalian cell lines from the milliliter-scale to the cubic-meter scale was demonstrated [35]. The power input in orbital-shaken systems can be determined with a torsion measurement at the outer shaker (when inner friction is neglected) based on Eq. (9.2) or when the liquid and air temperature in the vessel is measured [36–38]. The corresponding surface area for the calculation of the oxygen mass transfer can increase significantly when the reactor is orbital-shaken. The formation of a water spout leads to film formation on the vessel wall. The water film is saturated with dissolved oxygen very rapidly due to the high surface-to-volume ratio. Once the core of the liquid phase touches the film, it is mixed with the liquid core again. Thus, the oxygen dissolved in the film is contributing to a much higher $k_La$ value. The film formation depends primarily on the geometry, shaking motion, and surface tension of the wall material and the liquid phase, among various parameters. Hence, the oxygen mass transfer needs to be evaluated experimentally for each cultivation vessel. For the approximation of the $k_La$ value in shake flasks, the following equation was developed by Maier and Buchs [39]:

$$k_La \cong n^{0.84} V^{-0.84} d_0^{0.27} d_{m,SF}^{-1,25} \tag{9.7}$$

Maintenance of the mixing time among different scales has been described for orbital-shaken SUBs [40]. In order to achieve this, the inner vessel diameter to shaking diameter and the Froude number should be kept constant. This was achieved in different scales from the milliliter to the cubic-meter range.

Since film formation on the wall contributes greatly to the gas mass transfer, the characteristics of the surface of the bag material play an important role as well as the surface tension of the liquid phase. However, especially at small scale, further optimization of geometries and materials was performed in order to achieve comparably high gas mass transfer rates. First of all, the design was optimized for an improved oxygen mass transfer (e.g., flower plates, m2p-labs GmbH, Aachen, Germany). Cover technology had been introduced, which allows the application of high shaking speed without losing oxygen and carbon dioxide mass transfer capacities of the system by membrane wetting (Enzyscreen BV, Leiden, The Netherlands). When applying fluorescence sensor technology (PreSens), the pH, DOT, and biomass can be monitored *online*, which allows the adjustment of *in situ* substrate release systems (EnBase® from BioSilta Oy, Oulu, Finland and silicon pellets from Kuhner, Birsfelden, Switzerland). They enable the application of a nutrient-limited fed-batch mode in nearly any scale [41–43]. By released control with silicon pellets, the cultivation conditions of *Hansenula polymorpha* could be improved in shake flasks, which led to an increased concentration of the final biomass by 85% compared with batch cultures [44].

As reported [45], the BioLector technology (m2p-labs) can be coupled with a microfluidic control device to run fed-batch cultivations at the microwell plate scale. Constant and exponential feeding may be applied in parallel bioreactor systems driven by micropumps or liquid-handling systems [46]. The BioLector technology further enables the measurement of the biomass concentration, metabolites, and reporter proteins by measuring the intensities of fluorescence and scattered light *on line* in up to 96 wells of a disposable plate [47]. The $\mu$-24 miniature bioreactor system (Pall Life Sciences Inc., NY) is equipped with sensors for DO, temperature, and the pH value measurement [48]. A high well-to-well reproducibility, which is in some way crucial when applying cultivation methods at the microliter and low-milliliter scale, was proven for *Saccharomyces cerevisiae* cultivations [49]. The apparent volumetric oxygen mass transfer achievable in such a system ranged between 3 and 22 $h^{-1}$ for headspace aeration, and between 4 and 53 $h^{-1}$ for direct gas sparging, respectively [50]. Direct gas sparging reduced mixing times by a factor of up to 19.

A further contribution to fully operated disposable bioreactors at the small scale is their integration in automated liquid-handling systems. Several reports are describing applications with individual process control, feed profiles, and induction procedures in parallel cultures on liquid-handling platforms [33, 51]. Thus, together with platform-independent technologies such as EnBase [52] and feed-in-time (FIT) medium [53], microwell plates and shake flasks represent the initial scale at which fully equivalent orbital-shaken bioreactor systems are applicable nowadays.

#### 9.2.4.3
#### Scalability of Wave-Mixed Single-Use Bioreactors

In order to estimate the power input at wave-mixed SUBs, the momentum of one bidirectional movement between the corresponding angles $-\varphi_{max}$ and $\varphi_{max}$ is determined [10]:

$$P = \int_{-\varphi_{max}}^{\varphi_{max}} Mk\,d\varphi \tag{9.8}$$

The momentum can be analyzed from the observation of the fluid flow behavior and the corresponding change of the point of gravity and the spatial distribution of the fluid surface area in the bag. In order to be able to establish a classical relation of power input and the Reynolds number, a modified Reynolds number for wave-mixed processes was introduced [10]. Therefore, the characteristic length $l_C$ of the Reynolds number was assumed to be derived from the cross-section based on the liquid height $h$ and the width of the bag $w_b$:

$$l_C = \frac{2A_C}{l_B} \tag{9.9}$$

In Eq. (9.8), $l_B$ represents the true length (greatest length) of the fluid cross-section.

The cross-section $A_C$ depends on the geometry of the bag and is case-sensitive. When assuming that the volume has to be turned twice partially along the length of the bag, its velocity is assumed to be

$$w = \frac{2VnC}{A_C} \tag{9.10}$$

The constant factor $C$ depends on the part of the volume that passes the rotation point, thus depending on the bag geometry, filling level, and rocking angle. The modified Reynolds number for bag reactors can then be defined as [10]:

$$Re = \frac{w\,l_c}{\nu}\;;\; Re_{mod} = \frac{4\,V\,u\,C}{l_B\,\nu} \tag{9.11}$$

Transition between laminar and turbulent flow was found in the area between $Re = 400$ and 1000. Usually, turbulence can be achieved in wave-mixed SUBs, if the filling volume was not very high or very low.

The mixing time of wave-mixed bioreactors is in a similar range of those obtained in stirred SUBs [10]. As reported, distinct regions from 20% toward 30% or even up to 50% of filling volume in relation to the bag volume were found to yield the lowest mixing numbers. This is in agreement with the measurements of the gas mass transfer coefficient $k_La$ in wave-mixed SUBs, in which a maximum is achieved at a distinct filling volume of 40% of the bag volume [11]. An increase of the modified Reynolds numbers above 1000 did not lead to a further reduction of the mixing time [10].

Scalability of wave-mixed SUBs can be achieved, if a part of the bag is blocked for operation, for example, through clamps or channel blocks as shown in Figure 9.3

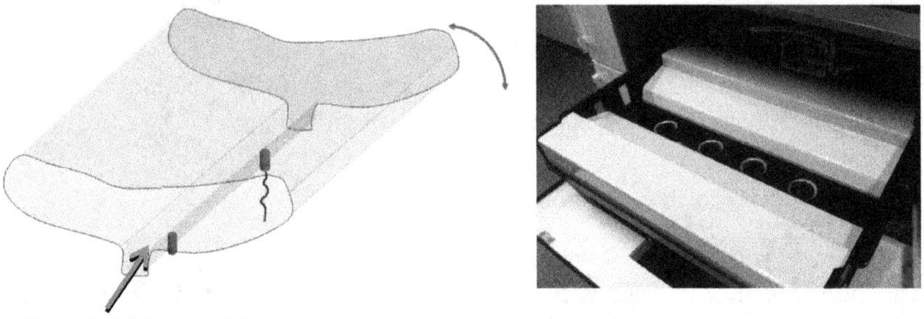


Figure 9.3 Removable bodies creating a narrow channel in the CELL-tainer SUB, ensuring a scalability from the milliliter to the liter range without the necessity to change the bag.

(Celltainer Biotech BV, Winterswijk, The Netherlands). Since the liquid height, rocking angle, and characteristic length rather remain constant, similar power inputs are maintained on the same rocking platform for a broad range of working volumes.

### 9.2.4.4 Recent Advances in the Description of the Mass Transfer in SUBs

In order to evaluate SUBs and scale-up/scale-down strategies, further engineering parameters need to be considered. Computational fluid dynamics (CFD) can contribute to broaden the knowledge about the fluid dynamics caused by the different agitation and shaking technologies applied and SUBs. Fluid flow characteristics in SUBs were investigated by means of particle image and tracking velocimetry [43], laser-Doppler anemometry [54], and hot-film anemometry [55]. CFD, including the information of the experimental determination of fluid flow, was applied in different stirred [15, 54] and wave-mixed bioreactors [10, 55]. The valuable result of CFD applications is the estimation of local fluid flow velocities, gas bubble distribution, and thus the local energy dissipation and shear stress. Shear stress is crucial, since many SUBs are employed for shear-sensitive cell lines. Maximum shear rates might occur at bubbles collapsing at the surface. However, shear rates at the stirrer tip speed can also harm cells due to the friction that occurs when a stirrer blade hits the cell. Including CFD in scale-up studies allows the consideration of the development of shear stress at the larger scale from a very early reactor design stage on. The impact of different bottom edge geometries due to constraints of production was investigated at a conventional and single-use version of the 2 l UniVessel® by means of CFD. It was found that the changes on the secondary flow field were of minor importance in this case for relevant engineering parameters, when a Rushton turbine and a segmented blade impeller were used [56]. The same approach is suitable to estimate the most suitable power input for cell culture applications at which there should be a compromise between the achievement of a sufficient gas mass transfer and not very high shear forces. In a CFD study, the Mobius® 3 l CellReady, which consists of an upward-pumping marine scoping impeller, was investigated with particle image velocimetry, combined with a study of the impact of the fluid flow characteristics on a cell culture performance [57]. Fluid compartmentalization and turbulence were observed at various tip speeds and working volumes. A GS-CHO cell line producing an IgG antibody was used for cultivation. Impacts on cellular growth and viability at the applied conditions (80–350 rpm and 1–2.4 l working volume) were hardly observable, although a significant reduction in recombinant protein productivity was detected at 350 rpm and 1 l working volume. Under these conditions, highest Reynolds numbers were achieved, likely causing locally very high shear forces in this case.

The challenges in scale-up of bioprocesses in SUBs are similar to those encountered in conventional reactors. However, the various designs, which are on the market, require various individual approaches. Since the geometries of stirred SUBs are similar to conventional stirred-tank reactors, scale-up methodologies can be adopted, thus making scale-up considerations feasible. However, the limitation of mechanical stability, the greater effort for the production of bags for directly stirred systems, the higher energy consumption, and comparably higher shear forces (especially due to direct sparging) do not allow denoting such stirred SUBs as a method of choice in every case. Orbital-shaken systems have a similar footprint, but totally different gas mass transfer features without the need of sparging. Wave-mixed systems, although characterized by a higher footprint, exhibit low shear forces and in some cases excellent gas mass transfer coefficients, which makes them suitable for microbial cultivations. A most efficient utilization of the space they need has to be achieved.

## 9.3 Cell Culture Application

### 9.3.1 Wave-Mixed Bioreactors

The introduction of the "wave" bioreactor by Singh in 1999 [2], being the first real SUB technology, was fully focused on the cultivation of animal, insect, and plant cells in suspension culture. The "wave" bioreactor (Figure 9.4) provides agitation by gentle movement of the bag (placed in a rocker-tray), and cells can be cultivated without being damaged by the shear of a stirrer or coalescing bubbles. However, the "wave" SUB is clearly restricted in oxygen mass transfer capacity. Singh [2] reports an oxygen transfer coefficient, $k_L a$, in the range of 1–5 $h^{-1}$, supporting cell growth of $3 \times 10^{-6}$ cells $ml^{-1}$ for NS0 cells and $2.7 \times 10^{-6}$ cells $ml^{-1}$ for HEK293 cells. More recent studies in "wave"-type SUBs report about cultivations in which a cell density of up to $1.5 \times 10^{-7}$ cells $ml^{-1}$ (CHO cells) was achieved in a perfusion process even up to 500 l of working volume [58].

Variants of rocking bioreactors (Appliflex, Applikon Biotechnology BV; XRS20 bioreactor, Pall Life Sciences Inc.; CELL-tainer® bioreactor, Celltainer Biotech BV) have been introduced in the market, where the XRS20 bioreactor as well as

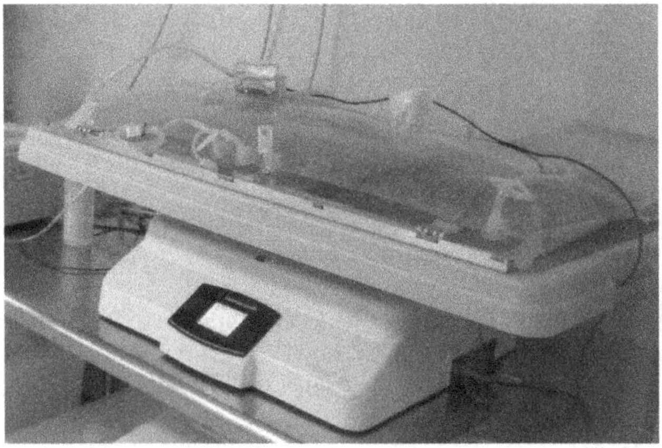

**Figure 9.4** Lab-scale wave-mixed single-use bioreactor.

**Table 9.3** Comparison of various types of shaking single-use bioreactors for cell culture application in comparison with a single-use stirred bioreactor.

| Property | BioWave | CELL-tainer | XRS20 | Stirred SUB |
|---|---|---|---|---|
| Mixing | Single rocking | 2D rocking | Orbital shaking | Stirrer |
| Bags | Simple | Simple | Simple | Complex |
| Working volumes | 1–200 l (various equipment) | 0.2–25 l<br>10–200 l | 2–20 l | 50–2000 l (various equipments) |
| Working volume range factor (in one bag) | 5–10 | 100–150 | 10 | 5 |
| Temperature control | Bottom (heater) | Incubator/ bottom | Incubator | Jacket |
| $k_La$ – Cell line versions | <50 $h^{-1}$ | >100 $h^{-1}$ | <75 $h^{-1}$ | <30 $h^{-1}$ |
| $t_m$ (20 l) | 50–100 s | 10–20 s | 16 s | 10–20 s |
| Sensor position | From top | Bottom – cup | From top | Side |
| Sensor type | Optical | Non-disposable electrochemical probes | Optical | Optical/reusable |

the CELL-tainer bioreactor show significantly higher $k_La$-values compared with "wave"-type bioreactors (Table 9.3).

### 9.3.2
### Stirred Single-Use Bioreactors

During the last decade, stirred SUBs have been introduced up to working volumes of 2000 l. These systems (Figure 9.5) consist of a disposable bag, which includes a

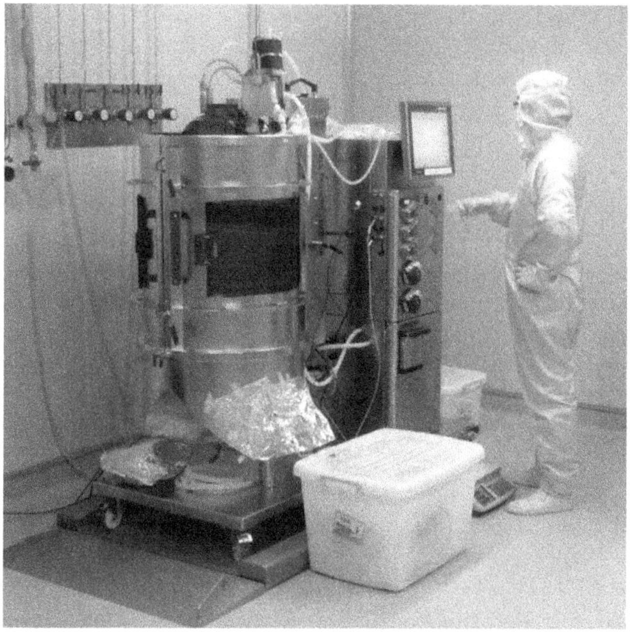

**Figure 9.5** Stirred-tank single-use bioreactor.

stirring device, comparable to stainless steel bioreactor designs. The bag is placed into a stainless steel tank-liner, which also contains the heat transfer device (electrically heated jacket or a jacket connected to a thermocirculator). The geometry of these single-use stirred reactors is comparable to traditional stainless steel stirred bioreactors for cell culture, with a height/diameter aspect ratio of 1 : 1 to 2 : 1. Some of the designs have an eccentrically located stirrer. Either the stirrer is connected by a magnetic drive (GE Healthcare Life Sciences – XCellerex™) or by a single-use sealing/bearing assembly (Sartorius Stedim Biotech SA; Thermo-Fisher Scientific Inc. – Hyclone™) as a part of the bag itself. The latter design requires a more laborious installation and connection of the shaft and motor, but results in a simpler bag design. The stirrer itself is made or molded from polyethylene or polycarbonate, both USP-class VI materials. Usually, for cell culture one or two 45° pitched blade impellers are applied.

Bags are provided with ports for media and base addition, and for sampling, respectively. Gassing is provided either with a microsparger or with a sparger ring, a porous plate, or open tube underneath the stirrer. The inlet and outlet gas lines are equipped with presterilized filters. The outlet gas filter is usually heated to prevent condensation, and later a blocking of the filter pores.

Typical oxygen mass transfer coefficients of these SUBs are comparable with traditional reusable cell culture bioreactors and range from 4 to 30 $h^{-1}$, depending on the stirrer speed and blade geometry [34]. The power input in these reactors range from 2 to 70 $W\,m^{-3}$.

The scalability of these SUBs is comparable to classical stirred-tank bioreactors [59]. Data by Genentech and by DSM Biologics [60] gained at CHO and PER.C6® cell cultures, respectively, indicate proper scalability in stirred SUBs, as well. CHO cells have been grown in a 250 l working volume (Thermo-Fisher Scientific Inc. – Hyclone). The cultivation was comparable with those in 2 l and 2000 l cultivations in glass and steel reactors, respectively. The PER.C6® cell cultures were grown in fed-batch mode in a working volume of 50 l and 250 l in the Hyclone SUB. These cultures show comparable results in terms of growth and productivity in these reactors compared with a 4 l cultivation in a glass-made stirred-tank lab bioreactor, as well. In the so-called "XD-process" of DSM, a cell density of $1.2-1.4 \times 10^8$ cells $ml^{-1}$ was achieved in a 50 l SUB, resulting in product titers of an IgG-type antibody of >10 $gl^{-1}$ [61, 62]. A remodeling of the sparger design of a Hyclone SUB from a frit to a ring sparger yielded similar results in terms of the gas mass transfer as achieved in the XDR™ [63]. Transferability between different SUBs in a scale of 1000 l was demonstrated by Minow *et al.* [64]. The transfer of a Chinese hamster ovary cells fed-batch process from an XDR to a Hyclone SUB was based on either maintaining the oxygen mass transfer or the power input. Finally, an operation range of the volumetric power input of 10–31 W $m^{-3}$ for the Hyclone SUB was identified. This was 35% higher than that in the XDR. Recently, the comparability of the cultivation of *Spodoptera frugiperda*-9 cells in conjunction with the baculovirus expression vector system, expressing a secreted alkaline phosphatase, was investigated in the wave-mixed BIOSTAT CultiBag RM and the stirred UniVessel SU [65]. Comparable growth rates, substrate consumptions, and maximum phosphatase activities were found in both bioreactors.

### 9.3.3 Orbital-Shaken Single-Use Bioreactors

Besides wave-mixed bioreactors and the stirred SUBs, several other types of disposable bioreactors have been introduced for cell culture application during the last 10–15 years.

Kuhner (Kuhner AG, Birsfelden, Switzerland) introduced the aformentioned orbital shaking bioreactor SBX [16], which is in fact a scaled version of a shake flask. In a 200 l shaken bioreactor system, cell counts for CHO cells of $7 \times 10^6$ cells $ml^{-1}$ were achieved. A reasonable scalability was found in scaling up from 50 ml tubes of 10 ml working volume up to 200 l. However, $k_La$ values are also restricted to less than 25 $h^{-1}$ in this system.

### 9.3.4 Mass Transfer Requirements for Cell Culture

On an average, mammalian cell cultures are characterized by a specific oxygen uptake rate of 5–10 pmol $cell^{-1}$ $day^{-1}$ (CHO cells) [66]. When assuming a required cell density of $10 \times 10^6$ cells $ml^{-1}$, this translates into a required oxygen transfer coefficient of 15–30 $h^{-1}$. Hence, most SUBs are able to support such cell densities. However, the transfer of carbon dioxide ($CO_2$) seems to be more critical. The specific $CO_2$ production rate for CHO cells is 4–6 pmol $cell^{-1}$ $day^{-1}$ [66]. The mass transfer rate for $CO_2$ is 89% of that of oxygen [67]. The excess of $CO_2$ in the off-gas (due to the production of the cells) can be 20–40%. Therefore, the required $CO_2$ transfer capacity will be lower than expected:

$$\mathrm{CTR} = k_\mathrm{L}a^{\mathrm{CO_2}}\left(\mathrm{cCO}_{2,\mathrm{L}} - \mathrm{pCO}_2^{\mathrm{gas}}/\mathrm{H}^{\mathrm{CO_2}}\right) \quad (9.12)$$

As a result of Eq. (9.12), the required mass transfer coefficient for $CO_2$ needs to be at least 1.5 times that of oxygen to ensure a proper removal. Apart from this, sufficient air is needed to strip the $CO_2$ from the gas phase (especially for air-overlay systems). From this, it can be concluded that the $CO_2$ transfer needs a $k_\mathrm{L}a^{\mathrm{CO_2}}$ of at least 40–50 $h^{-1}$ at a cell density of $10^{-7}$ cells $ml^{-1}$.

Especially, at the later stage of the process, the partial pressure of $CO_2$ in the liquid can rise to values of 100–150 mm Hg [68]. A high partial pressure of $CO_2$ leads to an increase of the $HCO_3^{-1}$ concentration in the bioreactor, and thus to a higher alkaline consumption. As a result, the osmolality in the culture can rise such that it starts to have a decreasing effect on the cells' viability. Therefore, control strategies are needed, which includes the maintenance of the gas composition by the removal of surplus $CO_2$, and pH control. It is obvious that in the case of perfusion processes, where cell densities of $10^8$ cells $ml^{-1}$ are obtained [62], special requirements are needed to prevent oxygen limitation and inhibition by $CO_2$ or $HCO_3^{-1}$.

In many large-scale cell culture processes, "wave"-type bioreactors are used for expansion (Figure 9.6a, [69]). These processes are rather laborious, and at least 6–8 steps are required to seed a large-scale bioreactor (200 l and more). In a co-development of Shire HGT and CELLution Biotech BV (now Celltainer Biotech BV), a simplified seeding process was developed, where several steps can be avoided, resulting in less labor needed and a lower risk for contamination.

By applying removable plastic bodies, which can be placed underneath the cell culture bag, a narrow channel is created, diminishing the starting volume of the culture in a 25 l bag (Figure 9.3). The applicable starting volume in this bag is reduced to 150 ml. When removing the plastic bodies, the bag further inflates to its final shape. A stepwise volume increase becomes feasible with this technique. The application of these "channel blocks" reduces the number of seed steps by 3–4 (Figure 9.3, [69]). A patent is pending on these "channel blocks" [26]. Stepwise medium addition can be performed automatically (based on the cell density) or on a daily basis. A medium container/bag only has to be coupled one time and cells need not be transferred from shaker flasks into the bag or from bag to bag. Due to the flexibility of the bags, the volume in one bag can be increased with a factor of 100–150. A procedure of volume increase by that factor is not feasible for stirred SUBs, because otherwise the impeller would not be touching the liquid phase. In general, the power input would change due to the different attachments of stirrers to the liquid phase. Results of a stepwise expansion in a CHO culture with channel blocks are shown in Figures 9.6b and 9.7.

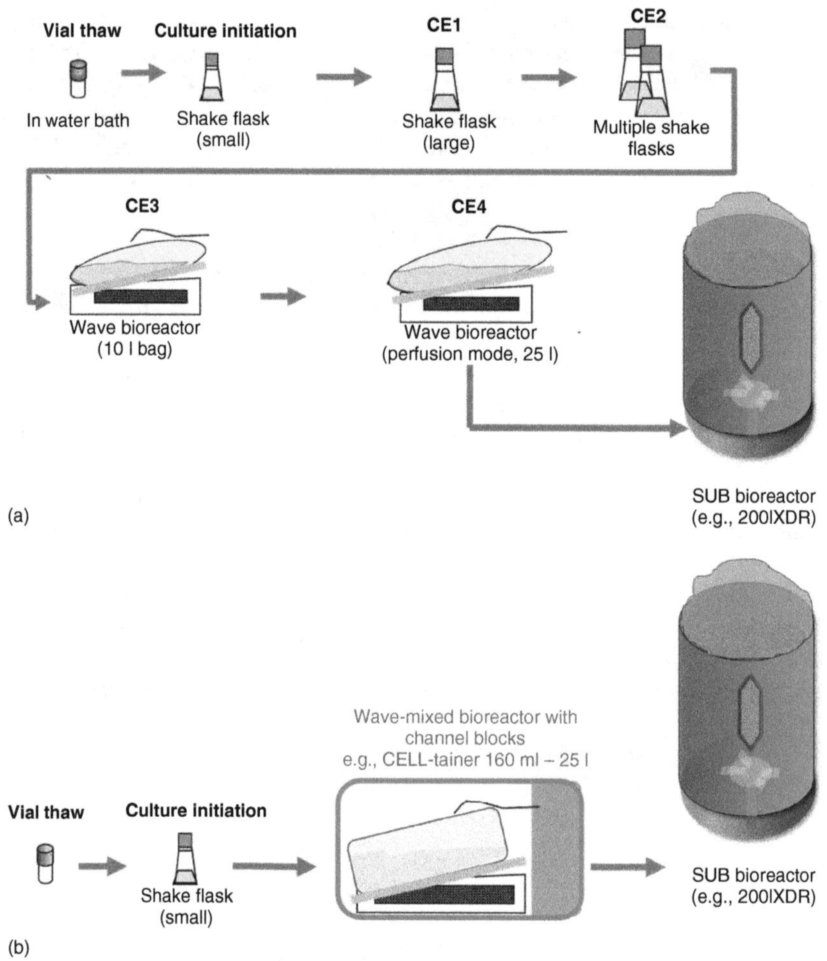


**Figure 9.6** Cell culture expansion process including the application of perfusion and wave-mixed single-use bioreactors. (a) Steps necessary involving different scales of wave-mixed single-use bioreactors. (b) Steps when using the scalability of one single-use bioreactor with devices that allow a noninvasive modulation of the cultivation volume such as channel blocks.

### 9.3.5
### Perfusion Processes in Single-Use Equipment

Many types of perfusion devices have been introduced over the last decades. Starting with a so-called spin filter (rotating mesh on the stirrer shaft), devices like hydrocyclones, settlers, centrifuges, filtration devices, hollow-fiber systems, acoustic separation and alternating tangential flow systems have been applied

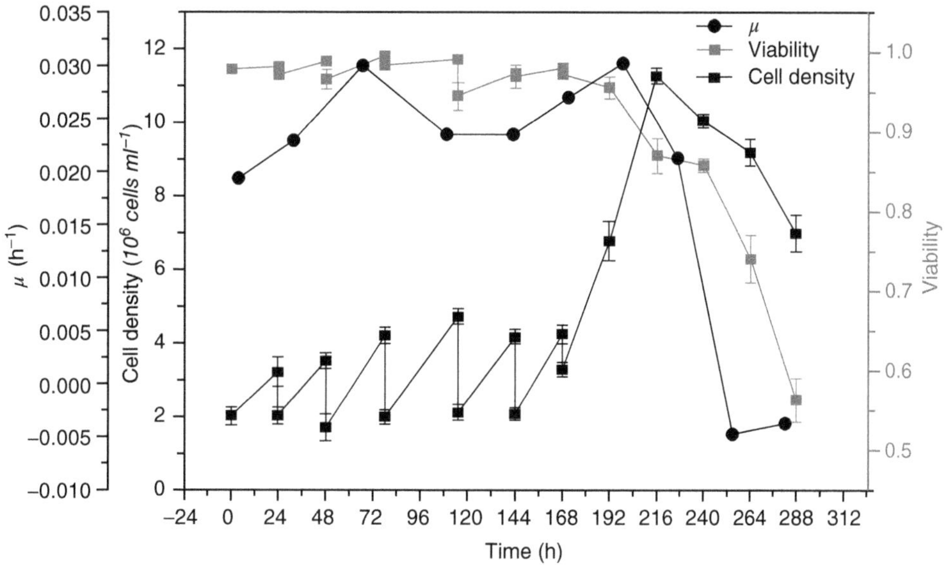


**Figure 9.7** Cultivation performance of a stepwise expansion process when channel blocks were applied [69].

more or less successfully [70]. Although there has been a continuous debate between regulatory bodies and process development, more and more attention is paid to perfusion-type processes in the last 5 years [1]. The advantages of perfusion-type processes or repeated batch processes are obvious. Due to a significantly higher volumetric productivity, costs of production are reduced [71]. In addition, product quality can be more uniform: cells stay continuously under the same environmental conditions.

Today, especially the ATF system (www.refinetech.com) becomes more and more accepted in pharmaceutical bioprocessing, mainly due to its positive properties in terms of reduction of fouling of the filters and the ability to define the filter cutoff. This allows not only the concentration of cells but also the increase of the concentration of the product [61]. Although the ATF is applied in SUBs [72], the housing of the filter remains to be autoclaved. Furthermore, sterile connections to the bioreactor have to be assured. The process control of the ATF is rather complex though. An internal filter device is available for rocking bags (GE Healthcare). This filter floats on the surface of the liquid [73]. The disadvantage of such a floating filter is that the filter reduces the surface for mass transfer. Fouling of these filters seems to be problematic. Recently, a single-use hollow-fiber bioreactor was evaluated for the production of high-titer influenza A virus [74]. MDCK cells infected with influenza virus H1N1 were cultivated in the extracapillary space to titers of $1.8 \times 10^{10}$ virions $ml^{-1}$. Influenza virus was collected by performing multiple harvests during the course of the cultivation.

### 9.3.6 Plant, Phototrophic Algae and Hairy Root Cell Cultivation in Single-Use Bioreactors

Although plant cell cultures are often grown as suspension culture at higher cell density at a broader pH-value range and lower temperature, most of the bioreactors suitable for growth of animal cell cultures are also suitable for the growth of plant cells. The high surface-to-volume ratio in wave-mixed SUBs favors illumination into the culture. In general, the low shear forces obtained in many SUBs also support the growth of plant cells. Reports describe the application of the BioWave reactor at tobacco, grape, and apple cell cultivations up to a volume of 10 l [6, 75]. A maximum fresh weight of 26 g $l^{-1}$ $d^{-1}$ was obtained in a batch culture of a tobacco cell line. A grape cell line *Vitis vinifera* was cultivated at a growth rate of 40 g $l^{-1}$ $d^{-1}$ of fresh weight in a wave-mixed bioreactor, the highest growth rate reported for plant cells in a SUB [75]. Cultivations of immobilized suspension cells of species of the Taxaceae family yielded very high product concentrations of nearly 21 mg $l^{-1}$ paclitaxel [10]. Since orbitally shaken SUBs can be scaled-up easily, a recent study investigated the cultivation performance of a tobacco BY-2 cell line in the SB200-X bioreactor (Kuhner) for the production of the human monoclonal antibody M12 [76]. Cell growth and recombinant protein accumulation were comparable to standard shake-flask cultivation in the 200-fold up-scaled cultivation. Final cell fresh weights of 300–387 g $l^{-1}$ and M12 yields of 20 mg $l^{-1}$ were achieved.

The low shear force and high surface area also leads to the feasibility of the cultivation of cells with a high demand of light, like phototrophic algae. Recent studies described the application of LED-based light rods at three different SUBs for diatom algae propagation [77]. The wave-mixed BIOSTAT CultiBag RM 20 (equipped with a 2 l bag) and the AppliFlex 10 l SUB were illuminated with LEDs either mounted at the bottom or top of the BIOSTAT. The Multitron Cell platform was shaken orbitally and operated with a CultiBag RM bag with a volume of 2 l. It was illuminated with fluorescent tubes unlike the other two systems. The volumetric mass transfer coefficients for all systems were above 8 $h^{-1}$, as mentioned by the authors. Identical growth curves for the *Phaeodactylum tricornutum* cultivations were obtained in the BIOSTAT and AppliFlex reactor at a maximum growth rate of $\mu_{max} = 0.773\ d^{-1}$. The cultivation in the Multitron Cell platform showed the highest growth rates of $\mu_{max} = 0.982\ d^{-1}$. The difference in the outcomes are likely due to the different light sources of which the fluorescent light emits more energy in the red and light spectra, which can be used by the cells via red and blue light–absorbing phytochromes. A bag was surrounded completely by 16 LED light rods in a BioWave SUB equipped with a bag for a working volume of 1 l [78]. The green algae *Chlorella vulgaris* and *Euglena gracilis* for paramylon production, a polysaccharide, were cultivated. Due to the heat generation of the LED rods, which were directly located at the bag's surface, cooling capacity was needed. A growth rate of 0.33 g $l^{-1}$ $d^{-1}$ was obtained, which seems typical for *C. vulgaris* cultivations. The feasibility was also proven for the *E. gracilis* cultivation. However, the requirements on the temperature control might exceed the ability of installed devices at several SUBs. Issues arising from this can be counteracted by an appropriate optimization.

Besides plant and algal cells, wave-mixed SUBs have been used to cultivate hairy root cells [10]. A BioWave SUB was operated in a mode that allowed temporary immersion ("ebb-and-flow"). The illumination was performed with external fluorescent light. The hairy roots cultivation of *Hyoscyamus muticus* and *Panax ginseng* was performed at a working volume between 0.1 and 0.5 l for 28 and 56 days, respectively. The yields were up to threefold higher than in glass spray reactors. In the same study, the successful application of bags with integrated mesh supporting root immobilization is described.

## 9.4 Microbial Application of Single-Use Bioreactors

The aforementioned limitation in the achievable $k_{L}a$ value in SUBs leads to restrictions in the application for microbial cultivations. However, an increasing number of examples are described for the successful implementation of single-use technology for the cultivation of bacteria, yeast, and fungi. An early report states the application of a WAVE SUB for the cultivation of the bakers' yeast *S. cerevisiae* [79]. Dissolved oxygen limitation was not determined before a biomass concentration of $8\,g\,l^{-1}$ was achieved at a working volume of 5 l. The inlet gas was blended with oxygen in order to achieve a sufficient oxygen mass transfer. Due to the necessity for a defined gas composition at cell line cultivations, installations for gas blending are usually employed in SUBs, and thus no further installations are needed.

Several applications of bacterial cultivation in wave-mixed bioreactors are described in the literature, as well. However, oxygen depletion occurred at a final biomass concentration of less than $2\,g\,l^{-1}$ in a BIOSTAT CultiBag RM system when cultivating *E. coli* in batch mode before oxygen depletion occurred [42]. In order to control, and thereby reduce the volumetric oxygen uptake rate of the cultivation, the enzymatic substrate release method EnBase (BioSilta) was applied. No mechanical feeding was necessary in this case, which reduces the contamination risks. Due to substrate limitation and the concomitant reduction of the oxygen demand per cell, an optical density of $OD_{600} = 30$ ($10\,g\,l^{-1}$ of dry biomass) is now achievable in the same system. Heterologous expression of the alcohol dehydrogenase (ADH) enzyme was proven. This approach is especially very suitable when monitoring systems are not reliable enough for *online* feeding control. Hence, some drawbacks of current monitoring systems at SUBs can be circumvented. The application of SUBs for the growth of *Corynebacterium diphtheriae* for vaccine production was investigated by Ullah *et al.* In the BIOSTAT CultiBag RM reactor, a final cell density of $OD_{590} = 5$ (approximately $2\,g\,l^{-1}$ of dry biomass) compared with $OD_{590} = 7.3$ in an aerated stirred-tank bioreactor was achieved [9]. Hitchcock described the cultivation of a recombinant *Listeria monocytogenes* for vaccine production. In a BIOSTAT CultiBag RM with a working volume 5 l, a final optical density of $OD_{600} = 12$ was achieved [80].

A study on a BioWave SUB of 5 l working volume described the growth of an *E. coli* culture to an $OD_{600}$ of 15 without any oxygen limitation [81].

Due to the larger $k_L a$ values at the CELL-tainer in comparison to other wave-mixed bioreactors (Table 9.1), it is well-suited for microbial cultivation. In a study of Sanofi-Pasteur, growth and protein production in a 7 l cultivation has been shown to be comparable to data obtained in a conventional stirred-tank reactor [3, 11]. A final optical density of $OD_{600} = 60$ was achieved in the batch mode and $OD_{600} = 90$ in the fed-batch mode. An investigation proved the potential of growing *E. coli* cultures in a scale-up of the CELL-tainer in a volume of 120 l [11]. The cultivation was operated in a nutrient-limiting fed-batch mode. At a working volume of 12 l and only after 32 h, a final $OD_{600}$ of more than 130 was reached. At the tenfold scale, a similar biomass yield was obtained (Figure 9.8).

As already described for cell cultivation, a suitable application of SUBs is the utilization for seed cultivations. This application is beneficial while a smaller cell density is usually targeted than in production processes. In addition, sterility is of basic importance and critical during the manifold transitions between different

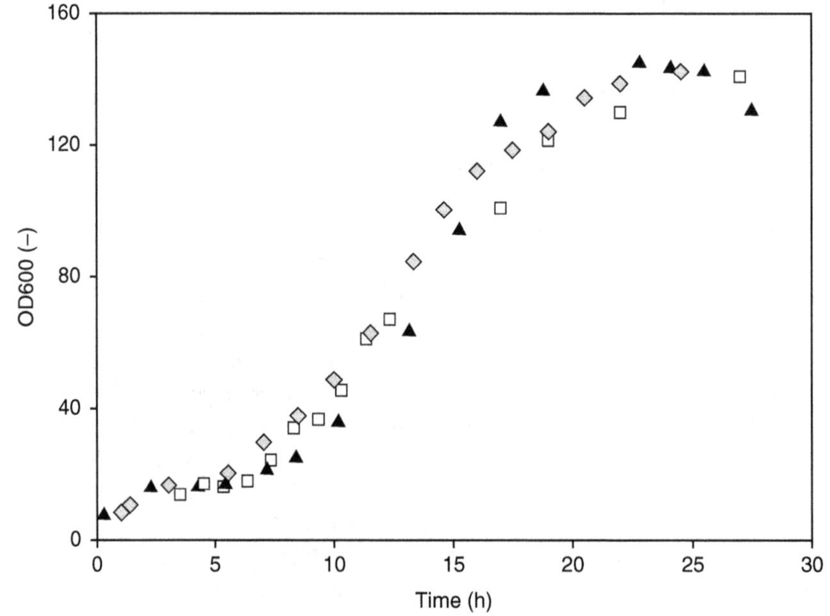


**Figure 9.8** Comparison of the growth performance of an *Escherichia coli* BL 21 cultivation [11] (fed-batch phase) performed in a stirred-tank reactor with a working volume of 2 l (unfilled squares), in a wave-mixed CELL-tainer CT 20 with a working volume of 12 l (gray diamonds), and in a wave-mixed CELL-tainer CT 200 with a working volume of 120 l (black triangles). The exponential feed rate was increased by a factor of $0.25\ h^{-1}$ t. After the induction of the expression of a maltogenic amylase at $t_{fed\text{-}batch} = 19\ h$, the feed rate was kept constant. No oxygen blending was applied.

cultivation vessels. The CELL-tainer SUB was applied in a seed cultivation for the production of lysine by *Corynebacterium glutamicum*. In this case, the cultivation from few milliliters to 15 l in the same bag was achievable with the application of channel blocks (Figure 9.3). This represents a good alternative to the common shake-flask approach. Pooling of and variability in growth in various flasks can have an impact on the process performance. In contrast, the application of one single bioreactor reduces the time needed for preparation and allows the monitoring and control of the pH value and dissolved oxygen throughout the seed cultivation process.

Microbial cultivations have also been performed successfully in orbital and stirred SUBs. Galliher describes an *E. coli* cultivation in an XDR SUB of a working volume of 50l [82, 83]. A cell density of $OD_{600} = 120$, which corresponds to a dry cell weight of approximately 40 g $l^{-1}$, was achieved. *Pseudomonas fluorescens* was cultivated in the XDR, in which a final dry cell weight of more than 100 g $l^{-1}$ was finally obtained. The cultivations were similar to process performances, which were obtained in traditional stirred-tank bioreactors [82]. A linear feed in a wave-mixed BIOSTAT CultiBag RM and an exponential feed for a stirred BIOSTAT CultiBag STR was applied in *E. coli* cultivations, in which a final optical density of up to $OD_{600} = 140$ was achieved [12, 13]. The maximum growth rate, which is approximately half of that reported for the CELL-tainer [11], corresponds to the ratio of the aformentioned maximum oxygen mass transfer rates for both systems. Another report described the successful application of a disposable stirred-tank reactor of a working volume of 250 ml for therapeutic protein and monoclonal antibody production using CHO cells, *Pichia pastoris*, and *E. coli* [84]. The cultivation performance concerning growth, cell viability, and product titer was similar to the conventional bioreactor runs. High cell densities above 400 g $l^{-1}$ of wet cell weight could be achieved in the bacterial cultivation.

SUBs were also applied for anaerobic processes and fungi with low growth rates. The shear stress–sensitive basidiomycetes *Flammulina velutipes* and *Pleurotus sapidus* were cultivated in a 5 l scale.

Differences concerning morphology, enzymatic activities, and growth in fungal cultivations were observed. In the bag reactor, growth in the form of small, independent pellets was observed while STR cultivations showed intense aggregation. *F. velutipes* cultures reached a higher biomass concentration and up to twofold higher peptidolytic activities in comparison to cultivations in a stirred-tank reactor [85]. Another reactor concept well-suited for the application in fungal cultivation is the Quorus bioreactor, which is equipped with hollow fiber modules that contain capillary membranes [86]. Growth occurs at the liquid–solid interface of a capillary membrane as cells accumulate at the outer membrane surface. Recombinant xylanase production in *Aspergillus niger* was 2–3 times higher in a 2 l scale than in a batch approach in a stirred-tank reactor. The successful application for $\beta$-lactamase production with *Lactococcus lactis* was proven.

Besides, the application of SUBs for marine cultures have benefits in comparison to the cultivation in steel stirred-tank reactors. Since the high salt content cannot be diminished in every case, the application of stainless steel reactors is crucial due to corrosion. However, alternatives such as glass reactors are usually restricted to a scale of 20 l. The CELL-tainer technology applied for the cultivation of the shear-sensitive heterotrophic microalga *Cryptecodinium cohnii* in marine media has been proven to be successful. Although nutrient limitation cannot be realized due to physiological constraints in this case, the gas transfer was sufficient to achieve a cell dry weight of more than $45\,g\,l^{-1}$ at maximum growth rates of $0.05\,h^{-1}$ [87, 88].

This application is a good example that single-use technology is not restricted to pharmaceutical processes and seed cultivation, but can provide a suitable alternative in many niche applications. Due to the easiness of installation, SUBs can be applied without much financial effort for initial process development. The alternative of an adopted traditional bioreactor system, which has to withstand the conditions of *in situ* sterilization requires usually higher investment costs, which cannot be covered at this stage of process development.

## 9.5 Outlook

In recent years, many different SUB systems were characterized, better understood, and scaled-up. However, the engineering characterization has not yet come to completion. An improved understanding of the fluid flow and mass transfer behavior is of great importance for a correct prediction of the dynamics at various cultivation conditions and – even more important – for an improvement of the design of SUBs. A similar degree of understanding of the fluid flow as it is the case for conventional stirred bioreactors will be necessary in order to achieve a similar degree of optimization and reliability of SUBs. The term “fully understood” in terms of power input, mixing, and mass transfer as it is valid for lab scale stirred-tank bioreactors is definitely not true for any scale and type of SUBs yet. However, the numerous applications of SUBs in industrial application and the broad distribution in the scientific community reflect the great potential of this technology in many areas of biotechnology. The various designs allow numerous suitable applications. While the achievable product titer could be increased for many products in recent years, the production scales of many bioprocesses were lowered. Nowadays, many products exist that are produced in batches of a few cubic meters per year, which makes the utilization of SUBs suitable for production purposes. However, besides the production and seed cultivation, SUBs are suited for the application in process research and development. In this case, scales are usually restricted to a few (hundred) liters. Cross-contamination might be a more crucial factor than at production because of nonstandardized and non-automated procedures. Besides, the environment might not be of the same sterile conditions similar to that in a production facility. In this case, presterilization and the feasibility to work in different scales as shown for wave-mixed reactors without the need of a reactor change reduce significantly the (cross-)contamination risks. The increased use of SUBs in contract manufacturing for this reason, for example, in the production scale for clinical trials, is ongoing.

Besides these benefits, the low consumption of energy, mainly due to the absence of steam sterilization, makes the application of SUBs more beneficial in terms of environmental impacts [89]. In addition, smart ways of waste treatment (recycling) of the plastic and sensor material will be found as was the case in other areas of plastic materials. The integration of SUBs into automated systems and the tailored combination with downstream single-use components will further support the penetration of the single-use technology while using the benefits of *ex situ* sterilization, design flexibility, and disposability.

## List of Abbreviations

| | |
|---|---|
| $A_C$ | characteristic cross-sectional area between gas and liquid [$m^2$] |
| $c_{CO_2,L}$ | concentration of dissolved $CO_2$ [$g\,L^{-1}$] |
| $c_{O_2}$ | dissolved oxygen concentration [$g\,L^{-1}$] |
| $c^*_{O_2}$ | dissolved oxygen saturation concentration [$g\,L^{-1}$] |
| $C$ | geometrical constant [–] |
| CTR | volumetric carbon dioxide transfer rate [$Mol\,L^{-1}\,h^{-1}$] |
| $d$ | stirrer diameter [m] |
| $d_0$ | shaking diameter [m] |
| $d_{m,SF}$ | maximal shake-flask diameter [m] |
| $D$ | reactor (bag) diameter [m] |
| $H$ | reactor (bag) height [m] |
| $H^{CO_2}$ | Henry coefficient of $CO_2$ [Pa] |
| $k$ | rocking rate [$s^{-1}$] |
| $k_La$ | volumetric oxygen gas mass transfer coefficient [$h^{-1}$] |
| $l_B$ | true length of the interfacial area between gas and liquid [m] |
| $l_C$ | characteristic length of the modified Reynolds number [m] |
| $M, M_0$ | torsion, torsion of reference state [N m] |
| $n$ | stirring speed [$s^{-1}$] |
| $Ne$ | Newton number [–] |
| OUR | volumetric oxygen uptake rate [$Mol\,L^{-1}\,h^{-1}$] |
| $p^{gas}_{CO_2}$ | partial pressure of $CO_2$ in the gas phase [Pa] |
| $P$ | power input [W] |
| $Re$, $Re_{mod}$ | Reynolds number, modified Reynolds number [–] |
| $u$ | stirrer speed [$m\,s^{-1}$] |
| $V$ | fluid volume [L] |
| $w$ | fluid velocity [$m\,s^{-1}$] |
| $\varphi$ | angle at rocking motions in wave-mixed bioreactors [°] |
| $\nu$ | kinetic viscosity [$m^2\,s^{-1}$] |
| $\rho_L$ | fluid density [$kg\,m^{-3}$] |
| $\tau_m$ | mixing time [s] |

## References


1. Langer, E.S. (2012) Single-use bioreactors Get Nod. *Genet. Eng. Biotechnol. News*, **32** (14), 16.
2. Singh, V. (1999) Disposable bioreactor for cell culture using wave-induced agitation. *Cytotechnology*, **30** (1–3), 149–158.
3. Oosterhuis, N.M.G., Neubauer, P., and Junne, S. (2013) Single-use bioreactors for microbial cultivation. *Pharm. Bioprocess.*, **1** (2), 167–177.
4. Eibl, R., Löffelholz, C., and Eibl, D. (2011) in *Single-Use Technology in Biopharmaceutical Manufacture* (eds R. Eibl and D. Eibl), John Wiley & Sons, Inc., Weinheim, Hoboken, NJ, pp. 33–52.
5. Klockner, W., Diederichs, S., and Büchs, J. (2014) Orbitally shaken single-use bioreactors. *Adv. Biochem. Eng. Biotechnol.*, **138**, 45–60.
6. Eibl, R., Kaiser, S., Lombriser, R., and Eibl, D. (2010) Disposable bioreactors: the current state-of-the-art and recommended applications in biotechnology. *Appl. Microbiol. Biotechnol.*, **86** (1), 41–49.
7. Neubauer, P., Cruz, N., Glauche, F., Junne, S., Knepper, A., and Raven, M. (2013) Consistent development of bioprocesses from microliter cultures to the industrial scale. *Eng. Life Sci.*, **13** (3), 224–238.
8. Löffelholz, C., Kaiser, S., Krause, M., Eibl, R., and Eibl, D. (2014) in *Disposable Bioreactors II*, Advances in Biochemical Engineering/Biotechnology (eds R. Eibl and D. Eibl), Springer-Verlag, Berlin, Heidelberg, pp. 1–44.
9. Ullah, M., Burns, T., Bhalla, A., Beltz, H.W., Greller, G., and Adams, T. (2008) Disposable bioreactors for cells and microbes-productivities similar to those achieved with stirred tanks can be achieved with disposable bioreactors. *BioPharm Int.*, November 2, 44.
10. Eibl, R., Werner, S., and Eibl, D. (2010) Bag bioreactor based on wave-induced motion: characteristics and applications. *Adv. Biochem. Eng. Biotechnol.*, **115**, 55–87.
11. Junne, S., Solymosi, T., Oosterhuis, N., and Neubauer, P. (2013) Cultivation of cells and microorganisms in wave-mixed disposable bag bioreactors at different scales. *Chem. Ing. Tech.*, **85** (1–2), 57–66.
12. Dreher, T., Husemann, U., Zahnow, C., de Wilde, D., Adams, T., and Greller, G. (2013) High cell density *Escherichia coli* cultivation in different single-use bioreactor systems. *Chem. Ing. Tech.*, **85** (1–2), 162–171.
13. Dreher, T., Walcarius, B., Husemann, U., Klingenberg, F., Zahnow, C., Adams, T. *et al.* (2014) Microbial high cell density fermentations in a stirred single-use bioreactor. *Adv. Biochem. Eng. Biotechnol.*, **138**, 127–147.
14. Eibl, D. and Eibl, R. (2009) Bioreactors for mammalian cells: general overview, in *Cell and Tissue Reaction Engineering* (eds R. Eibl, D. Eibl, R. Pörtner, G. Catapano, and P. Czermak), Springer-Verlag, Berlin.
15. Kaiser, S.C., Eibl, R., and Eibl, D. (2011) Engineering characteristics of a single-use stirred bioreactor at bench-scale: the Mobius CellReady 3L bioreactor as a case study. *Eng. Life Sci.*, **11** (4), 359–368.
16. Anderlei, T., Cesena, C., Bürki, C., De Jesus, M., Kühner, M., Wurm, F. *et al.* (2009) Shaken bioreactors provide culture alternative. *Genet. Eng. Biotechnol. News*, **29**, 44.
17. Vanhamel, S. and Masy, C. (2011) in *Single-Use Technology in Biopharmaceutical Manufacture* (eds R. Eibl and D. Eibl), John Wiley & Sons, Inc., Weinheim, Hoboken, NJ, pp. 113–134.
18. Bestwick, D. and Colton, R. (2009) Extractables and leachables from single-use disposables. *BioProcess Int.*, **7** (Suppl. 1), 88–94.
19. Uettwiller, I. (2006) Testing and validation of disposable systems. *Genet. Eng. News*, **26** (3), 54–57.
20. Glindkamp, A., Riechers, D., Rehbock, C., Hitzmann, B., Scheper, T., and Reardon, K.F. (2009) Sensors in disposable bioreactors status and trends. *Adv. Biochem. Eng. Biotechnol.*, **115**, 145–169.
21. Lindner, P., Endres, C., Bluma, A., Höpfner, T., Glindkamp, A., Haake,

C. *et al.* (2011) in *Single-Use Technology in Biopharmaceutical Manufacture* (eds R. Eibl and D. Eibl), John Wiley & Sons, Inc., Weinheim, Hoboken, NJ, pp. 67–81.

22. Bernard, F., Chevalier, E., Cappia, J.-M., Heule, M., and Paust, T. (2009) Disposable pH sensors. *BioProcess Int.*, 7 (Suppl. 1), 32–36.
23. Spichiger, S. and Sprichiger-Keller, U.E. (2011) in *Single-Use Technology in Biopharmaceutical Manufacture* (eds R. Eibl and D. Eibl), John Wiley & Sons, Inc., Hoboken, NJ, pp. 295–299.
24. Potyrailo, R.A. (2011) Passive multivariable temperature and conductivity RFID sensors for single-use biopharmaceutical manufacturing components. *Biotechnol. Progr.*, **27** (3), 875–884.
25. Nacke, T., Barthel, A., Frense, D., Meister, M., and Cahill, B.P. (2013) Application of high frequency sensors for contactless monitoring in disposable bioreactors. *Chem. Ing. Tech.*, **85** (1–2), 179–185.
26. Oosterhuis, N.M.G. and Tromper, A. (2014) Method and apparatus for cultivating cells. EP12704557
27. Rothe, S. and Eibl, D. (2011) in *Single-Use Technology in Biopharmaceutical Manufacture* (eds R. Eibl and D. Eibl), John Wiley & Sons, Inc., Weinheim, Hoboken, NJ, pp. 53–65.
28. Hartlep, M. and Künnecke, W. (2014) Online-Glucoseregelung für die Kultivierung von Mikroorganismen und Zelllinien. *Chem. Ing. Tech.*, **86**, 1586–1587.
29. Shukla, A.A. and Gottschalk, U. (2013) Single-use disposable technologies for biopharmaceutical manufacturing. *Trends Biotechnol.*, **31** (3), 147–154.
30. Yu, L.X. (2008) Pharmaceutical quality by design: product and process development, understanding, and control. *Pharm. Res.*, **25** (4), 781–791.
31. Nienow, A.W. (2006) Reactor engineering in large scale animal cell culture. *Cytotechnology*, **50** (1–3), 9–33.
32. Zhang, Z., Perozziello, G., Boccazzi, P., Sinskey, A., Geschke, O., and Jensen, K. (2007) Microbioreactors for bioprocess development. *J. Assoc. Lab. Autom.*, **12** (3), 143–151.
33. Knorr, B., Schlieker, H., Hohmann, H.-P., and Weuster-Botz, D. (2007) Scale-down and parallel operation of the riboflavin production process with *Bacillus subtilis*. *Biochem. Eng. J.*, **33** (3), 263–274.
34. Noack, U., De Wilde, D., Verhoeye, F., Balbirnie, E., Kahlert, W., Adams, T., Greller, G., and Reif, O.-W. (2011) in *Single-Use Technology in Biopharmaceutical Manufacture* (eds R. Eibl and D. Eibl), John Wiley & Sons, Inc., pp. 225–240.
35. Zhang, X., Stettler, M., De Sanctis, D., Perrone, M., Parolini, N., Discacciati, M. *et al.* (2010) Use of orbital shaken disposable bioreactors for mammalian cell cultures from the milliliter-scale to the 1,000-liter scale. *Adv. Biochem. Eng. Biotechnol.*, **115**, 33–53.
36. Buchs, J., Maier, U., Milbradt, C., and Zoels, B. (2000) Power consumption in shaking flasks on rotary shaking machines: I. Power consumption measurement in unbaffled flasks at low liquid viscosity. *Biotechnol. Bioeng.*, **68** (6), 589–593.
37. Klockner, W., Tissot, S., Wurm, F., and Buchs, J. (2012) Power input correlation to characterize the hydrodynamics of cylindrical orbitally shaken bioreactors. *Biochem. Eng. J.*, **65**, 63–69.
38. Raval, K., Kato, Y., and Buchs, J. (2007) Comparison of torque method and temperature method for determination of power consumption in disposable shaken bioreactors. *Biochem. Eng. J.*, **34** (3), 224–227.
39. Maier, U. and Buchs, J. (2001) Characterisation of the gas-liquid mass transfer in shaking bioreactors. *Biochem. Eng. J.*, 7 (2), 99–106.
40. Tissot, S., Farhat, M., Hacker, D.L., Anderlei, T., Kuhner, M., Comninellis, C. *et al.* (2010) Determination of a scale-up factor from mixing time studies in orbitally shaken bioreactors. *Biochem. Eng. J.*, **52** (2–3), 181–186.
41. Glazyrina, J., Krause, M., Junne, S., Glauche, F., Storm, D., and Neubauer, P. (2012) Glucose-limited high cell density cultivations from small to pilot plant scale using an enzyme-controlled glucose delivery system. *New Biotechnol.*, **29** (2), 235–242.

42. Glazyrina, J., Materne, E.M., Dreher, T., Storm, D., Junne, S., Adams, T. *et al.* (2010) High cell density cultivation and recombinant protein production with Escherichia coli in a rocking-motion-type bioreactor. *Microb. Cell Fact.*, **9**, 42.
43. Venkat, R.V., Stock, L.R., and Chalmers, J.J. (1996) Study of hydrodynamics in microcarrier culture spinner vessels: a particle tracking velocimetry approach. *Biotechnol. Bioeng.*, **49** (4), 456–466.
44. Jeude, M., Dittrich, B., Niederschulte, H., Anderlei, T., Knocke, C., Klee, D. *et al.* (2006) Fed-batch mode in shake flasks by slow-release technique. *Biotechnol. Bioeng.*, **95** (3), 433–445.
45. Funke, M., Buchenauer, A., Schnakenberg, U., Mokwa, W., Diederichs, S., Mertens, A. *et al.* (2010) Microfluidic BioLector-microfluidic bioprocess control in microtiter plates. *Biotechnol. Bioeng.*, **107** (3), 497–505.
46. Gebhardt, G., Hortsch, R., Kaufmann, K., Arnold, M., and Weuster-Botz, D. (2011) A new microfluidic concept for parallel operated milliliter-scale stirred tank bioreactors. *Biotechnol. Progr.*, **27** (3), 684–690.
47. Kensy, F., Zang, E., Faulhammer, C., Tan, R.-K., and Buechs, J. (2009) Validation of a high-throughput fermentation system based on online monitoring of biomass and fluorescence in continuously shaken microtiter plates. *Microb. Cell Fact.*, **4**, 8.
48. Tang, Y.J., Laidlaw, D., Gani, K., and Keasling, J.D. (2006) Evaluation of the effects of various culture conditions on Cr(VI) reduction by *Shewanella oneidensis* MR-1 in a novel high-throughput mini-bioreactor. *Biotechnol. Bioeng.*, **95** (1), 176–184.
49. Isett, K., George, H., Herber, W., and Amanullah, A. (2007) Twenty-four-well plate miniature bioreactor high-throughput system: assessment for microbial cultivations. *Biotechnol. Bioeng.*, **98** (5), 1017–1028.
50. Betts, J.P.J., Warr, S.R.C., Finka, G.B., Uden, M., Town, M., Janda, J.M. *et al.* (2014) Impact of aeration strategies on fed-batch cell culture kinetics in a single-use 24-well miniature bioreactor. *Biochem. Eng. J.*, **82**, 105–116.
51. Huber, R., Ritter, D., Hering, T., Hillmer, A.-K., Kensy, F., Mueller, C. *et al.* (2009) Robo-Lector - a novel platform for automated high-throughput cultivations in microtiter plates with high information content. *Microb. Cell Fact.*, **1**, 8.
52. Panula-Perälä, J., Šiurkus, J., Vasala, A., Wilmanowski, R., Casteleijn, M.G., and Neubauer, P. (2008) Enzyme controlled glucose auto-delivery for high cell density cultivations in microplates and shake flasks. *Microb. Cell Fact.*, **7** (1), 31.
53. Hemmerich, J., Wenk, P., Lütkepohl, T., and Kensy, F. (2011) Fed-batch cultivation in baffled shake-flasks. *Genet. Eng. Biotechnol. News*, **31** (14), 52–54.
54. Löffelholz, C., Kaiser, S.C., Werner, S., and Eibl, D. (2011) CFD as a tool to characterize single-use bioreactors, in *Single-Use Technology in Biopharmaceutical Manufacture* (eds R. Eibl and D. Eibl), John Wiley & Sons, Inc., Hoboken, NJ.
55. Oncul, A.A., Kalmbach, A., Genzel, Y., Reichl, U., and Thevenin, D. (2010) Characterization of flow conditions in 2 L and 20 L wave bioreactors (R) using computational fluid dynamics. *Biotechnol. Progr.*, **26** (1), 101–110.
56. Kaiser, S.C., Löffelholz, C., Werner, S., and Eibl, D. (2011) in *Computational Fluid Dynamics Technologies and Applications* (ed I. Minin), InTech, pp. 97–122.
57. Odeleye, A.O.O., Marsh, D.T.J., Osborne, M.D., Lye, G.J., and Micheletti, M. (2014) On the fluid dynamics of a laboratory scale single-use stirred bioreactor. *Chem. Eng. Sci.*, **111**, 299–312.
58. Pierce, L. (2004) Scalability of a disposable bioreactor from 25 L–500 L in perfusion mode with a CHO-based cell line: a tech review. *Bioprocess. J.*, **3**, 51–56.
59. Mardirosian, D., Guertin, P., Crowell, J., Yetz-Aldape, J., Hall, M., Hodge, G. *et al.* (2009) Scaling up a CHO produced hormone-protein fusion product. *Bioprocess Int.*, **7** (4), 30–35.
60. Oosterhuis, N.M.G., Hudson, T., D'Avino, A., Zijlstra, G.M., and Amanullah, A. (2011) in *Comprehensive Biotechnology*, 2nd edn (ed M.

Moo-Young), Elsevier, Amsterdam, pp. 249–261.
61. Zijstra, G.M., Hof, R.P., and Schilder, J. (2008) Improved process for culturing cells WO2008/006494A1.
62. Douwenga R. (2009) Moving to the next level of manufacturing. BioProcess International Conference, Raleigh, NC.
63. Minow, B., Seidemann, J., Tschoepe, S., Gloeckner, A., and Neubauer, P. (2014) Harmonization and characterization of different single-use bioreactors adopting a new sparger design. *Eng. Life Sci.*, **14** (3), 272–282.
64. Minow, B., Tschoepe, S., Regner, A., Populin, M., Reiser, S., Noack, C. *et al.* (2014) Biological performance of two different 1000 L single-use bioreactors applying a simple transfer approach. *Eng. Life Sci.*, **14** (3), 283–291.
65. Imseng, N., Steiger, N., Frasson, D., Sievers, M., Tappe, A., Greller, G. *et al.* (2014) Single-use wave-mixed versus stirred bioreactors for insect-cell/BEVS-based protein expression at benchtop scale. *Eng. Life Sci.*, **14** (3), 264–271.
66. Goudar, C.T., Piret, J.M., and Konstantinov, K.B. (2011) Estimating cell specific oxygen uptake and carbon dioxide production rates for mammalian cells in perfusion culture. *Biotechnol. Progr.*, **27** (5), 1347–1357.
67. Royce, P.N.C. and Thornhill, N.F. (1991) Estimation of dissolved carbon-dioxide concentrations in aerobic fermentations. *AIChE J.*, **37** (11), 1680–1686.
68. Nyberg, G., Jimenez, F., Li, F., Spasoff, A., Olsen, G., and Grampp, G. (2006) Predictive scaling of cell culture aeration in large-scale bioreactors. Cell Culture Engineering X, Whistler.
69. Frohlich, B., Bedard, C., Gagliardi, T., and Oosterhuis, N.M.G. (eds) (2012) Co-development of a new 2-D rocking single-use bioreactor to streamline cell expansion processes. IBC BioProcess International Conference, Rhode Island.
70. Castilho, L.R. and Medronho, R.A. (2002) Cell retention devices for suspended-cell perfusion cultures. *Adv. Biochem. Eng. Biotechnol.*, **74**, 129–169.
71. Sinclair, A. and Brown, A. (2013) in *Continuous Bioprocessing, Current Practice and Future Potential* (ed J. Bonham-Carter), Refine Technology, pp. 73–81.
72. Bonham-Carter, J., Weegar, J., Nieminen, A., Shevitz, J., and Eliezer, E. (2011) The use of the ATF system to culture Chinese hamster ovary cells in a concentrated fed-batch system. *BioPharm Int.*, **24** (6), 42.
73. Singh, V. (2003) Disposable perfusion bioreactor for cell culture, US2003/036192A1.
74. Tapia, F., Vogel, T., Genzel, Y., Behrendt, I., Hirschel, M., Gangemi, J.D. *et al.* (2014) Production of high-titer human influenza A virus with adherent and suspension MDCK cells cultured in a single-use hollow fiber bioreactor. *Vaccine*, **32** (8), 1003–1011.
75. Eibl, R. and Eibl, D. (2008) Design of bioreactors suitable for plant cell and tissue cultures. *Phytochem. Rev.*, **7** (3), 593–598.
76. Raven, N., Rasche, S., Kuehn, C., Anderlei, T., Klockner, W., Schuster, F. *et al.* (2015) Scaled-up manufacturing of recombinant antibodies produced by plant cells in a 200-L orbitally-shaken disposable bioreactor. *Biotechnol. Bioeng.* **112** (2), 308–321
77. Lehmann, N., Rischer, H., Eibl, D., and Eibl, R. (2013) Wave-mixed and orbitally shaken single-use photobioreactors for diatom algae propagation. *Chem. Ing. Tech.*, **85** (1–2), 197–201.
78. Hahne, T., Schwarze, B., Kramer, M., and Frahm, B. (2014) Disposable algae cultivation for high-value products using all around LED-illumination directly on the bags. *J. Algal Biomass Util.*, **5** (2), 66–73.
79. Mikola, M., Seto, J., and Amanullah, A. (2007) Evaluation of a novel Wave Bioreactor cellbag for aerobic yeast cultivation. *Bioprocess Biosyst. Eng.*, **30** (4), 231–241.
80. Hitchcock, T. (2009) Production of recombinant whole-cell vaccines with disposable manufacturing systems. *BioProcess Int.*, **5**, 36–45.
81. Mahajan, E., Matthews, T., Hamilton, R., and Laird, M.W. (2010) Use of disposable reactors to generate inoculum

cultures for *E. coli* production fermentations. *Biotechnol. Progr.*, **26** (4), 1200–1203.

82. Galliher, P.M. (2010) *Single-Use Bioreactor Platform for Microbial Fermentation*, John Wiley & Sons, Inc., Hoboken, NJ, pp. 241–250.
83. Galliher, P.M. (2008) Achieving high-efficiency production with microbial technology in a single-use bioreactor platform. *BioProcess Int.*, **6** (11), 60–65.
84. Bareither, R., Bargh, N., Oakeshott, R., Watts, K., and Pollard, D. (2013) Automated disposable small scale reactor for high throughput bioprocess development: a proof of concept study. *Biotechnol. Bioeng.*, **110** (12), 3126–3138.
85. Jonczyk, P., Takenberg, M., Hartwig, S., Beutel, S., Berger, R.G., and Scheper, T. (2013) Cultivation of shear stress sensitive microorganisms in disposable bag reactor systems. *J. Biotechnol.*, **167** (4), 370–376.
86. Fraser, S.J. and Endres, C. (2014) Quorus bioreactor: a new perfusion-based technology for microbial cultivation. *Adv. Biochem. Eng. Biotechnol.*, **138**, 149–177.
87. Hillig, F., Annemuller, S., Chmielewska, M., Pilarek, M., Junne, S., and Neubauer, P. (2013) Bioprocess development in single-use systems for heterotrophic marine microalgae. *Chem. Ing. Tech.*, **85** (1–2), 153–161.
88. Hillig, F., Porscha, N., Junne, S., and Neubauer, P. (2014) Growth and docosahexaenoic acid production performance of the heterotrophic marine microalgae *Crypthecodinium cohnii* in the wave-mixed single-use reactor CELL-tainer. *Eng. Life Sci.*, **14** (3), 254–263.
89. Pietrzykowski, M., Flanagan, W., Pizzi, V., Brown, A., Sinclair, A., and Monge, M. (2013) An environmental life cycle assessment comparison of single-use and conventional process technology for the production of monoclonal antibodies. *J. Cleaner Prod.*, **41**, 150–162.

# 10
# Computational Fluid Dynamics for Bioreactor Design

*Anurag S. Rathore, Lalita Kanwar Shekhawat, and Varun Loomba*

## 10.1
## Introduction

Most unit operations that are used in bioprocessing involve multiphase flows. This is particularly true for continuously stirred-tank reactors (CSTR) that are commonly used as bioreactors or fermenters. In these reactors, gas is sparged into a medium consisting of liquid media and solid cells. Back-mixing and gas-by-pass caused by poor mixing may result in stagnant or dead zones. These in turn create gradients within the working medium and result in inadequate supply of nutrients and oxygen to the cells. Knowledge of hydrodynamics of bioreactors is thus essential for understanding nonideal flow within the bioreactor [1].

When dealing with single-phase flows, the dynamics of the fluid can be easily determined by simultaneously solving the continuity equation and the Navier–Stokes equation. In case of multiphase flows, however, the complexity increases remarkably and a lot of factors need to be taken into account for reaching accurate conclusions. These include momentum exchange between phases, mass transfer between the phases, and absorption of one phase in others.

Proper mixing and gas hold-up in a reactor play a major role in determining its performance. Perfect mixing is a condition where there is uniformity in the concentration of every species across the reactor. Most reactors are designed based on the assumption of perfect mixing, which results in errors between the predictions and actual measurements, with the errors increasing with the size of the reactor [2]. In reality, the flow field inside the reactor is inherently unsteady and turbulent eddies that are formed in the reactor are nonuniform, thereby resulting in unsteady flow profiles. Considering the need for uniform mixing in a reactor, considerable efforts are made by the industry in moving as close to the ideal state as possible. Good mixing results in efficient mass transfer of dispersed phase into the continuous phase and also uniform concentration of all the species in the reactor [3]. In general, the bioreactors are equipped with impellers that can be operated at different impeller speeds and thus provide better contact and high interfacial area between the phases. This results in higher mass transfer coefficient between the phases in the reactor [4–6].

*Bioreactors: Design, Operation and Novel Applications,* First Edition. Edited by Carl-Fredrik Mandenius.

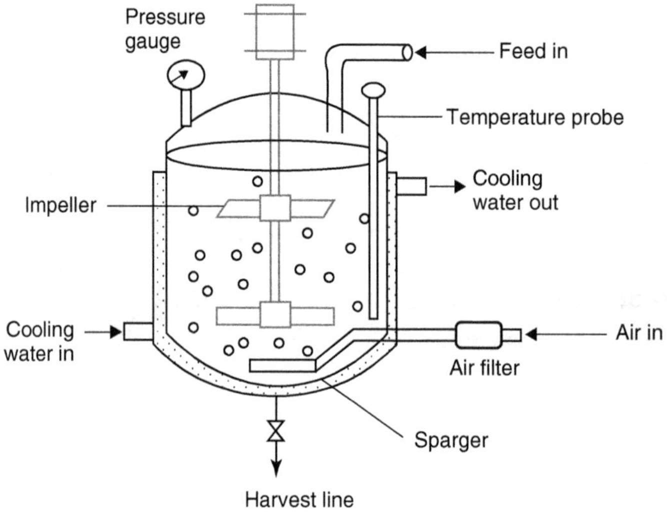


**Figure 10.1** Schematic of an aerated bioreactor.

In the biotech industry, aerated stirred-tank reactors are commonly used for growing microbial and mammalian cells (Figure 10.1). Oxygen is purged through the medium using a sparger placed at the bottom of the reactor. Product formation is often dependent on the rate at which the oxygen is transferred from one phase to another. To measure oxygen transfer between the phases, parameters such as gas hold-up and volumetric mass transfer coefficient are commonly used. Bubble size distribution is also an important factor as it directly affects the interfacial area between the different phases [4].

Computational fluid dynamics (CFD) is widely used for simulating the hydrodynamics within the bioreactor by solving the continuity equation and the Navier–Stokes equation simultaneously. The system volume is divided into smaller control volumes (called cells) and the momentum equations are numerically solved for each of the control volumes [6]. In order to avoid mathematical complexities, many researchers assume a single bubble size for the gas phase in the reactor. This makes the simulations computationally simple and thus faster. In a real process, however, the bubble size does change as the gas phase moves inside the reactor. Bubbles exit the sparger and then get broken down into smaller bubbles as they go through the high shear zone around the agitator. The interfacial area per unit volume of the reactor increases dramatically when this happens and this is in fact responsible for increase in the mass transfer rate in the reactor. A population balance model is, therefore, required when performing CFD simulations for a bioreactor [4].

Several operating parameters are known to affect mass transfer in a bioreactor. These include impeller speed, gas inlet flow rate, and gas hold-up. Figures 10.2a1 and a2 illustrate the effect of impeller speed and gas inlet flow rate, respectively. Figure 10.2a1 shows simulations performed at different impeller speeds on a

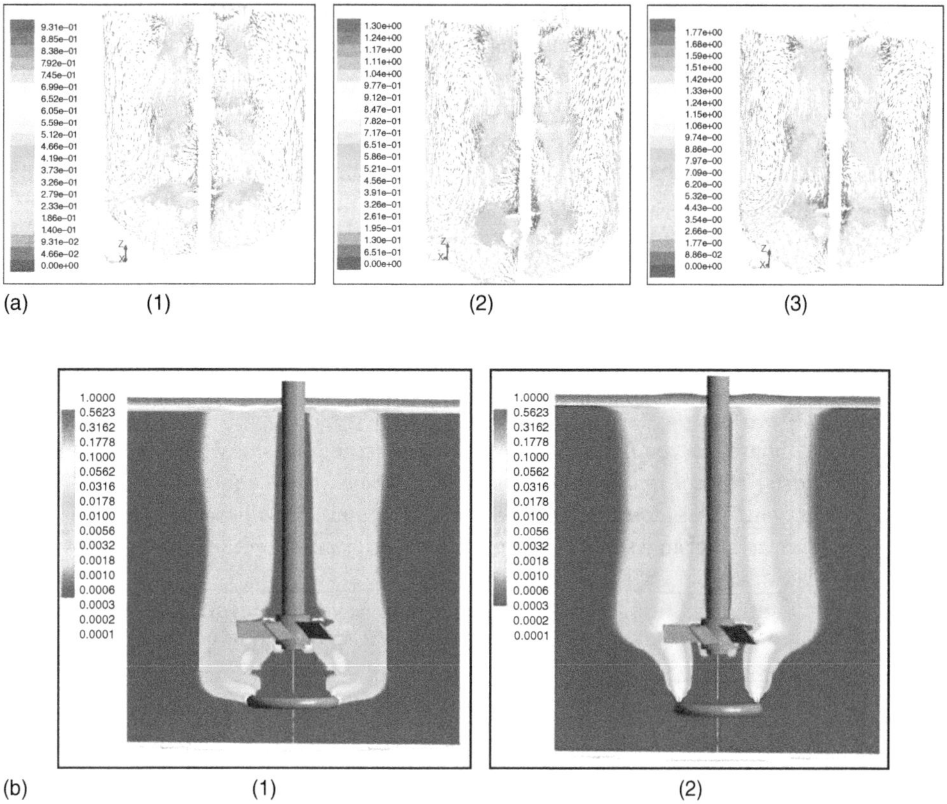


**Figure 10.2** (a) Velocity vector variation at different impeller speeds of (1) 100 RPM, (2) 150 RPM, and (3) 200 RPM. Three impellers out of which one is a Rushton turbine while the other two are 3-blade propeller-type impellers are mounted on shaft. The reactor capacity is of 67 l.(b) Gas phase volume fraction at gas inlet flow rate of (1) 1 LPM and (2) 4 LPM for 67 l capacity bioreactor.

bioreactor containing two 3-blade propeller-type impellers and one Rushton turbine impeller. It is seen that as the impeller speed is increased, turbulence inside the reactor also increases. At higher impeller speed (Figure 10.2a3), the mixing intensity and gas phase distribution are higher than at lower speed (Figure 10.2a1). However, shear and power consumption also increases. Thus, for the optimal operation of a bioreactor, it is critical to identify the impeller speed that yields the required mixing with minimal shear and power consumption. It is well known that Rushton turbine creates more turbulence as compared with propeller-type impellers [7].

Figure 10.2b shows the variation in gas phase volume fraction with gas inlet flow rates at constant impeller speed. On comparing Figures 10.2b1 and b2, it is observed that as the gas flow rate increases, the gas phase distribution also increases and more regions are seen to have high gas phase fractions (more regions are yellow than green). This increases the gas hold-up and improves the contact between the phases, thereby improving the performance of the reactor [8].

Hence, CFD is a useful tool to study the effects of various parameters on the performance of the bioreactors, including design of the bioreactor, agitation speed, gas-flow rate, and hydrodynamic properties such as velocity gradient distribution, circulation time distribution, gas hold-up, volume fractions, mass transfer coefficient, power dissipation, and shear stress. In Table 10.1, a summary of published CFD applications in the area of microbial fermentation or mammalian cell culture is listed [9–23].

## 10.2 Multiphase Flows

A "phase" can be defined as the thermodynamic state in which any component exists. Hence, multiphase flows refer to flows when more than one phase is present. Most industrial reactors have multiphase flow processes. In such cases, even small changes in the design of the reactor can result in large changes in the hydrodynamics of the reactor [24].

Multiphase flows can be classified into several categories either based on thermodynamics (e.g., gas–solid, gas–liquid, liquid–liquid) or on flow regimes. Examples of latter include **dispersed flow** in which only one phase is continuous while all others are dispersed, **multiphase flows** such as flow of gas bubbles in liquid, solid particles in liquid or gas, and liquid droplets in gas, **separated flows** in which all the phases are semi-continuous mode and have interfaces to differentiate them such as annular flow and jet flow, and finally, **mixed flows** in which both separated and dispersed flows are present, some phases are discrete and some are semi-continuous such as bubbly annular flow and droplet annular flow.

CFD modeling of multiphase flow involves the definition of a phase/flow regime, formulation of the governing equations, and finally, solution of the governing equations. Approaches that are commonly used include

1) Eulerian–Lagrangian approach that involves the application of the Eulerian approach for continuous phase and Lagrangian framework for dispersed phase
2) Eulerian–Eulerian approach that involves the application of the Eulerian approach for both phases without explicitly accounting for the interface
3) volume of fluid approach that involves the application of the Eulerian framework for both phases with reformation of the interface on volume basis.

### 10.2.1 Eulerian–Lagrangian Approach

In this approach, motion of the continuous phase is modeled in the Eulerian framework while the dispersed phase is modeled in the Lagrangian framework.

**Table 10.1** Summary of CFD applications in fermentation and cell culture bioprocess.

| Type of reactors | Application | Results |
|---|---|---|
| Wave reactor | Wave reactors for mammalian cell culture | In wave reactors (2 l and 20 l disposable polyethylene terephthalate bags), the flow is in the laminar regime leading to lower shear stress generation [9]. |
| Bioreactors | 3D simulation of mixing in tissue engineering applications | CFD is used to model the flow hydrodynamics within the bioreactors to optimize design, flow, nutritional, and metabolic requirements of cells and TECs without having to perform numerous and expensive bioreactor experiments saving time and resources [10]. |
| Bioreactors | Circulation time distribution (CTD) prediction | CFD used to estimate the circulation time distribution for three single-impeller mixing tanks based on the position of the particle instead of multiple triggering, which resulted into unimodal CTDs for tanks with similar geometries [11]. |
| Bioreactors | Cell culture: heterogeneous two-phase system | CFD model used for characterization of two-phase flow and hydrodynamic shear by solving momentum and mass transfer that occurs between the phases [12]. |
| Multi-impeller bioreactors | Cell culture: heterogeneous two-phase system | Two-fluid model along with MUSIF model (for polydispersed gas flow) implemented to predict the various flow regimes and hydrodynamic parameters in dual-turbine stirred bioreactor. Model successfully captured the flow regimes as observed during experiments. Effect of gas flow rate and impeller speed on gas hold-up and power consumption has been investigated. Gas hold-up increases with an increase in stirring speed for different gas flow rates. At higher impeller speeds, very small bubbles are generated and spend more time in the reactor [13]. |
| Bioreactors | Fermentation broth | CFD models revealed rheological analysis, that is, change in bioreactor mixing changes biomass concentration. Contour plots showed reduction in fluid viscosity in high shear stress regions and a direct dependence of temperature and concentration of biomass in the bioreactor on the apparent viscosity of fermentation broths. Loading concentration of solids more than 20% w/v affects fluid transport, mixing, and solids distribution negatively [14]. |

(*continued overleaf*)

**Table 10.1** (*continued*)

| Type of reactors | Application | Results |
|---|---|---|
| Bioreactors | Cell suspension cultures | High shear rate damages cell. Model helps to characterize the local shear flow by relating the shear stress ratio (a ratio of maximum to average shear stresses) and the velocity. The simulation of 2D flow field showed a strong agreement with the empirical results. Significant cell damage confined within two-stirrer diameters downstream of the stirrer, where the most intense energy dissipation rates were observed. Cell damage was found to correlate well with both the bulk energy dissipation and local energy dissipation rate. Significant cell damage occurs when the maximum energy dissipation is $50\,W\,kg^{-1}$ and the total energy dissipation is $104\,J\,kg^{-1}$ [15]. |
| Bioreactors | Enzymatic cell culture | CFD formulates a coupled hydrodynamics-reaction kinetics model to simulate a gas–liquid–solid three-phase heterogeneous system. Hydrodynamics model predicts of flow field, whereas reaction kinetics model portrays the reaction conversion process. Model is suitable for continuous flow systems but not for batch systems [16]. |
| Centrifugal impeller bioreactor | Cell culture | CFD is used to estimate the velocity profile, shear stress, liquid circulation, and circulation time distribution in a novel bioreactor. CFD defines design principles and characteristics of the CIB aiding in further optimization and scale-up/scale-down [17]. |
| Bioreactors | Fermentation broth | Bioprocess requires understanding and modeling the complex interactions between biological reaction and hydrodynamics. CFD is used to model the hydrodynamics behavior of bioreactors of different size and its interaction with the biological reaction. Hydrodynamics coupled with reactions is computationally intensive. Hence, to minimize the computational time, compartment or multizone model is used. Manual zoning algorithm and LBL2 automatic zoning algorithm based on the three velocity components explains the mixing of an inert tracer very well [18]. |
| Laminar flow bioreactor | Cell culture | CFD is used to simulate flow through a porous scaffold and recommends changes in design. Three different designs simulated conducted with different permeability values of scaffold and tissue. Bioreactor design optimized by understanding the flow patterns in combination with pressure distribution within the bioreactor [19]. |

**Table 10.1** (*continued*)

| Type of reactors | Application | Results |
|---|---|---|
| Stirred airlift bioreactor | Fermentation broth | CFD used to predict hydrodynamics of stirred airlift bioreactor. Simulated values of average gas hold-up and average mass transfer coefficient obtained from CFD showed good agreement with experimental data. The impeller effect is prevalent at low gas flow rates as it propels the liquid against the walls of the vessel, without recirculation, resulting in low values of gas hold-up and $k_L a$. Liquid recirculation is increased with increased air flow. Simulation carried out with constant bubble diameter showed 10% approximately difference between experimental and simulation values [20]. |
| Bioreactors | Fermentation broth | CFD model characterizes of 3D flow fields in bioreactors with simple and complex geometries [21]. |
| Membrane bioreactor | Biological waste water treatment | Membrane bioreactors are used for biological waste water treatment of effective solid–liquid separation. The optimization of MBRs requires knowledge of the membrane fouling, mixing, and biokinetics. CFD-based model developed to optimize mixing energy in MBRs. Results for RTD predicted from CFD model showed good agreement with experimental measured RTD data [22]. |
| Bioreactors | Fermentation broth | Pure CFD models are not feasible for studying bioreactors for transient operation and detailed population balances requiring large numbers of quantities in each grid cell. Hence, hybrid multizonal approach along with CFD is used as a tool to model aerobic bioreactors to capture close interaction between fluid flow and biological reactions [23]. |

This approach is preferred when the aim is to model particle motion including translational, rotational, and particle–particle collisions that occur in processes such as chemical reactions or heat and mass transfer between the dispersed phase and the gas phase at the individual particle scale.

In the Eulerian–Lagrangian approach, the dispersed phase is simulated by solving the equation of motion while the continuous phase is simulated by solving the Navier–Stokes equation. Equation of motion for a single particle in dispersed phase is written as

$$m_p \frac{dU_p}{dt} = F_p + F_D + F_{vm} + F_L + F_g + F_H \tag{10.1}$$

where $m_p$ and $U_p$ are the mass and mean velocity of the particle, respectively. The right-hand side of the equation represents the summation of all the forces acting on the particle, while the left-hand side of the equation denotes the rate of change of momentum.

The force, $F_p$, is the continuous-phase pressure gradient force and $F_g$ is the gravitational force acting on the particle. Both forces can be combined and written as

$$F_p + F_g = V_p \nabla p - \rho_p V_p g \tag{10.2}$$

where $V_p$ is the volume of the particle and $p$ is the pressure of the continuous phase.

The drag force on the particle due to the relative velocity between the continuous phase and the dispersed phase, $F_D$, at higher Reynolds number [25], can be stated as

$$F_D = \frac{\pi}{8} C_D \rho_c D_p^2 |U_p - U_c| \left(U_p - U_c\right) \tag{10.3}$$

where the subscript "c" denotes the continuous phase and "p" denotes the particulate phase. $C_D$ is the drag coefficient and depends on the flow regime whether the flow is laminar or turbulent and its value changes accordingly.

Other forces that can be included in Eq. 10.3 are lift force, added mass forces, and Basset forces that act on the particle. Vorticity or shear in the continuous phase also applies force on the dispersed phase particle. This force is included in the lift force. When the dispersed phase is moving up, it takes some part of the continuous phase with it. This adds up to the mass of the dispersed phase particles. As dispersed phase and continuous phase cannot occupy the same physical space simultaneously, added mass is modeled as a volume of fluid moving with the object for mathematical simplicity. This additional mass to the dispersed phase contributes to the "virtual mass force" or "added mass force." Other forces that may play a role include Basset forces that account for viscous effects and delay in boundary layer development as the relative velocity changes with time. However, the magnitude is very less as compared with the drag forces and so these are often ignored during simulation.

The primary forces that need to be accounted include drag force and pressure force acting on the dispersed phase. For resolving the flow of continuous phase, the velocity of the continuous phase is required. The latter can be calculated by simultaneously solving the equation of continuity and the Navier–Stokes equation for the continuous phase

$$\frac{\partial(\alpha_c \rho_c)}{\partial t} + \nabla \cdot (\alpha_c \rho_c U_c) = S_c \tag{10.4}$$

$$\frac{\partial(\alpha_c \rho_c U_c)}{\partial t} + \nabla \cdot (\alpha_c \rho_c U_c U_c) = -\alpha_c \nabla p - \nabla \cdot (\alpha_c \tau_c) + \alpha_c \rho_c g + S_{cm} \tag{10.5}$$

where $S_c$ and $S_{cm}$ are the source terms for mass and momentum, respectively. These terms represent the exchange between the continuous phase and the dispersed phase. When the dispersed phase volume fraction is too low, the effect of dispersed phase on the continuous phase can be ignored, thus reducing the continuous phase equations to those of the single phase.

The interaction between the dispersed phase and the continuous phase depends on the degree of coupling between the phases. Coupling can be one way when the dispersed phase volume fraction is too low and then it can be assumed that it will not affect the continuous phase but will be affected by the continuous phase. In a two-way coupling, the dispersed phase affects the continuous phase and vice versa, while in four-way coupling the particle–particle interactions are also taken into consideration [26].

As the number of particles to be simulated increases, the computational time required for performing the simulation following the Eulerian–Lagrangian approach also increases and so use of this approach is generally limited to cases where the dispersed phase volume fraction does not exceed 10%. For cases where the volume fraction of dispersed phase is higher or for cases where dispersed phase particle–particle interactions are strong, the Euler–Euler approach is more efficient and is discussed in the next section.

### 10.2.2 Euler–Euler Approach

In this approach, all the phases are assumed to be in the Eulerian framework, that is, the reference frame is assumed to be stationary and the fluid passes through the control volume, and the governing equations such as momentum balance, mass and heat balance are solved for the flowing fluid that is continuous phase [27, 42]. All phases are treated as separate continuum and share the same domain. As mentioned earlier, this approach is preferred for cases where the dispersed phase volume fraction is greater than 10%. Examples where this approach is preferred include modeling of fluidized bed, multiphase stirred reactor, and bubble column reactors.

As the phases can also interpenetrate each other, these interactions need to be accounted for and this is achieved by using the interphase momentum exchange term. Furthermore, averaging of the governing equations is performed on the per unit volume basis with the assumption

$$\sum_{k=1}^{n} \alpha_k = 1 \tag{10.6}$$

where $\alpha_k$ is the volume fraction of a particular phase and "$n$" is the total number of phases. The continuity equation then becomes

$$\frac{\partial(\alpha_k \rho_k)}{\partial t} + \nabla\cdot(\alpha_k \rho_k U_k) = \sum_{p=1,p\neq k}^{n} S_{pk} \tag{10.7}$$

where $\rho_k$ is the density and $U_k$ is the mean velocity of the phase "$k$," and $S_{pk}$ is the rate of mass transfer between phases "$p$" and "$k$." The equation of momentum can then be written as

$$\frac{\partial(\alpha_k \rho_k U_k)}{\partial t} + \nabla\cdot(\alpha_k \rho_k U_k U_k) = -\alpha_k \nabla p - \nabla\cdot(\alpha_k \tau_k) + (\alpha_k \rho_k g) + F_k + F_g \quad (10.8)$$

where $F_k$ is the interphase momentum exchange term between phase "$k$" and all other phases. It should be noted that pressure in Eq. 10.2 is common for all phases and hence it does not have any subscript in the equation. The interphase momentum exchange term is responsible for all the above-mentioned interactions between the phases and is a key differentiator between multiphases versus single-phase flows.

### 10.2.3 Volume of Fluid Approach (VOF)

In this approach, all the phases are modeled in the Eulerian framework and hence modeling involves solving the equations of continuity, momentum, and mass and energy transport simultaneously for all the phases. In case of two-phase systems, the dispersed phase (such as bubbles or drops) is separated from the continuous phase by a boundary. This boundary represents a solid geometry through which the fluid flows, and it executes motion under the influence of flow around it. When the interface is not stationary, applying a boundary condition becomes difficult. Hence, equations are solved for all the phases and motion of all the phases is captured, thereby capturing the motion of the interface indirectly. VOF is primarily used to simulate the interfacial properties such as wall adhesion and surface tension. This approach is useful for modeling only few dispersed phase particles because it requires significantly large computational resources.

Since all the fluids share a single set of equations, the model can be expressed as

$$\frac{\partial \rho}{\partial t} + \nabla\cdot(\rho U) = \sum S_k \quad (10.9)$$

$$\frac{\partial(\rho U)}{\partial t} + \nabla\cdot(\rho U U) = -\nabla \pi + \rho g + F \quad (10.10)$$

Front tracking method is used for solving these equations )numerically and is computationally cumbersome. Since a control volume contains more than one fluid, the above-mentioned equations are solved using mixture properties.

$$\rho = \sum \alpha_k \rho_k \quad (10.11)$$

$$C_p = \frac{\sum \alpha_k \rho_k C_{pk}}{\sum \alpha_k \rho_k} \quad (10.12)$$

where $\alpha_k$ is the volume fraction of the phase "$k$" in the control volume. The value of the volume fraction is calculated by solving the continuity equation of volume fraction for $N-1$ phases.

$$\frac{\partial \alpha_k}{\partial t} + (U_k\cdot\nabla)\alpha_k = S_k \quad (10.13)$$

It must be highlighted that volume fraction cannot determine the exact interface as different interface configurations can lead to same volume fractions. Hence,) several different techniques are applied to calculate the exact interface [28, 29].

## 10.3 Turbulent Flow

Turbulent flow has been defined by Hinze [30] as "an irregular condition of flow in which various quantities show a random variation with time and space coordinates, so that statistically distinct average values can be discerned." It is witnessed in the flowing fluid as the Reynolds number increases. Shear is the primary source of energy that introduces turbulence in laminar flow. Hence, it can be stated that if shear is reduced then turbulent flow would also reduce to laminar and the Reynolds number would decrease. Till now, the transition from laminar to turbulent flow has not been modeled. Existing theories can predict the transition for small disturbances but cannot do it for large disturbances. All of these theories work well at high Reynolds number but fail to work at low Reynolds number.

The Navier–Stokes equation cannot be applied to turbulent flows directly as the flow is very complex for determining fluid dynamics at each point. Hence, new terms need to be added to these equations to account for the fluctuations caused by turbulence. If averaging is done for these equations over space or time, then the flow can be analyzed to a greater extent. Thus, for resolving the turbulent flow, the instantaneous variable (velocity, temperature, etc.) is divided into two parts: the mean variable and the fluctuation variable. The mean variables can be determined experimentally and the fluctuation variables can be modeled to achieve complete visualization of the flow. Modeling of the fluctuations requires conversion of the unknown fluctuations into known variables that can be easily evaluated. These equations are then averaged over time. This method is called Reynolds averaging [31].

Richardson–Kolmogorov–Taylor phenomenology of energy cascade says that turbulence contains eddies of various sizes [31]. The large eddies are unstable and so they break up and give energy to smaller eddies. This process is continued until the energy is dissipated in the form of viscous dissipation.

### 10.3.1 Reynolds Stress Model

This model calculates the fluctuations occurring in the flow and thus allows for flow visualization. The variables in the governing equations are decomposed into a mean variable and a fluctuating variable. Time or ensemble averaging is performed on these equations. The fluctuating part goes to zero as the average of the fluctuations is zero and equations are left only in terms of mean variables. The mean variable equation is subtracted from the unaveraged decomposed equation to get the equation only in terms of fluctuations. This equation is multiplied by other fluctuating components to get the equations for Reynolds stress:

$$\frac{\partial \overline{v'_i v'_k}}{\partial t} + V_j \frac{\partial \overline{v'_i v'_k}}{\partial x_j} = -\overline{v'_i v'_j} \frac{\partial V_k}{\partial x_j} - \overline{v'_k v'_j} \frac{\partial V_j}{\partial x_j} - \frac{\partial \overline{v'_i v'_k v'_j}}{\partial x_j}$$
$$- \frac{1}{\rho} \left[ \frac{\partial \overline{p' v'_i}}{\partial x_k} + \frac{\partial \overline{p' v'_k}}{\partial x_i} - p' \left( \frac{\partial V_i}{\partial x_k} + \frac{\partial v'_k}{\partial x_i} \right) \right]$$
$$+ \vartheta \frac{\partial^2 \overline{v'_i v'_k}}{\partial x_j^2} - 2\vartheta \frac{\overline{\partial v'_i}}{\partial x_j} \frac{\overline{\partial v'_k}}{\partial x_j} \tag{10.14}$$

where $v'$ represents the fluctuating velocity, "$V$" represents the mean velocity, and the overbar represents the averaging that is performed over time or ensemble. The terms $v'_i v'_k$, $v'_i v'_j$, and $v'_k v'_j$ are called the Reynolds stresses [31]. By solving these equations, the fluctuations can be modeled and the flow profile can be estimated.

## 10.3.2
## *k*–*ε* Model

This model also uses a procedure that is similar to the Reynolds stress model (RSM) in order to get the equation in terms of fluctuating variable. The fluctuating variable is transformed to the turbulent kinetic energy "$k$" and the turbulent energy dissipation rate "$\varepsilon$." The latter is the rate at which a large eddy gives energy to smaller eddies. The final equation can be written as

$$\frac{\partial k}{\partial t} + V_j \frac{\partial k}{\partial x_j} = \vartheta_t \left( \frac{\partial V_i}{\partial x_j} + \frac{\partial V_j}{\partial x_i} \right) \frac{\partial V_i}{\partial x_j} + \frac{\partial \left[ \left( \vartheta + \frac{\vartheta_t}{\sigma_k} \right) \frac{\partial k}{\partial x_j} \right]}{\partial x_j} - \varepsilon \tag{10.15}$$

where $\vartheta_t$ is the turbulent kinematic viscosity and $\sigma_k$ is the empirical constant.

$$\frac{\partial \varepsilon}{\partial t} + V_j \frac{\partial \varepsilon}{\partial x_j} = -C_{\varepsilon 1} \frac{\varepsilon}{k} \overline{v'_i v'_j} \frac{dV_i}{\partial x_j} + \frac{\partial \left[ \left( \frac{\vartheta_t}{\sigma_\varepsilon} \right) \frac{\partial \varepsilon}{\partial x_j} \right]}{\partial x_j} - \frac{C_{\varepsilon 2} \varepsilon^2}{k} \tag{10.16}$$

where $C_{\varepsilon 1}$, $\sigma_\varepsilon$, and $C_{\varepsilon 2}$ are the empirical constants.

RSM gives better results than $k$–$\varepsilon$ model as it solves and estimates the fluctuating component directly but has a drawback that the number of equations to be solved is significantly more as compared with $k$–$\varepsilon$ model, thus resulting in cumbersome and time-consuming computation. Although there are many more models to model turbulence in the flow, the two mentioned here are the most commonly used.

## 10.3.3
## Population Balance Model

In typical bioreactors, a size distribution exists for the bubbles inside the bioreactor. This distribution occurs due to phenomena such as nucleation, aggregation, and breakage. Bubble breakup and coalescence occur because of interactions with the turbulent eddies. Breakup occurs when the surface energy of the bubble increases and reaches a critical value. The increase in surface energy is due to interaction with the turbulent eddies [6]. Eddies that cause breakup are either of the same size as of the bubble or of smaller size. Bigger eddies do not cause breakage; they simply carry the bubble with them [4]. Bubble coalescence takes place due to turbulence, buoyancy, and laminar shear. Bubble coalescence depends on the collision frequency and collision efficiency [32]. Thus, population balance needs to be incorporated into the momentum, mass and energy transport equations to account for the changing size of the dispersed phase. Since the size distribution determines the total available interfacial area, it also significantly impacts the total mass transfer between the phases.

The bubble distribution is calculated by PBM using the following equations:

$$\frac{\partial(\rho_G n_i)}{\partial t} + \nabla\cdot(\rho_G n_i G_i) = \rho_G(B_{iC} - D_{iC} + B_{iB} - D_{iB}) \tag{10.17}$$

where $B_{iC}$ and $B_{iB}$ are the birth rates of the particles and $D_{iC}$ and $D_{iB}$ are the death rates of the particles due to coalescence and breakage, respectively. The subscript "$i$" denotes the particular bubble class being considered. The birth rates and death rates can be calculated as

$$B_{iB} = \frac{1}{2}\int_0^V a(V - V', V')n(V - V', t)n(V', t)dV' \tag{10.18}$$

$$D_{iC} = \int_0^\alpha a(V, V')n(V, t)n(V', t)dV' \tag{10.19}$$

$$B_{iB} = \int_0^\alpha m(V')b(V')p(V, V')n(V', t)dV' \tag{10.20}$$

$$D_{iB} = b(V)n(V, t) \tag{10.21}$$

where $a(V,V')$ signifies the rate of coalescence between bubbles with volumes $V$ and $V'$, and $b(V')$ is the rate of breakage of bubble with volume $V'$. The term $m(V')$ is the number of daughter bubbles formed due to breakage of bubbles of volume $V'$, and $n(V,t)$ is the number of bubbles of volume $V$ at time $t$. The term $p(V,V')$ is the probability density function for bubbles of volume $V$ generated from bubbles of volume $V'$.

The volume fraction of the bubble size $\alpha_i$, of component "$i$," can then be defined as

$$\alpha_i = n_i V_i \tag{10.22}$$

There are several ways to solve the above-mentioned equations [33]:

1) **Discrete method**: The particle size is discretized into finite number of size intervals, that is, the dispersed phase can have only discrete sizes corresponding to size intervals. This approach can be used to directly define particle size distribution.
2) **Inhomogeneous discrete method**: Different groups of bins move at different velocities. This is a better representation of what actually occurs in the reactor (different-sized particles have different momentums).

3) **Standard method of moments**: Population balance equation (PBE) is changed into transport equations for moments of the distribution. This is computationally simpler than the discrete method, but has the limitation that aggregation and breakage are not included.
4) **Quadrature method of moments**: This method can be applied to a broad range of areas as it does not suffer from limitations of the standard method of moments.

## 10.4 CFD Simulations

### 10.4.1 Creation of Bioreactor Geometry

The first and arguably the most critical step in performing CFD simulations is creation of the geometry of the bioreactor. A lot of commercial software packages are available for creating geometry and meshing. These include Gambit, ICEM, Design Modeler, and so on. It is known that small changes in design can result in big variations in the results.

The first step is typically to create the design followed by meshing. The latter is a process of dividing the whole geometry into smaller cells such that each cell can be considered as a small control volume. Next, the governing equations are solved for each control volume and the flow profile is calculated. The local values of the parameters are then averaged over the whole domain to get an overall value. All commercial software provide the user with a variety of standard shapes that can be used to create any geometry as per requirement.

### 10.4.2 Meshing of Solution Domain

Meshing can be structured or unstructured depending on the geometry. If the geometry is simple and symmetric then structured mesh can be used, but if the geometry is complicated (as in most cases) then unstructured mesh has to be used. Structured mesh typically gives satisfactory results with less number of cells as compared with unstructured mesh. In the structured mesh, all cells are uniformly distributed while the unstructured mesh can have nonuniformity in distribution (Figure 10.3). Cell is defined as a control volume in which all discretized conservation equations are solved. Structured meshes typically have hexahedral cell (in 3D) and quadrilateral cells (in 2D). Unstructured meshes usually have tetrahedral cells (in 3D) and triangular cells (in 2D).

In case of 2D domains, the cells are specified by area, whereas 3D domains are specified by volume. The accuracy of the simulation results depended on the quality of the grid.

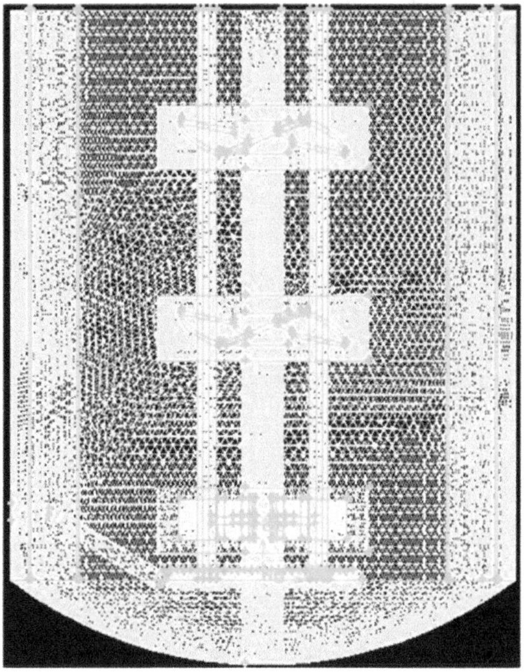

**Figure 10.3** Front view of meshed 67 l capacity bioreactor where tetrahedral and unstructured meshing used. Bioreactor equipped with three impellers consisting of one Rushton turbine and two 3-blade propeller-type impellers with a central shaft of diameter 0.025 m.

As the grid size is decreased, simulation results become independent of the grid size. To achieve grid size–independent solution, a grid independence study is done. In this study, grids of different sizes are generated and simulation is performed on each grid size. When the solution does not vary with grid size, grid independency is stated to have been achieved. With reduction in grid size for a given solution domain, the number of grid cells increases to achieve high accuracy at the expense of computational time. Hence, it is important to have a balance between grid size and computational time. Any parameter can be selected for visualization of grid independence.

After meshing is complete, the quality of mesh needs to be verified. For this purpose, skewness of each cell needs to be calculated. Skewness tells about how much the cell differs from a standard optimum cell shape, where the optimum cell size is the size of an equilateral cell with same circumradius [34].

$$\text{Skewness} = \frac{\text{optimum cell size} - \text{cell size}}{\text{optimum cell size}} \tag{10.23}$$

If skewness is zero then it is considered an excellent cell and when it becomes one, the cell is considered degenerate and is unacceptable. This check is done on every cell in the grid and if most of the cells have skewness below 0.5 then the mesh is

**Table 10.2** Cell quality based on skewness value [34].

| Value of skewness | Cell quality |
|---|---|
| 1 | Degenerate |
| 0.9–1 | Bad |
| 0.75–0.9 | Poor |
| 0.5–0.75 | Fair |
| 0.25–0.5 | Good |
| 0–0.25 | Excellent |
| 0 | Equilateral |

considered a good mesh. For structured mesh, the value is zero or close to zero, while for unstructured mesh any value can exist depending upon the geometrical complicacy (Table 10.2) [34].

### 10.4.3 Solver

Once meshing is complete, simulations need to be performed. Many solvers exist for this purpose including FLUENT, CFX, POLYFLOW, and STAR CCM. All solvers meet the basic requirements of incorporating the properties of the fluid flowing through the domain and applying the appropriate boundary conditions. However, they each have their unique advantages and disadvantages.

The governing equations that are to be solved numerically are (i) Navier–Stokes equation for momentum transport, (ii) continuity equation, and (iii) heat and mass transport equation. Since these equations are complex partial differential equations, it is not possible to solve them analytically and a numerical approach is typically required. Finite volume method is often used to discretize the partial differential equations to ordinary difference equations. Once these equations have been discretized to ordinary difference equations, numerical methods are used to solve them using the appropriate initial and boundary conditions.

Finite element method (FEM) is also widely used for numerical computations. In this method, the solution domain is divided into finite elements. Depending on the spatial problem, usually 1D (subinterval), 2D (triangles, quadrilaterals, curvilinear elements), and 3D (tetrahedrons, hexahedrons, prisms, curvilinear elements) element types are employed. Each element is solved numerically by discretizing the selected piecewise polynomial functions. At element interfaces, the functions have to fulfill certain continuity equation.

## 10.5 Case Studies for Application of CFD in Modeling of Bioreactors

The three case studies presented in this section are intended to showcase the practical applications of using CFD for modeling bioreactors. In the first case study, a unique combination of design of experiments (DOE) and CFD has been used for defining a design space with respect to achieving a targeted volume averaged mass transfer coefficient ($k_L a$). In the second case study, CFD tools such as parallel computing as well as adaptive unstructured meshes have been used to make efficient use of the computational power for calculating gas hold-up, Sauter mean bubble diameter, gas–liquid mass transfer coefficient, and flow structure using the MRF model with population balance. Here, the multiple reference frames model has been used to model the impeller region. In the third case study, predictive capability of the model is improved by explicitly including (i) the impeller geometry in the model, and (ii) the interfacial force averaged drag coefficient ($\overline{C_D}$) instead of the standard drag coefficient correlation (based on rise through a stagnant liquid). The resulting model is significantly better as the standard drag coefficient correlation does not consider the continual acceleration and decelerations experienced by the bubble due to turbulent eddies.

### 10.5.1
### Case Study 1: Use of CFD as a Tool for Establishing Process Design Space for Mixing in a Bioreactor

Optimal mixing in a bioreactor is a necessity so as to achieve efficient mass transfer of oxygen to the cells and maintain uniformity of concentration of oxygen and other nutrients across the reactor. However, excess mixing can also damage shear sensitive cells [35].

CFD has often been used as a tool to model local hydrodynamics, mixing, agitation, and mass transfer in the bioreactor. Design space has been defined as "multidimensional combination and interaction of input variables and process parameters that have been demonstrated to provide assurance of quality" [36]. In a recent study, authors used an Eulerian–Eulerian multiphase approach for modeling the gas–liquid two-phase system [3]. Navier–Stokes equation was solved for both phases including the interphase momentum exchange term in each equation. For turbulence, a standard $k-\varepsilon$ two-equation model was used. The turbulence model was used only for continuous phase, and the dispersed phase was treated as laminar. For the continuous phase, turbulent kinetic energy "$k$" and turbulent energy dissipation rate "$\varepsilon$" were solved. This assumption was based on the fact that the volume fraction of the secondary phase is too low as compared with the continuous phase and hence the density difference between the two phases is quite high.

A population balance model in discrete form was included to account for coalescence and breakage of bubbles in the reactor. Since in the discrete form, the bubble size distribution is required as input, a trial-and-error method was applied for estimating the bubble size distribution so that the simulated results match the experimental results. The overall mass transfer coefficient was then modeled as a function of impeller speed, gas flow rate, and the liquid level in the reactor. Gambit 2.4.3 software was used to design and mesh the reactor. Care was taken for mesh quality such that the skewness factor for most of the cells was below 0.5 and aspect ratio was between 1 and 2. CFD simulations were performed using Fluent 6.3.26.

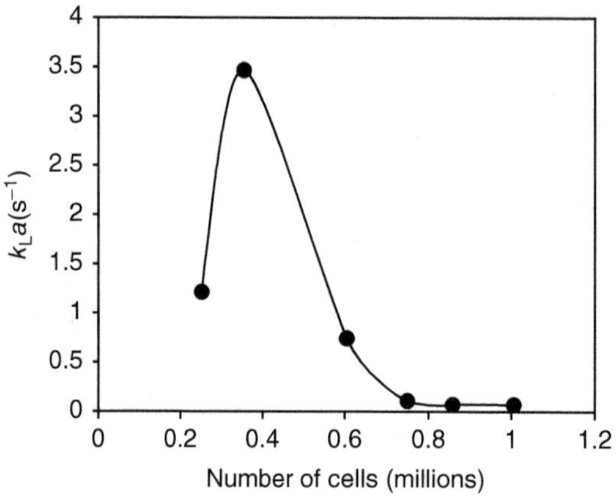


**Figure 10.4** Grid independence test: plot of variation of $k_La$ with grid size (data taken from [3]).

For the region near the impeller, a rotating frame of reference was used for solving the governing equations, while for the bulk region a stationary reference frame was used.

Initially, a grid independence study was performed to find out the minimum number of cells that should be used in simulations. It was found that a minimum of 0.838 million cells were required. Simulations were performed for different rotational speeds of the impeller and at different gas inlet flow rates. Figure 10.4 shows the results from the grid independence study and Figure 10.5 shows the variation in mass transfer coefficient inside the reactor.

As expected, the value of the mass transfer coefficient was found to be maximum near the impeller region while the value decreased as one moved away from the impeller region. This agrees with the general understanding that more turbulence is present near the impeller and hence efficient contact between the phases is possible in this region. Also large eddies are present, which increase the collision frequency in this region. Another observation from the simulations was the dominance of coalescence over breakage in the reactor. The volume averaged diameter from the bubble size distribution data was found to be 0.0036 m, which is higher than that of the bubble with 3 mm diameter formed at the sparger.

The volume averaged mass transfer coefficient $k_La$ data was modeled as a function of liquid level in the reactor, impeller rotating speed, and gas inlet flow rate using the JMP software. An expression showing the effect of these parameters on $k_La$ value was obtained as follows:

$$\begin{aligned}k_La = &-0.04095 + (0.0073667 \times \text{flow rate}) + (2.85e^{-5} \times \text{RPM})\\ &+ (0.17361 \times \text{Liquid Level}) + 8.125e^{-6}((\text{Flow Rate} - 1.5) \times (\text{RPM} - 600))\\ &+ 0.11833((\text{Flow Rate} - 1.5) \times (\text{Liquid Level} - 0.16))\\ &+ 3.625e^{-4}((\text{RPM} - 600) \times (\text{Liquid Level} - 0.16))\end{aligned} \tag{10.24}$$

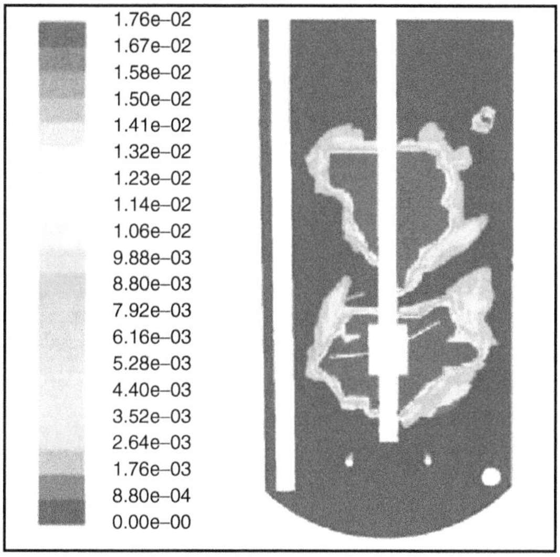


**Figure 10.5** $k_L a$ Variation in reactor along $x = 0.0625$ plane with PBE model (figure adapted from [3] with copyright permission from John Wiley and Sons).

The authors pointed out that this being an empirical expression, its usefulness is restricted to the present reactor and within the window of operating conditions under consideration.

For establishing the design space, a DOE study was designed and executed using the above-mentioned parameters. Analysis of the results allowed the authors to identify the design space where one can operate and achieve $k_L a$ higher than the prescribed cutoff. Thus, the authors successfully demonstrated the usefulness of CFD in defining the design space for a bioreactor. The work also suggests how CFD simulations can reduce the need for large-scale experimentation for bioprocessing applications.

### 10.5.2 Case Study 2: Prediction of Two-Phase Mass Transfer Coefficient in Stirred Vessel

Commercial solvers are quite efficient in predicting flow dynamics in most cases, but using them to design a stirred bioreactor with complex flow profiles inside the reactor is still a big challenge for the researchers. In this case study, the authors used CFD to visualize mass transfer coefficient in the reactor [37]. Parallel computing was used for better utilization of resources. The stirred reactor was a laboratory scale, 3 l, BioFlo 110 bioreactor with a working volume of 2 l. It contained a single impeller with three blades pitched at an angle of 45°. Concentration of oxygen was detected by polarographic-membrane dissolved oxygen probe. The authors compared experimental and numerically simulated estimations of the overall mass transfer coefficient. Once again, the Eulerian–Eulerian multiphase approach was used. Continuity and momentum transport equations were solved separately for each phase. An interfacial momentum exchange term was included in the momentum transport equations to account for the drag force. A dispersed $k-\varepsilon$ model, two-equation model was used to calculate the turbulence in the reactor. The concentration of the secondary phase was very low hence a dispersed model was used.

Population balance model was used to capture the combined effect of bubble breakup and coalescence in the tank. Out of the several numerical approaches that are available for solving the PBE, the discrete methods of classes were used [38, 39]. The bubbles were assumed to be of a discrete size. The bubble volume was divided between the two classes such that the total volume remained the same and the overall mass was conserved. The authors explained that breakage occurs when an eddy having size smaller than the bubble collides with the bubble, and breakage was assumed to be binary and depended on the collision frequency and the probability of breakage.

Gambit 2.0 was used for designing and meshing of the reactor. Nearly 0.483 million cells were used for simulation. The mesh contained cells of various shapes such as tetrahedron, hexahedron, wedge, and pyramid. Skewness of all the cells was kept under 0.7, and FLUENT 6.2 solver was used to discretize the equations and solve them numerically. The reactor domain was divided into a number of partitions automatically and each partition was sent to a computational node for obtaining a solution. Different numbers of computational nodes were considered and the results were compared. It was observed that increasing the number of nodes increased the efficiency but after 16 nodes the effectiveness did not change significantly. Hence, 16 nodes were used for further simulations.

Parallel computing was performed for the four automatic partitions of the reactor domain. Each partition was sent to a separate node for computation, and the results were automatically combined. Different number of bins for input to method of classes was taken, and the results were compared. The four bubble size distribution classes, that is, Class 7, Class 9, Class 11, and Class 13 with PBE equation contained 7, 9, 11, and 13 number of bins, respectively, and each with bubble size range of 0.75–12.00.

The authors observed that on increasing the number of bins and keeping the minimum and maximum bubble diameter constant, the accuracy of numerical method in predicting the $k_La$ value increased while the computational time increased. A comparison of results for different number of bins is shown in Table 10.3.

It is evident in Table 10.3 that the overall mass transfer coefficient for the maximum number of bins is closest to the experimental value. From the Sauter mean diameter contour plots, it was clear that the bubble size increased as we move from bottom to the top of the reactor. Smaller bubbles are present at the bottom while larger bubbles are present near the impeller region. This was found to be in accordance with the experimental observations. Figure 10.6 represents the local volume averaged mass transfer coefficient $k_La$ contour plot, which is independent of the bubble Sauter diameter and the number of bins. It is observed that the simulated

**Table 10.3** Comparison of CFD predictions of mass transfer coefficient for different distribution of PBE with experimental result.

| | $a$ (m$^{-1}$) | $k_L$ (m s$^{-1}$) | $k_La$ (s$^{-1}$) |
|---|---|---|---|
| MRF($d$ = 3 mm) | 22 | 1.59E−03 | 0.035 |
| MRF + 7 Classes PBE | 12.3 | 1.59E−03 | 0.0195 |
| MRF + 9 Classes PBE | 11.6 | 1.59E−03 | 0.0185 |
| MRF + 11 Classes PBE | 11.2 | 1.59E−03 | 0.0178 |
| MRF + 13 Classes PBE | 10.7 | 1.59E−03 | 0.017 |
| Experimental | | | 0.0169 |

Source: Data taken from [37] with copyright permission from Elsevier.

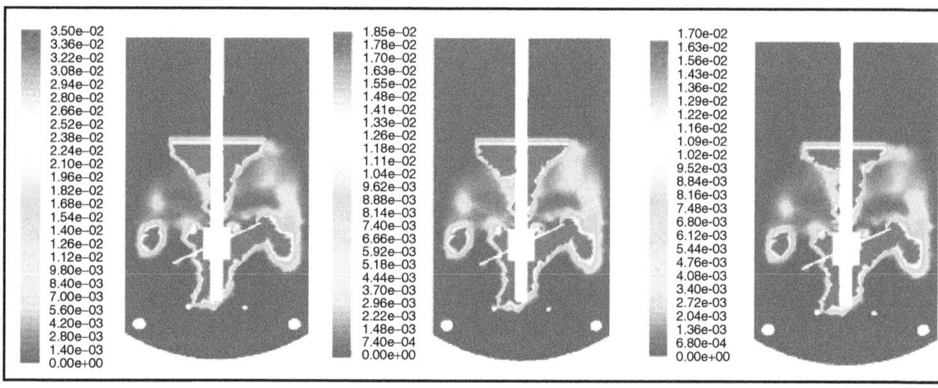


**Figure 10.6** Contour plot for mass transfer coefficient $k_La$ (s$^{-1}$) for the case: (a) with–without the PBE, (b) 9 classes PBE, and (c) 13 classes PBE (figure adapted from [37] with copyright permission from Elsevier).

overall mass transfer coefficient values were close to experimental values when the population balance model was used.

Thus, the authors successfully demonstrated the need of using the population balance model when modeling agitated dispersion in a reactor. Gas hold-up, dispersion, Sauter mean diameter, and overall mass transfer coefficient distributions were predicted correctly when MRF model along PBE was used with different bubble size distribution classes. The numerical and experimental results were compared and showed good agreement for 13 class bubble size distribution.

### 10.5.3 Case Study 3: Numerical Modeling of Gas–Liquid Flow in Stirred Tanks

The aim of this study was to create a model for gas–liquid flow in a stirred reactor using Eulerian two-fluid model that is independent of empirical data, so as to predict gas hold-up, bubble sizes, and flow patterns in reactors stirred by (i) a Rushton turbine and (ii) a Lightnin A315 impeller [40, 41]. Once again, an Eulerian–Eulerian approach was used for solving the governing equations for the gas and liquid phases. The interface momentum exchange term was assumed to be a combined effect of drag force, added mass force, lift force, and turbulent dispersion force. Since drag is dependent on the turbulence, the following correlation was proposed:

$$\frac{U_S}{U_T} = 1 - 1.4\left(\frac{\tau_P}{T_L}\right)^{0.7} \exp\left(-0.6\frac{\tau_P}{T_L}\right) \tag{10.25}$$

where $U_S$ and $U_T$ are the actual slip velocity and stagnant terminal velocity, respectively; and $\tau_P$ and $T_L$ are the bubble relaxation time and turbulent time scale, respectively.

Standard $k-\varepsilon$ model was applied for turbulence. An additional term was added to the expression for turbulent viscosity to account for bubble slip velocity. The gas phase was also considered to be turbulent in nature. Instead of considering the full bubble size distribution, which increases the usage of computational resources, model was used to predict the mean bubble diameter in the reactor based on the number density of the bubble. The model is given as follows:

$$\frac{\partial n(x,t)}{\partial t} + \nabla\cdot(nU_2) = \phi_{br} - \phi_{co} + \phi_{ph} \tag{10.26}$$

where $\varphi_{br}$, $\varphi_{co}$, and $\varphi_{ph}$ are the rates of break up, coalescence, and phase change, respectively.

The authors modeled the ventilated gas cavity formation near the impeller blades, where gas bubbles do not behave like dispersed phase. Modifications in the governing equations were made to incorporate these conditions in the same Eulerian framework. Use of a Rushton turbine as well as a Lightnin A315 impeller was examined. The operating conditions were maintained similar to that of the experimental operating conditions. With the Rushton turbine, only a 60° section of the tank, including a baffle and one blade of the impeller was modeled and the periodic boundary condition was applied (the flow dynamics in this section of the reactor was repeated in subsequent sections). About 59 000 cells were used for the simulation of this section. Multiple reference frame (MRF) model was used to model a rotating frame for the impeller region. For the Lightnin impeller, the full reactor needed to be simulated because of the impeller design, and hence 183 000 cells were used. The MRF method failed to give stable results for the latter impeller and hence the sliding mesh method was used. CFX 4 software was used for simulation purpose and the finite volume method was used for discretization.

Four different simulations were performed with the Rushton turbine by changing the gas inlet flow rate and the impeller rotating speed. Results obtained with both the standard method (as mentioned in CFX code) and improved models for drag and turbulent dispersion suggested by the authors were compared, and it was seen that the models suggested by the authors produced results closer to the experimental results (Table 10.4).

Figure 10.7 shows the bubble diameter variation and volume fraction variation in a section of the reactor (in a vertical plane through the center of the tank half way between baffles). It is observed that the gas bubbles accumulate in the

**Table 10.4** Comparison of simulated results for gas hold-up (%) with experimental gas hold-up (%) for different operating conditions of impeller speed, gas flow rate, gas flow number, and Froude number.

| Case number | Case 1 | Case 2 | Case 3 | Case 4 | Case 5 |
|---|---|---|---|---|---|
| Impeller type | Rushton | Rushton | Rushton | Rushton | Lightnin A315 |
| Impeller speed (RPM) | 180 | 250 | 180 | 285 | 600 |
| Gas flow rate ($m^3 s^{-1}$) | 0.00164 | 0.00164 | 0.00687 | 0.00687 | $5.5 \times 10^{-4}$ |
| Gas flow number | 0.015 | 0.011 | 0.062 | 0.039 | 0.010 |
| Froude number | 0.31 | 0.59 | 0.31 | 0.77 | 1.81 |
| Experimental gas hold-up (%) | 3 | 3.7 | 7.8 | 9.7 | 4.6 |
| Simulated gas hold-up (%) from revised model | 2.3 | 3.0 | 7.8 | 9.8 | 4.5 |
| Simulated gas hold-up (%) from standard model | 1.2 | 2.1 | 3.4 | 6.2 | 2.1 |

Source: Data taken from [40] with copyright permission from Elsevier.

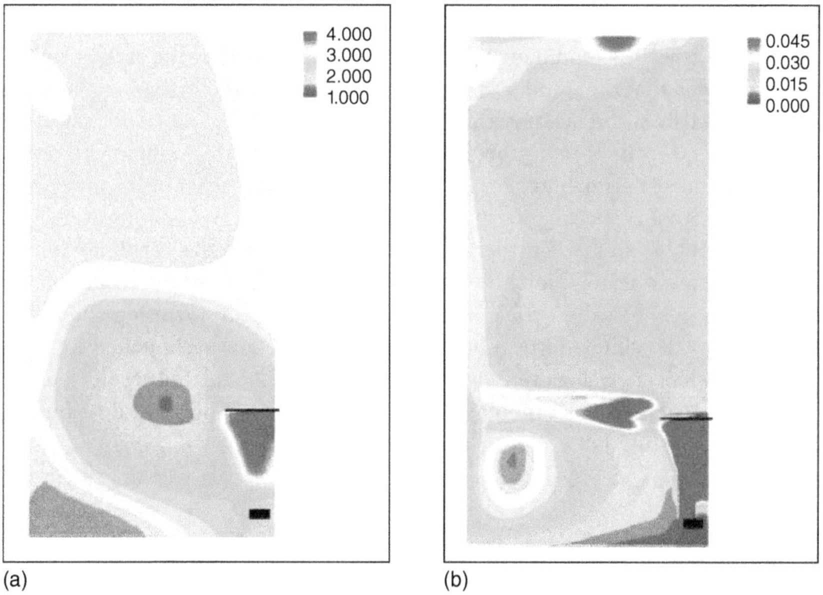


**Figure 10.7** Bioreactor with Rushton turbine with 180 RPM impeller speed, gas flow rate 0.00164 $m^3 s^{-1}$, Froude number = 0.31 and experimental gas hold-up 3%: (a) bubble diameter variation (mm) and (b) gas volume fraction variation in the section of tank (figure adapted from [40] with copyright permission from Elsevier).

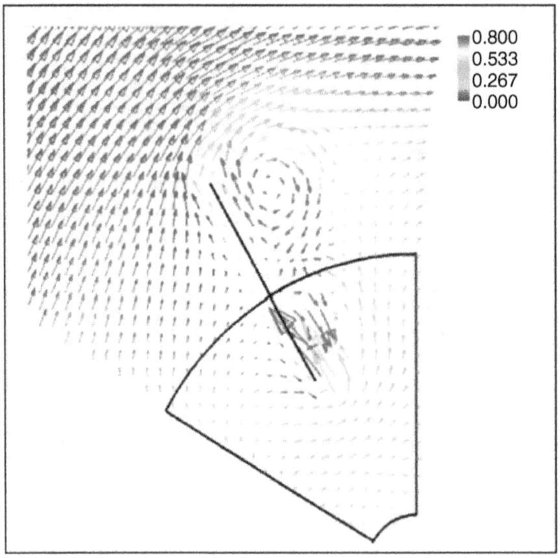


**Figure 10.8** Circulations produced in velocity illustrating the cavity formation (figure adapted from [40] with copyright permission from Elsevier).

region below the impeller. On increasing the flow rate, the accumulation increases. Figure 10.8 shows the gas phase circulating in the cavities.

Standard as well as modified models were also compared by the authors, and it was concluded that the standard models underpredicted the gas volume fraction values especially at the upper region (Table 10.4).

In Figure 10.9, the mean bubble diameter obtained from simulations is compared with the experimental data for Case 4 at different heights of the medium.

Thus, the authors successfully modeled the gas–liquid flow and improved upon the earlier published models, thereby improving the predictive capability for the model. A new model to predict the drag coefficient was created to account for the interactions between the bubble and the eddy. A range of operating conditions was used for the simulation of the reactor. The reactor was simulated for both Rushton turbine as well as the Lightnin impeller. The simulated results were in good agreement with experimental results of gas hold-up, bubble size distribution, and gas volume fraction.

## Summary

CFD is increasingly finding applications in bioprocessing, particularly in modeling flow distribution in bioreactors. It has emerged as a potent tool that can be effectively used to understand the effects of various parameters on the output responses, thereby reducing the need for performing experimentation in the large-scale reactors. It is evident that the knowledge of hydrodynamics can be of

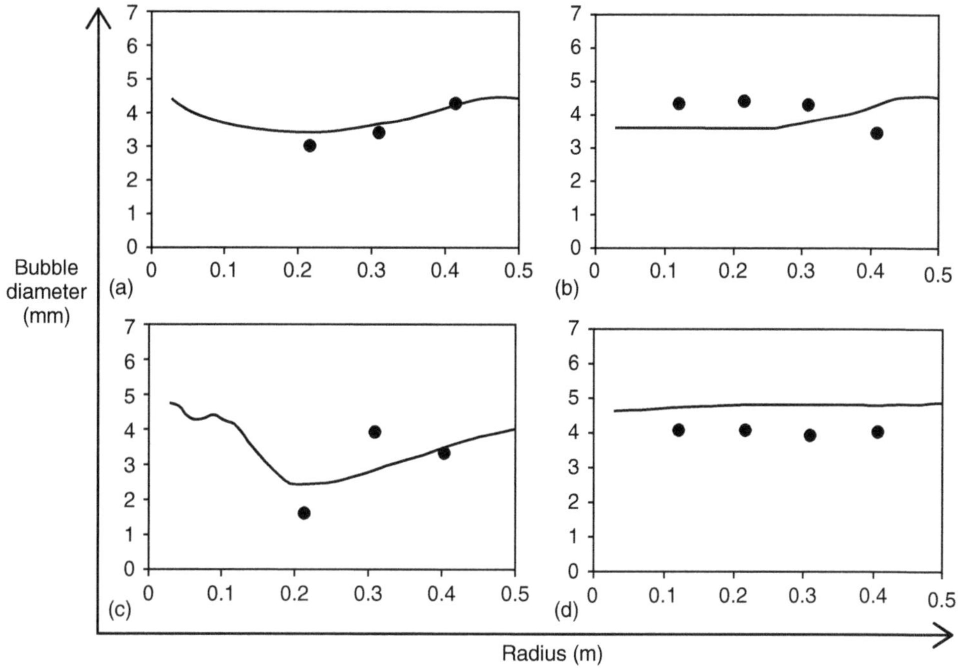


**Figure 10.9** Comparison of mean bubble diameter (mm) in Case 4 (——) with experimental measurements of (•) at a height (in m) of (a) 0.125, (b) 0.25, (c) 0.435, and (d) 0.805 in the tank stirred by Rushton turbine (data taken from [40]).

immense help to researchers looking (i) to improve the reactor design, (ii) performing optimization and scale-up of the process, and/or (iii) aiming to understand of interactions between the various significant input parameters (such as agitation speed, flow rate of gas). The future is likely to see a large-scale adoption of CFD amongst process developers.

## References


1. Choi, B.S., Wan, B., Philyaw, S., Dhanasekharan, K., and Ring, T.A. (2004) Residence time distribution in a stirred tank: comparison of CFD predictions with experiment. *Ind. Eng. Chem. Res.*, **43**, 6548–6556.
2. Guha, D., Dudukovic, M.P., and Ramachandran, P.A. (2006) CFD based compartmental modeling of single phased stirred tank reactors. *AIChE J.*, **52**, 1836–1846.
3. Rathore, A.S., Sharma, C., and Persad, A. (2012) Use of CFD as a tool for establishing process design space for mixing in a bioreactor. *Biotechnology*, **28**, 382–391.
4. Venneker, B.C.H. and Durkson, J.J. (2002) Population balance modelling of aerated stirred vessel based on CFD. *AIChE J.*, **48**, 673–685.
5. Dhanasekharan, K.M., Sanyal, J., Jain, A., and Haidari, A. (2005) A generalized approach to model oxygen transfer in bioreactor using population balance and CFD. *Chem. Eng. Sci.*, **60**, 213–218.

6. Sharma, C., Rathore, A.S., and Malhotra, D. (2011) Review of CFD applications in biotechnology processes. *Biotechnol. Progr.*, **27**, 1497–1510.
7. Nienow, A.W. (1996) Gas-liquid mixing studies: a comparison of Rushton turbine with some modern impellers. *Chem. Eng. Res. Des.*, **74**, 417–423.
8. Jahoda, M., Tomaskova, L., and Mostek, M. (2009) CFD prediction of liquid homogenisation in gas-liquid stirred tank. *Chem. Eng. Res. Des.*, **87**, 460–467.
9. Oncul, A.A., Kalmbach, A., Genzel, Y., Reichl, U., and Thevenin, D. (2010) Characterization of flow conditions in 2L and 20L wave bioreactors using computational fluid dynamics. *Biotechnol. Progr.*, **26**, 101–110.
10. Dietmar, W.H. and Harmeet, S. (2008) Computational fluid dynamics for improved bioreactor design and 3D culture. *Trends Biotechnol.*, **26**, 166–172.
11. Davidson, K.M., Sushil, S., Eggleton, C.D., and Marten, M.R. (2003) Using computational fluid dynamics software to estimate circulation time distributions in bioreactors. *Biotechnol. Progr.*, **19**, 1480–1486.
12. Kelly, W.J. (2008) Using computational fluid dynamics to characterize and improve bioreactor performance. *Biotechnol. Appl. Biochem.*, **49**, 225–238.
13. Ahmed, S.U., Ranganathan, P., Pandey, A., and Sivaraman, S. (2010) Computational fluid dynamics modeling of gas dispersion in multi impeller bioreactor. *Biosci. Bioeng.*, **109**, 588–597.
14. Um, B.-H. and Hanley, T.R. (2008) A comparison of simple rheological parameters and simulation data for *Zymomonas mobilis* fermentation broths with high substrate loading in a 3L bioreactor. *Appl. Biochem. Biotechnol.*, **145**, 29–38.
15. Sowana, D.D., Williams, D.R.G., Dunlop, E.H., Dally, B.B., O'Neill, B.K., and Fletcher, D.F. (2001) Turbulent shear stress effects on plant cell suspension cultures. *Chem. Eng. Res. Des.*, **79**, 867–875.
16. Wang, X., Ding, J., Guo, W.-Q., and Ren, N.-Q. (2010) A hydrodynamics-reaction kinetics coupled model for evaluating bioreactors derived from CFD simulation. *Bioresour. Technol.*, **101**, 9749–9757.
17. Xia, J.-Y., Wang, S.-J., Zhang, S.-L., and Zhong, J.-J. (2008) Computational investigation of fluid dynamics in a recently developed centrifugal impeller bioreactor. *Biochem. Eng. J.*, **38**, 406–413.
18. Delafosse, A., Delvigne, F., Collignon, M.L., Crine, M., Thonart, P., and Toye, D. (2010) Development of a compartment model based on CFD simulations for description of mixing in bioreactors. *Biotechnol. Agron. Soc. Environ.*, **14**, 517–522.
19. Israelowitz, M., Weyand, B., Rizvi, S., Vogt, P.M., and Schroeder, H.P.V. (2012) Development of a laminar flow bioreactor by computational fluid dynamics. *J. Healthcare Eng.*, **3**, 455–476.
20. Jesus, S.S.D., Martinez, E.L., Binelli, A.R.R., Santana, A., and Filho, R.M. (2013) CFD simulation of hydrodynamic behaviors and Gas-liquid mass transfer in a stirred Airlift bioreactor. *Int. J. Chem. Nucl. Metall. Mater. Eng.*, **17**, 680–684.
21. Emily, L.W.T., Kumar, P., and Samyudia, Y. (2009) CFD approach for non-ideally mixed bioreactor modeling. Curtin University of Technology Engineering and Science International Conference.
22. Brannock, M., Leslie, G., Wang, Y., and Buetehorn, S. (2010) Optimising mixing and nutrient removal in membrane bioreactors: CFD modelling and experimental validation. *J. Desalin.*, **250**, 815–818.
23. Bezzo, F., Macchietto, S., and Pantelides, C.C. (2003) General Hybrid Multizonal/CFD approach for bioreactor modeling. *AIChE J.*, **49**, 2133–2148.
24. Felix, G.O. and Emilio, G. (2009) Bioreactor scale-up and oxygen transfer rate in microbial processes: an overview. *Biotechnol. Adv.*, **27**, 153–176.
25. Christopher, E.B. (2005) *Fundamentals of Multiphase Flows*, Cambridge University Press.
26. Boris, V.B., Alex, C.H., Paweel, K., and Lee, D.R. (2010) Eulerian-Eulerian CFD model for the sedimentation of spherical particles in suspension with high particle concentrations. *Eng. Appl. Comput. Fluid Mech.*, **4**, 116–126.

27. Holbeach, J.W. and Davidson, M.R. (2009) An Eulerian-Eulerian model for the dispersion of a suspension of microscopic particles injected into a quiescent liquid. *Eng. Appl. Comput. Fluid Mech.*, **3**, 84–97.
28. Rider, W.J., Kothe, D.B., Mosso, S.J., Cerutti, J.H., and Hochstein, J.I. (1995) Accurate solution algorithms for incompressible multiphase flows. AIAA paper, Aerospace Science Meeting, Reno, Nevada. 95-0699.
29. Rudman, M. (1997) Volume tracking method for interstitial flow calculations. *Int. J. Numer. Methods Fluids*, **24**, 671–691.
30. Hinze, J.O. (1959) *Turbulence: An Introduction to Its Mechanism and Theory*, McGraw-Hills.
31. Wilcox, D.C. (1998) *Turbulence Modeling for CFD*, 2nd edn, DCW Industries, 3rd edn (2006)..
32. Prince, M.J. and Blanch, H.W. (1990) Bubble coalescence and break-up in air sparged bubble column. *AIChE J.*, **36**, 1485–1499.
33. ANSYS (2013) *ANSYS Fluent Population Balance Module Manual*, ANSYS, Inc.
34. Fluent Inc.(2007) *Gambit User Guide*, www.fluent.com.
35. LaRoche, Richard, D., Sanjib Das Sharma, and Srinivasa Mohan, L. (2006) "BIOT 102-Mammalian cell bioreactor scale-up using fluid mixing analysis and computational fluid dynamics." Abstracts of papers of the American chemical society. **232**. 1155 16th St, NW, Washington, DC 20036 USA: Amer chemical soc, 2006. (Website address:www.acs.org)
36. US Department of Health and Human Service, Food and Drug Administration (FDA) (2006) Guidance for Industry: Q8 Pharmaceutical Development, May 2006. Q8 Annex Pharmaceutical Development, Step 3, November 2007.
37. Fouzi, K., Bannari, A., Pierre, P., Rachid, B., Skrga, M., and Labrecque, Y. (2008) Two-phase mass transfer coefficient in stirred vessel with a CFD model. *Comput. Chem. Eng.*, **32**, 1943–1955.
38. Kumar, S. and Ramkrishna, D. (1996) On solution of population balance equations by discretization-I. A fixed pivot technique. *Chem. Eng. Sci.*, **51**, 1311–1332.
39. Ramkrishna, D. (2000) *Population Balances*, Academic Press, San Diego, CA.
40. Lane, G.L., Schwarz, M.P., and Evans, G.M. (2005) Numerical modeling of gas-liquid flow in stirred tanks. *Chem. Eng. Sci.*, **60**, 2203–2214.
41. Khopkar, A.R., Ranade, V.V., and Dudukovic, M.P. (2005) Gas-liquid flow generated by a Rushton turbine in a stirred vessel: CARPT/CT measurements and CFD simulations. *Chem. Eng. Sci.*, **60**, 2215–2229.
42. Balachandar, S., and John, K.E. (2010) Turbulent dispersed multiphase flow. *Annu. Rev. Fluid Mech.*, 42, DOI:10 .1146|annurev.fluid.010908.165243.

# 11
# Scale-Up and Scale-Down Methodologies for Bioreactors

*Peter Neubauer and Stefan Junne*

## 11.1
## Introduction

A knowledge-based scale-up and scale-down methodology is vital, regardless of the microorganism or target product, for the transfer of a bioprocess from the laboratory to the industrial production scale. In any case, the estimation of the sensitivity of specific parameters on the process performance in industrial scale is of basic interest.

Such process knowledge has to consider fluid dynamic aspects with regard to the specific type of bioreactor applied and the responses of the biological system to changes in physical and chemical process parameters. One major issue in the scale-up of fed-batch processes is the formation of concentration gradients at high cell densities. Heterogeneous conditions evolve near the feed addition, the gas inlet, and near the addition of any controlling agent such as base or acid (Figure 11.1). The occurrence of gradients is based on the global and local power input in the gas–liquid phase of the bioreactor.

While an efficient and successful bioprocess scale-up is very crucial for time-to-market considerations, also a robust scale-up is important for the transfer of a bioprocess from one production facility to another, especially if the type, size, or other construction parameters of the bioreactor are changed. However, scale-up cannot be reduced to classical engineering parameters such as power input, stirrer tip speed, dissolved oxygen concentration, or oxygen transfer coefficient, but it should include the physiological conditions inside the cell and the space–time dynamics of the environmental parameters that determine the specific responses and response time constants (kinetics) in the biological system. This is specific for each case, and thus needs specific studies of the fluid phase and the complex responses of cells to fast environmental perturbations at heterogeneous cultivation conditions.

*Bioreactors: Design, Operation and Novel Applications*, First Edition. Edited by Carl-Fredrik Mandenius.

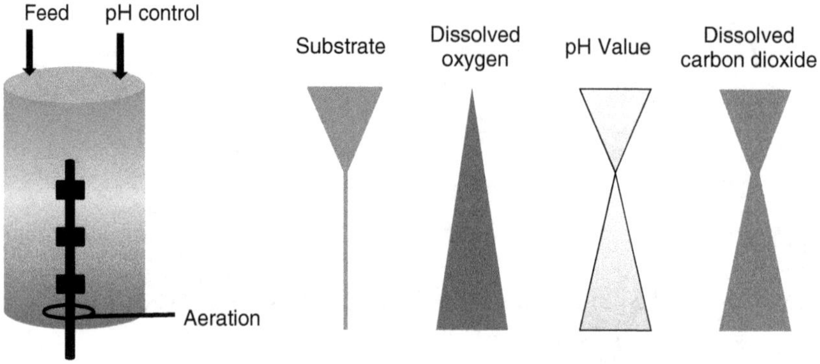


**Figure 11.1** Anticipated distribution of substrate, dissolved oxygen, pH value, and dissolved carbon dioxide in a top-fed industrial-scale high-cell-density cultivation process. Usually, the substrate nearly depletes in the middle and bottom parts. The altered metabolic activity of cells in the different regions due to varying substrate availability leads to an opposite dissolved oxygen concentration gradient. The pH value is affected by large mixing times of its controlling agents. The dissolved oxygen concentration is elevated in zones of high metabolic activity and in the bottom part due to hydrostatic pressure.

## 11.2 Bioprocess Scale-Down Approaches

### 11.2.1 A Historical View on the Development of Scale-Down Systems

While early industrial bioprocesses only relied on batch technologies, the strategy of the most common principle applied in industrial production, the fed-batch technology, was patented in 1915 in connection to the production of baker's yeast. This milestone paper in fermentation implied the controlled feeding of glucose in an aerated yeast process in order to avoid the formation of ethanol with a limited addition of concentrated wort [1] (the reader can refer the review [2] for further information). The fed-batch process is the basis for high-cell-density bioprocesses today, as it allows the restriction of the volumetric oxygen consumption of a culture with the aid of a simple feeding procedure. However, the large concentration difference between the highly concentrated sugar feed solution (up to $> 800\,\mathrm{g\,l^{-1}}$) and the low, growth rate controlling sugar concentration in the liquid bulk, which is in the milligrams per liter range, is the major reason for the formation of concentration gradients within the bioreactor.

Surprisingly, despite the wide application of the fed-batch technology in various bioprocesses, approaches of obtaining a better understanding of how living systems respond to the mixing in large-scale bioprocesses started not earlier than the early 1980s. The progress in this field of research and application and concomitantly the evolvement of scale-down bioreactors (Figure 11.2) is divided into three phases, which are detailed in the following sections.

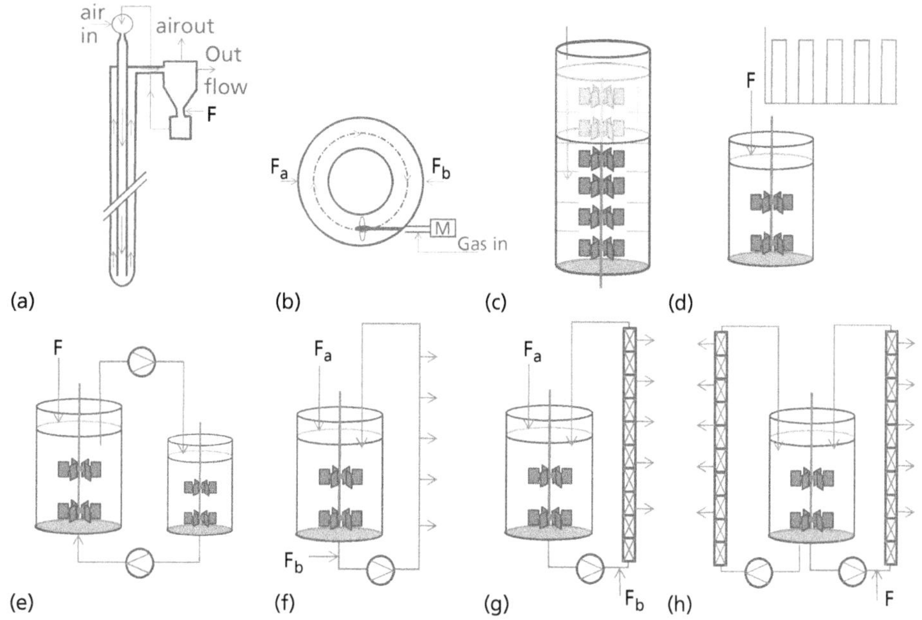


**Figure 11.2** Schematic presentation of different scale-down simulators. (a–d) One-reactor systems. (a) tubular closed-loop air-lift reactor [3], (b) single-loop bioreactor design by Gschwend *et al.* [4], (c) reactor with internal disks for increased mixing times [5], (d) cyclically changing feeding approaches. (e–g) More compartment reactors with (e) two stirred-tank reactors, (f) with a simple plug-flow reactor, (g) with a plug-flow reactor that contains static mixers and can be aerated, (h) three-compartment reactor system with two plug-flow reactors for simulation of different zones at the same time [6]. F = substrate feed.

#### 11.2.1.1
#### Phase 1: Initial Studies of Mixing Behavior and Spatial Distribution Phenomena

During this first studies, the basics of mixing and spatial distribution phenomena in bioreactors were investigated. Oosterhuis and Kossen characterized the impact on different engineering parameters such as the power input [7], the oxygen transfer by a two-compartment model [8] and a five-compartment model [9] for scale-up effects, respectively. The latter clearly showed that the oxygen transfer rate differs in various parts of a reactor. Data were provided on the dependency of the oxygen transfer rate, the gas hold-up, and the stirrer speed. The liquid circulation and mixing time were determined with a radio pill in the same 19 $m^3$ bioreactor in a later study [10, 11]. Based on the experiments, which were performed in water, a first model was developed that described the circulation of the liquid through different compartments of the bioreactor. The circulation time was strongly influenced by the aeration rate and differed in various regions of the bioreactor. These studies provided a good experimental and theoretical basis that microorganisms in large-scale cultivations could be exposed to changing environmental conditions. In parallel, Sweere *et al.* [12]

performed small-scale experiments, to which oxygen oscillations were applied. They provided evidence that microorganisms respond to such fluctuations with yeast as an example. These studies also provided first data about the response of yeast depending on the cycle length, that is, the time interval for the fluctuations for biomass production and several metabolites (ethanol, acetic acid, glycerol). Distinct responses were observed when the oxygen limitation phases were longer than 7.5 s with a total cycle time of 30–60 s. Sweere *et al.* [13] studied the effects of a glucose feed zone on the metabolism of yeast in a small-scale simulator. Therefore, a two-compartment system consisting of two stirred-tank reactors was used. The authors examined the response at various circulation times. Already a circulation time of 1 min reduced the biomass yield by 10% and showed clearly the accumulation of ethanol. These studies provoked the conclusion that the response and relaxation kinetics of the biological system would be a key parameter to describe the process behavior.

An important study of that time described by Vardar and Lilly showed that cycling oxygen concentrations reduce the product formation rate of penicillin in *Penicillium* sp. cultivations. If the respiration rate is reduced to 90% of the maximum value, no product is formed at all. The authors demonstrated that oscillations of the dissolved oxygen tension around the critical value and oxygen starvation for 16 min resulted in an irreversible reduction of the oxygen uptake rate (OUR) and the penicillin production [14].

These studies entailed detailed physiological studies, which investigated the effect of discontinuous or cyclic feed additions on the dynamics of microbial cultures. Such investigations were performed with most of the organisms, which are usually applied in large-scale production, including *Penicillium chrysogenum* [14, 15], *Streptomyces* strains [16], baker's yeast [13, 17–19], *Gluconobacter oxydans* for gluconic acid fermentation [20], and even butanol production by *Clostridium* sp. (Schoutens, referred in [21]). There was also an interesting study conducted on paraffin degradation by yeast [3] apart from desulfurization of coal and waste water treatment.

All the studies of this period describe preferably perturbations of aeration or feed supply. However, first ideas came up in using plug-flow-type or loop reactors to separate the different zones in a bioreactor system and to study the microbial response kinetics in detail. Besides Katinger [3], Gschwend *et al.* used a two-compartment model of a single-loop (100 l torus) bioreactor with an integrated agitator. Different concentrations of NADH were detectable by online fluorescence determination at 460 nm in a continuous chemostat culture in the two different parts of the scale-down reactor. Hence, an oscillatory physiological response of the strictly aerobic yeast *Trichosporon cutaneum* was proven [4]. Another circulating system was applied by Larson and Enfors [15], who used a stirred-tank reactor with an external loop. Their system was easier to implement in standard laboratory bioreactor systems. Therefore, it was applied in later studies, however, with several modifications.

In parallel to the question of how microorganisms respond to certain zones in a large bioreactor, the question occurred, which parameters could prove such a response. While the $NAD^+/NADH$ ratio was appropriate to follow responses to oxygen limitation, it was proposed that the feeding zone of large-scale bioreactors is likely to provoke high metabolic activities and consequently cause oxygen limitation. Cleland and Enfors proposed to measure $H_2$ as an anaerobic metabolite apart from the $NAD^+/NADH$ ratio in *Escherichia coli* cultivations [22].

#### 11.2.1.2
#### Phase 2: Evolvement of Scale-Down Systems Based on Computational Fluid Dynamics

The basic observations that microorganisms respond to cyclic perturbations and kinetic studies of microbial responses to single pulses provoked the development of further scale-down systems. The interactions between fluid dynamics in industrial-scale bioreactors and transient biokinetics came into the focus of research. Advancements originated initially from a larger Nordic European network based on the support of the Nordic Industrial Fund, which was directed to the "Development of methods to study the performance and characteristics of bioreactors" (1991–1993). The interdisciplinary activities by groups from Sweden, Norway, and Denmark were focused on the application of computational fluid dynamics (CFD), based on the analysis of industrial-scale bioreactors and the development of scale-down systems. The activities included extensive measurements in a 30 $m^3$ stirred-tank bioreactor at Statoil (Stavanger, Norway) and in a 200 $m^3$ bubble column at Jästbolaget (Stockholm, Sweden). The results, with the fermentations with *Saccharomyces cerevisiae* as example, proved that glucose gradients exist in large stirred-tank bioreactors and provided a better knowledge about the size of these gradients. The data also showed that the decision of whether to use a feeding from the top or from the bottom has a major impact on the glucose profile in the bioreactor. The authors applied a fast sampling technique for the first time and showed that zones with higher glucose concentration move through the bioreactor. The local variation of the glucose concentration depended on the turbulence within the zone, into which the feed was added, and on the distance between the sampling spot and the feed addition [23]. The interest in studying the kinetics of the microbial reaction in a feed zone provoked the development of special scale-down systems. The two-compartment reactor concept was further modified so that it contained a static mixer element in the loop. This allowed to achieve a certain gas mass transfer supported by direct aeration into the loop and to maintain the plug flow at the same time [24]. The reactor was used to separate the effects of high glucose concentration (feeding zone) and oxygen limitation, which would be a result of the high metabolic activity in this zone. The examiners succeeded in maintaining aerobic conditions at the entrance of the loop with a continuous addition of pure oxygen. In this case, ethanol production was simply related to the higher glucose concentration. The same system was applied by Neubauer *et al.* [25] to study acetate overflow metabolism in an aerobic PFR zone. The authors also compared cultivations with the supply of oxygen enriched and non-enriched air into the PFR to cultivations. While using the PFR loop as a starvation zone by feeding the glucose to the STR compartment, it was shown that short-time starvation induces cyclically the stringent response regulator guanosine tetraphosphate (ppGpp), that is, *E. coli* responds to cyclic starvation as it may occur in larger parts of large-scale bioreactors [26].

The results in the above-mentioned Nordic project were the basis for a larger interdisciplinary approach within the Framework 4 program on "Bioprocess scale-up strategy based on integration of microbial physiology and fluid dynamics" (1996–1999). This project, which was based on *E. coli* cultivations, included detailed analysis of the metabolic reactions in the two-compartment bioreactor system [27]. The results indicated that *E. coli* responds with the induction of a typical mixed-acid fermentation metabolism when cells are exposed to short-time glucose pulses and concomitant oxygen limitation. This anaerobic response was more clearly observed in *E. coli* compared with yeast cultivations, because the glucose affinity constant ($k_s$) is one order of magnitude lower in *E. coli*, which entails a lower mean substrate concentration during glucose limited fed-batch conditions. Cultivations with *E. coli* in the 22 $m^3$ scale, which had been applied previously for yeast, showed that the local glucose variation was higher and the gradient effects were more pronounced [28]. Most importantly, the microbial kinetics was not only studied in connection with the metabolic profiles of key end products of the carbon metabolism, but with quantitative mRNA analyses for a number of marker genes. A response at transcriptomic level was proven by Schweder *et al.* Stress-related mRNAs were increased within the residence of the cells in the PFR loop [29]. In addition, differential gene expression was detected for these genes in the large scale [28].

#### 11.2.1.3
#### Phase 3: Recent Approaches Considering Hybrid Models

The advancements in metabolic models, which have been developed on the basis of pulse experiments in chemostat cultures, for example, by Reuss *et al.* for yeast [30–34] and *E. coli* [35] in the context of systems biology, and the response behavior of cells described by these models were combined with fluid dynamic models in order to reflect and estimate gradients in every part of a bioreactor [36–38].

Currently, with respect to quality-by-design approaches, several issues are approached with scale-down research concepts:

1) Developments are driven from the interest in simulating oscillating conditions already at the screening stage for new strains. New techniques for achieving perturbations in microwell plates are currently developed. This is challenging as such approaches should be performed under nutrient-limiting conditions, that is, by applying the fed-batch technology. Recent developments such as microvalves that are coupled with microwells [39] or a biocatalyst-based feed systems [40, 41] make such studies feasible. The responses of microorganisms to oscillations are investigated in connection with the development of populations at the single-cell level. New microsystems allow following such developments in real time [42].
2) Single-cell analyses benefit from fluorescence-based gene reporter systems, which provide new insight into the time kinetics of the cellular responses with

a resolution over the population [43, 44]. More and more online tools are at the development stage and will become applicable in the following years. The combination with online tools that allow following the physiology and morphology of single cells does allow using other than the classical engineering parameters to design scale-down experiments. A similar cellular condition would be a proof that the surrounding environment is likely identical in the different scales.

3) The impact of additional perturbating environmental factors in large-scale bioreactors, on the cell physiology and population heterogeneity, and concomitantly tailored scale-down systems for their investigation, for example, for the consideration of carbon dioxide concentrations [45, 46].
4) Deeper comprehensive metabolic network analysis and flux determination will increase the overall understanding of the cellular response within the comprehensive regulatory network at perturbations.
5) Tailored scale-down reactor designs are needed, due to the fact that fluid flow and mixing conditions vary, as basically no standard bioreactor design, including the location of feed addition, exist for any kind of processes. For a recent paper, see Lemoine *et al.* [6] who performed studies in a three-compartment scale-down bioreactor, which comprised two plug flow reactors based on an industrial case (Figure 11.3). The reactor represents a

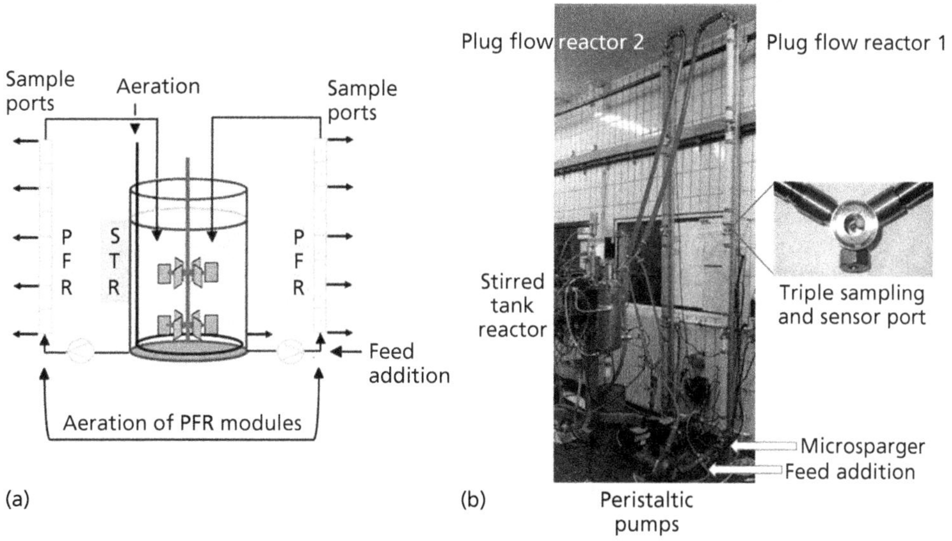


**Figure 11.3** (a) Schematic and (b) photographic presentation of a three-compartment scale-down bioreactor. It comprises a stirred-tank reactor TECHFORS-S with a working volume of 10 l (Infors AG, Switzerland) and two plug-flow reactor modules (total liquid volume of 1.4 l). The residence time can be set by peristaltic pumps at the bottom part between 30 s and 1.5 min. The plug flow reactors comprise four static mixer modules. They are connected with Tri-Clamps. Five triple ports for monitoring pH, DO, and for sampling are located between each static mixer. Aeration is conducted via a sparger from the bottom (if needed). All parts are steam-sterilizable.

further development step of a two-compartment reactor, which comprised only commercially available parts, and enabled monitoring and sampling at five positions along the height of a 4 m long plug flow module. This module was equipped with static mixers and had a volume of 15% of the total liquid volume [47].

6) For an appropriate scale-down design, new tools to characterize large-scale bioreactors are needed. This includes sensors and sampling systems, which can provide data from different parts of a bioreactor, especially from the liquid core phase. First approaches with mobile sensors and sampling systems have been tested in biogas processes, but their application in other bioprocesses can also be expected [48].
7) Studies of microbial responses to the large-scale environment with a broader range of microorganisms need to be performed. The continuous interest in steadily increasing reactor volumes in the biotech industry on one side and the cost pressure for products on the other side increase the relevance of understanding and handling of large-scale effects. Recently, for instance, the number of papers that study large-scale effects for *Corynebacterium glutamicum*, one of the most important microorganisms in view of number of products and scale, is rising [49, 50].

### 11.2.2
### Scale-Up of Bioreactors

Scale-up is often facing problems due to the restrictions in the maintenance of key process parameters, which cannot be held constant for mechanical and economic reasons. However, a consideration of the rate of change of these parameters from the lab to the industrial scale allows an estimation of the change of the degree of homogeneity, to which the cells are exposed to. As a rule of thumb, the power-to-volume ratio $P/V$ and the oxygen mass transfer coefficient $k_La$ is applied to either 30% as the dominant scale-up criteria in industry [21]. Hence, the aim is to keep these parameters as constant as possible among the different scales. However, the equality of the fluid flow velocity as described by the Reynolds number would account for an increase of the volumetric power input ratio ($P/V$) of $10^5$ from 10 l to 10 $m^3$, which is far above of what is usually reachable in large scale (Table 11.1).

The conventional stirred-tank reactor is the system, for which most correlations had been established and intensive research on scale-up phenomena had been performed. Therefore, this reactor design is comparably easy to describe with classical engineering approaches in any scale, which is an advantage for recently developed single-use versions of these reactor types [57]. However, scale-up is also described at the same set of process parameters for shaken bioreactors, for example, single-use orbital shakers [58, 59]. Scale-up studies for various other bioreactor designs have been performed, but are usually restricted to certain mechanical constraints or technical limitations.

A summary of the most relevant parameters, which are considered for measurements in pilot plant experiments or which are estimated by models for the

**Table 11.1** Process parameters, usually applied as scale-up factors in bioprocesses, their determination, and typical values in the cubic-meter scale.

| Scale-up criteria | Variable (equation) | Typical range in a scale of >20 m$^3$ | References |
|---|---|---|---|
| Geometric similarity | Height to diameter $H/D$ | up to 8 : 1 | [51] |
| Volumetric power input | $\frac{P}{V_l} = \frac{2\pi nM}{V_l}$ | 1.0–2.0 kW m$^{-3}$ | [52] |
| Volumetric oxygen mass transfer coefficient | $k_L a = c_0 \left(\frac{P}{V_l}\right)^{c_1} w_{gas}^{C_2}$ | ~400 h$^{-1}$ | [53] |
| Mixing number | $\Theta_{95} = n\tau$ | 100 | [9] |
| Stirrer tip speed | $u = 2\pi n d_i$ | <7 m s$^{-1}$ | [54] |
| Volumetric gas flow rate | $\frac{Q}{V_l}$ (or vvm) | 1 | [54] |
| Ratio of the local to the mean specific energy dissipation rate | $\frac{\varepsilon_T}{\bar{\varepsilon}_T} = \frac{P/V\rho}{P/V_l\rho_l}$ | 70 | [55] |
| Maximum shear stress | Shear rate $\dot{\gamma}$ | <1.5 N m$^{-2}$ (cell culture processes) | [56] |

purpose to predict their values in the large scale is provided in Table 11.1. These parameters are used to estimate the differences of the process conditions in the different scales. Depending on the scale, type of microorganism, process mode, and features of substrate and product, various other parameters are considered further, for example, the viscosity if polymer substances were produced.

#### 11.2.2.1 Dissolved Oxygen Concentration

The oxygen transfer is the most important parameter that should be maintained in a sufficient range during scale-up. The dissolved oxygen concentration in a suspension of aerobic microorganisms depends on the rate of oxygen transfer from the gas phase to the liquid, on the rate, at which oxygen is transported into media, and on the OUR by the microorganism for growth, maintenance, and production [60]. The transport from the gas into the liquid phase strongly depends on the gas–liquid surface area, which increases due to dispersion by the stirrers. Many attempts have been made to estimate the $k_L a$ value by empiric equations for any purposes and reactor designs, independent of the scale. Several comprehensive reviews have been conducted, which summarize estimations of the $k_L a$ value for different gassing rates, bioreactor geometries, and liquid viscosities in dependence on the power input and stirrer speed, for example, [53, 60].

Relevant for the achievement of a sufficient dissolved oxygen concentration in large scale is the relation between the volumetric power input $P/V$ and the $k_La$ value. Indeed, during scaling processes from the multibioreactor scale (typically between the microliter and milliliter range) to lab scale processes, the maintenance of the $k_La$ value was identified as being most important in order to achieve the highest degree of similarity. These results were obtained when the $k_La$ value was considered as scale-up factor in comparison to the gassing-power-to-volume ($Q/V$) and power-to-volume ratio ($P/V$) in a scale-up study from 100 ml to 2 l of working volume [61]. While maintaining a constant $k_La$ value, Alam *et al.* [62] discovered a scaling factor for the stirring speed and air flow rate of 0.28 and 3.1, when the two scales 16 and 150 l of working volume were compared. The reduction in the stirrer speed could be counteracted by a higher energy support due to the increased gassing rate. Ju *et al.* [52] proposed to replace the scale-up criterion of a constant $k_La$ value by the oxygen transfer rate at a predetermined value of dissolved oxygen. It can be maintained by applying different oxygen partial pressures in the influent gas streams for different scales. However, the achievement of various gas-phase oxygen partial pressures might be difficult at large-scale processes.

Another important parameter that affects the $k_La$ value is the local energy dissipation rate near the stirrer. The gradient between the zones near the tip of a Rushton turbine and in zones in between the stirrers is increased in a larger working volume [55]. A good indicator is the ratio of the local to the mean specific energy dissipation ratio (Table 11.1). The reduction of this ratio surely leads to reduced inhomogeneities in the liquid phase.

#### 11.2.2.2
#### Consideration of Similarities and Dimensionless Numbers

In order to determine the crucial process parameters, dimensionless numbers provide a useful estimation of similarity between different scales. Basically, the volumetric ratios of key parameters are held rather constant, if the dimensionless numbers are maintained during scale-up. Oosterhuis described the dissolved oxygen availability as a limiting component in gluconic acid production in a scale of 25 $m^3$, since the oxygen mass transfer and circulation times were assumed to be critical, the first being lower than the OUR of the culture [20]. Since the mixing time is the most crucial parameter during scale-up for aerobic fed-batch processes, numerous experiments were conducted in large scale to confirm the estimated increase in mixing time from the lab scale. One reason for the typical increase in mixing time in the large scale is the requirement of additional power input. It increases by 2/3 the power of the increase of liquid volume [63]. Usually, the power input is reduced during scale-up in industrial production scale. Chemical factors affect scale-up as well, as it is the partial pressure of gasses in the bottom part (e.g., dissolved carbon dioxide) and foam formation in the top part [54]. Although both cases cannot be circumvented in the large scale as they result from the increased liquid volume, the process design and strain engineering can be focused to target these typical problems of the industrial scale.

#### 11.2.2.3 Shear Rate

Since the stirrer tip speed is increased in the large scale in order to limit the reduction of the volumetric power input during scale-up, the shear rates are increased as well near the stirrers. As a consequence, the gradient between the shear forces far away from the stirrers and the ones at the tip of the stirrers is comparably large in industrial scale. However, the maintenance of the shear rate is especially important for shear-sensitive cells such as cell lines [56]. Thus, it represents a suitable scale-up criterion for these types of cultures.

#### 11.2.2.4 Cell Physiology

Since the classical engineering approach for scale-up is not always feasible, several studies proposed that a classical dimensionless analysis might not be sufficient for the description of the scale-up, and consequently for the scale-down, of biotechnological processes, since – unlike in chemical processes – the cellular behavior plays a major role in the success of the industrial production [21, 64]. Therefore, the authors propose that a mechanistic analysis should be added to the dimensionless analysis, which considers parameters of the cellular metabolism and physiological state. Altogether, a regime analysis of both sets of parameters is considered as a suitable approach for a general process analysis. Votruba and Sobotka state that the transfer of microbial technology from the laboratory to the industrial production level is critically affected, in contrast to chemical reactors, by the physiology of growth and production, that is, by the relationship between the potential production ability of selected microorganisms and the external conditions in the bioreactor [64]. Since restrictions of the gas mass transfer, gradients of nutrient supply, the pH value, and other process parameters can directly affect the physiological state of the cell, a similar state of it can most likely be linked to very similar cultivation conditions. As a consequence, the scale-up of a bioprocess should include a proper analysis of the physiological state, eventually by means of the currently rising possibilities to apply online sensors. They allow to achieve a high resolution of physiologic and morphologic pictures of a culture or even single cells in a high time resolution. The quantity and quality of data similar to that for traditional engineering parameters is achievable.

### 11.2.3 Most Severe Challenges During Scale-Up

Up to now, scale-up is usually performed by a randomly chosen set of experiments, which is performed according to the available infrastructure. It is performed rarely based on mathematical models, which describe the large-scale behavior of a process based on the lab-scale experiments (and probably suitable scale-down approaches, which were performed in advance). These empirical approaches are evolved iteratively, relying strongly on the experiences of the developers. If a certain degree of reproducibility between pilot and production scale is achieved, the scale-up is generally considered as being successful [65]. The lack of data within the scientific community about scale-up performances and the industrial scale itself hinders a detailed scientific approach. Since only very limited data are available, many approaches could not be evaluated for various cases. Naturally, the available models, which try to predict the behavior in the large scale, can only be as good as the data (in quantity and quality) on which they rely. Increasing efforts in sharing data between industry and science are needed when advances should be achieved. One example is the evolvement of refined and thus improved fluid flow models. While basic models have been derived already a decade ago, high resolution will be feasible only if sufficient data of the distribution of substrate, oxygen, and carbon dioxide including the gas bubble and cell size are achievable by suitable monitoring methods. The differences of Reynolds numbers at similar power input and stirrer speed in different scales are leading to variances in mass and heat transfer. However, the great differences due to the various biotechnological applications are of even greater importance. For example, a fungal cultivation needs a development of traditional engineering parameters that is different from a bacterial cultivation due to viscosity changes. Hence, the limited knowledge about the behavior in large scale is often not transferable from one investigation to the other. The issues of different behaviors of cells in the large scale due to inhomogeneities in the liquid phase are counteracting the benefits of scale-up. Any improvement relies on the correct definition of the conditions in the large scale in order to properly simulate these conditions in a smaller scale as an iterative approach that relies on quality-by-design principles and model-based process optimization in combination with strain engineering (Figure 11.4). The lack of data needs to be overcome in order to face the current problems of scale-up by most suitable scale-down systems design. A suitable set of models based on relevant data is a prerequisite to achieve this goal.

## 11.3
## Characterization of the Large Scale

In order to gain sufficient knowledge of conditions in the large scale, the collection of experimental data is absolutely necessary. However, the usually performed "good manufacturing practice" (GMP) conditions in the pharmaceutical industry do not allow for the installation of sophisticated equipment for the elucidation of gradients inside a bioreactor. This is, among the risk of losing confidential data of bioprocesses, the main reason for the limited data available of the large scale. Still not much is known about the distribution of concentration gradients. Suitable monitoring systems of the large scale should, therefore, possess the following features: (i) the disturbance of the fluid field should be barely recognizable, (ii) the

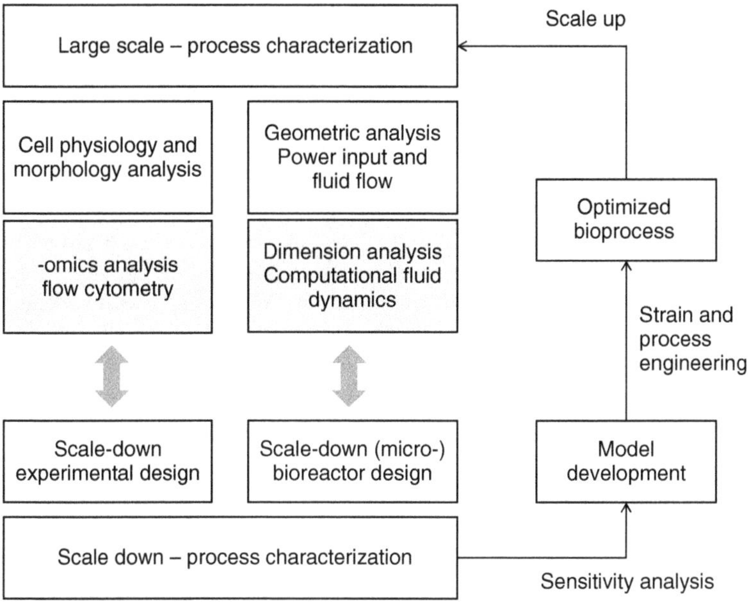


**Figure 11.4** The methodology of an iterative scale-up and scale-down approach – knowledge is gained from the large scale (left) and mimicked in the small scale. Lack of resolution and measurement methods is (partly) replaced by model approaches for the description of the conditions in large scale.

monitoring method should be noninvasive or at least mounted to the available ports, (iii) the sterility of the process should not be affected, and (iv) the monitoring method should be able to identify gradients of relevant parameters with a sufficient resolution. These features are hardly combined in any system. Therefore, a compromise has to be made. Especially, sterility cannot be assured, for example, when multiposition sensors are applied. However, the flexible position of a sensor is a pre-requisite for the determination of critical zones and gradients in between them. A way to circumvent this problem is to focus on processes with low requirements for sterility, at least non-GMP processes. Processes for the production of commodity materials, food, biofuel, and biogas might be suitable targets, since these processes do not usually require absolute sterility, at least not at production scale before termination, when gradients have evolved to the greatest extent.

A rather simple approach for the detection of gradients of dissolved oxygen is the mounting of typically applied DO sensors at different reactor heights. The dynamic method (oxygen is removed by nitrogen (gassing out) or oxygen is sparged into an anoxic liquid phase in a reactor (gassing in)) is a suitable method that provides data of the oxygen mass transfer when the response time of the DO electrode is smaller than the $k_La$ value. Similar approaches can be followed for the determination of the carbon dioxide dissolution. In this case, a pH electrode can be applied to follow the pH change caused by dissolved carbon dioxide. Several reviews describe the comparison of different methods and applications for $k_L a$ value determination, for example, [60]. The effect of insufficient mixing, that is insufficient power input with regard to the liquid volume, causes gradients of substrates, co-factors such as dissolved oxygen, and products. In order to investigate the supply of oxygen to a *Streptomyces* cultivation in the 112 $m^3$ scale, Manfredini *et al.* installed a vertically flexible DO and temperature probe that was mounted on a steel rig [66]. It could be fixed at every position in the bulk in the liquid phase. Static measurements of dissolved oxygen and temperature at various concentrations of dry biomass and thus viscosity yielded gradients along the height of the bioreactor. While a dissolved oxygen concentration of up to 65% could be maintained near the sparger throughout the cultivation, a clear reduction was seen at the top part, where only 30% of saturation could be measured. Relevant gradients of temperature were not detected. A similar approach to detect gradients of oxygen was performed by Oosterhuis [9]. Similar to the previously mentioned approach, a polarographic electrode was mounted along a cord, which itself was mounted to a pipe on the bottom and top in a bioreactor with a working volume of 19 $m^3$. Hence, the sensor could be moved freely in horizontal and vertical direction. Measurements were assessed under production conditions. A saturation concentration from >90% near the sparger toward <5% in between the impellers was measured. A fast indication of the mixing time by a pH-dependent (de-)coloring agent and video analysis was established by Cabaret *et al.* [67]. The method allows a global estimation of the mixing time. The method applied by Alves *et al.* relied on small capillaries in which the gas–liquid phase was sucked into at multiple positions. The bubbles were well-separated from each other by the liquid phase. Mixing time was estimated in a horizontal rotating tubular bioreactor using a temperature step method at various rotating speeds in a temperature-controlled room [68].

Due to the different optical features of the gas and liquid phase, a separation was performed with photo-optical methods [69–72]. Other studies describe the application of a flow-cell-based system in which the bubbles were photographed [69, 73]. The sensitivity of nonintrusive photo-optical techniques was improved so that a minimum bubble diameter of 40 μm could be detected [74].

An important process parameter, which affects the oxygen availability, is the power input *P*. While gradients do exist in the large scale, the power input can be regarded as gradient free when only the stirrer-based power input is considered. It can be measured with torsion dynamometers, or based on the heat balance in an insulated vessel independent of the scale [75]. Due to the easiness of measurements, the power input has been used as scale-up parameter in many studies [52, 76–78]. Own studies even revealed a valid correlation between the power input of the stirrer as measured at the stirrer shaft, which was necessary to keep the stirrer speed, and the viscosity of culture broth in an industrial-scale plug-flow-type bioreactor for biogas production [79]. Since the viscosity affects highly the separation of zones and high viscosity favors gradient formation, the power input needs to be adjusted if possible to avoid an increase in mixing time. However, this parameter can be determined easily and accurately even in large scale in a noninvasive way.

In order to investigate the occurrence of glucose gradients, Larsson *et al.* [23] conducted sampling at three parts (top/middle/bottom zone) of a 30 $m^3$ cultivation. The challenge in the measurement of defined gradients with sampling and offline analysis lies in the rapid inactivation of the metabolic activity of the cell. Many approaches originally invented and applied for gaining a high sample frequency after a glucose pulse for the purpose of systems biology studies can be transferred to follow gradient formation in the large scale.

Recently, a multiposition and multisensor device for the observation of gradients was established for the application in anaerobic fermentation processes. These consisted of commercially available sensor probes for sterilizable applications. Sensors were equipped with amplifiers already directly at the sensor, which allowed data transmittance over long distances. The sensors were mounted into a water-proof steel case that was connected to a cord that allowed a flexible movement in the liquid phase. While three sensors (e.g., pH, redox, and dissolved oxygen (in connection to temperature)) were combined in one single case of a diameter of 76 mm, solutions of the Kurt-Schwabe-Institute (Meinsberg, Germany) allowed the combination of six parameters with biochemical microsensors (both systems are depicted in Figure 11.5) [80]. These sensors were connected to a steel lance for the measurements of gradients in biogas reactors [48]. For the application in industrial-scale brewing tanks, the sensor heads were guided via a fishing line. Nienow *et al.* performed a study in brewing tanks of a minimum filling level of over 1600 $h l^{-1}$. Eight probes of biomass were taken at different heights of a beer fermenter, guided through a boom arm and a winch [81]. This application is feasible when no moving parts such as stirrers are installed. In beer fermentation, ultrasonic Doppler velocimetry (UDV) was applied at a scale of 270 l in order to determine the velocity fields with 128 measuring points. Measurements were taken in the turbid fluid with the existing yeast cells as tracer particles in it [82]. In general, the recent advances in optical sensors will allow the recognition of flow patterns of particles more easily in the future.

## 11.4 Computational Methods to Describe the Large Scale

Computational fluid dynamic tools were used to estimate the degree of mixing in a large-scale bioreactor. In general, a two-phase problem has to be considered, since the bubble size of the gas phase and the resulting interfacial area is crucial for the gas mass transfer between the liquid and the gas phase. Many approaches have been conducted to study and predict the spatial distribution of dissolved oxygen in stirred-tank reactors. Therefore, the correct estimation of the bubble size, and thus the surface area, is of basic importance. For a review, see, for example, [55, 83, 84].

Several approaches relying on a compartmentation of the liquid phase for an improved understanding of mixing in the large-scale stirred-tank bioreactors have been analyzed. Early studies described the simulation of the agitator performance and mass transfer in multiturbine bioreactors of industrial scale. The

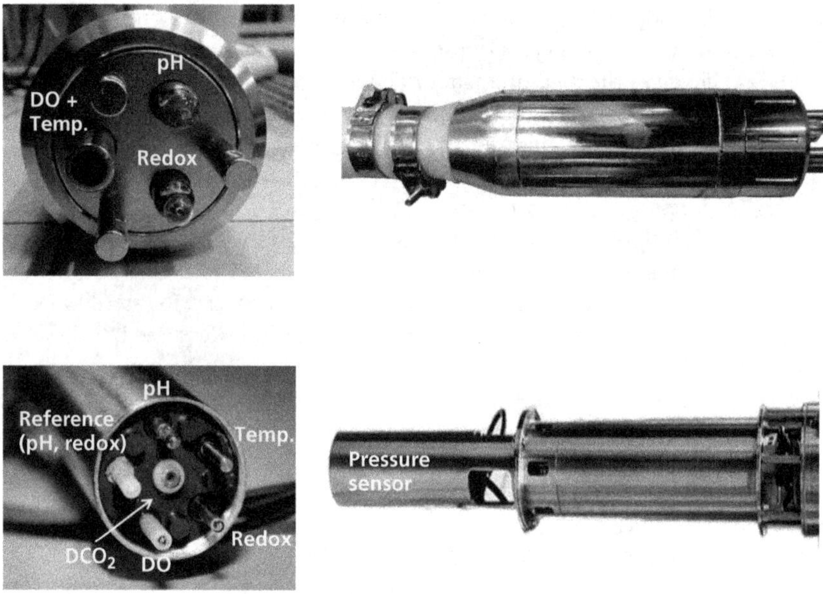


**Figure 11.5** Electrodes as applied in multi-sensor devices; upper left corner: Hamilton ARC sensors for the optical measurement of dissolved oxygen, redox potential, pH value (both glass electrodes), and temperature; lower left corner: microsensors of the Kurt-Schwabe-Institute for temperature measurement (Pt1000), redox (platinum electrode), pH value, and reference (both glass electrodes); diameter: maximum 4 mm, length: 40 mm; right: waterproof, movable built-in sensor heads with a diameter of 76 mm (top) and 45 mm (bottom).

application of this approach provides a method for determining axial dissolved oxygen profiles under conditions of known mass transfer rates as a function of agitation and aeration. The bioreactor is divided into several mixing cells, which allows the consideration of the mass transfer at each turbine individually. The structure of the modeling approach can serve as a basis for testing single-turbine correlations and adapting them to multiturbine systems [85]. CFD software was used to simulate flow fields related to the fed-batch conditions by solving the Navier–Stokes equations of a compartment model using the turbulence model. Tracer particles were released from a reference zone within the simulated flow fields, and the time these tracer particles needed to return to the reference zone was calculated [86]. Another method of estimating the mixing time in industrial Rushton turbine-agitated reactors under aerated conditions was presented and called the analogy mixing-time model. It is based on the hypothesis of the analogy of the mixing time at the same turbine agitation speed in gas–liquid phases. It was further demonstrated that the internal geometry of the reactor, the number of impellers, the distance between impellers, and the degree of homogeneity are less sensitive to the model outcome. Only the location at which pulses are injected is found to influence the model parameters. Concentration gradients in one of the reactors studied (12 $m^3$ of volume) were estimated at both top and bottom injection, both of which were located in the liquid phase [87]. Rhanganatan and Sivaraman applied multiphase CFD coupled with a population balance method in order to study the hydrodynamics and mass transfer in a gas–liquid dual-turbine stirred-tank reactor. A steady-state method of multiple frame of reference (MFR) approach is used to model the impeller and tank regions. The population balance for bubbles is considered using both homogeneous and inhomogeneous polydispersed flow equations to account for bubble size distribution due to breakup and coalescence of bubbles [88]. Findings previously described [69] were matched. Alves *et al.* found that for Rushton turbines, bubble size increases from the stirrer tip along the discharge stream. Bubble sizes in electrolyte solutions are smaller and more sensitive to power input than in water. Addition of surfactants results in a further decrease in bubble size and has to be considered for model-based prediction of the gas mass transfer [69].

Particularly crucial to successful modeling of such systems is the coupling of the physical transport phenomena and the biological activity in one model. The compartment model approach (CMA) was applied to generate a model that included kinetic considerations and was based on the general knowledge of the hydrodynamics of both unaerated and aerated stirred-tank reactors [89]. A model for the expression of cellulosic enzymes by the filamentous fungus *Trichoderma reesei* was established, in which the viscosity and the influence on the mass transfer had the most pronounced effect on the outcome of the model [90]. Morchain *et al.* simulated a lab-scale (70 l) and industrial-scale (70 $m^3$) aerated fermenter with an Euler–Euler approach for the liquid and gas phase [91]. It was coupled to a population balance model, which reflected the adaptation to gradients and a kinetic model for the estimation of biochemical turnover rates. The growth rate was chosen as the discriminating parameter between the two subpopulations. A distinct region of high substrate concentration and concomitant high substrate uptake was observed for either case: feeding from the top or from the bottom. Growth was found to be spatially rather independent. The authors denoted the estimated differences between the uptake rate and growth rate as a major source of disturbance. The evolvement of gradients has been studied by a biased random-walk model (BRWM) able to reproduce the displacement of microorganisms inside the nonmixed part of a scale-down reactor [92, 93]. The model is capable to simulate individual cells of a microbial population crossing the reactor. It has been previously used to characterize the dissolved oxygen and substrate gradients experienced by a single microbial cell in large-scale reactor [94]. The local production rate of butanediol was found to be a function of the local power input in a batch fermentation of *Bacillus subtilis* by coupling a mechanistic model to parameterized functions of the (gas) mass transfer in stirred bioreactors based on CFD studies [84]. A similar methodology was applied to estimate the availability of glucose and concomitant gradients in an *E. coli* fed-batch cultivation in the cubic-meter scale [36]. Especially, the outcomes of this report represent the basis for many scale-down experiments. Further case-specific hybrid modeling needs to be performed, since gradients can be quantified, which are then mimicked in lab scale and thus providing valuable data for the large scale.

## 11.5 Scale-Down Experiments and Physiological Responses

A comprehensive understanding of the cell physiology in industrial-scale inhomogeneous bioreactors is one of the very big challenges of applied biotechnology. The importance of response and relaxation times of biological systems was recognized earlier [21]. However, still there is a scientific gap between the fundamental efforts made to understand the molecular regulation of cellular networks and the implication of this knowledge at the dynamic conditions of bioprocesses.

The key problems of understanding the interacting responses to inhomogeneities of a complex cellular system are related to the following parameters:

1) **Limited understanding of reactor dynamics.** It may be impossible to map the dynamically varying chemical and physical parameters in a large-scale bioreactor over time and place. Current models and scale-down methods consider only a few parameters. However, the cell will respond to the distinct local conditions and also respond to small changes while moving through the reactor.
2) **Limited understanding of the kinetics of cellular response networks.** The response and adaptation of the cells to the local conditions occur in different time horizons, for example, by metabolic fluxes, by protein expression, or by reorganization of membrane systems. The movement of a cell through different zones of a reactor may be faster than the response. Thus, the response may be detected only at a site different from the site where it was triggered.
3) **Limited understanding of cell history effects.** The response of a cell depends on the specific growth rate, which itself is influenced by the local environmental conditions and the history of the experienced dynamic changing environmental conditions.
4) **Limited use of technical tools for measurement and sampling.** Currently, methods for collecting representative samples from a large-scale bioreactor from different sites in the bulk liquid of the reactor are rarely applied, although such probe devices are known for small-scale bioreactors. Special attention is necessary for the immediate stabilization of samples, which are collected from the core liquid phase of a large-scale bioreactor, because its state may change in the sampling system due to the fast metabolic responses, especially if long sampling paths are considered and if the cell densities are high. Also, the availability of online sensors, which can be placed freely in different locations of a large-scale bioreactor, is limited.

### 11.5.1 Scale-Down Experiments with *Escherichia coli* Cultures

*E. coli* responds to repeated glucose perturbations by the formation of side products of the glycolysis when cells are exposed to a high glucose concentration. Many experiments have been performed to investigate this response in various bioreactor designs. The response is affected by the strain and the specific growth rate, the glucose concentration at the feed point, that is, the intensity of the shift

[95], and the residence time in the feed zone. It seems that the cellular response to a glucose shift is at least biphasic [26, 96, 97], whereby the first phase connected to the filling of the glycolysis shunt is very fast – even shorter than 5 s [96, 98]. The capacity of the cells for the uptake of glucose depends not only on the actual specific growth rate, but it is also influenced by the history of the culture. The response is strongly affected by the availability of oxygen in the feed zone [25, 96]. It seems that *E. coli* cells respond differently to single pulses and repeated perturbations. While Lara *et al.* [96] detected a higher specific uptake capacity of cells, which were exposed to an anaerobic glucose pulse compared with cells exposed to an aerobic glucose pulse (31 vs. 15 mmol $g^{-1}$ $h^{-1}$) in a single-pulse experiment in the rapid sampling unit BioScope [98], Neubauer *et al.* [25] detected a similar high capacity for the glucose uptake in a two-compartment reactor system where the PFR compartment, which simulated the feed-zone, was at least partially characterized by oxygen limitation compared with fully aerobic conditions. However, this capacity declined with the decrease of the specific growth rate throughout the glucose-limited fed-batch cultivation. In general, the available data suggest that the glucose uptake capacity of *E. coli* can be more than one order of magnitude higher under glucose-limited conditions than the specific glucose uptake rate during a batch phase.

While the cycle time, during which cells were exposed to glucose pulses, seems to have no major effect on the central glucose flux of *E. coli* for the two intervals studied (7 and 18 min) [99], the cycle time can strongly influence other responses, and thus the outcome of a process [100]. In a study with regular feed rate changes, Lin *et al.* observed that the expression of a recombinant product and the overgrowth of a plasmid-free cell population were strongly affected by the length of a feed cycle in a range between 1 and 4 min.

The metabolic response of *E. coli* to repeated glucose pulses includes the formation of glycolytic overflow products or even a full mixed-acid fermentation if the culture was exposed to oxygen limitation simultaneously. This includes the accumulation of acetate, D-lactate, formate, succinate [25, 27, 99], and hydrogen [22, 101]. However, a metabolic flux model has not been established yet under these conditions. The accumulating metabolites are partially remetabolized in other parts of the reactor, in which glucose is limiting or depleted [27]. Thus, in industrial processes one cannot conclude that there are no feed-zone effects, if these metabolites were not found in the analysis. The only exception is formate, which is slowly remetabolized, and thus has been considered as an indicator for feed-zone effects. However, formate accumulation is strongly reduced if fermentation media contain the elements selenium, nickel, and molybdenum, which are needed for a functional formate–hydrogen lyase complex [102]. Then, formate is converted to hydrogen and thus is hardly detectable.

Besides the accumulation of typical mixed-acid fermentation products, simultaneous appearance of high glucose concentrations and oxygen limitation results in the accumulation of pyruvate. This in turn can increase the flux into the vicinal amino acid pathways and may result in an accumulation of alanine, valine, and even in the formation of the noncanonical amino acid norvaline [103, 104].

Oscillating conditions not only provoke metabolic responses, but also affect the cellular gene expression and protein composition. Experiments by Schweder *et al.* [29] prove that a residence time of 1 min in a glucose feed zone of a two-compartment reactor is long enough that various mRNAs are induced. The authors detected a manifold increase of mRNAs, which are typical for osmotic stress, general stress, and for oxygen limitation (shown for *proU*, *ackA*, *dnaK*, *clpB*, *pfl*, and *frd*). An increase was already visible at the first sampling point of the two-compartment reactor for some of these mRNAs (*proU*, *dnaK*, and *pfl*), that is, within 13 s after a glucose pulse. In a very recent study, Brognaux *et al.* [105] investigated the release of cellular proteins into the extracellular medium. While *E. coli* is usually known to possess very few proteins that are secreted, it is known that a release of proteins can occur from the periplasm in older cultures and after overexpression of proteins in the periplasm of the *E. coli*. Studies suggest that the appearance of proteins in the extracellular medium in *E. coli* high-cell-density cultivations is not necessarily related to cell lysis. Brognaux *et al.* showed that this release of intracellular proteins is connected to a change in the permeability of the cell membrane, and most importantly that this release is stronger under homogeneous glucose-limited fed-batch cultivations compared with oscillating cultures. This work strongly supports earlier observations by flow cytometric analyses by Hewitt and colleagues [28, 106–108], which show that the cells cultivated under large-scale conditions can exhibit a higher viability.

Further insight into the knowledge about cellular responses could be obtained by reporter gene fusions. Although they have been widely applied in the bacterial physiology, only recently they were applied to the area of the scale-down of bioprocesses by Delvigne *et al.* Using a *rpoS*:GFP reporter, they aimed to monitor the level of the $\sigma^S$-based general starvation response [43]. Interestingly, they observed (i) a lower accumulation of GFP and (ii) a higher degree of differentiation of the GFP expression level in the cell population of the heterogeneous cultures. More information was obtained with unstable GFP reporters (*gfp*AAV mutants), which were cloned behind different promoters [109]. In all cases (promoters tested: *rrnB*, *fis*, *rpoS*), the GFP levels were lower in cyclically perturbed cultivations compared with the cultures with continuous feeding, although it is still an open question whether the decreased GFP level is due to the decreased activity of the distinct promoter or due to a higher proteolytic activity under perturbed conditions [28, 107, 110].

Do inhomogeneous conditions lead to more population heterogeneity? This interesting question, which is highly related to process robustness, is currently studied from different sides. Delvigne *et al.* found, in experiments with cyclic perturbations, two populations of *E. coli* cells on the basis of the GFP reporter fluorescence. When collected and recultivated, these two populations turn into single population, which indicates that there is transient heterogeneity. The reason could be a phenotypic heterogeneity, which is enhanced under perturbing environmental conditions [44].

The concentration of carbon dioxide in the fermentation broth is a parameter, which is probably very important in large-scale cultivations, but has been rarely considered in scale-down investigations. Baez *et al.* [111] studied the effect of $CO_2$ in a two-compartment STR–STR scale-down simulator for concentrations up to 300 mbar while varying the mean circulation times between 50 and 375 s. As a result, they found a reduction of the specific growth rate, higher acetate formation, and a reduction of the recombinant product GFP by one-third. Gene expression was only affected by the highest $CO_2$ concentration. The mRNAs of glutamate decarboxylase (*gadA*) and glutamate/$\gamma$-aminobutyric acid antiporters (*gadC*) were elevated, and two components of the $\alpha$-ketoglutarate dehydrogenase complex (*sucA*, *sucB*) were downregulated. This may indicate an inhibition of the tricarbonic acid cycle activity, in a similar to that observed earlier in *E. coli* cultivations [112].

An interesting question is whether perturbations can have a positive effect on a large-scale process. Jazini and Herwig [113] recently found that a heterologous protein (alkaline phosphatase), which is expressed as a periplasmic product in *E. coli*, is produced at higher levels with a cyclic feed profile. Interestingly, in a further study, the authors found that more of the product is released into the cultivation medium (60% increase) in cyclically perturbed cultivations compared with well-mixed and homogeneous conditions. As the final optimal procedure, the authors propose a triangular feed profile, which is applied after induction [114]. These results are so far an exception. In most cases, the yield of a recombinant product is significantly decreased by the exposure of the culture to inhomogeneous conditions. This has been observed for an instable variant of protein A [110], where the lower yield was related to a higher proteolytic activity. Cyclic glucose feeding under fully aerobic conditions resulted in a high proteolytic instability of heterologously expressed yeast $\alpha$-glucosidase in *E. coli* [100]. Cycling between aerobic and anaerobic conditions leads to a significant decrease of recombinant insulin yield [115], as shown by a study with different circulation times in a two-compartment STR–STR system.

### 11.5.2
### Scale-Down Experiments with *Corynebacterium glutamicum* Cultures

One of the most important industrial microorganisms is *C. glutamicum*. Surprisingly, the first investigations about the robustness of *C. glutamicum* under large-scale conditions consider the development of a special scale-down bioreactor with a longer mixing time ($\Theta_{90} = 130$ s vs. 10 s), which was selected in consideration of the specific power input ($P/V$) of a 10 $m^3$ production reactor. This increase of the mixing time was reached by implementing five fixed disks between six Rushton turbines. As a result, cultivations in this scale-down reactor showed a reduced growth (−7%) connected to a lower sugar and ammonia consumption and a lower lysine production (−12%) [5]. However, *C. glutamicum* seems to be relatively robust. Neither extreme mechanical stress nor high aeration or low dissolved oxygen concentrations caused a significant physiological response [116].

This high robustness of *C. glutamicum*, which might be one of the reasons for its extraordinary industrial success, was also confirmed by the studies carried out by

Käß *et al.* [49, 50]. They studied the feed-zone impact (concomitant glucose access and oxygen limitation) in a two-compartment reactor system consisting of an aerated STR coupled to a nonaerated PFR module. The authors applied different strains, a wild-type strain and a lysine overproducer, and different residence times in the PFR-module up to 3 min [49, 50]. Interestingly only differences in the lactate accumulation were found, but no reduction of growth yield or productivity losses.

Recently, two- and three-compartment reactor cultivations with *C. glutamicum* were performed and compared with a standard STR cultivation. While the STR was aerated, the two coupled PFRs were nonaerated: one with and the second without feed addition [6]. The results showed that lactate and succinate as well as a number of amino acids (glycine, threonine, glutamate, glutamine), which derive from the central carbon metabolism, were accumulated up to twofold in the three-compartment reactor cultivation than in the two-compartment reactor cultivation. In contrast to the two-compartment reactor cultivation, no intracellular accumulation of pyruvate is observed in the three-compartment reactor cultivation, since the carbon fluxes are directed toward lactate. This adaptation of cells, which can be considered as successful, is revealed by a flow cytometric analysis of BOX-stained cells and a series of electro-optical *in line* measurements of the cell polarizability. Both methods indicate a higher polarizability of cells in the three-compartment reactor cultivation. PI-staining does not indicate any significant membrane damage or accelerated cell death in either system. However, although the strain shows robustness, the product yield of lysine was reduced in scale-down cultivations, which underlines the relevance of process optimization.

Recently, Buchholz *et al.* [45, 46] investigated the effect of oscillating $CO_2/HCO_3^-$ levels in a three-compartment cascade bioreactor on the biomass yield, as well as the metabolome and transcriptional response of a *C. glutamicum* wild-type strain and a lysine producer. The residence time in the $CO_2/HCO_3^-$ elevated environment was adjusted to 3.6 min at a volume ratio of 8.4%, which seemed to be relevant for the industrial production scale. Pressure gradients of $pCO_2$ of 75–315 mbar were applied. Although more than 60 genes were either up- or downregulated under these conditions depending on the pressure gradient and residence time, no remarkable changes were found in the biomass and product concentration compared with the reference culture. These interesting findings indicate not only a strong response to oscillating $CO_2$ availability, but also a high robustness of *C. glutamicum*, although detailed mechanisms remain to be resolved.

### 11.5.3
### Scale-Down Experiments with *Bacillus subtilis* Cultures

Although *B. subtilis* is one of the most important bacteria applied in large-scale industrial cultivations, so far very limited investigations on the response of *B. subtilis* to large-scale conditions have been published. In the 1980s, Moes *et al.*

[117] investigated the response of *B. subtilis* to transient oxygen depletion in a two-compartment reactor. The ratio between acetoin and 2,3-butanediol, which is very sensitive to the oxygen level [118], was shown in a two-compartment reactor to be shifted toward more 2,3-butanediol production. A kinetic model to predict this behavior depending on the mixing time was developed. These results were used as a basis to investigate the effect of pH oscillations [119]. The pH shift was established by addition of pH-controlling agent at the entrance of the PFR module of a two-compartment reactor, and the residence time was ≥60 s. 2,3-Butandiol synthesis rates were found to be sensitive to pH value shifts between 6.5 and 7.2. Changes in the product accumulation were probably caused by the sensitivity of the acetoin and 2,3-butanediol-forming enzymes to pH and dissociated acetate. A study with *B. megaterium* showed that substrate perturbations, which were caused by applying an insensitive proportional feed controller (P-controller) caused (i) starvation of amino acids and (ii) overflow of acetate. At the same time, the production of GFP as a product was reduced. The authors showed that this lower production could be reversed by an extra addition of the limiting amino acids [120].

### 11.5.4
### Scale-Down Experiments with Yeast Cultures

The metabolic responses of bacteria seem to be significantly faster compared with the response of lower eukaryotes [98], such as yeasts. Furthermore, also in *S. cerevisiae*, overflow metabolism can be considered in high glucose zones by accumulation of ethanol, even if this compartment is maintained aerobically [24]. Such scale-down cultivations with a circulation time of 60 s could imitate the conditions in a 215 $m^3$ bubble column very well. Most interestingly, in this process, the product quality, that is, the gassing power, of the baker's yeast was higher when the cultivation was performed under inhomogeneous conditions, either in the SDR or in the production bubble column. Interestingly, also the respiration capacity of the cells was increased under these oscillating conditions [121]. Recent data from our group show that cultivations in the SDR system also affect the sterol synthesis pathway when oxygen limitation occurs in the glucose-feeding zone [122].

Gradients in brewery processes have been intensively studied. Mixing can promote $CO_2$ evolution and decrease early settling of yeast into the conic part of the tank and thus shorten the fermentation time by 20%. Such an improvement is also achievable by the use of stirrers or jet base mixing devices in brewing tanks (for an extensive review see [81]).

Besides *S. cerevisiae*, a number of scaling effects have been described for the yeast *Yarrowia lipolytica*, an industrially important lipase producer. Different environmental fluctuations can negatively impact the lipase yield in *Y. lipolytica* [123]. In contrast, other parameters such as oscillations of the pH and the dispersion of the inducing substrate methyl oleate, lead to lower biomass yields, and thus affect the lipase concentration indirectly [125–127].

### 11.5.5
### Scale-Down Experiments with Cell Line Cultures

A lot of work on the impact of oscillations has been performed in the area of cell culture. The aim was mainly to evaluate the robustness of the cells and their product formation in the larger scale. Currently, cultivations are performed in up to 25 $m^3$ vessels in turbulent conditions. Considering the size and sensitivity of the cells to hydrodynamic conditions, the specific energy dissipation rate $\overline{\varepsilon}_T$ plays a special role and also $\Phi$, which describes the ratio between the maximum local specific energy dissipation rate and the average specific energy dissipation rate ($\varepsilon_T/\overline{\varepsilon}_T$). While $\overline{\varepsilon}_T$ seems to be in the order of 0.2 W $kg^{-1}$ in typical industrial setup, $\varepsilon_T$ may be ~30 times higher (for recent discussion see [128]). Recent data show that CHO and other industrially used cell lines are relatively robust when exposed to such hydrodynamic conditions. Interestingly, pH variations could influence the cell number significantly. Such pH deviations from the control point have been measured in cell culture bioreactors in various studies (see discussion at [129]). It is assumed that pH values of ~9.0 can be reached at set points of ~7.2. Also, in the later phase of the cultivations, alkali additions are frequent (<10 s interval). Since such additions are not fully dispersed, the pH is higher near the addition point than the set point. Osman *et al.* showed with GS-NS0 cultures that pH value oscillations can decrease the cell number in a two-compartment STR–STR system. However, the authors did not investigate the effect on product quality

## 11.6
## Outlook

The development of suitable scale-down systems is spreading toward microbioreactor cultivations. This enables a faster investigation of possible issues, which might appear during scale-up. However, a sufficient knowledge of the large scale remains a major bottleneck, which can only partly be circumvented with CFD approaches. However, the rapid development of (noninvasive) monitoring tools enables a scale-down based on the physiological and morphological characteristics of a cell population. Methods, which are capable of online monitoring and which at least partially replace instrumentation such as offline flow cytometry, can contribute a lot to an improved design of scale-down reactors. If these monitoring tools were miniaturized or coupled to automated parallel sampling methods, scale-down could be adjusted even in microbioreactors. The additional information also enables the refinement of population balance models and thus the refinement of hybrid models.

The various scale-down systems, when coupled to automated sampling systems, allow the consideration of systems biology approaches to a higher extent than currently performed. The exact determination of the physiological response under relevant oscillatory conditions will support the understanding of the regulatory cell and their engineering for more robustness. Hence, scale-up and scale-down of bioprocesses demands more than ever for a deep interdisciplinary approach between natural science and engineering disciplines.

## Nomenclature

| | |
|---|---|
| $c_1, c_2$ | Empirical constants |
| $c_o$ | Dissolved oxygen saturation concentration |
| $d_i$ | Impeller diameter |
| $D$ | Reactor vessel diameter |
| $H$ | Reactor vessel height |
| $K_L a$ | Volumetric oxygen mass transfer coefficient |
| $M$ | Torsion |
| $n$ | Impeller speed |
| $P$ | Power input |
| $Q$ | Gas flow rate |
| $u$ | Circumvental velocity |
| $V, V_l$ | Total liquid volume, local liquid volume |
| $w_{gas}^{O_2}$ | Oxygen velocity |
| $\gamma$ | Shear rate |
| $\varepsilon_T$ | Energy dissipation |
| $\theta_{95}$ | Mixing number |
| $\rho, \rho_l$ | Density, local density |
| $\tau$ | Mixing time |

## References


1. Hayduck, F. (1915) Verfahren der Hefefabrikation ohne oder mit nur geringer Alkohlolerzeugung, Germany German Patent no. 300662.
2. Gélinas, P. (2014) Fermentation control in baker's yeast production: mapping patents. *Compr. Rev. Food Sci. Food Saf.*, **13** (6), 1141–1164.
3. Katinger, H.W.D. (1976) Physiological-response of *Candida tropicalis* grown on N-paraffin to mixing in a tubular closed-loop fermenter. *Eur. J. Appl. Microbiol.*, **3** (2), 103–114.
4. Gschwend, K., Beyeler, W., and Fiechter, A. (1983) Detection of reactor nonhomogeneities by measuring culture fluorescence. *Biotechnol. Bioeng.*, **25** (11), 2789–2793.
5. Schilling, B.M., Pfefferle, W., Bachmann, B., Leuchtenberger, W., and Deckwer, W.D. (1999) A special reactor design for investigations of mixing time effects in a scaled-down industrial L-lysine fed-batch fermentation process. *Biotechnol. Bioeng.*, **64** (5), 599–606.
6. Lemoine, A., Maya Martínez-Iturralde, N., Spann, R., Neubauer, P., and Junne, S. (2014) Response of *Corynebacterium glutamicum* exposed to oscillations cultivation conditions in a two- and a novel three compartment scale down bioreactor. *Biotechnol. Bioeng.*, in press. **112** (6), 1220–1231.
7. Oosterhuis, N.M.G. and Kossen, N.W.F. (1981) Power input measurements in a production scale bioreactor. *Biotechnol. Lett.*, **3** (11), 645–650.
8. Oosterhuis, N.M.G. and Kossen, N.W.F. (1983) Oxygen-transfer in a production scale bioreactor. *Chem. Eng. Res. Des.*, **61** (5), 308–312.
9. Oosterhuis, N.M.G. (1984) Scale-down of bioreactors. PhD thesis. Delft University of Technology, The Netherlands.

10. Vanbarneveld, J., Smit, W., Oosterhuis, N.M.G., and Pragt, H.J. (1987) Measuring the liquid circulation time in a large gas–liquid contactor by means of a radio pill. 2. Circulation time distribution. *Ind. Eng. Chem. Res.*, **26** (11), 2192–2195.
11. Vanbarneveld, J., Smit, W., Oosterhuis, N.M.G., and Pragt, H.J. (1987) Measuring the liquid circulation time in a large gas–liquid contactor by means of a radio pill. 1. Flow pattern and mean circulation time. *Ind. Eng. Chem. Res.*, **26** (11), 2185–2192.
12. Sweere, A.P.J., Vandalen, J.P., Kishoni, E., Luyben, K.C.A.M., Kossen, N.W.F., and Renger, R.S. (1988) Theoretical-analysis of the baker's-yeast production – an experimental-verification at a laboratory scale. 1. Liquid-mixing and mass-transfer. *Bioprocess. Eng.*, **3** (4), 165–171.
13. Sweere, A.P.J., Matla, Y.A., Zandvliet, J., Luyben, K.C.A.M., and Kossen, N.W.F. (1988) Experimental simulation of glucose fluctuations – the influence of continually changing glucose-concentrations on the fed-batch bakers-yeast production. *Appl. Microbiol. Biotechnol.*, **28** (2), 109–115.
14. Vardar, F. and Lilly, M.D. (1982) The measurement of oxygen-transfer coefficients in fermentors by frequency response techniques. *Biotechnol. Bioeng.*, **24** (7), 1711–1719.
15. Larsson, G. and Enfors, S.O. (1988) Studies of insufficient mixing in bioreactors – effects of limiting oxygen concentrations and short-term oxygen starvation on *Penicillium chrysogenum*. *Bioprocess. Eng.*, **3** (3), 123–127.
16. Yegneswaran, P.K., Gray, M.R., and Thompson, B.G. (1991) Experimental simulation of dissolved-oxygen fluctuations in large fermenters – effect on *Streptomyces clavuligerus*. *Biotechnol. Bioeng.*, **38** (10), 1203–1209.
17. Heinzle, E., Moes, J., and Dunn, I.J. (1985) The influence of cyclic glucose feeding on a continuous bakers' yeast culture. *Biotechnol. Lett.*, **7**, 235–240.
18. Sweere, A.P.J., Giesselbach, J., Barendse, R., Dekrieger, R., Honderd, G., and Luyben, K.C.A.M. (1988) Modeling the dynamic behavior of *Saccharomyces cerevisiae* and its application in control experiments. *Appl. Microbiol. Biotechnol.*, **28** (2), 116–127.
19. Sweere, A.P.J., Janse, L., Luyben, K.C.A.M., and Kossen, N.W.F. (1988) Experimental simulation of oxygen profiles and their influence on baker yeast production. 2. 2-Fermentor system. *Biotechnol. Bioeng.*, **31** (6), 579–586.
20. Oosterhuis, N.M., Kossen, N.W., Olivier, A.P., and Schenk, E.S. (1985) Scale-down and optimization studies of the gluconic acid fermentation by *Gluconobacter oxydans*. *Biotechnol. Bioeng.*, **27** (5), 711–720.
21. Sweere, A.P.J., Luyben, K.C.A.M., and Kossen, N.W.F. (1987) Regime analysis and scale-down – tools to investigate the performance of bioreactors. *Enzyme Microb. Technol.*, **9** (7), 386–398.
22. Cleland, N. and Enfors, S.O. (1987) A biological system for studies on mixing in bioreactors. *Bioprocess. Eng.*, **2** (3), 115–120.
23. Larsson, G., Tornkvist, M., Wernersson, E.S., Tragardh, C., Noorman, H., and Enfors, S.O. (1996) Substrate gradients in bioreactors: origin and consequences. *Bioprocess. Eng.*, **14** (6), 281–289.
24. George, S., Larsson, G., and Enfors, S.O. (1993) A scale-down two-compartment reactor with controlled substrate oscillations: metabolic response of *Saccharomyces cerevisiae*. *Bioprocess. Eng.*, **9** (6), 249–257.
25. Neubauer, P., Häggström, L., and Enfors, S.O. (1995) Influence of substrate oscillations on acetate formation and growth yield in *Escherichia coli* glucose limited fed-batch cultivations. *Biotechnol. Bioeng.*, **47** (2), 139–146.
26. Neubauer, P., Ahman, M., Törnkvist, M., Larsson, G., and Enfors, S.O. (1995) Response of guanosine tetraphosphate to glucose fluctuations in fed-batch cultivations of *Escherichia coli*. *J. Biotechnol.*, **43** (3), 195–204.
27. Xu, B., Jahic, M., Blomsten, G., and Enfors, S.O. (1999) Glucose overflow metabolism and mixed-acid fermentation in aerobic large-scale fed-batch

processes with *Escherichia coli*. *Appl. Microbiol. Biotechnol.*, **51** (5), 564–571.

28. Enfors, S.O., Jahic, M., Rozkov, A., Xu, B., Hecker, M., Jurgen, B. *et al.* (2001) Physiological responses to mixing in large scale bioreactors. *J. Biotechnol.*, **85** (2), 175–185.
29. Schweder, T., Kruger, E., Xu, B., Jurgen, B., Blomsten, G., Enfors, S.O. *et al.* (1999) Monitoring of genes that respond to process-related stress in large-scale bioprocesses. *Biotechnol. Bioeng.*, **65** (2), 151–159.
30. Rizzi, M., Baltes, M., Theobald, U., and Reuss, M. (1997) In vivo analysis of metabolic dynamics in *Saccharomyces cerevisiae*: II. Mathematical model. *Biotechnol. Bioeng.*, **55** (4), 592–608.
31. Theobald, U., Mailinger, W., Baltes, M., Rizzi, M., and Reuss, M. (1997) In vivo analysis of metabolic dynamics in *Saccharomyces cerevisiae*: I. Experimental observations. *Biotechnol. Bioeng.*, **55** (2), 305–316.
32. Vaseghi, S., Baumeister, A., Rizzi, M., and Reuss, M. (1999) In vivo dynamics of the pentose phosphate pathway in *Saccharomyces cerevisiae*. *Metab. Eng.*, **1** (2), 128–140.
33. Visser, D., van der Heijden, R., Mauch, K., Reuss, M., and Heijnen, S. (2000) Tendency modeling: a new approach to obtain simplified kinetic models of metabolism applied to *Saccharomyces cerevisiae*. *Metab. Eng.*, **2** (3), 252–275.
34. Muller, D., Exler, S., Aguilera-Vazquez, L., Guerrero-Martin, E., and Reuss, M. (2003) Cyclic AMP mediates the cell cycle dynamics of energy metabolism in *Saccharomyces cerevisiae*. *Yeast*, **20** (4), 351–367.
35. Chassagnole, C., Noisommit-Rizzi, N., Schmid, J.W., Mauch, K., and Reuss, M. (2002) Dynamic modeling of the central carbon metabolism of *Escherichia coli*. *Biotechnol. Bioeng.*, **79** (1), 53–73.
36. Lapin, A., Schmid, J., and Reuss, M. (2006) Modeling the dynamics of *E. coli* populations in the three-dimensional turbulent field of a stirred-tank bioreactor—A structured–segregated approach. *Chem. Eng. Sci.*, **61** (14), 4783–4797.
37. Lapin, A., Klann, M., and Reuss, M. (2010) Multi-scale spatio-temporal modeling: lifelines of microorganisms in bioreactors and tracking molecules in cells. *Adv. Biochem. Eng. Biotechnol.*, **121**, 23–43.
38. Lapin, A., Muller, D., and Reuss, M. (2004) Dynamic behavior of microbial populations in stirred bioreactors simulated with Euler-Lagrange methods: traveling along the lifelines of single cells. *Ind. Eng. Chem. Res.*, **43** (16), 4647–4656.
39. Wilming, A., Bahr, C., Kamerke, C., and Buchs, J. (2014) Fed-batch operation in special microtiter plates: a new method for screening under production conditions. *J. Ind. Microbiol. Biotechnol.*, **41** (3), 513–525.
40. Panula-Perälä, J., Siurkus, J., Vasala, A., Wilmanowski, R., Casteleijn, M.G., and Neubauer, P. (2008) Enzyme controlled glucose auto-delivery for high cell density cultivations in microplates and shake flasks. *Microb. Cell Fact.*, **7**, 31.
41. Siurkus, J., Panula-Perälä, J., Horn, U., Kraft, M., Rimseliene, R., and Neubauer, P. (2010) Novel approach of high cell density recombinant bioprocess development: optimisation and scale-up from microliter to pilot scales while maintaining the fed-batch cultivation mode of *E. coli* cultures. *Microb. Cell Fact.*, **9**, 35.
42. Grünberger, A., van Ooyen, J., Paczia, N., Rohe, P., Schiendzielorz, G., Eggeling, L. *et al.* (2013) Beyond growth rate 0.6: *Corynebacterium glutamicum* cultivated in highly diluted environments. *Biotechnol. Bioeng.*, **110** (1), 220–228.
43. Delvigne, F., Boxus, M., Ingels, S., and Thonart, P. (2009) Bioreactor mixing efficiency modulates the activity of a prpoS::GFP reporter gene in *E. coli*. *Microb. Cell Fact.*, **8**, 15.
44. Delvigne, F. and Goffin, P. (2014) Microbial heterogeneity affects bioprocess robustness: dynamic single-cell analysis contributes to understanding of microbial populations. *Biotechnol. J.*, **9** (1), 61–72.

45. Blombach, B., Buchholz, J., Busche, T., Kalinowski, J., and Takors, R. (2013) Impact of different $CO_2/HCO_{3-}$ levels on metabolism and regulation in *Corynebacterium glutamicum*. *J. Biotechnol.*, **168** (4), 331–340.
46. Buchholz, J., Graf, M., Freund, A., Busche, T., Kalinowski, J., Blombach, B. *et al.* (2014) $CO_2/HCO_3$ (–) perturbations of simulated large scale gradients in a scale-down device cause fast transcriptional responses in *Corynebacterium glutamicum*. *Appl. Microbiol. Biotechnol.*, **98** (20), 8563–8572.
47. Junne, S., Klingner, A., Kabisch, J., Schweder, T., and Neubauer, P. (2011) A two-compartment bioreactor system made of commercial parts for bioprocess scale-down studies: impact of oscillations on *Bacillus subtilis* fed-batch cultivations. *Biotechnol. J.*, **6** (8), 1009–1017.
48. Kielhorn, E., Sachse, S., Moench-Tegeder, M., Naegele, HJ, Haelsig, C., Oechsner, H., Vonau, W., Neubauer, P., Junne, S. (2015) Multiposition Sensor Technology and Lance-Based Sampling for Improved Monitoring of the Liquid Phase in Biogas Processes. Energy & Fuels, **29** (7), 4038–4045
49. Käß, F., Hariskos, I., Michel, A., Brandt, H.J., Spann, R., Junne, S. *et al.* (2014) Assessment of robustness against dissolved oxygen/substrate oscillations for *C. glutamicum* DM1933 in two-compartment bioreactor. *Bioprocess. Biosyst. Eng.*, **37** (6), 1151–1162.
50. Käß, F., Junne, S., Neubauer, P., Wiechert, W., and Oldiges, M. (2014) Process inhomogeneity leads to rapid side product turnover in cultivation of *Corynebacterium glutamicum*. *Microb. Cell Fact.*, **13**, 6.
51. Najafpour, G.D. (2007) *Biochemical Engineering and Biotechnology*, 1st edn, Elsevier, Amsterdam, pp. 142–169.
52. Ju, L.K. and Chase, G.G. (1992) Improved scale-up strategies of bioreactors. *Bioprocess. Eng.*, **8** (1–2), 49–53.
53. Yawalkar, A.A., Heesink, A.B.M., Versteeg, G.F., and Pangarkar, V.G. (2002) Gas–liquid mass transfer coefficient in stirred tank reactors. *Can. J. Chem. Eng.*, **80** (5), 840–848.
54. Junker, B.H. (2004) Scale-up methodologies for *Escherichia coli* and yeast fermentation processes. *J. Biosci. Bioeng.*, **97** (6), 347–364.
55. Nienow, A.W. (1998) Hydrodynamics of stirred bioreactors. *Appl. Mech. Rev.*, **51**, 3–32.
56. Maranga, L., Cunhaa, A., Clemente, J., Cruz, P., and Carrondo, M. (2004) Scale-up of virus-like particles production: effects of sparging, agitation and bioreactor scale on cell growth, infection kinetics and productivity. *J. Biotechnol.*, **107**, 55–64.
57. Löffelholz, C., Kaiser, S., Krause, M., Eibl, R., and Eibl, D. (2014) in *Advances in Biochemical Engineering/Biotechnology* (eds R. Eibl and D. Eibl), Springer-Verlag, Berlin, Heidelberg, pp. 1–44.
58. Klockner, W., Diederichs, S., and Büchs, J. (2014) Orbitally shaken single-use bioreactors. *Adv. Biochem. Eng. Biotechnol.*, **138**, 45–60.
59. Klockner, W., Gacem, R., Anderlei, T., Raven, N., Schillberg, S., Lattermann, C. *et al.* (2013) Correlation between mass transfer coefficient $k_La$ and relevant operating parameters in cylindrical disposable shaken bioreactors on a bench-to-pilot scale. *J. Biol. Eng.*, **7** (1), 28.
60. Garcia-Ochoa, F., Gomez, E., Santos, V.E., and Merchuk, J.C. (2010) Oxygen uptake rate in microbial processes: an overview. *Biochem. Eng. J.*, **49** (3), 289–307.
61. Gill, N.K., Appleton, M., Baganz, F., and Lye, G.J. (2008) Quantification of power consumption and oxygen transfer characteristics of a stirred miniature bioreactor for predictive fermentation scale-up. *Biotechnol. Bioeng.*, **100** (6), 1144–1155.
62. Alam, M.N.H.Z. and Razali, F. (2005) Scale-up of stirred and aerated bioengineering bioreactor based on constant mass transfer coefficient. *J. Teknol.*, **43** (F), 95–110.
63. Uhl, V.W. and von Essen, J.A. (1986) in *Mixing: Theory and Practice* (eds V.W.

Uhl and J.B. Gray), Academic Press, New York, pp. 155–167.
64. Votruba, J. and Sobotka, M. (1992) Physiological similarity and bioreactor scale-up. *Folia Microbiol.*, **37** (5), 331–345.
65. Gernaey, K.V., Nørregaard, A., Bolic, A., Hernandez, D.Q., Hagemann, T., Larsson, H. *et al.* (2014) Challenges in industrial fermentation technology research. *Biotechnol. J.*, **9** (6), 727–738.
66. Manfredini, R., Cavallera, V., Marini, L., and Donati, G. (1983) Mixing and oxygen-transfer in conventional stirred fermenters. *Biotechnol. Bioeng.*, **25** (12), 3115–3131.
67. Cabaret, F., Bonnot, S., Fradette, L., and Tanguy, P.A. (2007) Mixing time analysis using colorimetric methods and image processing. *Ind. Eng. Chem. Res.*, **46** (14), 5032–5042.
68. Santek, B., Horvat, P., Novak, S., Moser, A., and Maric, V. (1998) Studies on mixing in horizontal rotating tubular bioreactor part III: influence of liquid level and distance between the partition walls on prediction systems for adjustable model parameters. *Bioprocess. Eng.*, **19** (2), 91–102.
69. Alves, S.S., Maia, C.I., Vasconcelos, J.M.T., and Serralheiro, A.J. (2002) Bubble size in aerated stirred tanks. *Chem. Eng. J.*, **89** (1–3), 109–117.
70. Greaves, M. and Kobbacy, K.A.H. (1984) Measurement of bubble size distribution in turbulent gas–liquid dispersions. *Trans. IChemE., Part A, Chem. Eng. Res. Des.*, **62**, 3–12.
71. Barigou, M. and Greaves, M. (1991) A capillary suction probe for bubble size measurement. *Meas. Sci. Technol.*, **2**, 318–326.
72. Barigou, M. and Greaves, M. (1992) Bubble size distributions in a mechanically agitated gas–liquid contactor. *Chem. Eng. Sci.*, **47**, 2009–2025.
73. Parthasarathy, R., Jameson, G.J., and Ahmed, N. (1991) Bubble breakup in stirred vessels—predicting the Sauter mean diameter. *Trans. Instrum. Chem. Eng.*, **69**, 295–301.
74. Machon, V., Pacek, A.W., and Nienow, A.W. (1997) Bubble sizes in electrolyte and alcohol solutions in a turbulent stirred vessel. *Trans. Instrum. Chem. Eng.*, **75**, 339–348.
75. Vilaca, P.R., Badino, A.C., Facciotti, M.C.R., and Schmidell, W. (2000) Determination of power consumption and volumetric oxygen transfer coefficient in bioreactors. *Bioprocess. Eng.*, **22** (3), 261–265.
76. Gamboa-Suasnavart, R.A., Marin-Palacio, L.D., Martinez-Sotelo, J.A., Espitia, C., Servin-Gonzalez, L., Valdez-Cruz, N.A. *et al.* (2013) Scale-up from shake flasks to bioreactor, based on power input and *Streptomyces lividans* morphology, for the production of recombinant APA (45/47 kDa protein) from *Mycobacterium tuberculosis*. *World J. Microbiol. Biotechnol.*, **29** (8), 1421–1429.
77. Mockel, H.O., Wolleschensky, E., Drewas, S., and Rahner, H.J. (1990) Modeling of the calculation of the power input for aerated single-stage and multistage impellers with special respect to scale-up. *Acta Biotechnol.*, **10** (3), 215–224.
78. Rocha-Valadez, J.A., Estrada, M., Galindo, E., and Serrano-Carreon, L. (2006) From shake flasks to stirred fermentors: scale-up of an extractive fermentation process for 6-pentyl-alpha-pyrone production by *Trichoderma harzianum* using volumetric power input. *Process Biochem.*, **41** (6), 1347–1352.
79. Schimpf, U., Hanreich, A., Mähnert, P., Unmack, T., Junne, S., Renpenning, J. *et al.* (2013) Improving the efficiency of large-scale biogas processes: pectinolytic enzymes accelerate the lignocellulose degradation. *J. Sustainable Energy Environ.*, **4**, 53–60.
80. Sachse, S., Bockisch, A., Enseleit, U., Gerlach, F., Ahlborn, K., Kuhnke, T., Rother, U., Kielhorn, E., Neubauer, P., Junne, S., Vonau, W. (2015) To the use of electrochemical multi-sensors in biologically charged media. *J Sens Sens Syst*, **4**, 295–303.

81. Nienow, A.W., Nordkvist, M., and Boulton, C.A. (2011) Scale-down/scale-up studies leading to improved commercial beer fermentation. *Biotechnol. J.*, **6** (8), 911–925.
82. Bottcher, K. and Meironke, H. (2012) Determining of velocity fields during real and experimental simulated beer fermentations by UDV and LDA. *Proc. Appl. Math. Mech.*, **12**, 559–560.
83. Ochieng, A., Onyango, M., and Kiriamiti, K. (2009) Experimental measurement and computational fluid dynamics simulation of mixing in a stirred tank: a review. *S. Afr. J. Sci.*, **105** (11–12), 421–426.
84. Schmalzriedt, S., Jenne, M., Mauch, K., and Reuss, M. (2003) Integration of physiology and fluid dynamics. *Adv. Biochem. Eng./Biotechnol.*, **80**, 19–68.
85. Bader, F.G. (1987) Modeling mass-transfer and agitator performance in multiturbine fermenters. *Biotechnol. Bioeng.*, **30** (1), 37–51.
86. Davidson, K.M., Sushil, S., Eggleton, C.D., and Marten, M.R. (2003) Using computational fluid dynamics software to estimate circulation time distributions in bioreactors. *Biotechnol. Progr.*, **19** (5), 1480–1486.
87. Guillard, F. and Tragardh, C. (2003) Mixing in industrial Rushton turbine-agitated reactors under aerated conditions. *Chem. Eng. Process.*, **42** (5), 373–386.
88. Ranganathan, P. and Sivaraman, S. (2011) Investigations on hydrodynamics and mass transfer in gas–liquid stirred reactor using computational fluid dynamics. *Chem. Eng. Sci.*, **66** (14), 3108–3124.
89. Vrabel, P., van der Lans, R.G.J.M., van der Schot, F.N., Luyben, K.C.A.M., Xu, B., and Enfors, S.O. (2001) CMA: integration of fluid dynamics and microbial kinetics in modelling of large-scale fermentations. *Chem. Eng. J.*, **84** (3), 463–474.
90. Albaek, M.O., Gernaey, K.V., Hansen, M.S., and Stocks, S.M. (2012) Evaluation of the energy efficiency of enzyme fermentation by mechanistic modeling. *Biotechnol. Bioeng.*, [Evaluation Studies], **109** (4), 950–961.
91. Morchain, J., Gabelle, J.C., and Cockx, A. (2014) A coupled population balance model and CFD approach for the simulation of mixing issues in lab-scale and industrial bioreactors. *AIChE J.*, **60** (1), 27–40.
92. Delvigne, F., Destain, J., and Thonart, P. (2006) Toward a stochastic formulation of microbial growth in relation to bioreactor performances: case study of an *E. coli* fed-batch process. *Biotechnol. Progr.*, **22** (4), 1114–1124.
93. Delvigne, F., Destain, J., and Thonart, P. (2006) A methodology for the design of scale-down bioreactors by the use of mixing and circulation stochastic models. *Biochem. Eng. J.*, **28** (3), 256–268.
94. Vlaev, D., Mann, R., Lossev, V., Vlaev, S.D., Zahradnik, J., and Seichter, P. (2000) Macro-mixing and *Streptomyces fradiae*: modelling oxygen and nutrient segregation in an industrial bioreactor. *Trans. Instrum. Chem. Eng.*, **78**, 354–362.
95. Sunya, S., Delvigne, F., Uribelarrea, J.L., Molina-Jouve, C., and Gorret, N. (2012) Comparison of the transient responses of *Escherichia coli* to a glucose pulse of various intensities. *Appl. Microbiol. Biotechnol.*, **95** (4), 1021–1034.
96. Lara, A.R., Taymaz-Nikerel, H., Mashego, M.R., van Gulik, W.M., Heijnen, J.J., Ramirez, O.T. *et al.* (2009) Fast dynamic response of the fermentative metabolism of *Escherichia coli* to aerobic and anaerobic glucose pulses. *Biotechnol. Bioeng.*, **104** (6), 1153–1161.
97. Schaub, J. and Reuss, M. (2008) In vivo dynamics of glycolysis in *Escherichia coli* shows need for growth-rate dependent metabolome analysis. *Biotechnol. Progr.*, **24** (6), 1402–1407.
98. De Mey, M., Taymaz-Nikerel, H., Baart, G., Waegeman, H., Maertens, J., Heijnen, J.J. *et al.* (2010) Catching prompt metabolite dynamics in *Escherichia coli* with the BioScope at oxygen rich conditions. *Metab. Eng.*, **12** (5), 477–487.
99. Sunya, S., Bideaux, C., Molina-Jouve, C., and Gorret, N. (2013) Short-term dynamic behavior of *Escherichia coli* in

response to successive glucose pulses on glucose-limited chemostat cultures. *J. Biotechnol.*, **164** (4), 531–542.
100. Lin, H.Y. and Neubauer, P. (2000) Influence of controlled glucose oscillations on a fed-batch process of recombinant *Escherichia coli*. *J. Biotechnol.*, **79** (1), 27–37.
101. Larsson, G. and Enfors, S.O. (1993) Kinetics of *Escherichia coli* hydrogen-production during short-term repeated aerobic-anaerobic fluctuations. *Bioprocess. Eng.*, **9** (4), 167–172.
102. Soini, J., Ukkonen, K., and Neubauer, P. (2008) High cell density media for *Escherichia coli* are generally designed for aerobic cultivations – consequences for large-scale bioprocesses and shake flask cultures. *Microb. Cell Fact.*, **7**, 26.
103. Soini, J., Falschlehner, C., Liedert, C., Bernhardt, J., Vuoristo, J., and Neubauer, P. (2008) Norvaline is accumulated after a down-shift of oxygen in *Escherichia coli* W3110. *Microb. Cell Fact.*, **7**, 30.
104. Soini, J., Ukkonen, K., and Neubauer, P. (2011) Accumulation of amino acids deriving from pyruvate in *Escherichia coli* W3110 during fed-batch cultivation in a two-compartment scale-down bioreactor. *Adv. Biosci. Biotechnol.*, **2**, 336–339.
105. Brognaux, A., Francis, F., Twizere, J.C., Thonart, P., and Delvigne, F. (2014) Scale-down effect on the extracellular proteome of *Escherichia coli*: correlation with membrane permeability and modulation according to substrate heterogeneities. *Bioprocess. Biosyst. Eng.*, **37** (8), 1469–1485.
106. Hewitt, C.J., Boon, L.A., McFarlane, C.M., and Nienow, A.W. (1998) The use of flow cytometry to study the impact of fluid mechanical stress on *Escherichia coli* W3110 during continuous cultivation in an agitated bioreactor. *Biotechnol. Bioeng.*, **59** (5), 612–620.
107. Hewitt, C.J., Nebe-Von Caron, G., Axelsson, B., McFarlane, C.M., and Nienow, A.W. (2000) Studies related to the scale-up of high-cell-density *E. coli* fed-batch fermentations using multiparameter flow cytometry: effect of a changing microenvironment with respect to glucose and dissolved oxygen concentration. *Biotechnol. Bioeng.*, **70** (4), 381–390.
108. Hewitt, C.J., Nebe-von Caron, G., Nienow, A.W., and McFarlane, C.M. (1999) The use of multi-parameter flow cytometry to compare the physiological response of *Escherichia coli* W3110 to glucose limitation during batch, fed-batch and continuous culture cultivations. *J. Biotechnol.*, **75** (2–3), 251–264.
109. Han, S., Delvigne, F., Brognaux, A., Charbon, G.E., and Sorensen, S.J. (2013) Design of growth-dependent biosensors based on destabilized GFP for the detection of physiological behavior of *Escherichia coli* in heterogeneous bioreactors. *Biotechnol. Progr.*, **29** (2), 553–563.
110. Rozkov, A. (2001) Control of proteolysis of recombinant proteins in *Escherichia coli*. PhD thesis. Royal Institute of Technology, Stockholm.
111. Baez, A., Flores, N., Bolivar, F., and Ramirez, O.T. (2011) Simulation of dissolved $CO_2$ gradients in a scale-down system: a metabolic and transcriptional study of recombinant *Escherichia coli*. *Biotechnol. J.*, **6** (8), 959–967.
112. Baez, A., Flores, N., Bolivar, F., and Ramirez, O.T. (2009) Metabolic and transcriptional response of recombinant *Escherichia coli* to elevated dissolved carbon dioxide concentrations. *Biotechnol. Bioeng.*, **104** (1), 102–110.
113. Jazini, M. and Herwig, C. (2011) Effect of post-induction substrate oscillation on recombinant alkaline phosphatase production expressed in *Escherichia coli*. *J. Biosci. Bioeng.*, **112** (6), 606–610.
114. Jazini, M. and Herwig, C. (2014) Substrate oscillations boost recombinant protein release from *Escherichia coli*. *Bioprocess. Biosyst. Eng.*, **37** (5), 881–890.
115. Sandoval-Basurto, E.A., Gosset, G., Bolivar, F., and Ramirez, O.T. (2005) Culture of *Escherichia coli* under dissolved oxygen gradients simulated in a two-compartment scale-down system: metabolic response and production

of recombinant protein. *Biotechnol. Bioeng.*, **89** (4), 453–463.
116. Chamsartra, S., Hewitt, C.J., and Nienow, A.W. (2005) The impact of fluid mechanical stress on *Corynebacterium glutamicum* during continuous cultivation in an agitated bioreactor. *Biotechnol. Lett.*, **27** (10), 693–700.
117. Moes, J., Griot, M., Heinzle, E., Dunn, I.J., and Bourne, J.R. (1986) A microbial culture as an oxygen sensor for reactor mixing effects. *Ann. N.Y. Acad. Sci.*, **469**, 118–130.
118. Moes, J., Griot, M., Keller, J., Heinzle, E., Dunn, I.J., and Bourne, J.R. (1985) A microbial culture with oxygen-sensitive product distribution as a potential tool for characterizing bioreactor oxygen transport. *Biotechnol. Bioeng.*, **27** (4), 482–489.
119. Amanullah, A., McFarlane, C.M., Emery, A.N., and Nienow, A.W. (2001) Scale-down model to simulate spatial pH variations in large-scale bioreactors. *Biotechnol. Bioeng.*, **73** (5), 390–399.
120. Korneli, C., Bolten, C.J., Godard, T., Franco-Lara, E., and Wittmann, C. (2012) Debottlenecking recombinant protein production in *Bacillus megaterium* under large-scale conditions – targeted precursor feeding designed from metabolomics. *Biotechnol. Bioeng.*, **109** (6), 1538–1550.
121. George, S., Larsson, G., Olsson, K., and Enfors, S.O. (1998) Comparison of the baker's yeast process performance in laboratory and production scale. *Bioprocess. Eng.*, **18** (2), 135–142.
122. Marba, A., Bockisch, A., Emmerich, J., Maaß, S., Rojahn, J., Päßler, S. *et al.* (2014) Process analytical tools for the characterization of industrial-scale bioprocesses and scale-down experiments. 10th European Symposium on Biochemical Engineering Sciences and 6th International Forum on Industrial Bioprocesses, Lille, France, September 7–10, 2014.
123. Kar, T., Delvigne, F., Masson, M., Destain, J., and Thonart, P. (2008) Investigation of the effect of different extracellular factors on the lipase production by *Yarrowia lipolityca* on the basis of a scale-down approach. *J. Ind. Microbiol. Biotechnol.*, **35** (9), 1053–1059.
124. Kar, T., Destain, J., Thonart, P., and Delvigne, F. (2010) Impact of scaled-down on dissolved oxygen fluctuations at different levels of the lipase synthesis pathway of *Yarrowia lipolytica. Biotechnol. Agron. Soc. Environ.*, **14**, 523–529.
125. Kar, T., Delvigne, F., Destain, J., and Thonart, P. (2011) Bioreactor scale-up and design on the basis of physiologically relevant parameters: application to the production of lipase by *Yarrowia lipolytica. Biotechnol. Agron. Soc. Environ.*, **15** (4), 585–595.
126. Kar, T., Destain, J., Thonart, P., and Delvigne, F. (2012) Physical and physiological impacts of different foam control strategies during a process involving hydrophobic substrate for the lipase production by *Yarrowia lipolytica. Bioprocess. Biosyst. Eng.*, **35** (4), 483–492.
127. Kar, T., Destain, J., Thonart, P., and Delvigne, F. (2012) Scale-down assessment of the sensitivity of *Yarrowia lipolytica* to oxygen transfer and foam management in bioreactors: investigation of the underlying physiological mechanisms. *J. Ind. Microbiol. Biotechnol.*, **39** (2), 337–346.
128. Nienow, A.W. (2014) Re "Development of a scale-down model of hydrodynamic stress to study the performance of an industrial CHO cell line under simulated production scale bioreactor conditions" [Sieck, J.B., Cordes, T., Budach, W.E., Rhiel, M.H., Suemeghy, Z., Leist, C., Villiger, T.K., Morbidelli, M., and Soos, M. (2013). J. Biotechnol., **164**, 41–49]. *J. Biotechnol.*, **171**, 82–84.
129. Osman, J.J., Birch, J., and Varley, J. (2002) The response of GS-NS0 myeloma cells to single and multiple pH perturbations. *Biotechnol. Bioeng.*, **79** (4), 398–407.

# 12
# Integration of Bioreactors with Downstream Steps

*Ajoy Velayudhan and Nigel Titchener-Hooker*

## 12.1
## Introduction

A characteristic feature of biotechnological processes is the lack of detailed information about the impurity profile in the final product . This incomplete information has been an important reason for the empirical approach to bioprocess design embodied in the dictum "the process is the product". Fortunately, the recent trend toward fundamental understanding of critical events in manufacturing, driven by the "quality by design" initiative of the regulatory agencies, and the increasing value being placed on robust bioprocess models within industry, have led to more detailed analyses of *individual* bioprocess unit operations. However, there is still the urgent need to combine unit operations, and seek integrated models and understanding of the *whole* bioprocess. In that context, this chapter describes interactions between the bioreactor step and those that follow it.

It is now widely recognized in bioprocessing that the many parameters chosen to increase the efficiency of the cell-culture step can have important consequences for the unit operations downstream. For instance, the use of dextran sulfate to reduce cell clumping has been found to promote changes in the structure of product proteins, often accelerating their aggregation and reducing their shelf life. Another example is extending the duration of a production bioreactor to increase productivity: in some cases, such extended reactor runs, with final cell viabilities less than 80% seem to lead to changes in the folding of product proteins. However, such bioprocess-wide interactions have not received significant quantitative attention in the literature, in spite of their importance. This chapter summarizes the interactions between cell culture and subsequent isolation, recovery, and purification steps.

The schematic in Figure 12.1 facilitates the characterization of interactions between the bioreactor and steps downstream. Class A, which describes the relevant metrics within the bioreactor, includes the usual metrics for bioreactor productivity: the growth rate represents the effectiveness with which cells are produced, and the cell-specific productivity represents the effectiveness of a single (average) cell in generating product. However, there are other parameters

*Bioreactors: Design, Operation and Novel Applications*, First Edition. Edited by Carl-Fredrik Mandenius.

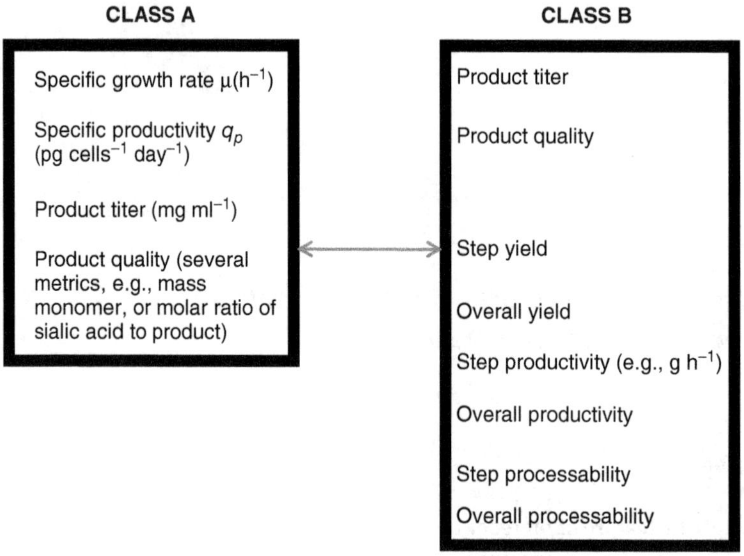


**Figure 12.1** Interactions between the cell culture step and other steps in a typical bioprocess. Class A pertains to the bioreactor, and Class B to all the steps downstream of the bioreactor. Class A includes the usual nonproduct-based as well as product-based metrics for bioreactor efficiency. Class B describes all the metrics describing impacts of the bioreactor step on steps downstream of it, whether a single step (unit operation) or multiple steps are affected. Any interaction between parameters of different classes represents a trade-off: either within the bioreactor, or between the bioreactor step and the steps downstream.

that are important in characterizing the quality of the product made in the bioreactor, which are also listed in Class A. If the product can be found in several multimeric forms, the fraction of the desired form(s) is an important metric of quality; often the monomer is the desired variant, and various aggregated forms (dimer, trimer, etc.) are undesirable. Similarly, if the product is appreciably glycosylated, the extent of glycosylation could be critical. One approximate measure of glycosylation that is often used in industry is the molar ratio of sialic acid (often the NANA variant) to the product. As is discussed later, such a molar ratio may be misleading; it is more often important to characterize the *extent* of glycosylation; in particular, the aglycosylated form may be so rapidly scavenged when the product is introduced into a patient's bloodstream that it should not be counted as a useful product variant. Interactions between parameters within Class A represent one form of trade-off in optimizing bioreactor operation.

In the same way, all the product quality metrics for all steps downstream of the bioreactor are listed in Class B. This could not only include product quality parameters of the kind already discussed but also bulk measures of efficiency such as yield or throughput, which can be affected by bioreactor conditions. For instance, if the bioreactor is run to low final viability in order to maximize product formation, the subsequent impact of the lysed cells on recovery steps such

**Table 12.1** Representative interactions within cell culture (Class A interactions) and between cell culture and downstream steps (Class A–Class B interactions).

| Step(s) | Interaction | Source |
|---|---|---|
| Class A–Class A interactions | | |
| Cell culture | Growth rate and cell-specific productivity | [1] |
| Cell culture | Overall metabolism and product formation rate | [2] |
| Class A–Class B interactions | | |
| Cell culture–centrifugation | Cell state and ease of homogenization | [3] |
| Cell culture–filtration | High-density cell culture, filterability, and improvement by interposition of chromatography | [4] |
| Cell culture–chromatography | Bioreactor production rate facilitated by product capture by chromatographic resin | [5, 6] |

The examples shown here provide a sense of the range of possible interactions; they are discussed in more detail in subsequent sections.

as centrifugation or filtration could be substantial. If the cost of increasing the product amount in the bioreactor is that the filtration or centrifugation step takes twice as long, and possibly with a substantial reduction in step yield, this may not represent the globally optimal bioprocess. An interaction between any metric in Class B and any metric in Class A represents another form of trade-off in optimizing bioreactor operation. (There is a third form of trade-off, involving two or more metrics in Class B, but that is outside the scope of this chapter.)

The list of possible interactions is large and complex. Representative examples, which are discussed in more detail later, are summarized in Table 12.1. It can be seen that multiple interactions of many kinds have been reported in the literature. More quantitative data of this kind are needed to evaluate how global the design process should be, especially for products at early stages of the development process, for example, Phase I or II clinical trials.

To facilitate the structure of this chapter, interactions are discussed on the basis of unit operations. Each important unit-operation step – the bioreactor itself, centrifugation, filtration, chromatography – is considered in turn, and examples of how characteristic trade-offs can occur are provided. These trade-offs elucidate that a bioprocess should be designed globally: optimizing each step individually is unlikely to lead to a globally optimal set of operating conditions. One very poor step can compromise the efficiency of the whole bioprocess; it is, therefore, better to accept some decrease in performance in other steps in order to improve substantially the performance of this poor step. There are exceptions to this interactivity among steps; for instance, [7] (discussed later) showed that cell engineering could be improved without negative impacts downstream. But this lack of interactivity must be demonstrated, which still requires an integrated view of proposed changes. The next two sections of the chapter discuss integrated processes, and integrated models. Finally, some conclusions on the potential value of further work on integrated processes are offered.

## 12.2 Improvements in Cell-Culture

Traditionally, cell-culture media were supplemented by the use of serum, or by the use of serum proteins such as transferrin, to improve growth rates and provide some protection against shear stress. However, these media were highly variable and complicated downstream processing [8]. The use of chemically defined media has decreased variability in the entire bioprocess. Nevertheless, some additives, such as antifoaming agents, can still have deleterious effects on downstream processing, for example, in filtration [9] and chromatography.

The trade-off between increased specific growth rate and rate and quality of product is a topic of great current importance [1, 2]. Srivastava *et al.* [1] found experimentally that high specific productivities and high glycosylation rates could be maintained for several cell lines through optimization of feeding strategies, that is, they did not find that they had to reduce specific productivities significantly in order to maintain high levels of glycosylation. This is an extremely interesting result, since the conventional wisdom is that such a loss is to be expected. Mathematically, one would expect ultimately to reach part of a Pareto surface, over which improving one component of a multicriterion objective function inevitably leads to a worsening of at least one other component. (It is possible that these experimental results had not yet reached a Pareto surface, in which case further improvements could be expected until such a surface was reached.) It is to be hoped that more results of this kind will be published, allowing the field to reach a more settled understanding of this critical trade-off. Another indication of such a trade-off was recently published by Nocon *et al.* [2]. They constructed a genome-level model for the central metabolism of *Pichia pastoris*, and used it to evaluate the impact of overproduction of heterologous proteins on primary metabolism. The results showed overexpression targets in the pentose pathway and the TCA cycle, and knockout targets at several branch points in the glycolysis pathway. Making the appropriate changes led to increased production of the heterologous protein (cytosolic human superoxide dismutase) in five out of nine targets. It would be fascinating to evaluate the additional impact of incorporating protein variants, such as glycoforms, to discuss the possible trade-off between product quality and quantity discussed earlier.

An interesting example of cell engineering is the overexpression of the twin arginine translocation (Tat) pathway in *Escherichia coli* [7]. The Tat pathway, which can export potentially valuable proteins that cannot be exported by the Sec pathway, has a low capacity in the native form. Engineering the cells to overexpress the Tat pathway led to a 25% improvement in growth rate and a 40-fold increase in periplasmic accumulation of the desired protein. In addition, the integrity of the membrane was unaffected in tests with increasing levels of shear. A scale-down model of continuous disk-stack centrifugation predicted that overexpressing the

Tat pathway did not affect recovery steps. This is an instance where optimizing one step did not have any deleterious consequences on subsequent steps.

## 12.3 Interactions with Centrifugation Steps

Centrifugation is a critical harvesting step in the biotechnology industry, particularly at manufacturing scale. An important assessment of the importance of the operational parameters for typical centrifugation processing was contributed by King *et al.* [10], who carried out a global sensitivity analysis (GSA) of the centrifugation of mammalian and high-density yeast cell broths. They showed that the global analysis showed important differences, because of interactions, than the more traditional single-parameter sensitivity calculations. They also demonstrated (Figure 12.2) that the critical particle size was the most important parameter in processing mammalian cell broth, and that the flow rate was the most important parameter in processing yeast cell broth. Naturally such conclusions are limited to the ranges across which the GSA was performed; these ranges are provided in Table 12.2. Such results are important in evaluating the combined effect of the bioreactor and the centrifuge: the operating flow rate of the centrifuge and its performance (typically the clarification level) must be chosen based on the *interactive* dependence on the relevant cell-broth parameters.

Li *et al.* [3] presented an instructive example of the cumulative impact of fermentation and homogenization on centrifugation in an ultra-scale-down (USD) format. Fermentation conditions are known to affect homogenization,

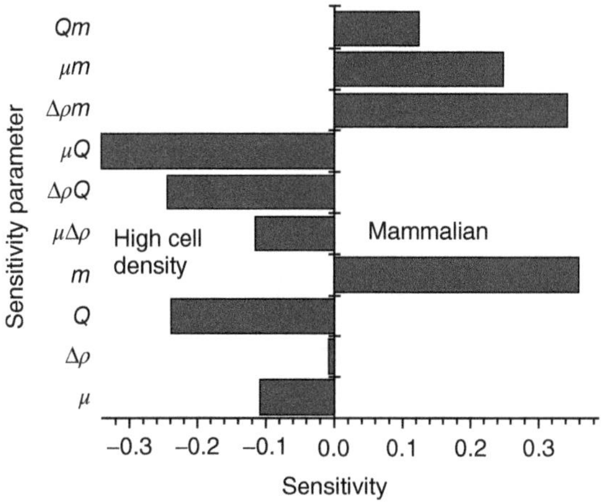


**Figure 12.2** Global sensitivities of operational parameters for typical centrifugation of mammalian and yeast culture broth [10]. The mammalian culture results are on the right, and the yeast results (referred to as "high cell density") are on the left of the diagram. The operational parameters are: $m$, the mean particle size; $\mu$, the broth viscosity; $\Delta\rho$, the density difference between the particles and the liquid; and $Q$, the flow rate. The mammalian centrifugation is dominated by $m$, but the yeast centrifugation is dominated by $Q$.

**Table 12.2** Operational parameters for centrifugation of mammalian and yeast broths (adapted from [10]).

| Parameter | Mammalian broth | Yeast broth |
|---|---|---|
| Flow rate $Q$ ($\mathrm{l\,h^{-1}}$) | 170–190 | 350–400 |
| Viscosity $\mu$ ($\mathrm{kg\,(m\,s)^{-1}}$) | 0.0015–0.0018 | 0.0020–0.0021 |
| Density difference $\Delta\rho$ ($\mathrm{kg\,m^{-3}}$) | 90–110 | 160–170 |
| Particle size $m$ (μm) | 9–11 | 4.9–5.1 |

for example, cells harvested in the exponential phase are easier to homogenize than cells harvested in the stationary phase [11]. Homogenization, in turn, affects centrifugation as well as steps further downstream, such as chromatography [12]. It is, therefore, extremely important to mimic the bench- or manufacturing-scale process quantitatively in the USD format. Li *et al.* (op. cit.) were able to show that the critical parameters were quantitatively preserved upon using focused acoustics to replace homogenization.

## 12.4 Interactions with Filtration Steps

A variety of dead-end and tangential flow filtration steps are commonly found in bioprocess sequences today; Reis and Zydney [13] provide a useful summary.

Crucell (now a subsidiary of Johnson & Johnson) presented an interesting approach [4] to cell separation at extremely high densities (approximately 120 million PER.C6 cells per ml). Depth filtration cannot cope with such high loadings; while centrifugation can, it is likely to become expensive and slow and result in reduced recovery. Flocculation can also be effective, but then presents an additional separation challenge because of the added flocculants. In this work, Schirmer *et al.* [4] used ion-exchange (IEX) resins as a pretreatment step to induce cell settling and thereby reduce the loading onto a subsequent depth filtration step. Comparison of settling rates (which are proportional to the supernatant volumes reported as a function of time) is provided in Figure 12.3 for a variety of resins; the best of these, silicon-based polytheyleneimine or Si-PEI, was also shown to provide 40% reduction in host-cell protein (HCP) levels by electrostatic adsorption. This is an unusual approach to solving a cell-culture issue – of extremely high cell density, which of course corresponds to very high product generation rates – by incorporating a step traditionally used for polishing much further downstream into the recovery phase, to address the high cell density while still providing HCP clearance. Subsequent studies with varying amounts of Si-PEI showed (cf. Figure 12.4) that increasing the amount of the resin decreased the level of cells while maintaining product recovery.

Felo *et al.* [14] analyzed a two-step harvest process, the first step being either centrifugation or depth filtration, and the second step being secondary clarification by depth filtration. The base-case was the harvesting of a 5000 l

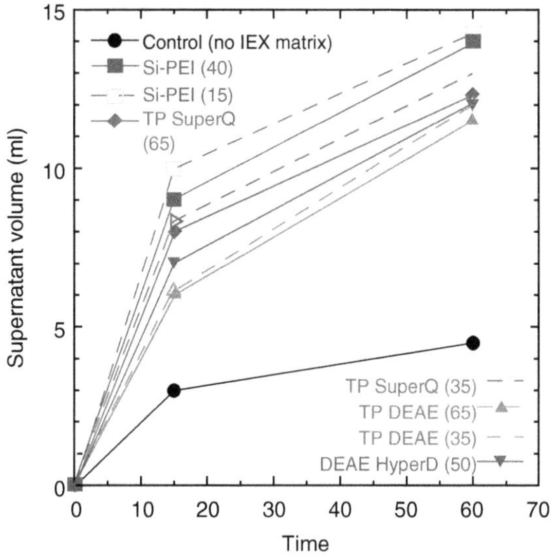


**Figure 12.3** Supernatant volumes collected as a function of time when cell-culture broth was exposed to a variety of ion-exchange resins [4]. The number in parentheses next to each resin is the average particle size of the resin beads. The Si-PEI provided the most efficient collection of supernatant volume, and was chosen for further study (see text for more details).

**Table 12.3** Base-case data for the comparison of centrifugation and depth filtration for mammalian cell culture [14], and associated range studied during sensitivity analysis.

| Model parameter | Base-case value | Range for sensitivity analysis |
|---|---|---|
| Bioreactor volume (l) | 5 000 | 500–12 000 |
| Facility utilization rate (%) | 80 | 20–100 |
| Product titer at the time of clarification ($g\,l^{-1}$) | 3.0 | 0.5–10.0 |
| Capacity of primary depth filter ($l\,m^{-2}$) | 69 | 35–500 |

bioreactor running a CHO cell line to produce a mAb. Table 12.3 describes the range of values considered in their sensitivity analysis. The essential result provided in Figure 12.5 shows that centrifugation is slightly better in this study for bioreactor volumes above 5000 l. However, this cost advantage diminished from 6.2% to 2.4% as the cell titer increased from 0.5 to 10 $g\,l^{-1}$.

## 12.5 Interactions with Chromatographic Steps

*Adsorption to improve product recovery:* Process integration has long been recognized as a promising path toward reducing costs and improving performance in the production of biologics [15].

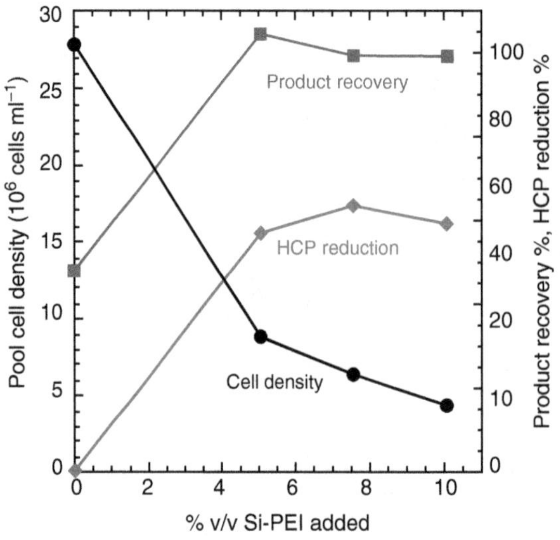


**Figure 12.4** Addition of Si-PEI ion-exchange resin to reduce pool cell density and host-cell protein (HCP) content while recovering the product [4]. Increasing the amount of resin, through 10%, decreased the pool cell density and HCP while maintaining product recovery (see text for more details).

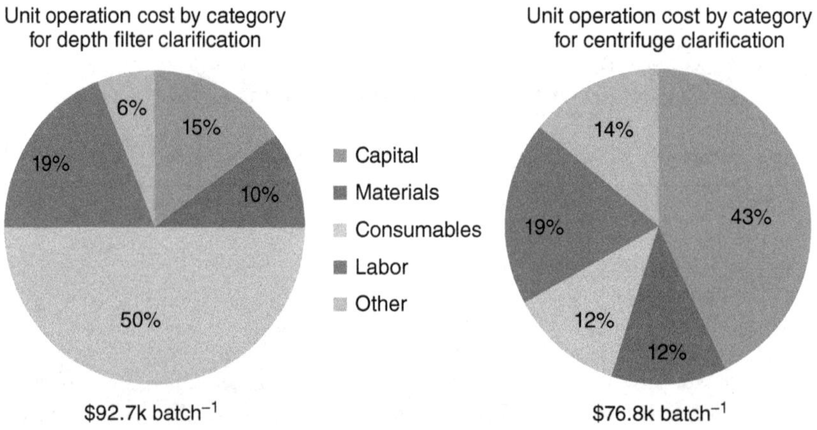


**Figure 12.5** Comparison of centrifugation and depth filtration for 5000 l bioreactor [14] (see text for more details).

Kumar *et al.* [6] presented an integration of the bioreactor and product capture step for the production of urokinase from animal cell culture. In this study, a continuous super-macroporous polyacrylamide cryogel column was used to provide a scaffold for the growth and proliferation of anchorage-dependent cells and simultaneously as an adsorbent for the capture of the urokinase product. HT1080 human fibroscarcoma and HCT116 human colon cancer cell lines were used. An indication of the value of the dual role played by the cryogel column is afforded by the head-to-head comparison the authors made of the current system with a standard bioreactor followed by a chromatographic column. The latter was found to foul quickly with cell debris, while the former worked effectively without the need of an intervening cell-harvesting step.

The use of such immediate product-harvesting strategies is quite common in plant-cell cultures; a typical example is furnished by Gao *et al.* [5], who reported significant improvements in the recovery of the taxoid product taxiyunnanine C from suspension cultures of *Taxus chinensis* through adsorption *in situ*. They combined this adsorption with repeated elicitation with a jasmonate analogue and an optimized sucrose-feeding strategy. The best results achieved, approximately $1.7\,g\,l^{-1}$, amounted to more than double the highest value previously reported in the literature [16]. It was found that more than 90% of the product generated adsorbed to the XAD resins used, thereby also simplifying the subsequent downstream processing.

Expanded-bed adsorption (EBA) has been used in the past, most commonly with ion-exchange functionality, to attempt to feed cell broth directly into a chromatographic column, thereby obviating clarification steps [17, 18]. Kelly *et al.* [19] have recently investigated next-generation high-particle-density EBA adsorbents, including two mixed-mode resins, for the direct capture of a recombinant protein expressed in yeast at high cell densities. Using the most retentive resin, Fastline® MabDirect, competitive binding of nontarget proteins was found, and adsorption from a cell broth containing 5–10% cells reduced the equilibrium binding capacity by 30% to $50\,mg\,ml^{-1}$ settled bed. The process was roughly comparable at low flow rates to a conventional process using packed beds in terms of step yield and HCP levels, but cleared substantially less DNA (about 50× less than the conventional process). As cell densities and the structural complexity of proteinaceous products increase, clarification may become more time-consuming and expensive; EBA may then become economically attractive.

*Genuine multistep interactions:* There is a dearth of experimental data in the literature on the combined evaluation of cell culture and chromatographic polishing steps, no doubt partly because of the complexity of such experiments. However, such data would help to evaluate whether the global optimum (e.g., for overall productivity) for the sequence is substantially different from the combination of the individual optima for each step. We have recently simulated this situation for a product with five sialylated species for the sequence of fed-batch reactor and two polishing chromatographic columns [20]. (It was assumed that the intervening steps of clarification and capture did not to affect the glycoform distribution substantially.) The cell culture was assumed to produce either a uniform or a symmetric distribution of sialyl forms, as a first approximation. The order of product elution from the cation-exchange (CX) and hydrophobic interaction (HIC) polishing columns were, therefore, reversed: the most sialylated species was the most retained on the CX column, being the most negatively charged, and it was the least retained on the HIC column for the same reason. (It is assumed that these sialyl charges play a critical role in the binding configuration assumed by the protein on both adsorbents.) Detailed simulations of the process led to the comparison of the HIC–CX and CX–HIC sequences for the different sialyl distributions for four operating parameters in each chromatographic step: loading, flow rate, gradient slope, and initial modulator level. (In fact, all of the pooled material from the first column was fed into the second column, thereby removing the loading parameter as an independent variable for the second column; this resulted in seven independent variables for the two-column polishing sequence.) The interactions were found to be quite complex, and led to complex operating windows in the operating parameters. Various objective functions were considered – yield, purity, amount of acceptably pure product recovered, throughput, and the sialic acid molar ratio (ratio of sialic acid to ratio of protein) – and Pareto plots were constructed for the multicriteria optimization. These results were compared with the analogous results for scalarized objective functions, for example, by combining many or all of the individual objective functions into a single weighted product (an approach that is quite common as a base-case calculation in multiobjective optimization). The central result was, as might have been expected, that the quality of the product was as important as its quantity, and that there were conditions under which it would be advantageous to reduce the rate of product formation in order to improve its quality.

## 12.6
## Integrated Processes

In an extreme but impressive example, Caparon *et al.* [21] reported the purification challenges that Pfizer faced in purifying an apolipoprotein variant produced in *E. coli*. It appeared that several HCPs were co-eluting with the product, and that a truncated form of the product was being generated. A purification train involving five chromatographic steps (three IEX and two HIC) was insufficient to provide robust clearance of HCPs. They used multiple analytical methods, including 1D and 2D gel electrophoresis, Western blot, protein deglycosylation, mass spectrometry (MS), Edman degradation, and a sandwich ELISA. Two-dimensional Western blot analysis of the purified ApoA-1M samples detected two major HCPs along with 6–7 minor HCP species. As shown in Figure 12.6b, these two major HCPs have molecular weights around 60 kDa, each exhibiting multiple forms on the 2D gel of a purified protein sample. Both molecular weight and isoelectric point of these HCPs are thus quite close to those of product dimer (56 kDa, p*I*: 5.1–5.3), which may explain why they were co-purified with the product. This allowed these HCPs to be located on the corresponding stained gel at high loading (Figure 12.6a), and then isolated by 2D electrophoresis for identification by mass spectrometry. Three HCPs were identified: a dipeptide-binding protein and an oligopeptide-binding protein with sizes and charges close to those of the product dimer, both of which are found in abundance in *E. coli* periplasm; a maltose-binding periplasmic protein with slightly smaller size. Furthermore, a bacterial protease was suspected as the cause of the product cleavage reaction. It was decided that a second-generation process be carried out by sequentially knocking out the genes for these four compounds, using a method described by

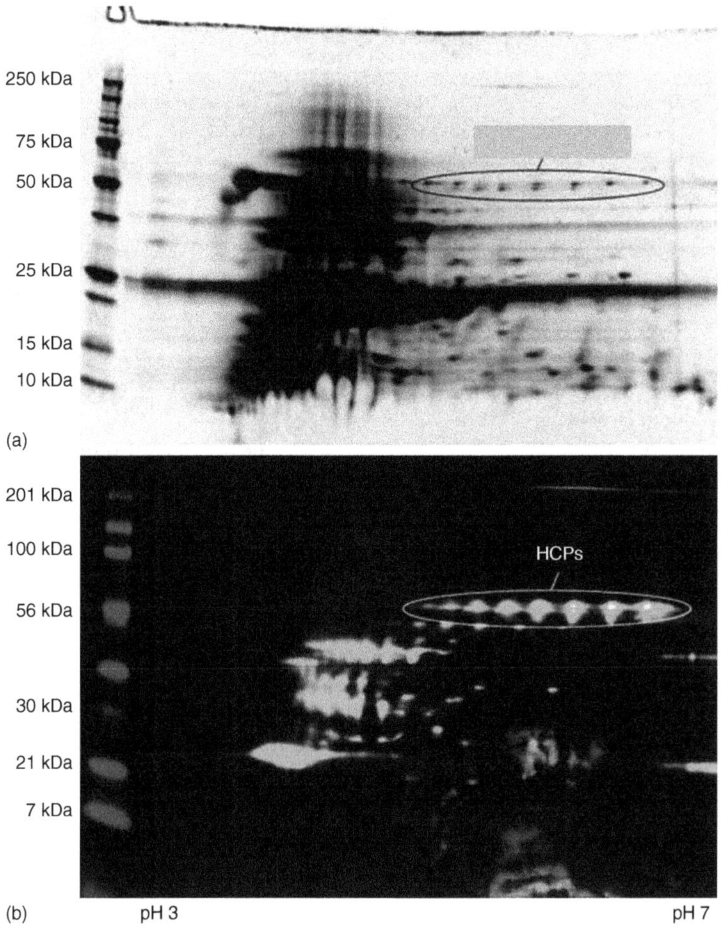


**Figure 12.6** Two-dimensional gel electrophoresis pictures [21] of an intermediate process sample from the apolipoprotein purification process: (a) total protein stain; (b) *E. coli* HCP Western blot of the same gel. Total protein loading was 400 μg (see text for more details).

Link *et al.* [22]. The results are shown in Table 12.4, where the 2-deletion refers to the removal of the dipeptide- and oligopeptide-binding proteins, the 3-deletion refers to the additional removal of the maltose-binding protein, and the 4-deletion refers to the additional removal of the protease. It can be seen from the table that these identifications were successful: the HCP level has been dramatically reduced (the corresponding 2D Western blots – data not shown – do not present any trace of the HCPs seen earlier). Furthermore, the percentage of the truncated product form is well controlled, indicating that the protease was in fact the cause of the truncation. The purification process for this next-generation process was then dramatically simplified, with only two polishing columns needed to achieve the desired purity. This is an unusual example, both in the level of difficulty of the

**Table 12.4** Effects of HCP deletion on process performance (summarized from [21]; see text for more details).

| | HCP (ng HCP/mg product) | | Truncated product species (%) | |
|---|---|---|---|---|
| | 2-Deletions | 3-Deletions | 3-Deletions | 4-Deletions |
| Capture pool | 22 800 | 660 | 1.9 | 0.1 |
| DEAE pool | 75 | <3 | 1.7 | 0.2 |
| DEAE postpeak pool | – | – | 7.0 | 1.4 |
| Selected DEAE postpeak fraction | – | – | 14 | 1.6 |

purification process, and in the degree of success that genetic knockouts were able to provide. Nevertheless, it shows the value of making fundamental changes when the problem is sufficiently difficult.

Another important trend in integrated processing is the use of continuous unit-operations. While classical engineering suggests that processes operated continuously at steady state are more efficient than batch processes, the biotechnological industry has historically been dominated by batch processing. Although some products were commercially manufactured by perfusion cell culture, they were exceptions; and no commercial biologic has been manufactured using continuous chromatography so far. Nevertheless, both perfusion culture and continuous chromatography are being evaluated critically by many companies in collaboration with many academicians, and the near future may decide whether a dramatic change in bioprocessing may be in the offing. Genzyme has made a substantial contribution toward demonstrating the robustness of such continuous processes for complex biologics: their recent publications have shown that process stability and product quality have been maintained for many weeks in a single run of a continuous process. Their recent paper [23] summarized their findings. They use perfusion bioreactors and four-column periodic countercurrent chromatography (PCC) to produce and purify several exemplar therapeutics, including a mAb and a recombinant human enzyme. A quasi-steady state was achieved in the bioreactor at about 50–60 million cells $ml^{-1}$ for more than 60 days. The fully automated PCC system ran for more than 30 days, with product quality comparable to that of a traditional batch purification process. This integration was achieved with the removal of several hold steps that would be necessary in traditional batch operation. The literature would benefit from other reports of this kind, describing continuous processes operated over long periods.

## 12.7 Integrated Models

Integrated process models are widely used in the chemical process industry, where continuous unit operations are typically run at steady state. More recently, corresponding approaches have been used in the pharmaceutical industry, where

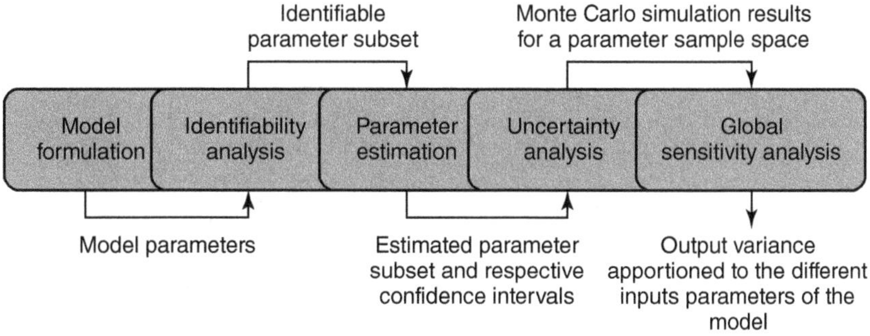


**Figure 12.7** Schematic workflow for model analysis [26] (see text for more details).

small molecule products allow for complete characterization of impurity profiles [24, 25]. However, such models are more difficult to contemplate for bioprocesses, partly because of the incomplete analytical information on impurity profiles and partly because of the intrinsically higher variability of bioprocesses. Nevertheless, the wider use of such models would have great value in the biotechnology industry, not least in showing the sources of variability that are most important in practice. There has been a trend toward such holistic bioprocess models [26, 27]. We briefly address some representative work.

Taras and Woinaroschy [28] have combined Matlab (as the numerical tool) and SuperPro Designer (as the process simulator) to provide a setting for the interactive multicriterion optimization of whole bioprocesses. Appropriate linking software was created in order to allow for a simple user interface. The method was applied to the production of lysine, and incorporated environmental objectives in addition to the more common process and economic objectives.

Fernandes *et al.* [26] have reviewed the mechanistic modeling approach in the context of bioprocess development. The iterative process of model selection, parameter estimation, and validation is summarized in Figure 12.7. They applied the approach to a yeast model [29]. They combined their approach with sensitivity and uncertainty analysis, and suggested that the approach can be readily extended to other, more complex situations.

## 12.8 Conclusions

While there is a dearth of published studies quantifying the interactions between the bioreactor and the steps downstream, it is clear that such studies are urgently needed to increase the overall process efficiency, and to reduce avoidable mismatches among unit operations deriving from the "silo-based" approach to process development, where each unit operation is developed in isolation. The results reported here are an indication that the search for process understanding will eventually result in seamless descriptions of whole bioprocesses.

## References


1. Srivastava, R., Rao, L., Shukla, K., Prabhu, S., Desan, S., Baskar, D., Bhatnagar, A., Goel, A., and Iyer, H. (2011) *BMC Proc.*, **5** (Suppl. 8), P95.
2. Nocon, J. *et al.* (2014) *Metab. Eng.*, **24**, 129–138.
3. Li, Q., Mannall, G.J., Ali, S., and Hoare, M. (2013) *Biotechnol. Bioeng.*, **110**, 2150–2160.
4. Schirmer, E.B., Kuczewski, M., Golden, K., Lain, B., Bragg, C., Chon, J., Cacciuttolo, M., and Zarbis-Papastoitsis, G. (2010) *Bioprocess Int.*, **8** (1), 32–39.
5. Gao, M.-B., Zhang, W., and Ruan, C.-J. (2011) *World J. Microbiol. Biotechnol.*, **27**, 2271–2279.
6. Kumar, A., Bansal, V., Nandakumar, K.S., Galaev, I.Y., Roychoudhury, P.K., Holmdahl, and Mattiasson, B. (2006) *Biotechnol. Bioeng.*, **93**, 636–646.
7. Branston, S.D., Matos, C.F.R.O., Freedman, R.B., Robinson, C., and Keshavarz-Moore, E. (2012) *Biotechnol. Bioeng.*, **109**, 983–991.
8. Fassnacht, D., Rossing, S., Ghaussy, N., and Portner, R. (1997) *Biotechnol. Lett.*, **19**, 35–38.
9. Aunins, J.G. and Henzler, H.J. (1993) in *Biotechnology*, vol. 3 (eds H.J. Rehm and G. Reed), Wiley-VCH Verlag GmbH, pp. 219–281.
10. King, J.M.P., Titchener-Hooker, N.J., and Zhou, Y. (2007) *Bioprocess Biosyst. Eng.*, **30**, 123–134.
11. Casadei, M.A., Manas, P., Niven, G., Needs, E., and Mackey, B.M. (2002) *Appl. Environ. Microbiol.*, **68**, 5965–5972.
12. Hubbuch, J.J., Brixius, P.J., Lin, D.Q., Mollerup, I., and Kula, M.R. (2006) *Biotechnol. Bioeng.*, **94**, 543–553.
13. van Reis, R. and Zydney, A. (2007) *J. Membr. Sci.*, **297**, 16–50.
14. Felo, M., Christensen, B., and Higgins, J. (2013) *Biotechnol. Prog.*, **29**, 1239–1245.
15. Chisti, Y. and Moo-Young, M. (1996) *Trans. Inst. Chem. Eng.*, **74**, 575–583.
16. Qian, Z.G., Zhao, Z.J., Xu, Y.F., Qian, X.H., and Zhong, J.J. (2005) *Biotechnol. Bioeng.*, **90**, 516–521.
17. Hubbuch, J., Thommes, J., and Kula, M.-R. (2005) *Adv. Biochem. Eng./Biotechnol.*, **92**, 101–123.
18. Mullick, A. and Flickinger, M. (1999) *Biotechnol. Bioeng.*, **65**, 282–290.
19. Kelly, W., Garcia, P., McDermott, S., Mullen, P., Kamguia, G., Jones, G., Ubiera, A., and Goklen, K. (2013) *Biochem. Appl. Biotechnol.*, **60**, 510–520.
20. Konstantinidis, S., Field, N., Jurlewicz, K., and Velayudhan, A. (2014) Whole-process evaluation of operating windows and critical process parameters. Presentation at Recovery of Biological Products XVI, Rostock, Germany, July 27–31, 2014.
21. Caparon, M.H., Rust, K.J., Hunter, A.K., McLaughlin, J.K., Thomas, K.E., Herberg, J.T., Shell, R.E., Lanter, P.B., Bishop, B.F., Dufield, R.L., Wang, X., and Ho, S.V. (2010) *Biotechnol. Bioeng.*, **105**, 239–249.
22. Link, A.J., Phillips, D., and Church, G.M. (1997) *J. Bacteriol.*, **179**, 6228–6237.
23. Warikoo, V., Godawat, R., Brower, K., Jain, S., Cummings, D., Simons, E., Johnson, T., Walther, J., Yu, M., Wright, B., McLarty, J., Karey, K.P., Hwang, C., Zhou, W., Riske, F., and Konstantinov, K. (2012) *Biotechnol. Bioeng.*, **109**, 3018–3029.
24. Schaber, S.D., Gerogiorgis, D.I., Ramachandran, R., Evans, J.M., Barton, P.I., and Trout, B.L. (2011) *Ind. Eng. Chem. Res.*, **50**, 10083–10092.
25. Troup, G.M. and Georgakis, C. (2013) *Comput. Chem. Eng.*, **51**, 157–171.
26. Fernandes, R.L., Bodla, V.K., Carlquist, M., Heins, A.-H., Lantz, A.E., Sin, G., and Gernaey, K.V. (2012) *Adv. Biochem. Eng./Biotechnol.*, **132**, 137–166.
27. Velayudhan, A. (2014) *Curr. Opin. Chem. Eng.*, **6**, 83–89.
28. Taras, S. and Woinaroschy, A. (2012) *Comput. Chem. Eng.*, **43**, 10–22.
29. Sonnleitner, A. and Kappeli, E. (1986) *Biotechnol. Bioeng.*, **28**, 927–937.

# 13
# Multivariate Modeling for Bioreactor Monitoring and Control

*Jarka Glassey*

## 13.1
## Introduction

Over the last decade, the importance of bioprocess monitoring, modeling, and control rapidly gained prominence in the scientific community and industry, partly due to the emphasis placed on this aspect of processing by the quality-by-design (QbD) and process analytical technologies (PAT) guidelines documents [1]. Another important driver raising the prominence of this discipline is the impact of effective bioprocess monitoring and control upon the business performance of the manufacturing industry. In addition, modeling can significantly speed up bioprocess development, improving the market competitiveness of new products and emerging branches of bioprocess industry, including cell therapies, stem cell technologies, and artificial organ production described earlier in this book (see Chapters 9, 11–13).

A plethora of scientific articles, reviews, and books dealing with aspects of bioprocess monitoring, modeling, and control exists currently. A quick search of Web of Science for these key words indicates 3214 entries over all years (1864–2015), although as Figure 13.1 illustrates, the use of these terms emerged in the late 1970s and expanded at a phenomenal rate since then (note: the literature sources shown in Figure 13.1 were not checked for duplication between the three categories).

However, this scientific field continues to grow and evolve to encompass the latest developments in analytical methods, as well as in single-use bioreactors (see Chapter 6), scale-down (see Chapter 10) and microbioreactors, and high-throughput technology (see Chapter 16), used in bioprocess industries more and more extensively. These aspects introduce additional opportunities and challenges for monitoring, modeling, and control of typically highly complex processes within this tightly regulated sphere. For example, the reports of the expert panel meetings of the modeling, monitoring, measurement, and control ($M^3C$) section of the European Society of Biochemical Engineering Sciences (ESBES) on hybrid modeling [2], cell culture processing [3], microbioreactors [4], and soft sensors [5] in bioprocess industries summarize the state-of-the-art in

*Bioreactors: Design, Operation and Novel Applications*, First Edition. Edited by Carl-Fredrik Mandenius.

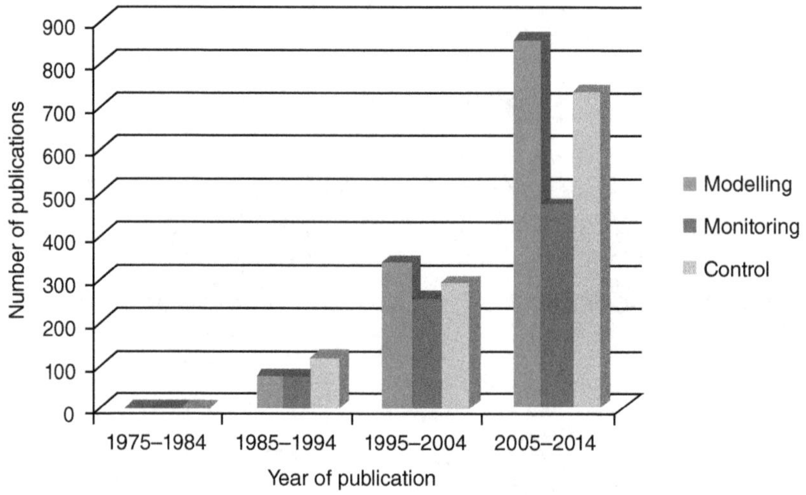


**Figure 13.1** Overview of publication rate in bioprocess modeling, monitoring, and control area between 1975 and 2014, based on a search of Web of Science.

respective scientific areas and propose recommendations for future development in each area with relevance to monitoring and control in bioindustries. A recurring theme through these reports is a need for more accurate analytical techniques, providing reliable, timely, and representative measurements of critical process parameters (CPPs) even at very low concentrations of analytes and further development and successful industrial application of mathematical representations of the bioprocesses.

This chapter, therefore, gives an overview of the measurement methods currently applied in bioprocess measurement and monitoring and then reviews multivariate data analysis methods used for bioprocess data exploration and modeling for the purpose of monitoring and control. Bioprocess modeling based on first-principles understanding of the underlying biological processes is outside the scope of this publication, but the readers are referred to various reviews and specific articles dealing with this topic for their respective processes (e.g., [6–8]).

## 13.2
## Analytical Measurement Methods for Bioreactor Monitoring

Timely monitoring of process behavior is not only required by the PAT guidelines [1], but forms a critical basis of effective process control. A detailed review of specific process instrumentation for critical parameter measurement is beyond the scope of this chapter, and a number of authors offer a useful review of measurement techniques useful to bioprocesses in general (e.g., [9, 10]) or concentrating either on specific types of bioprocesses (e.g., [11] in cell culture) or types of sensors (e.g., [12–14]). Regardless of the bioprocess considered, it is important to consider carefully the characteristics of the sensors before their application in a monitoring and control scheme. Glassey discusses the general sensor characteristics in more detail in [15]. These include considerations of accuracy and resolution, precision, sensitivity, reliability, response time, practicality, as well as the cost. Vojinovic *et al.* [14] include additional requirements on *in situ* bioprocess sensors, such as the ability to function without fouling over prolonged periods of time, endure extreme conditions during sterilization, to have a dynamic range covering the expected variation in the measured process variable. Further considerations include the ability to generate multianalyte data without analyte consumption and noninterference with culture metabolism [16].

Given the extensive literature resource relating to sensors, the frequency of the measurement, and the placement of the sensors already available, only a brief overview of sensors most relevant to the bioprocesses considered in this book are described in the following sections.

### 13.2.1
### Traditional Measurement Methods

Sonnleitner offers a detailed description of various methods of monitoring physical and chemical variables, including measurements of temperature, flow rates, pH, $pO_2$, $pCO_2$, and gas composition [9]. A comprehensive and up-to-date review of methods of biomass and bioactivity is also included, highlighting the benefits and the challenges associated with various methods detailed in the review.

A further review of the historical progress of monitoring and computer control, with a particular emphasis on monitoring and control of fermentation processes over the last four decades is provided by Junker and Wang's [10] review commemorating the seventieth birthday of Daniel I.C. Wang whose research significantly shaped the directions of bioprocess computer control and monitoring. The early development concentrated on the application of engineering principles, such as mass and energy balancing, in conjunction with existing traditional measurement methods. Such approaches are still extensively used in bioprocess monitoring and form a substantial component of advanced monitoring systems utilizing more advanced concepts of soft sensors (see Chapter 15, [5]).

In the context of multivariate modeling for bioprocess monitoring, it is important to note that the traditional measurement methods typically provide extensive data sets with relatively high sampling frequency often requiring data preprocessing and reduction as discussed in Section 13.2.3. In industrial applications, the sampling frequencies are thus often set to exception logging with values from sensors recorded only when specific criteria for a change from the set point values are satisfied.

Regardless of the established nature of the traditional measurement methods in bioprocessing, the scale-down and high-throughput [17–19] approaches increasingly used in bioprocess development introduce additional challenges even in this area. Real-time monitoring of relatively straightforward physical and chemical process variables in microbioreactors can increase the monitoring challenges substantially. Most commercially available platforms for microprocessing and high-throughput bioprocessing provide the fundamental monitoring capabilities (for a useful review of such processing for animal cell culture processes see, e.g., Kim *et al.* [18]). These typically cover temperature, pH, and in some cases dissolved oxygen levels, with biologically relevant variables, such as cell number or activities, measured offline. At smaller scales, the measurements and the control of pH and dissolved oxygen (and/or dissolved carbon dioxide in the case of cell cultures) frequently relies on the quenching properties of fluorescent dyes [20–22]. This introduces additional challenges to monitoring in terms of range and sensitivity. Further discussion on the challenges, particularly in terms of data preprocessing and multivariate analysisintroduced by such scale-down and/or high-throughput processing are discussed in more detail in Section 13.2.3.

### 13.2.2
### Advanced Measurement Methods

With the rapid development of the analytical techniques, the categorization of methods into traditional and advanced is becoming more and more arbitrary. A wide range of techniques considered advanced (at least from an industrial application point of view) are now routinely used in bioprocessing. A number of these techniques are still used predominantly offline, that is, requiring manual or automatic sampling with a sample transfer to dedicated laboratory for analysis. However, there are techniques already applied online, frequently *in situ*, with a probe placed inside a bioreactor/unit operation from which the measurement is taken. The most striking progress in bioprocess monitoring was unsurprisingly registered in the upstream (bioreactor) stage. Arguably, this stage is the most challenging part of the bioprocess from the monitoring and control point of view, given the complexity of the process and its potential impact upon the product quality and quantity.

#### 13.2.2.1
#### Spectral Methods

Recent advances particularly in the optical fiber technology and the development of spectrometers and robust *in situ* probes make the spectral methods of bioprocess monitoring especially attractive. In the bioindustries sector, spectroscopy, particularly near-infrared (NIR) spectroscopy, was traditionally widely used in food industry [23–26]. In the case of NIR spectroscopy, high-throughput, short response time, fast and nondestructive basis, remote sampling capability, little or no sample preparation, and the ability to provide simultaneous determination of multiple components per measurement in real time were highlighted as particularly beneficial for the food industry [23]. These characteristics are equally beneficial in other areas of bioprocessing, as discussed later.

Various methods relying on the absorption of radiation in specific regions of the electromagnetic spectrum, from ultraviolet (UV) through visible to radio waves, are available and are used to varying extents in the bioindustries. Tamburini *et al.* [27] outline various techniques and analytes measured by these methods with respect to the wavelength in which they operate. Lourenco *et al.* [16] provide a comprehensive overview of the spectroscopic analysis (UV/vis, NIR, mid-infrared (MIR), Raman, and fluorescence) in the broad bioprocess industry including environmental, agro and food, biofuels, and pharmaceuticals. For each application, the scale of the production, details of the spectroscopic system used as well as the chemometric data analysis method and the mode of application (qualitative/quantitative monitoring and control) are shown. Although UV/vis spectroscopy, fluorometry, Raman, and NMR spectroscopy have been shown to provide useful monitoring of biomass, NADH, ATP, glucose, and other metabolites (e.g., [28–30]), infrared spectroscopy, particularly NIR, found the most diverse application both in research and industry over the past few decades.

#### Near- and Mid-Infrared Spectroscopy

The ability to detect groups containing –CH (whether aliphatic, aromatic, or alkene), –NH, and –OH bonds enables rapid fingerprinting of bioprocess samples in NIR (700–2500 nm) or MIR (2500–40000 nm) region of the spectrum. A typical NIR fingerprint from a bioprocess can be seen in Figure 13.2, indicating the changing composition with time (different colored spectra) over a range of wavelengths scanned.

NIR was reported to be successfully applied routinely in biomanufacturing, particularly in raw material and final product quality testing [31, 32] and more widely in bioprocessing in monitoring a range of analytes in upstream processing (e.g., [16, 33] for overview) or the quality of product downstream [34]. The latter

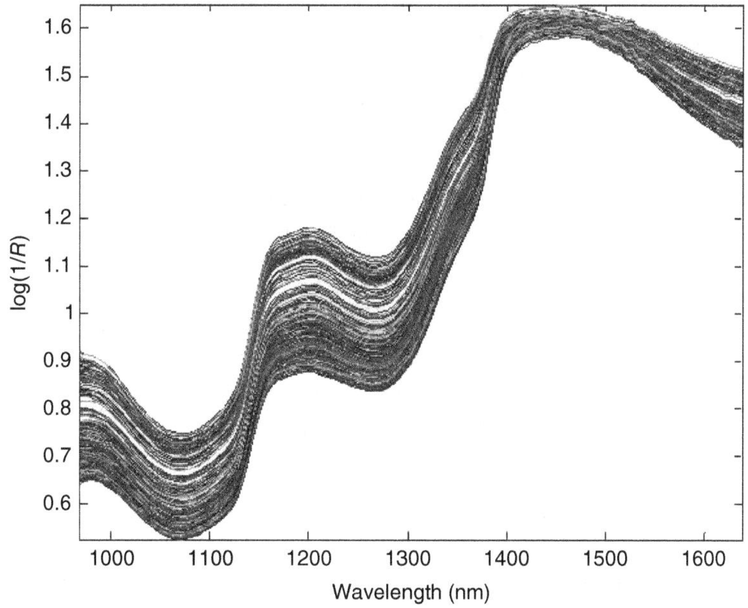


**Figure 13.2** Example of an NIR fingerprint from bioprocess monitoring.

publication was particularly useful from bioprocess monitoring point of view, as it demonstrated the ability of MIR spectroscopy to quantify soluble antibody aggregates (two different antibodies tested) post-protein A elution as well as in two different clarified CHO cell culture supernatants. The authors were able to demonstrate satisfactory prediction ability down to 1% aggregates compared with overall antibody amount with coefficients of variation for most of the samples below 20% [34].

However, the success of the application is often influenced by a number of interfering factors, such as agitation, gas bubbles, temperature changes, feeding, and changes in media composition [16]. Appropriate data preprocessing and multivariate data analysis methods are, therefore, particularly important in such cases and these are discussed in subsequent sections.

#### 13.2.2.2
#### Other Fingerprinting Methods

Alternative multianalyte measurement methods, such as electronic noses and electronic tongues, have also been successfully used in bioprocess monitoring [35–37]. These types of approaches are multisensory arrays, frequently utilizing a variety of transduction principles, such as electrochemical, gravimetric, and optical, as Rudnitskaya and Legin [35] highlight together with providing an overview of the respective sensing materials and analytes/applications. A particular advantage of electronic nose sensor, often highlighted in literature, is their ability to measure volatile analytes directly in the headspace/exit gas without requiring direct contact with the cultivation medium. In addition, there could be time delays, particularly for low-concentration analytes, and the sensors may be affected by drifting and water vapor interference [35]. Although the electronic tongue may require some sample preparation to remove any suspended solids in order for the liquid medium to be contacted with the array of sensors.

Various "omics" methods in essence also produce a fingerprint of either the characteristics of the organism or that of the environment in which they are cultured and could arguably classed as fingerprinting monitoring methods. Examples of genomics, transcriptomics, proteomics, and metabolomics tools used for bioprocess monitoring and increased understanding demonstrate the added value these methods can offer in bioprocess development and operation [38–41]. Although detailed analysis of individual "omics" methods is beyond the scope of this chapter, their application in bioprocess monitoring significantly increases the challenges in effective data preprocessing and multivariate data analysis as discussed in Section 13.2.3.

### 13.2.3
### Data Characteristics and Challenges for Modeling

The diversity of bioprocesses leads to a significant variability in the quality and the quantity of process data measured. As argued by Vojinovic *et al.* [14], the process variables measured, the frequency, precision, and time delay are affected by the

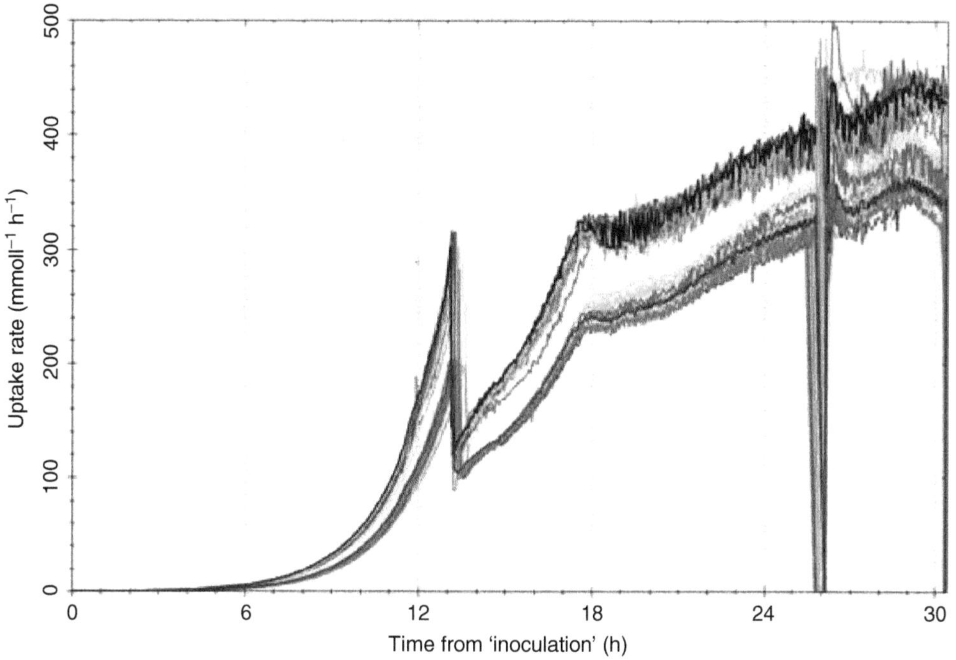


**Figure 13.3** An example of gas uptake measurements from high-throughput monitoring experiments (courtesy of Fujifilm Diosynth Biotechnologies, Billingham, UK, 2015).

manufacturing process that is being monitored. In general, however, the multitude of analytical methods used in bioprocess modeling (see previous sections) results in a substantial amount of continuous and discrete data of varying quality and recorded at various frequencies.

This issue of varying frequency and large amount of data is exacerbated by the use of high-throughput technologies, where the process development bottleneck is moved from the resource availability to conduct the required experiments to data analysis of the resulting large data sets (see Figure 13.3, for the example of a high-throughput monitoring of gas uptake rates). Significant data preprocessing is thus required in order for the data to be used in the development of bioprocess models for monitoring and control purposes. This preprocessing and data reconciliation may include simple time-alignment, in-filling missing data by interpolation, or the use of more sophisticated methods [42].

Spectroscopic data in particular have been shown to benefit from appropriate preprocessing in order to reduce the sources of variability and noise in the data and to enhance the informational content of the resulting data set. Such preprocessing methods frequently include mean-centering (since the spectra are measured in the same units, this standardization is often satisfactory), the use of derivatives to eliminate baseline offset or slope variation or various methods of smoothing and scatter correction. These include smoothing noisy data using a

Savitzky–Golay filter, multiplicative scatter correction (MSC), and standard normal variate (SNV) [16].

One of the major challenges for effective multivariate modeling of bioprocess data for monitoring and control purposes is the fusion of data from various sources and with various characteristics (e.g., [43]). Various mathematical representations of the data are possible to capture the salient characteristics of each stage of the bioprocess and methodologies to combine these into an effective framework capable of capturing the whole process behavior are still not standardized sufficiently. Section 13.3 will overview the most frequently used modeling approaches, highlighting their applicability in bioprocess monitoring and control as well as raising areas for further development.

## 13.3 Multivariate Modeling Approaches

The challenges introduced by data collected from the bioprocesses, particularly the high dimensionality of the resulting data sets make the multivariate approaches to bioprocess data modeling an essential requirement. The use of chemometric tools is well established in other scientific fields, particularly chemistry, although their wide acceptance in bioprocess modeling is evidenced by a large number of publications reporting the routine use of these tools both in research and in industrial applications (e.g., [11, 16, 42]). Various multivariate modeling approaches, both linear and nonlinear, are briefly introduced in the following with specific emphasis on their applicability to bioprocess monitoring and control.

Although the argument of using methods based on the fundamental assumption of linear relationships between the process variables is frequently cited as a major limitation of linear methods, they remain widely used and accepted in bioprocess data analysis and modeling. Despite the nonlinear character of bioprocess, it has been shown that various modifications and data preprocessing enable linear methods to effectively capture the underlying characteristics of the bioprocesses modeled (e.g., [44]).

### 13.3.1 Feature Extraction and Classification

One of the most frequent exploratory data analysis/feature extraction method described in the chemometrics literature is principal component analysis (PCA) [45, 46]. This technique is frequently employed as a data reduction approach, particularly before further regression analysis. This data reduction ability results from the decomposition of the original data matrix of process measurements ($X$) into a set of uncorrelated variables (principal components – PCs). The resulting orthogonal PCs are a linear combination of the original process variables with the first PC capturing most of the variance in the original data and subsequent PCs capturing

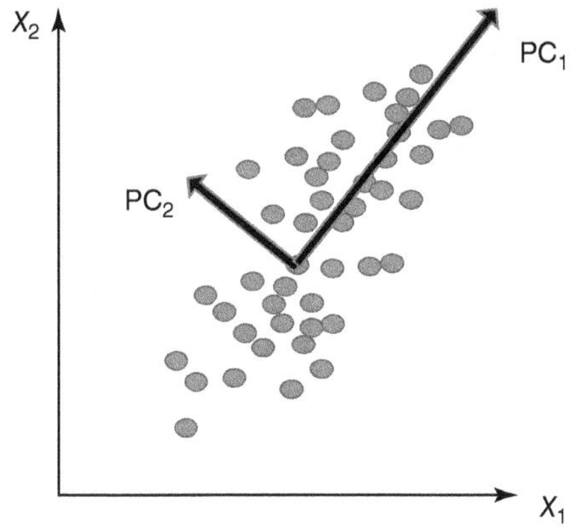


**Figure 13.4** Visual representation of a PCA transformation of process measurements.

the decreasing proportion of variance, respectively (see Figure 13.4 for visual representation of this decomposition).

Equation 13.1 represents this decomposition in terms of the resulting scores (***T***) and loadings (***P***) matrices and a residual error matrix ***E***:

$$X = \boldsymbol{TP}^{T} + \boldsymbol{E} \tag{13.1}$$

Since the PCs are ordered in the order of decreasing variance, it is possible to capture the underlying features in the original data using fewer PCs, thus reducing the dimensionality of the original data. This makes PCA particularly suitable for analysis of highly dimensional data produced by fingerprinting and multianalyte methods, such as spectroscopic techniques, electronic noses (see case study in Section 13.4) and tongues as well as the output of various "omics" measurement as numerous literature reports confirm (e.g., [47, 48])

Literature sources on PCA application within bioprocesses data analysis cover a broad range of bioprocessing aspects from raw material, seed cultivation, production batch, or downstream process quality monitoring (e.g., [44, 49, 50]). The PCA functionality of process data dimensionality reduction is typically utilized in all these applications. The resulting set of principal components is typically subsequently used as inputs in monitoring process models within multivariate statistical process control (MSPC) schemes.

In a range of applications, PCA and similar techniques, such as linear/quadratic/regularized discriminant analysis, *k*-nearest neighbors, or hierarchical clustering [51], are used as "classifiers," that is, assigning samples/objects into one of the possible classes on the basis of the measurements collected for the given sample/object as compared with a library of similar samples collected over a period of time. Such applications are particularly useful in identifying similarities, for example, in gene expression data collected through a range of DNA/RNA microarrays or pattern in metabolomics data (e.g., [48, 52]). In such applications, the important issue is specifying criteria to discriminate between various classes. The dangers of arbitrary decisions in this respect are discussed by Glassey [42], where a case study on using PCA to discriminate between high- and low-producing batches of recombinant monoclonal antibody fraction process using *Escherichia coli* expression system is presented. Although initially the clustering of principal components appeared to provide a reasonable classification for the required discrimination, a more in-depth analysis revealed that the initial arbitrary clustering was more closely related to the feed composition than the product titer.

In multivariate quality control, typically in food industry as well as potentially very important in biopharmaceutical industry in final product quality assurance, the above-mentioned techniques are usually less applicable [53]. In such cases, it is important to determine without any doubt whether the product belongs to a particularly category (e.g., unadulterated food products from a specific region). Forina *et al.* [53] argue that this task is most suitable for the so-called class modeling techniques (CMT), such as UNEQ, SIMCA, POTFUN (potential functions modeling), and MRM (multivariate range modeling). These methods are then discussed in more detail in terms of their discriminatory power and sensitivity to noise in the data.

To counteract the argument of the limitation of linear feature extraction and classification techniques, a large number of nonlinear methods were introduced. These include not only nonlinear variants of linear techniques described earlier but also artificial neural network–based techniques described in more detail Marini [54] or Glassey [42]. The latter also provides examples of bioprocess applications of such techniques for identifying physiological state of the culture and thus potentially improving the estimation of important process parameters.

### 13.3.2 Regression Models

There is a plethora of challenges in bioprocess monitoring and control where exploratory data analysis, clustering and classification do not fulfill the requirements of the specific application. For example, in soft sensors (see Chapter 15), regression methods enabling the prediction of a desired critical quality attribute (CQA) or CPP from process data are required. In an ideal world, such data would be obtained through timely, accurate, and robust analytical measurements. However, the lack of appropriate techniques to measure CQAs, particularly when present in very low quantities in complex media, raises the need for inferential estimation through soft sensor technology (see Chapter 15) of such attributes and CPPs using process measurements that are readily available. This part of the chapter, therefore, briefly reviews established regression methods, highlighting applications in bioreactor modeling particularly.

The most widely used linear regression technique used to correlate causal ($X$) and output ($Y$) variables for a range of processes, is the partial least squares (PLS) method [55] and its various modifications. The PLS algorithm is based on the projection of the input ($X$) and output ($Y$) variables onto a number of latent variables and identifying the least-squares correlation between these new variables by single-input–single-output linear regression (Eqs. 13.2 and 13.3):

$$X = \sum_{k=1}^{np<nx} t_k p_k^T + \boldsymbol{E} \quad \text{and} \quad Y = \sum_{k=1}^{np<nx} u_k q_k^T + \boldsymbol{F}^* \tag{13.2}$$

where $\boldsymbol{E}$ and $\boldsymbol{F}$ are residual matrices, $np$ is the number of inner components that are used in the model, and $nx$ is the number of causal variables.

$$u_k = b_k t_k + \varepsilon_k \tag{13.3}$$

where $b_k$ is a regression coefficient and $\varepsilon_k$ refers to the prediction error.

The challenge associated with large data sets collected using multianalyte measurement methods (e.g., spectroscopic of "omics" methods) is partly addressed by the ability of PLS to reduce the data dimensions. However, as discussed in [42], evidence shows that preselecting variables for PLS modeling can significantly improve the model performance. For example, the orthogonalized PLS (O-PLS) eliminates orthogonal variation with respect to the output variables $Y$ from the input variable set $X$. Using this method, Yang *et al.* [56] showed that analyte-irrelevant optical variability in NIR spectra of human tissues enabled them to account for tissue overlying and thus a more accurate representation of the analyte concentrations within a specific tissue.

The reported limitation due to the assumption of linear relationships between the variables is addressed in a range of alternative PLS methods. These include incorporating polynomial relationships into the PLS structure [57], using artificial neural networks as inner PLS models [58] or hybrid structures incorporating mass balance equations based on a first-principles understanding of the process [59]. Just as with the feature extraction and classification methods, neural networks represent an alternative nonlinear regression modeling approach with an extensive range of examples of application in bioprocess modeling and monitoring (see Ref. [42] for a more detailed overview).

## 13.4 Case Studies

### 13.4.1 Feature Extraction Using PCA

An example of the application of feature extraction technique to identify the differences in bioreactor performance under various conditions using electronic nose measurements is presented to demonstrate the utility of such approaches. For this case study, a series of six recombinant *E. coli* fermentations were carried out and used to investigate the effect of induction regimes on the production of $\beta$-galactosidase. A commercial hybrid electronic nose from the former Nordic Sensor Technologies A.B, Linköping (subsequently Applied Sensors), NST 3320, was used during these cultivations. The formation of $\beta$-galactosidase in all the batches was induced with isopropyl-$\beta$-d-thiogalactopyranoside (IPTG). Two of the batches, referred to as 1A and 1B, were induced early (during exponential growth phase at 19 log hours). Another two batches, referred as 2A and 2B, were induced during the stationary phase at 26–27 log hours. The *E. coli* strain with the lac UV promoter operated expression vector was used in all four of these batches. The two remaining cultivations, 3A and 3B, were carried out as a control. Batch 3A was induced early at 20 log hours and batch 3B was induced late at 27 log hours. In these two batches, a strain that carried an empty plasmid was used, resulting in no $\beta$-galactosidase production. Extensive online and offline data were measured, including temperature, pressure, pH, gas flow rate, percentage of $O_2$ and $CO_2$ in the off gas and the percentage of dissolved oxygen in the broth, concentrations of glucose, acetic acid, and formate, optical density (600 nm), and dry weight.

Five of the electronic nose sensors, namely those sensitive to hydrogen, methane, and humidity, were removed from the data set before analysis due to their irrelevance in this application. PCA was carried out on the Response part of the remaining sensor signals and the scores plots for the raw Response data are shown for all six batches in Figure 13.5a. This is compared with the same analysis

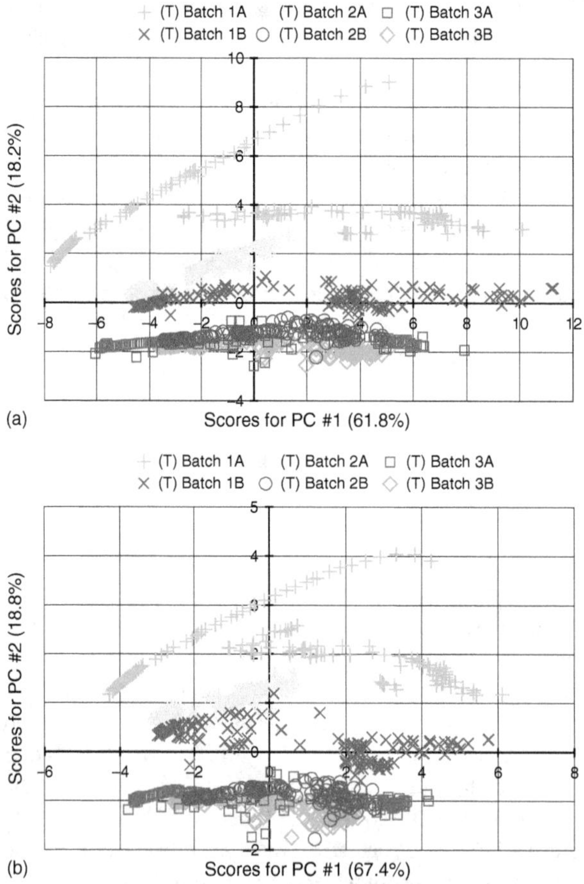


**Figure 13.5** (a) Scores plot of sensor RESPONSE parameter for all sensors, (b) scores plot for sensors RESPONSE parameter after the removal of insignificant and highly correlated variables.

following the removal of insignificant variables (reducing the data dimensionality from 120 variables to 49 variables), shown in Figure 13.5b.

In these figures, the scores for each of the batches are shown in different colors using different symbols (see legend). Clearly, the variable reduction has not affected the ability of the PCA to discriminate between the individual batches. Figure 13.5 also clearly shows the differences between the batches, although batches producing $\beta$-galactosidase (batches 2A, 2B, 3A, and 3B) are rather compressed in this figure. Batches that stand out as having a different scores profile, that is, 1A, 2A, and to some extent 1B were, therefore, analyzed in a pair-wise comparison to elucidate the differences observed. Figure 13.6 shows

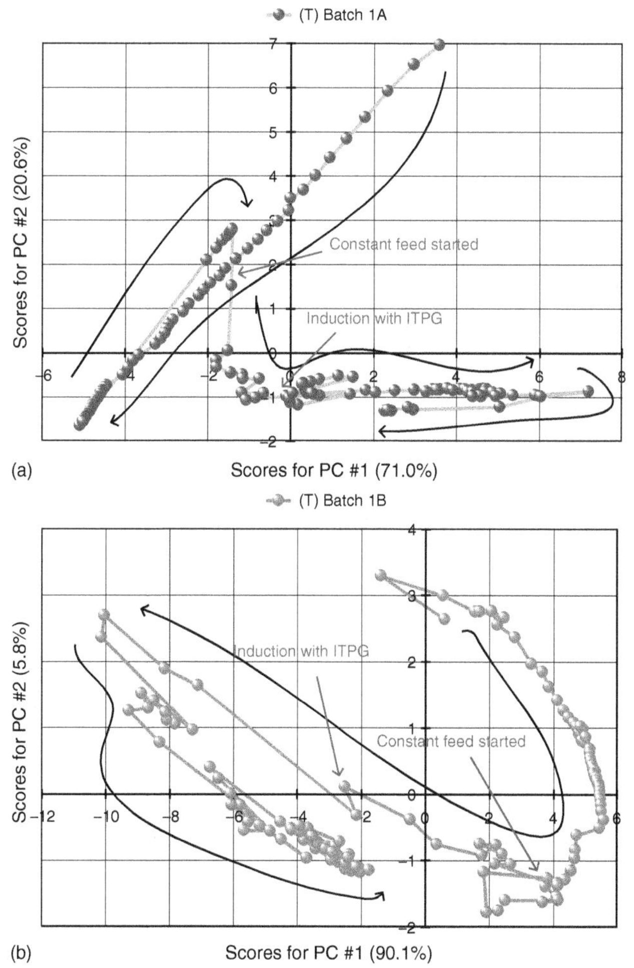


**Figure 13.6** (a) Scores plot for batch 1A response parameter only, (b) scores plot for batch 1B response parameter only.

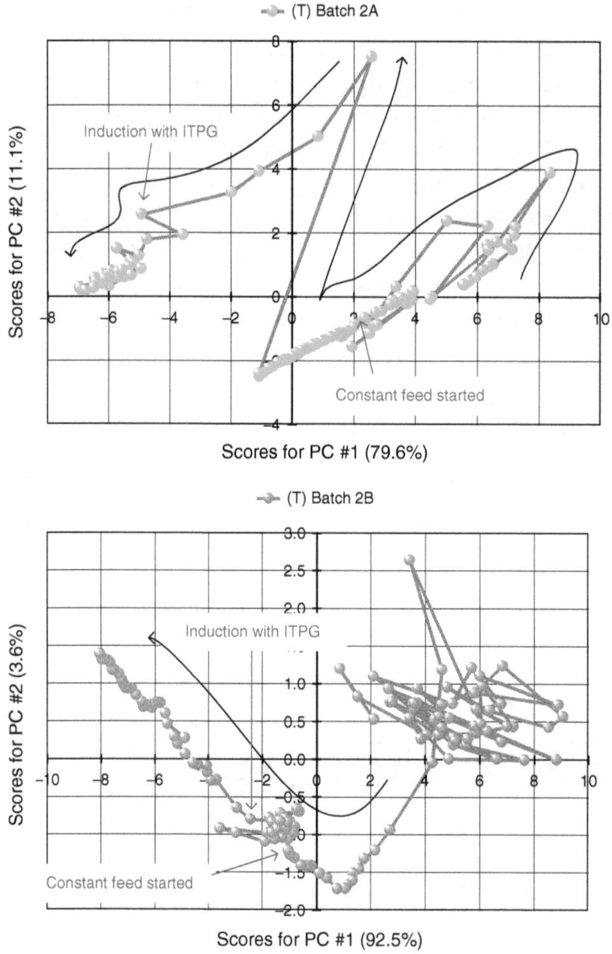


**Figure 13.7** (a) Scores plot for batch 2A response parameter only, (b) scores plot for batch 2B response parameter only.

the scores plots for batches 1A and 1B based on the Response signal from the electronic nose.

Clearly the behavior of the two batches, as reflected by the time trajectory of score plots (indicated by the black arrows), is different. This could be expected as, although both are control batches, IPTG addition in each batch took place at a different time (for batch 1A it was early and for batch 1B later on). However, a dissimilarity in the scores plots is also observed in batches 2A and 2B, in which the $\beta$-galactosidase production was induced early on (Figure 13.7).

Although the scores highlight the differences between the batches, further analyses (and potentially experiments) are required to establish the reasons for these differences. An example of a loadings plot for the first two principal components

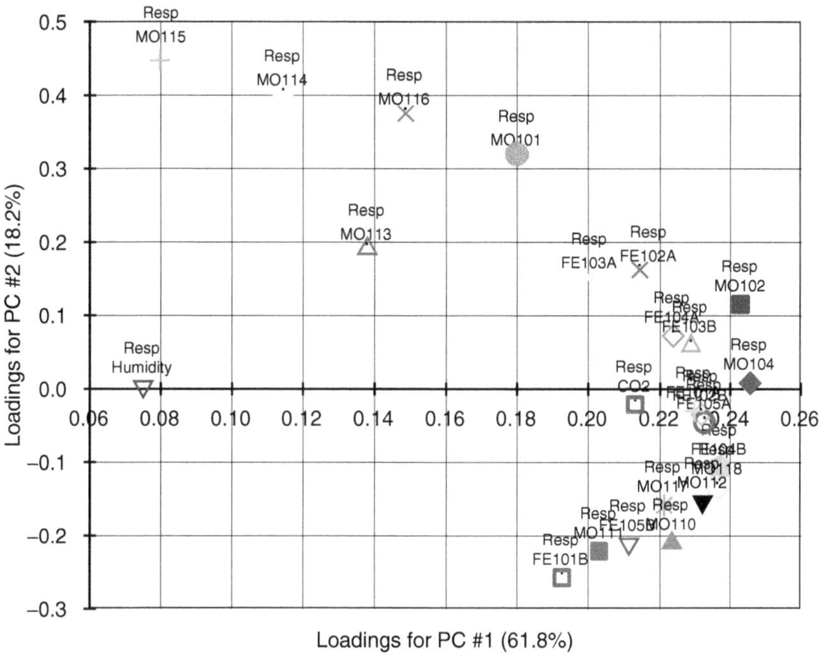


**Figure 13.8** PCA loading plot for all sensor responses

(PCs) is shown in Figure 13.8, and it indicates the ability to identify the sensor response values resulting in the clustering of scores within the PC1 and PC2 space.

While such a loading plot can point to particular volatile components detected by relevant sensors as the potential source of the difference between the batches, additional measurements, for example, the offline measurements of metabolites, biomass, and product, may be needed in order to reveal the biological processes leading to the release of these volatiles. In this case, the total number of cultivations carried out was insufficient for reliable conclusions to be drawn. However, the potential benefits of using multivariate data analysis in conjunction with multianalyte measurements are clearly evident.

### 13.4.2
### Prediction of CQAs

As highlighted in Section 13.3.2, the use of regression multivariate methods is particularly useful in the estimation of CQAs and CPPs that are not easily measured in real time. The following case study demonstrates the use of PLS in the estimation of amino acid concentrations and of the glycosylation form in the production monoclonal antibody produced using hybridoma cell culture. Green and Glassey [60] reported that various multivariate data analysis feature extraction methods, such as PARAFAC, PCA, and multiway PCA, were able to capture the differences between batches carried out under different operating

**Table 13.1** Summary of PLS models for amino acid concentration estimation based on online data collected during the hybridoma cell culture cultivation .

| Model | *X* matrix | *Y* matrix | LVs | Validation RMSE – Run 5; Run 13 |
|---|---|---|---|---|
| A | DO, $O_2$, $CO_2$, pH, and base | All amino acid data | 3 | 0.12; 0.23 |
| B | DO, $O_2$, and $CO_2$ | All amino acid data | 3 | 0.22; 0.23 |
| C | DO, $O_2$, $CO_2$, pH, and base | All amino acid data as separate models | Variable | 0.19; 0.24 |
| D | DO, $O_2$, and $CO_2$ | All amino acid data as separate models | Variable | 0.24; 0.24 |

LVs, latent variables; RMSE, root mean square error.

conditions, highlighting particularly the negative impact of high values of dissolved oxygen and pH. These observations confirmed the earlier observations of Ivarsson *et al.* [61]. Subsequently, PLS models were developed using online and offline data in order to estimate important process parameters (such as the amino acid concentrations during the cultivations) and a CQA of the monoclonal antibody – its glycosylation forms. A range of models were developed for each output variable, as highlighted in Tables 13.1 and 13.2. Table 13.1 summarized the amino acid estimation models, the input and output variables for each of the models together with the number if latent variables and the performance of the models on the validation data sets in terms of root mean square error [60].

Figure 13.9 shows graphically the performance of each of the models on the amino acid concentration profiles for the two validation runs. Clearly, the online data were sufficient to estimate the concentration of some amino acids with sufficient accuracy although for other amino acids (e.g., ALA, GLU, GLY, and VAL – number 1, 7, 8, and 20) the accuracy is lower. Nevertheless, the benefits

**Table 13.2** Summary of PLS models for glycan concentration estimation based on online and offline data collected during the hybridoma cell culture cultivation.

| Model | *X* matrix | *Y* matrix | LVs | Validation RMSE – Run 5; Run 13 |
|---|---|---|---|---|
| 15 | DO, $O_2$, $CO_2$, pH, and base | Glycosylation profile | 5 | 0.79; 1.06 |
| 16 | DO, $O_2$, and $CO_2$ | Glycosylation profile | 6 | 0.69; 1.16 |
| 17 | DO, $O_2$, $CO_2$, pH, and base, viable cell count, % viability, glucose conc., lactate conc., and product titer | Glycosylation profile | 5 | 0.75; 1.03 |
| 18 | Amino acid profile | Glycosylation profile | 5 | 0.83; 0.85 |

LVs, latent variables; RMSE, root mean square error.

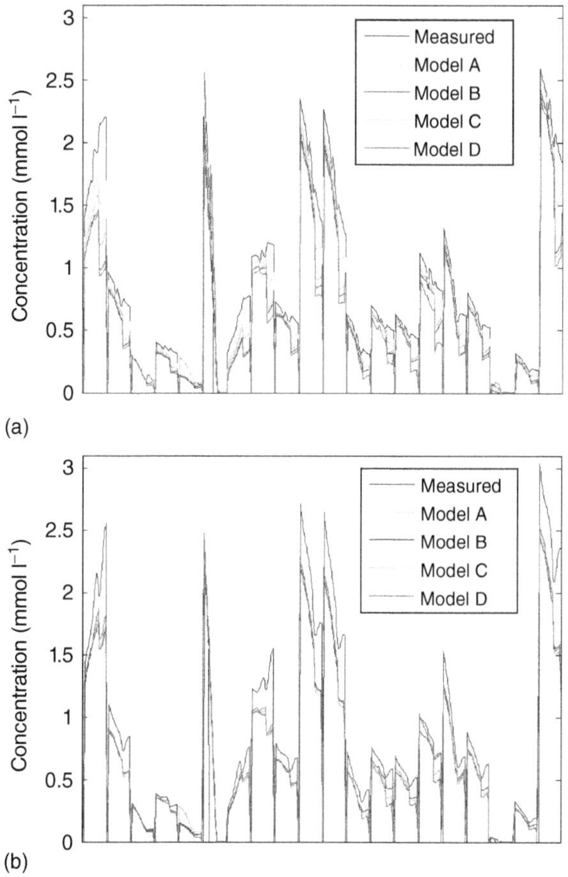


**Figure 13.9** Amino acid concentration predictions for (a) Run 5 and (b) Run 13. The details of each model in terms of input and output data, model structure and the RMSE values of each model for each validation run are provided in Table 13.1.

in understanding the behavior of the cell culture and the opportunity to obtain the estimates of important metabolic parameters in real time is undoubtedly a significant benefit in understanding the metabolic state of the culture.

However, it would be even more beneficial to be able to estimate CQAs, such as the concentration of the glycoforms produced by the cell culture. A set of PLS models was developed to predict this important attribute, as summarized in Table 13.2.

Figure 13.10 demonstrates graphically the ability of these models to predict the concentration of the nine glycans as the end of the cultivation (measured as described in Ivarsson *et al.* [61]). This figure clearly shows that the 'predictions' for most of the glycans are satisfactory, with Model 18 – based on amino acid profile as input to the model – providing the lowest RMSE values for validation run 13 and at the same time providing satisfactory performance for run 5. The

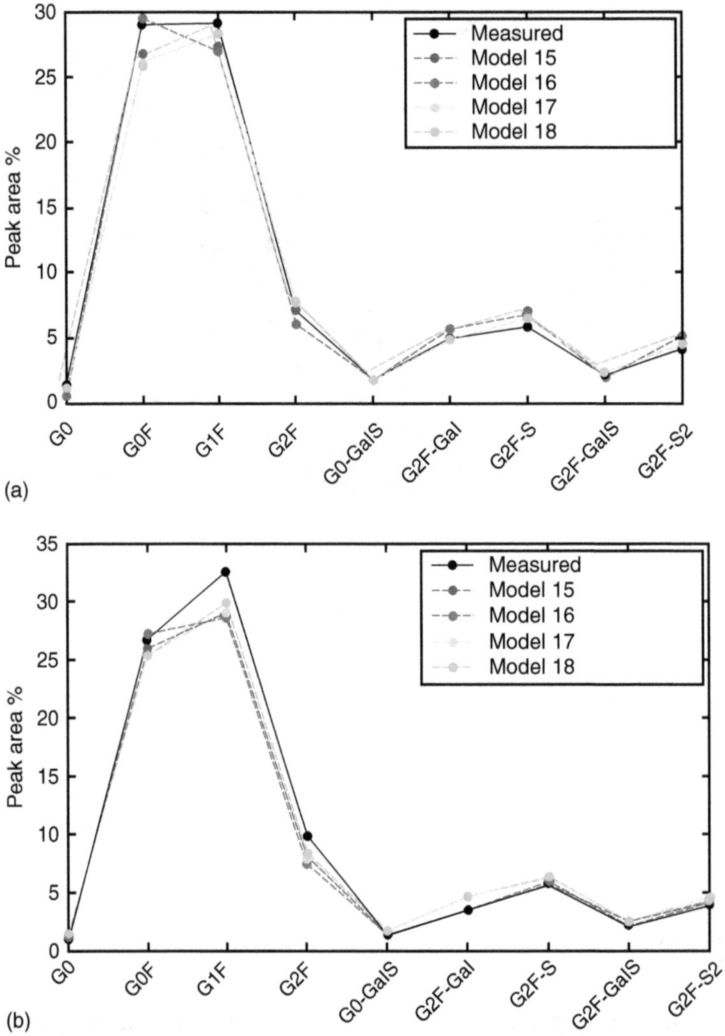


**Figure 13.10** Glycoform predictions for (a) Run 5 and (b) Run 13. The details of each model in terms of input and output data, model structure, and the RMSE values of each model for each validation run are provided in Table 13.2.

least accurate predictions were for glycans G0F and G1F. The reasons for this are currently being investigated further.

## 13.5 Conclusions

This chapter reviews the various sensors and analytical techniques used in bioprocess measurement, both in traditional bioprocessing as well as in high-throughput and scale-down experiments. The challenges introduced by varying frequency of the measurements and the large number of measurements from various sensor technologies (particularly the multianalyte methods, such as spectral and "omics" methods) were discussed. Specific methods of data preprocessing, particularly in the case of spectral data, were reviewed. A brief outline of PCA and PLS as examples of feature extraction/classification and regression methods was provided. Nonlinear variants of both classification and regression methods were outlined as an alternative to PCA and PLS. Finally, two bioprocess case studies were presented. In the first case study, PCA was used to identify the differences between batches of recombinant *E. coli* $\beta$-galactosidase production based on electronic nose data monitored during the cultivation. In the second case study, PLS was used to predict amino acid concentration and glycosylation profiles from online and offline data collected during monoclonal antibody production using hybridoma cell culture. These case studies highlighted both the benefits of such approaches as well as the challenges faced in their application, particularly when limited data are available from the processes and the fundamental understanding of the underlying processes is limited.

## Acknowledgments

The contribution of Nicola Dawes and Amy Green to the case studies in this chapter, carried out as part of their Master and Doctoral studies ad Newcastle University, is gratefully acknowledged.

## References

1. US Food and Drug Administration (2004) *Guidance for Industry: PAT — A Framework for Innovative Pharmaceutical Development, Manufacturing, and Quality Assurance*, FDA, Rockville, MD, http://www.fda.gov/downloads/Drugs/GuidanceComplianceRegulatoryInformation/Guidances/UCM070305.pdf (accessed 02 November 2015).
2. von Stosch, M., Davy, S., Francois, K., Galvanauskas, V., Hamelink, J.M., Luebbert, A., Mayer, M., Oliveira, R., O'Kennedy, R., Rice, P., and Glassey, J. (2014) Hybrid modelling for quality by design and PAT – benefits and challenges of applications in biopharmaceutical industry. *Biotechnol. J.*, **9** (6), 719–726.
3. Carrondo, M.J.T., Alves, P.M., Carinhas, N., Glassey, J., Hesse, F., Merten, O.W., Micheletti, M., Noll, T., Oliveira, R., Reichl, U., Staby, A., Teixeira, A.P., Weichert, H., and Mandenius, C.F. (2012) How can measurement, monitoring, modeling and control advance cell culture in industrial biotechnology? *J. Biotechnol.*, **7** (12), 1522–1529.
4. Gernaey, K.V., Baganz, F., Franco-Lara, E., Kensy, F., Kruhne, U., Luebberstedt, M., Marx, U., Palmqvist, E., Schmidt, A., Schubert, F., and Mandenius, C.F. (2012) Monitoring and control of microbioreactors: an expert opinion on development needs. *Biotechnol. J.*, **7**, 1308–1314.
5. Luttmann, R., Bracewell, D.G., Cornelissen, C., Gernaey, K.V., Glassey, J., Hass, V.C., Kaiser, C., Lindström, I.M., Preusse, C., Striedner, G., and Mandenius, C.M. (2012) Soft sensors in bioprocessing: a status report and recommendations. *Biotechnol. J.*, **7** (8), 1040–1047.

6. Gernaey, K.V., Eliasson Lantz, A., Tufvesson, P., Woodley, J.M., and Sin, G. (2010) Application of mechanistic models to fermentation and biocatalysis for next generation processes. *Trends Biotechnol.*, **28**, 346–354.
7. Nolan, R. and Lee, K. (2011) Dynamic model of CHO cell metabolism. *Metab. Eng.*, **13** (1), 108–124.
8. Maschke, R.W., Geipel, K., and Bley, T. (2015) Modeling of plant in vitro cultures: overview and estimation of biotechnological processes. *Biotechnol. Bioeng.*, **112** (1), 1–12.
9. Sonnleintner, B. (2013) Automated measurement, monitoring and estimation of bioprocesses — key elements of the $M^3C$ strategy. *Adv. Biochem. Eng. Biotechnol.*, **132**, 1–33.
10. Junker, B.H. and Wang, H.Y. (2006) Bioprocess monitoring and computer control: key roots of the current PAT initiative. *Biotechnol. Bioeng.*, **95** (2), 226–261.
11. Teixeira, A.P., Oliveira, R., Alves, P.M., and Carrondo, M.J.T. (2009) Advances in on-line monitoring and control of mammalian cell cultures: Supporting the PAT initiative. *Biotechnol. Adv.*, **27**, 726–732.
12. Beutel, S. and Henkel, S. (2011) In situ sensor techniques in modern bioprocess monitoring. *Appl. Microbiol. Biotechnol.*, **91**, 1493–1505.
13. Broger, T., Odermatt, R.P., Huber, P., and Sonnleitner, B. (2011) Real-time on-line flow cytometry for bioprocess monitoring. *J. Biotechnol.*, **154** (4), 240–247.
14. Vojinovic, V., Cabral, J.M.S., and Fonseca, L.P. (2006) Real-time bioprocess monitoring. Part I: in situ sensors. *Sens. Actuators, B*, **114** (2), 1083–1091.
15. Gernaey, K., Glassey, J., Skogestad, S., Kramer, S., Weiss, A., Engell, S., Pistikopoulos, E.N., and Cameron, D.B. (2012) Process systems engineering, 5. Process dynamics, control, monitoring and identification, in *Ullmann's Encyclopedia of Industrial Chemistry*, Wiley-VCH Verlag GmbH, Online ISBN: 9783527306732, doi: 10.1002/14356007.
16. Lourenco, N.D., Lopes, J.A., Almeida, C.F., Sarraguca, M.C., and Pinheiro, H.M. (2012) Bioreactor monitoring with spectroscopy and chemometrics: a review. *Anal. Bioanal. Chem.*, **404**, 1211–1237.
17. Bhambure, R., Kumar, K., and Tathore, A.S. (2010) High-throughput process development for biopharmaceutical drug substances. *Trends Biotechnol.*, **29** (3), 127–135.
18. Kim, B.J., Diao, J., and Shuler, M.L. (2012) Mini-scale bioprocessing systems for highly parallel cell cultures. *Biotechnol. Progr.*, **28** (3), 595–607.
19. Tsang, V.L., Wang, A.X., Yusuf-Makagiansar, H., and Ryll, T. (2014) Development of a scale down cell culture model using multivariate analysis as a qualification tool. *Biotechnol. Progr.*, **30** (1), 152–160.
20. Wu, M.H., Huang, S.B., and Lee, G.B. (2010) Microfluidic cell culture systems for drug research. *Lab Chip.*, **10**, 939–956.
21. Zhang, H., Lamping, S.R., Pickering, S.C.R., Lye, G.J., and Shamlou, P.A. (2008) Engineering characterisation of a single well from 24-well and 96-well microtitre plates. *Biochem. Eng. J.*, **40**, 138–149.
22. Lamping, S.R., Zhang, H., Allen, B., and Shamlou, P.A. (2003) Design of a prototype miniature bioreactor for high throughput automated bioprocessing. *Chem. Eng. Sci.*, **58**, 747–758.
23. Woodcock, T., Downey, G., and O'Donnell, C.P. (2008) Better quality food and beverages: the role of near infrared spectroscopy. *J. Near Infrared Spectrosc.*, **16** (1), 1–29.
24. Ranzan, C., Strohm, A., Ranzan, L., Trierweiler, L., Hitzmann, B., and Trierweiler, J.O. (2014) Wheat flour characterization using NIR and spectral filter based on Ant Colony Optimization. *Chemom. Intell. Lab. Syst.*, **132**, 133–140.
25. Wu, D. and Sun, D.W. (2013) Advanced applications of hyperspectral imaging technology for foor quality and safety analysis and assessment: a review – Part I: fundamentals. *Innovative Food Sci. Emerg. Technol.*, **19**, 1–14.
26. Li, Y.S. and Church, J.S. (2014) Raman spectroscopy in the analysis of food and

pharmaceutical nanomaterials. *J. Food Drug Anal.*, **22** (1), 29–48.
27. Tamburini, E., Marchetti, M.G., and Pedrini, P. (2014) Monitoring key parameters in bioprocesses using near-infrared technology. *Sensors*, **14**, 18941–18959.
28. van den Broeke, J., Langergraber, G., and Weingartner, A. (2006) On-line and in situ UV/vis spectroscopy for multiparameter measurements: a brief review. *Spectrosc. Eur.*, **18**, 15–18.
29. Surribas, A., Geissler, D., Gierse, A., Scheper, T., Hitzmann, B., Montesinos, J.L., and Valero, F. (2006) State variables monitoring by in situ multi-wavelength fluorescence spectroscopy in heterologous protein production by *Pichia pastoris*. *J. Biotechnol.*, **124**, 412–419.
30. Picard, A., Daniel, I., Montagnac, G., and Oger, P. (2007) In situ monitoring by quantitative Raman spectroscopy of alcoholic fermentation by *Saccharomyces cerevisiae* under high pressure. *Extremophiles.*, **11**, 445–452.
31. Skibsted, E. (2006) Near infrared spectroscopy: the workhorse in the PAT toolbox. *Spectrosc. Eur.*, **18**, 14–17.
32. Forcinio, H. (2003) Pharmaceutical industry embraces NIR technology. *Spectroscopy*, **18**, 16–19.
33. Qiu, J., Arnold, M.A., and Murhammer, D.W. (2014) On-line near infrared bioreactor monitoring of cell density and concentrations of glucose and lactate during insect cell cultivation. *J. Biotechnol.*, **173**, 106–111.
34. Capito, F., Skudas, R., Kolmar, H., and Hunzinger, C. (2013) Mid-infrared spectroscopy-based antibody aggregate quantification in cell culture fluids. *Biotechnol. J.*, **8** (8), 912–917.
35. Rudnitskaya, A. and Legin, A. (2008) Sensor systems, electronic tongues and electronic noses for the monitoring of biotechnological processes. *J. Ind. Microbiol. Biotechnol.*, **35**, 443–451.
36. Rosi, P.E., Miscoria, S.A., Bernik, D.L., and Negri, R.M. (2012) Customized design of electronic noses placed on top of air-lift bioreactors for in situ monitoring the off gas patterns. *Bioprocess Biosyst. Eng.*, **35**, 835–842.
37. Navratil, M., Norberg, A., Lembren, L., and Mandenius, C.F. (2005) On-line multi-analyzer monitoring of biomass, glucose and acetate for growth rate control of a *Vibrio cholerae* fed-batch cultivation. *J. Biotechnol.*, **115** (1), 67–79.
38. Shen, D., Kiehl, T.R., Khattak, S.F., Li, Z.J., He, A., Kayne, P.S., Patel, V., Neuhaus, I.M., and Sharfstein, S.T. (2010) Transcriptomic responses to sodium chloride-induced osmotic stress: a study of industrial fed-batch CHO cell cultures. *Biotechnol. Progr.*, **26** (4), 1104–1115.
39. Chrysanthopoulos, P.K., Goudar, C.T., and Klapa, M. (2012) Metabolomics for high-resolution monitoring of the cellular physiological state in cell culture engineering. *Metab. Eng.*, **12**, 212–222.
40. Oldiges, M.L., Lutz, S., Pflug, S., Schroer, K., Stein, N., and Wiendah, C. (2007) Metabolomics: current state and evolving methodologies and tools. *Appl. Microbiol. Biotechnol.*, **76** (3), 495–511.
41. Huang, E.L., Orsat, V., Shah, M.B., Hettich, R.L., VerBerkmoes, N.C., and Lefsrud, M.G. (2012) The temporal analysis of yeast exponential phase using shotgun proteomics as a fermentation monitoring technique. *J. Proteomics.*, **75** (17), 5206–5214.
42. Glassey, J. (2013) Multivariate data analysis for advancing the interpretation of bioprocess measurement and monitoring data. *Adv. Biochem. Eng. Biotechnol.*, **132**, 167–191.
43. Regis, M., Doncescu, A., and Desachy, J. (2007) Use of evidence theory for the fusion and the estimation of relevance of data sources: application to an alcoholic bioprocess. *Trait. Signal*, **24** (2), 115–132.
44. Cunha, C.C.F., Glassey, J., Montague, G.A., Albert, S., and Mohan, P. (2002) An assessment of seed quality and its influence on productivity estimation in an industrial antibiotic fermentation. *Biotechnol. Bioeng.*, **78**, 658–669.
45. Brereton, R. (2009) *Chemometrics for Pattern Recognition*, John Willey & Sons, Ltd, Chichester.
46. Chatfield, C. and Collins, A.J. (1980) *Introduction to Multivariate Analysis*, Chapman & Hall, London.

47. Wang, Y., Shi, M., Niu, X., Zhang, X., Gao, L., Chen, L., Wang, J., and Zhang, W. (2014) Metabolomic basis of laboratory evolution of butanol tolerance in photosynthetic *Synechocystis* sp. PCC 6803. *Microb. Cell Fact.*, **13**, 151.
48. Chong, W.P.K., Goh, L.T., Reddy, S.G., Yusufi, F.N.K., Lee, D.P., Wong, N.S.C., Heng, C.K., Yang, M.G.S., and Ho, Y.E. (2009) Metabolomics profiling of extracellular metabolites in recombinant Chinese hamster ovary fed-batch culture. *Rapid Commun. Mass Spectrom.*, **23** (23), 3763–3771.
49. Arnold, S.A., Crowley, J., Woods, N., Harvey, L.M., and McNeil, B. (2003) In-situ near infrared spectroscopy to monitor key analytes in mammalian cell cultivation. *Biotechnol. Bioeng.*, **84** (1), 13–19.
50. Pate, M.E., Turner, M.K., Thornhill, N.F., and Titchener-Hooker, N.J. (2004) Principal component analysis of nonlinear chromatography. *Biotechnol. Progr.*, **20**, 215–222.
51. Balabin, R.M., Safieva, R.Z., and Lomakina, E.I. (2010) Gasoline classification using near infrared (NIR) spectroscopy data: comparison of multivariate techniques. *Anal. Chim. Acta.*, **671** (1–2), 27–35.
52. Dawes, N.L. and Glassey, J. (2007) Normalisation of multicondition cDNA macroarray data. *Comp. Funct. Genomics*. doi: 10.1155/2007/90578, Article ID 90578, 12.
53. Forina, M., Oliveri, P., Bagnasco, L., Simonetti, R., Casolino, MC., Griffi, FN, Casale, M., , , , , 2015) Artificial nose, NIR and UV-visible spectroscopy for the characterisation of the PDO Chianti Classico olive oil. Talanta, **144**, 1070–1078.
54. Marini, F. (2009) Artificial neural networks in foodstuff analyses: trends and perspectives a review. *Anal. Chim. Acta.*, **635**, 121–131.
55. Geladi, P. and Kowalski, B.R. (1986) Partial least-squares regression: a tutorial. *Anal. Chim. Acta.*, **185**, 1–17.
56. Yang, Y., Shoer, L., Soyemi, O.O., Landry, M.R., and Soller, B.R. (2006) Removal of analyte-irrelevant variations in near-infrared tissue spectra. *Appl. Spectrosc.*, **60**, 1070–1077.
57. Wold, S., Trygg, J., Berglund, A., and Antti, H. (2001) Some recent developments in PLS modelling. *Chemom. Intell. Lab. Syst.*, **58**, 131–150.
58. Lee, D.S., Lee, M.W., Woo, S.H., Kim, Y.J., and Park, J.M. (2006) Nonlinear dynamic partial least squares modeling of a full-scale biological wastewater treatment plant. *Process Biochem.*, **41**, 2050–2057.
59. von Stosch, M., Oliveira, R., Peres, J., and Feyo de Azevedo, S. (2011) A novel identification method for hybrid (N) PLS dynamical systems with application to bioprocesses. *Expert Syst. Appl.*, **38** (9), 10862–10874.
60. Green, A. and Glassey, J. (2014) Multivariate analysis of the effect of operating conditions on hybridoma cell metabolism and glycosylation of produced antibody. *J. Chem. Technol. Biotechnol.*, **90**, 303–313.
61. Ivarsson, M., Villiger, T.K., Morbidelli, M., and Soos, M. (2014) Evaluating the impact of cell culture process parameters on monoclonal antibody N-glycosylation. *J. Biotechnol.*, **188C**, 88–96. doi: 10.1016/j.jbiotec.2014.08.026

# 14
# Soft Sensor Design for Bioreactor Monitoring and Control

*Carl-Fredrik Mandenius and Robert Gustavsson*

## 14.1
## Introduction

The term *soft sensor* or *software sensor* is frequently used to describe the combination of "hardware" sensors with software-implemented models where the latter use the signals from the hardware to derive new information. The soft sensor principle is illustrated in Figure 14.1, showing sensors measuring a process where the sensor signals are processed by the model into new variables, signals, or other information. The process may be a single bioreactor or, of course, a whole bioprocess including both upstream and downstream processing. As the figure illustrates, the soft sensor operates online to the process, in real time and the collected signals are usually more than one [3–8].

The soft sensor principle opens up for a more extensive exploitation of existing sensors and sensor signals as well as taking advantage of prior knowledge of mechanisms of process systems.

By that, soft sensors become useful for monitoring and control applications in the process industry but may also, due to the generality of its principle, be used in any continuous process or transformation where sensor signals are available for modeling. In the more than 1000 soft sensor reports appearing in the engineering literature applications to commodity chemicals [1, 2, 9–27] and metallurgical processes [1, 2, 9–27], unit operation control [1, 2, 9–27], energy [1, 2, 9–27], and environment [27, 28] dominate, while examples of use in biotechnology are relatively few.

To be attractive for use in industry, in particular for bioreactor applications, soft sensors should be based on sensors and models that are robust and easy to operate, and the resulting soft sensor setup should be possible to validate convincingly. This calls for simplicity both for chosen hardware and software – for devices and computation methods. By that, a soft sensor should provide a less complicated analytical system, it should reduce the need for extensive operational surveillance of the analytical system itself, it should reduce operational and maintenance work, and it should also lead to decreased equipment cost (Table 14.1).

*Bioreactors: Design, Operation and Novel Applications*, First Edition. Edited by Carl-Fredrik Mandenius.

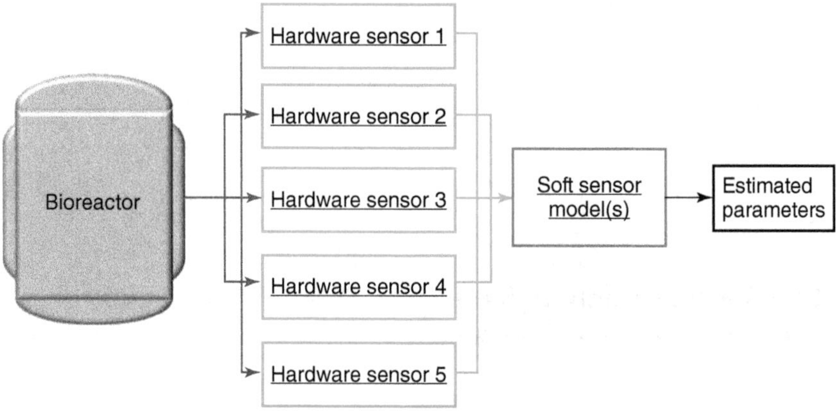


**Figure 14.1** The soft sensor principle as defined in Refs. [1, 2]. The figure shows one hardware sensor and one estimator. In reality, these can be multiplied.

**Table 14.1** Important constrains and requirements for efficient use of soft sensors.

| Constraint/requirement | Reasons |
|---|---|
| Cost for sensor components in the soft sensor should be low | Important for products with low margin, especially if other manufacturer goals follow that |
| Readouts from sensors should be accessible as analogue or digital signals | Allowing signal processing in an existing process control system |
| The soft sensor system should be able to automate or robotize | Allow continuous use Reduce interaction by technicians/operators |
| Algorithms in the soft sensor should be stable and allow validation | Model algorithms should by that be fairly robust and transparent for the user to handle Recalibration of the soft sensor should be minimized |
| Analytical principle applied in the soft sensor should be general | Being user-friendly for technicians and operators to handle and maintain Allow the soft sensor to be “clone” into similar applications |
| Sensor parts should preferably be disposable and/or exchangeable | Reduce calibration needs Requires realistic cost level of disposables |

## 14.2 The Process Analytical Technology Perspective on Soft Sensors

If the criteria outlined in Table 14.1 are realised the soft sensors may also be a valuable tool in the light of the process analytical technology (PAT) initiative [29]. In particular, the PAT outlines a set of tools and methods for making pharmaceutical manufacturing processes more reliable and efficient. The way of reasoning in PAT has been supported by several regulatory bodies [30, 31] and the academic research has been active in pursuing these directions in a variety of aspects such as quality-by-design, online sensor development, and statistical experimental design [32, 33]. Moreover, the PAT and QbD principles can be extended to all biotechnological manufacturing processes including proteins, foods, and other chemicals [117].

US FDA's PAT guidance [29] mentions six goals that should be accomplished by using the PAT tools: (i) reducing production cycling time, (ii) preventing rejection of batches, (iii) enabling real-time release, (iv) increasing automation, (v) improving efficiency of energy and material use, and (vi) facilitating continuous processing. These goals are expected to be applicable to all pharmaceutical production, including manufacture of biotherapeutic protein-based drugs (proteins, antibodies) [34], gene therapy vectors as well as cell therapy products [35].

Of these goals, the soft sensors may potentially contribute especially to increased automation, due to the possibility of using signals for feedback control of the soft sensor–derived variables, and to enabling real-time release, due to immediate provision of estimates of critical quality parameters. It is possible to assume that this may have significant impact on reducing the cycle time, facilitating continuous processing, and improving material balances, due to provision of early real-time information for regulating manufacturing conditions. Table 14.2 provides a few concrete examples of how this can be accomplished with the soft sensors.

Although the soft sensors are applied in many industrial manufacturing areas, applications related to manufacturing processes, bioprocess applications, to

**Table 14.2** Examples of existing or potential soft sensor applications relevant for biotherapeutic manufacture.

| Hardware signals | Soft sensor model action | Examples of estimates |
|---|---|---|
| Online sensor probes generating real-time spectra with peaks or other patterns related to components in measured media | The model interprets spectral patterns by multivariate analysis or neural network resulting in prediction of the components | Quantitation of raw materials, products and by-products in process media. These may be directly related to CQA and economical attributes of the production process |
| A single online sensor measuring specifically and directly a component of the medium | The model estimates volumetric rate from signal | Estimated volumetric rates for biomass, media components, products and by-products. |
| Two or more online sensors measure components and biomass in the medium | The model estimates specific rate signals | Estimates are used as control variables for optimal values |
| Online gas effluent analysis by mass spectrometry or other gas specific sensors/instruments | The model estimates volumetric uptake or evolution rates or ratios of these | Estimated volumetric rates for biomass, media components, products, and by-products |

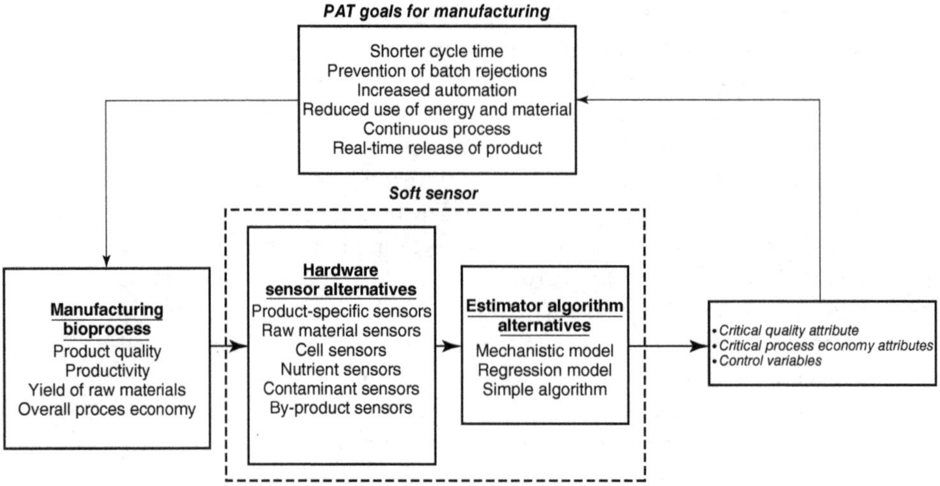


**Figure 14.2** The information flow in a soft sensor with examples from typical bioreactor applications.

bioreactors, or other bioprocess operations, are mainly limited to demonstrations performed in academic research laboratories. The reasons for this may be that the biological process parameters of interest are analytically less stable and reproducible than parameters appearing in applications in the mechanical, electrical, and chemical industry. Figure 14.2 provides an overview of how soft sensor in principle can be applied to bioreactors for supporting the PAT goals. The figure is further detailed in Table 14.2, which exemplifies how these principles can be accomplished.

## 14.3 Conceptual Design of Soft Sensors for Bioreactors

The design of process analytical systems, from a PAT perspective or any other viewpoint, does seldom start with a thorough analysis of the manufacturer's actual needs. This should, however, be the first step to take as commonly done with other engineering design tasks [36–38].

The manufacturer needs for analytical data are basically tied to the basic production goals: (i) high or defined quality of the product and (ii) to ascertain the lowest possible production cost. As a consequence, the needs of the process analytical systems are guided by these two goals. For the same reasons, the PAT goals do to a large extent overlap with these.

With this as a starting point, it is most likely that the manufacturer will mention the analysis of the purity of the product and be able to measure side products and other impurities that may occur as high-ranked needs; but also, at what stage in the manufacturing, how accurate, how fast, how often, and with what effort these are analyzed [38].

It is also important to clarify how the manufacturer wants to use the analytical information. It could be for withdrawal of a batch of too low quality, to make decisions of how or when to continue processing in a subsequent process step, or to use the analytical information for automatic control of the process.

A very important issue for the manufacturer is that the analytical equipment can be trusted. Mostly, that relies on precision, accuracy, and repeatability of analytical data. If that is attained, the manufacturer is willing to accept higher cost for the instrumentation, at least until it does not drastically increase the production cost. Normally, the expenditure for analytics is not that high in comparison to the total production cost. However, too extensive analytical procedures in a process may overshadow other operational routines in the manufacture. Due to that, analytics that have lesser logistic demands are favored.

Other manufacturers' needs are directly related to the process economy and processing time. This could include analysis of the utilization of the raw materials to ensure that optimal conditions are prevailing in the upstream process steps or that product yields are not lost in downstream processing steps. For such analytical information, the purpose is often to be able to adjust process conditions immediately by control actions. Although the reasons behind these needs are related to the process economy, PAT motives are indirectly possible to derive, such as savings of materials and energy [29].

At a certain level, the manufacturer needs to become product specific and must be directly connected to the actual product case. Production of, for example, a biotherapeutic protein in a mammalian cell culture is different from a microbial production process and, therefore, has different analytical requirements [38]. Production of cells for cell therapy differs considerably from a cell culture process for a biotherapeutic protein [39]. Production of gene therapy vectors has a process layout that deviate from other processes in a number of ways.

Table 14.4 maps common manufacturer needs and compares these for microbial, cell culture, and cell therapy products. The table highlights the differences in needs between the products. The intention is to illuminate general differences; in a real production case, the mapping of needs must be more detailed and specific to be useful [22]. When the analytical needs listed in Table 14.4 are provided with target values a much more precise map of the manufacturer needs can be stated [16]. This allows the ranking of the importance of analytical needs which would facilitate design of the analytical system [37]. With the support of mapping, the boundaries for the design of the analytical system are set. Once this is done, soft sensor approaches can seriously be considered.

## 14.4 "Hardware Sensor" Alternatives

Obviously, the detecting sensor devices (in a soft sensor often referred to as the "hardware sensor") delivering the signals to the algorithms are the decisive components of the soft sensor; if these do not meet the necessary criteria,

**Table 14.3** Reaching PAT goals by soft sensors for biotherapeutic manufacture.

| PAT goal | Application principle | Examples of soft sensor solution |
|---|---|---|
| Automation | The soft sensor is a part of a feedback control system where the soft sensor estimate is controlled | 1) The soft sensor variable is a quality variable defined by QbD<br>2) The soft sensor estimates a variable for start/stop of induction. |
| Real-time release | The soft sensor provides quality parameters identical to QC parameters normally analyzed off-line. | Allowing signal processing in an existing process control system |
| Facilitate recycling | The soft sensor provides signals that should work automatically without operator actions to allow continuous use. | Measurement of CQAs for recycling |
| Decrease use of material and energy | The soft sensor provides signals that ensure that raw materials are left unused and not converted to product | 1) Signals from multivariate spectral data are in a model used to estimate raw materials in the medium<br>2) Signals from multivariate spectral data are in a model used to estimate products and material-consuming by-products |
| Facilitate continuous processing | The fact that the soft sensor provides information decisive for subsequent process steps online makes continuity of processing possible | Allowing the soft sensor to be "clone" into another application |
| Prevent rejection of batches | The soft sensor provides signals that secure that optimal state is reached faster and maintained | Provided low cost, easy to discharge, in GMP/GLP |

especially sensitivity and robustness, the soft sensor will fail to fulfill its purpose (cf. Table 14.1). The hardware sensors should be able to cope with the PAT goals (Table 14.3) and when combined with a soft sensor algorithm provide analytical data of sufficiently good quality. Attaining the manufacture needs in Table 14.4 to a large extent relies on the capacity and performance of the hardware. The analytical tasks this leads to are typically found in the purposes mentioned in Table 14.5.

The soft sensors also presuppose real-time sensing, normally realized by online or inline devices, either invasively or noninvasively. If this is not accomplishable, and that is often the case for critical biological parameters, other solutions and strategies must be considered.

**Table 14.4** Manufacturer needs on process analytics.

| Needs | Purpose/motivation | Microbial process | Cell culture process | Cell therapy process |
|---|---|---|---|---|
| Robustness | Adapt to process environment | +++ | +++ | +++ |
| Low investment cost | Expensive instrumentation burdens total cost | ++ | + | + |
| Low operational cost | Avoiding increasing personnel costs | +++ | ++ | + |
| Reliability | Consistency of operation, minor troubleshooting, few stops | +++ | +++ | +++ |
| Precision | Satisfying regulator demands, quality criteria | +++ | +++ | +++ |
| Simplicity | Short start up, not requiring special competence, easy to use | +++ | +++ | ++ |
| Trainability | Reduced effort for new operation personnel | ++ | ++ | ++ |
| Flexibility | Can be used with more than one product | + | + | + |
| Transferability | Adapts easily to a new product (CMC) | + | + | + |
| Durability over time | Robust but more referring to construction, spare parts | +++ | +++ | ++ |
| Self-controlled | Inbuilt routines, automatic performance | +++ | +++ | ++ |
| Measure impurities | To check quality for release or further purification | +++ | +++ | +++ |
| Measure main product | Protein structure identity, cell type | +++ | +++ | +++ |
| Measure side-products | Adverse protein forms, overflow metabolites, other cell types | +++ | +++ | +++ |
| Measure impurities | From cells, gels, culture media | +++ | +++ | +++ |
| Measure substrate | Nutrient sources in media | ++ | ++ | + |
| Measure gases | For gas balance calculations, respiration state | ++ | + | + |

**Table 14.5** Analytical purposes with hardware sensors.

| Purpose | When? |
|---|---|
| Intracellular capacity | If decisive parameter resides inside the cell. Is difficult when concentrations are very low |
| Complex extracellular media | If components are excreted to media, if media components are exhausted |
| Biological activity | Viability of cells, proliferation, growth |
| Biomolecular activity | Protein activity, enzyme activity, number of proteins |

The analytical criteria and targets highlighted in Table 14.4 – easy-use, disposable, response time (just-in-time), accuracy, and price – should be compared, evaluated, and ranked. If this becomes undoable, for example, due to the kind of biological data that must be at hand to infer useful information for process decisions (feedback control, sequential actions, etc.), a refined monitoring strategy is needed.

**Table 14.6** Hardware sensor alternatives: Price levels and utility of sensors and instrumentation.

| Price level | Method | Biological relevance | Just-in time | References |
|---|---|---|---|---|
| 0.01–1 kEUR | Temperature and pressure sensors | • | ••• | [40] |
| | Specific electrodes (pH, $pO_2$) | • | ••• | [40] |
| | Gas analyzers for $O_2$ and $CO_2$ | • | ••• | [41–46] |
| | Fluorometric optodes | • | ••• | |
| | Mass flowmeters | • | ••• | [46] |
| | Balance for tanks | • | ••• | [43, 44] |
| | Dipsticks for glucose and antigens | ••• | •• | [47, 48] |
| 1–10kEUR | Capacitance biomass probes | •• | ••• | [49, 50] |
| | Capacitive immunosensors | •• | •• | [51, 52] |
| | Quadrupole mass spectrometry | •• | ••• | [53] |
| | Membrane gas sensors | • | •• | [54] |
| | Flow injection electrodes | •• | •• | [55–58] |
| | Metabolic heat sensing | ••• | • | [59–62] |
| | Enzyme calorimetric sensors | •• | •• | [63, 64] |
| | Electronic noses | •• | ••• | [65–68] |
| | Electronic tongues | •• | ••• | [69] |
| | Photoacoustic spectroscopy online | • | ••• | [70] |
| 10–100 kEUR | HPLC online | •• | •• | [71–76] |
| | Advanced mass spectrometry | •• | •• | [77, 78] |
| | Multiwavelength fluorescence online | •• | ••• | [79–82] |
| | Near infrared (NIR) spectroscopy online | •• | ••• | [83, 84] |
| | Mid-infrared (MIR) spectroscopy online | •• | ••• | [85–87] |
| | Raman spectroscopy | •• | ••• | [88] |
| | One-channel surface plasmon resonance | •• | •• | [89–91] |
| | Capillary electrophoresis online | •• | •• | [92] |
| | In situ microscopy online | ••• | •• | [93, 94] |
| 100–1000 kEUR | Flow cytometry online | ••• | •• | [95–98] |
| | Surface plasmon resonance imaging (iSPR) | ••• | • | [99] |
| | Nuclear magnetic resonance online | ••• | • | [100] |

The gray zone represents hardware sensors presently rarely used in biomanufacturing.

Table 14.6 provides an overview of hardware sensor alternatives for soft sensors and rank these according to their utility, price, biological relevance, and just-in-time performance.

Several sensor alternatives reach the criteria of low cost as well as robustness and reliability. Unfortunately, these sensors do not provide the most critical biological process parameters. However, they are very essential for maintaining physical and chemical conditions in the bioprocess, such as temperature, pressure, pH, and $pO_2$ electrodes [40]. Moreover, they can provide signals that are valuable to integrate into soft sensor algorithms.

Several of the sensors are standard in existing process equipment [40, 46]. Partly, this can be ascribed to reliability, durability, stability, and low price of these sensors. Other common low-cost alternatives are gas analyzers [41–45] and dip-stick methods for metabolites and nutrient components [47, 48]. The latter are so far not commercially available for online use but may technically be realized.

A few of the low-cost sensors are ideal for soft sensors. Several sensors also show good-to-very-good analytical performance with respect to sensitivity and stability, while others, such as electrodes may require recalibration due to fouling and other drift effects.

Online high-performance-liquid-chromatography [71–76], and online mass spectrometry [53, 77, 78] are examples of more expensive stand-alone instruments with the capacity to provide online data for proteins, metabolites, or other low-molecular-weight compounds. These methods are more costly to set up in combination with soft sensors, they require significant support and operator surveillance, but could, in many cases, supply reliable and valuable process data with short duration. In this category, we also find other online alternatives, such as capacitive probes [49, 50], calorimetric sensors [59–64], electronic noses [65–68], and electronic tongues [69]. These instruments are demanding alternatives for soft sensors, sometimes accurate enough but less robust to use.

Online spectrometric methods, such as NIR [83, 84], MIR [85–87], Raman [88], and multiwave fluorescence spectrometry [79–82], require extensive calibrations and validation of soft sensor setups for the accurate prediction of components in the process media. However, these methods have the advantage of fast response. If the process media are not too complex, pure signals and good predictions can be expected especially for small molecules. This may be difficult in bioreactor media but in downstream processing, for example, NIR and fluorescence can show high reproducibility and are, therefore, viable alternatives also at large scale [71].

Cell-specific sensors may address important process parameters, such as morphology, cell type, cell surface biomarkers, are usually difficult to realize in an online fashion. This includes microscopy, normally requiring trained microscope technicians, but where automated microscopes are under development for *in situ* use in bioreactors [93, 94]. Also, flow cytometers [95–98] have shown interesting possibilities to be applied online with PLS soft sensor models. Importantly, these hardware alternatives are in particular important to consider for the monitoring of manufacture of stem cell or for cell therapy products [39].

More exclusive methods such as online NMR [100] and imaging SPR [99] are here considered far too expensive and unrealistic to justify at their present stage for soft sensor applications.

The PAT applicability of the hardware alternatives in Table 14.6 decreases in the high cost categories although biological relevance of analytical data may be good. Further sensor development would, however, contribute to expand the PAT applicable zone in the table. Especially desirable is to qualify those hardware sensors with a high biological relevance for potential PAT use.

## 14.5
## The Modeling Part of Soft Sensors

The algorithms in the soft sensors use more or less complex models and computation methods. The most common approaches are models based on multivariate data analysis (MVDA). Interpretation of analytical data or signals from NIR and MIR spectrometry [83, 85], fluorometry [79, 101], mass spectrometers [46], and electronic noses [102] as well as combination of data from different analytical techniques [103, 121] are examples of hardware instrumentations where MVDA methods are successfully applied. Spectrometric instruments where the data are superimposed in continuously updated spectra and where the spectral signals are deconvoluted into sets of analytical state variables are especially attractive as soft sensors.

An alternative and popular model approach to MVDA is artificial neural network (ANNs).

Extensive calibrations and tuning are required for the tuning of parameters in the MVDA and ANN models to the actual measurement situation, for example, the culture, the medium environment, and the analytes. Nevertheless, MVDA and ANN techniques are flexible in use and could furnish many alternatives for predicting data relevant for the PAT goals [79, 101–108].

Algorithms based on mechanistic models are sometimes the preferred choice due to their adherence to actual biological events and processes and to the fact that the analytical data are easier and mostly possible to relate to accepted theoretical models in bioengineering [42, 43, 109, 110, 118, 119]. Typically, the algorithms are derived from mass balances and kinetic equations for the biological systems that can easily be implemented in the soft sensor with rather limited work. The most demanding efforts will be to determine values of parameters in the models from experimental data. Often the users of soft sensors do not want to be hampered by this and, therefore, prefer the nondeterministic models by linking inputs and outputs according to weight functions or polynomial factors. Unfortunately, this may result in that critical metabolic events or pathways may be hidden in a set of numerical factors only derived from calibration and training experiments.

Nevertheless, partial-least-square (PLS) and ANN methods are common modeling methods that have frequently been applied in soft sensors. This is the case with, for example, electronic noses and electronic tongues where gas sensors and ion-sensitive electrode signals are used as inputs to the PLS or ANN models linked to outputs such as media analyte concentrations, biomass, or expressed proteins. The same has frequently been attained with success with mass and optical spectra.

Combinations of modeling approaches are also interesting alternatives, such as hybrid models [111, 112], as well as other alternatives, such as genetic algorithms [113] and inference methods.

Sometimes the computation tasks for the soft sensors are uncomplicated and straightforward, for example, two signal values added, subtracted, multiplied, or divided, as for example when converting volumetric rates to specific rates, computing yield factors, or filtering noise. Although quite simple to implement, these information can be of great value for operator communication, decision-making, and as control set-points.

Validation of the soft sensor performance is a very critical step for the evaluation of its usefulness in manufacturing [120]. Typically, calibration data sets are separated from validation data and statistical methods, and performance criteria are applied. It is up to the user to decide these criteria and at the level of the soft sensor estimates that is suffice for its purpose. In PLS and ANN models, standard procedures for validation are comprehensively described and normally followed (e.g., [79, 102]). In mechanistic models, correlations to reliable reference methods for the estimated variables are pivotal. Process validation guidelines from regulatory authorities are helpful and applicable to validation of soft sensors [31]. Self-validation procedures built in into the soft sensor have recently been suggested could also be considered [27].

In Table 14.7, the commonly used modeling approaches are summarized. The table indicates pros and cons with the models in soft sensor applications.

**Table 14.7** Modeling alternatives in soft sensors.

| Model type | Benefits/drawbacks(+/−) | References |
|---|---|---|
| Unstructured mechanistic model | + Modls based on first principles and may by that provide understanding<br>− Models do not mirror the full biological complexity<br>− Parameterization required<br>− Parameters must fit the design space | [42, 60, 61, 74, 75, 102, 114] |
| Structured mechanistic model | + Has opportunity to render enlarged understanding of the process<br>− Extensive biological knowledge about the biological complexity required | [42–45, 50, 93, 114] |
| Partial-least-square regression model | + The methodology fairly easy to set up<br>− Extensive validation of parameters required<br>− Nonexplanatory<br>− Must be recalibrated more frequently | [24, 69, 79, 81, 101, 103] |
| Artificial neural network model | + Easto set up in existing software with established ANN methodology<br>− Require training for setting weight parameters<br>− Must be retrained frequently | [16, 18, 23, 25, 102, 107, 108] |
| Hybrid model | + Takadvantage of best of mechanistic and regression models<br>− May be difficult to grasp and validate | [13, 15, 112, 115] |
| Transfer function | + Strit forward, easy to implement<br>+ Trasparent<br>+ Norally few parameters | [75, 102, 110] |

A general comment is that it is up to the user's own preferences to choose an approach. It may be that to a large extent the process knowledge required to set up a mechanistic model is available and that the model parameters are previously determined for the actual conditions or design or control space. Then, it may be attractive to use that model in the soft sensor but very little preknowledge exists, which could be the case, for example, cell therapy products, or other complex biological systems. This would favor the use of PLS or ANN models in the soft sensor.

## 14.6
## Strategy for Using Soft Sensors

A strategy of using soft sensors as PAT tools should be based on the trade-off between the user needs and soft sensor constrains as discussed earlier (Tables 14.2–14.4).

A ranking (e.g., as suggested by Mandenius and Björkman [35, 37]) of analytical performance versus user requirements might be the efficient way to select the most relevant soft sensor configuration. This should also at some level involve an economical assessment [37].

Figure 14.3 shows the examples of principal soft sensor configurations. Hardware sensors are placed *in situ* or at-line in a bioprocess unit operation. In the configurations illustrated in (a)–(c), the unit operation is a bioreactor with a growing culture producing a biomolecular or cellular product. The sensors monitor the cells' state, their number or viability, the surrounding metabolites or proteins produced by the cells, components of the added growth medium either in the liquid or gaseous phase of the reactor. In some cases, monitoring is motivated both before and after the unit, in others only after or at the unit. In (d), the example with three sequential units, the stepwise conversion in the processing is monitored. The strategy of using soft sensors can be extended to downstream configurations as well; the configurations illustrated in (e)–(g) show how the two effluent streams reflect the efficiency of the separation accomplished in the unit.

The main objectives of using the sensor signals in these examples are to mirror the purity of the main product on a cell-specific level, to provide signals for control actions, and to monitor the efficiency of the process to make use of the raw materials, or to alert when a control action or change in the process needs to be initiated.

The utility of these process parameters are obvious when it comes to decision-making. This can be founded on reflecting a metabolic or physiological state, which is critical for changing process operations and which has a direct effect on productivity, quality, or efficiency of using the raw materials.

Previous studies with soft sensors have exemplified and demonstrated a variety of alternatives. These alternatives should be evaluated and compared with a strategic perspective and versus application criteria.

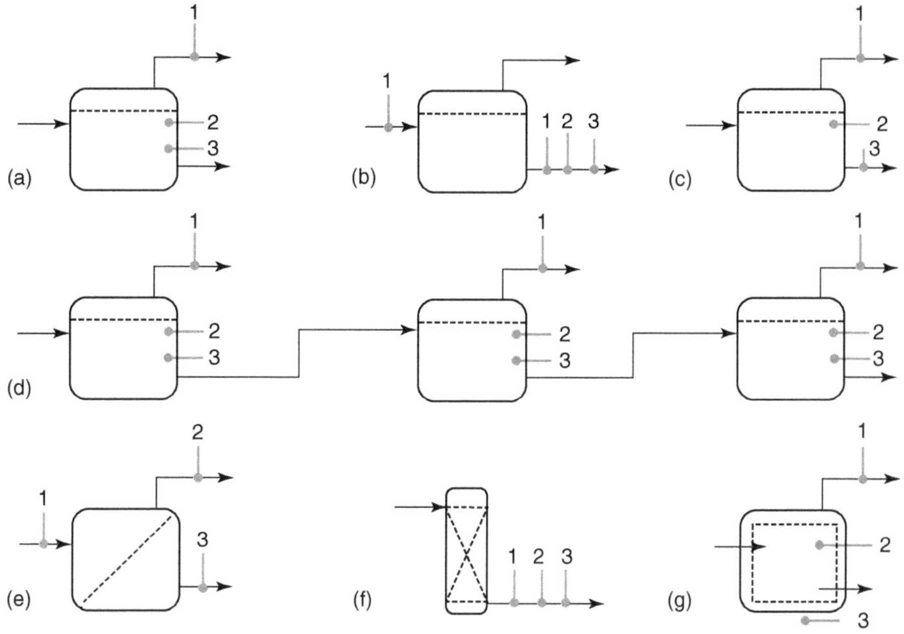


**Figure 14.3** Alternative soft sensor configurations : (a) a bioreactor unit with media inlet and outlet and gas vent equipped with three online sensor sensors (e.g., 1, $CO_2$ gas analysis; 2, cell density sensor; 3, dissolved oxygen electrode); (b) sensors placed inline in the bioreactor outlet (e.g., 1, HPLC for metabolites; 2, immunosensor for a recombinant protein; 3, NIR sensor for biomass); (c) sensors placed *in situ* and inline (e.g., 1, mass spectrometer for volatile metabolites; 2, capacitive sensor for cell viability; 3, HPLC for substrate); (d) bioreactor sequence where a product is formed (e.g., 1, gas sensor volatile product; 2, cell density sensor; 3, substrate sensor); (e) separation unit where component in an input flow is separated (e.g., 1, cell density sensor; 2,3, protein sensors); (f) a column separation where the outlet stream is monitored (e.g., 1, UV sensor, 2, conductivity sensor, 3, immunosensor); (g) a downstream unit where a product is dried (e.g., 1, moisture, 2, temperature, 3, weight).

## 14.7
## Applications of Soft Sensors in Bioreactors

A typical bioreactor setup at the laboratory scale provides real-time signals for a number of operational parameters such temperature, pressure, aeration rate, pH, and $pO_2$ (Figure 14.4). A well-equipped bioreactor may also have an out-gas analyzer and a cell density sensor as standard.

Figures 14.5–14.9 show five real examples of soft sensor applications to bioreactor monitoring and control collected from recent academic reports [60, 74, 75, 81, 102, 110]. The configurations of the soft sensors in these examples promote the strategy of optimizing productivity, enhance reproducibility, and reduce quality variation. And it is done by combining online sensors and scientific knowledge.

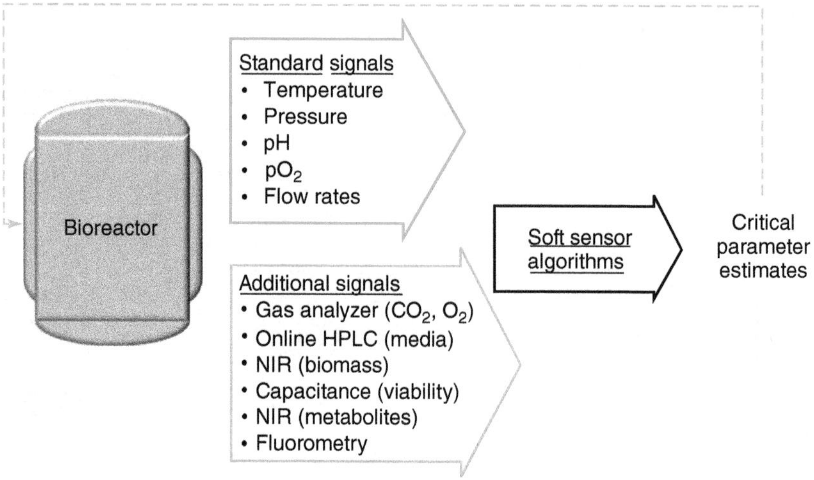


**Figure 14.4** A general bioreactor soft sensor setup showing typical standard sensor signals supplemented with additional hardware sensor signals.

### 14.7.1
### Online Fluorescence Spectrometry for Estimating Media Components in a Bioreactor

In the example shown in Figure 14.5, a multiwavelength fluorescence spectrometer probe monitors a yeast cultivation online [81]. Concentrations of biomass, glucose, and ethanol are predicted from scanned spectra analyzed by segmented PLS models. The prediction models were calibrated on standard online process data and displayed low root mean square error of prediction (RMSEP) for ethanol and glucose, corresponding to 4%, 2%, and 2% of the respective concentration intervals.

These batch bioreactor cultivations were done using *Saccharomyces cerevisiae* at high (with 190–305 $g\,l^{-1}$ glucose) and low (with 21–25 $g\,l^{-1}$ glucose) biomass concentrations and the multiwavelength fluorescence method and standard monitoring sensors. Partial least squares models were calibrated for the prediction of cell dry weight, ethanol and consumed glucose, using the two data types separately. The low-density cultivations consisted of two diauxic phases (glucose uptake with ethanol production followed by ethanol uptake after glucose depletion), which required different models for the two phases, but improved the predictions significantly. The prediction models calibrated on standard online process data displayed similar or lower RMSEPs compared with the fluorescence models. The best prediction models for high-density cultivations had RMSEPs of 1.0 $g\,l^{-1}$ CDW, 1.8 $g\,l^{-1}$ ethanol, and 5.0 $g\,l^{-1}$ consumed glucose, corresponding to 4%, 2%, and 2% of the concentrations, while low-density models had 0.3 $g\,l^{-1}$ CDW, 0.7 $g\,l^{-1}$ ethanol, and 1.0 $g\,l^{-1}$ consumed glucose, corresponding to 4%, 8%, and 4% of the concentrations.

This work demonstrates that online prediction of important bioreactor analytes such as biomass, glucose, and ethanol can be achieved by modeling

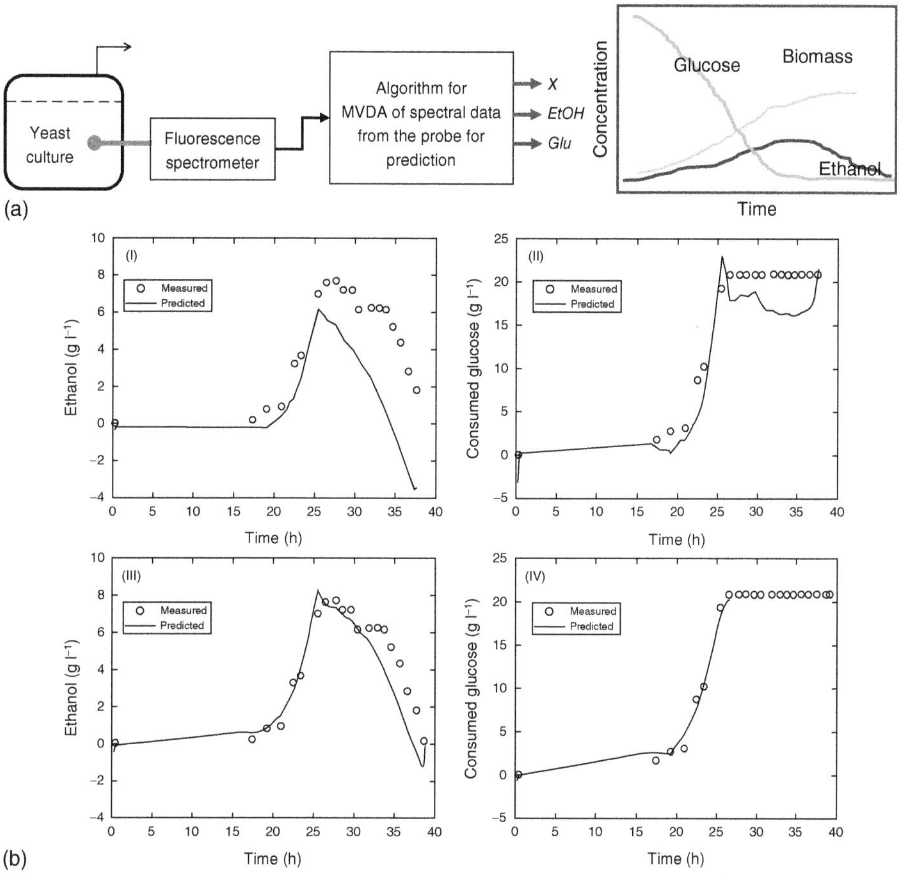


**Figure 14.5** (a) A soft sensor setup where an online fluorescence probe predicts concentrations of glucose, biomass, and ethanol using an MVDA algorithm in yeast cultivation. (b) Four examples of predictions of analytes from the multiwavelength fluorescence spectrometer. Time profiles of (○) measured and (—) predicted (I) ethanol and (II) consumed glucose using one TRAD PLS model for the whole cultivation. (III) and (IV) show the corresponding values using segmented models (with permission from [81]).

multiwavelength fluorescence data as well as by using soft sensors consisting of PLS models based on standard online bioreactor data.

## 14.7.2 Temperature Sensors for Growth Rate Estimation of a Fed-Batch Bioreactor

In the example shown in Figure 14.6, two sensors measure the temperature of the influent and effluent cooling medium of the bioreactor with an *Escherichia coli* culture producing a recombinant protein [60, 110]. A third sensor measures the culture temperature inside the bioreactor. The signals are used in a set of

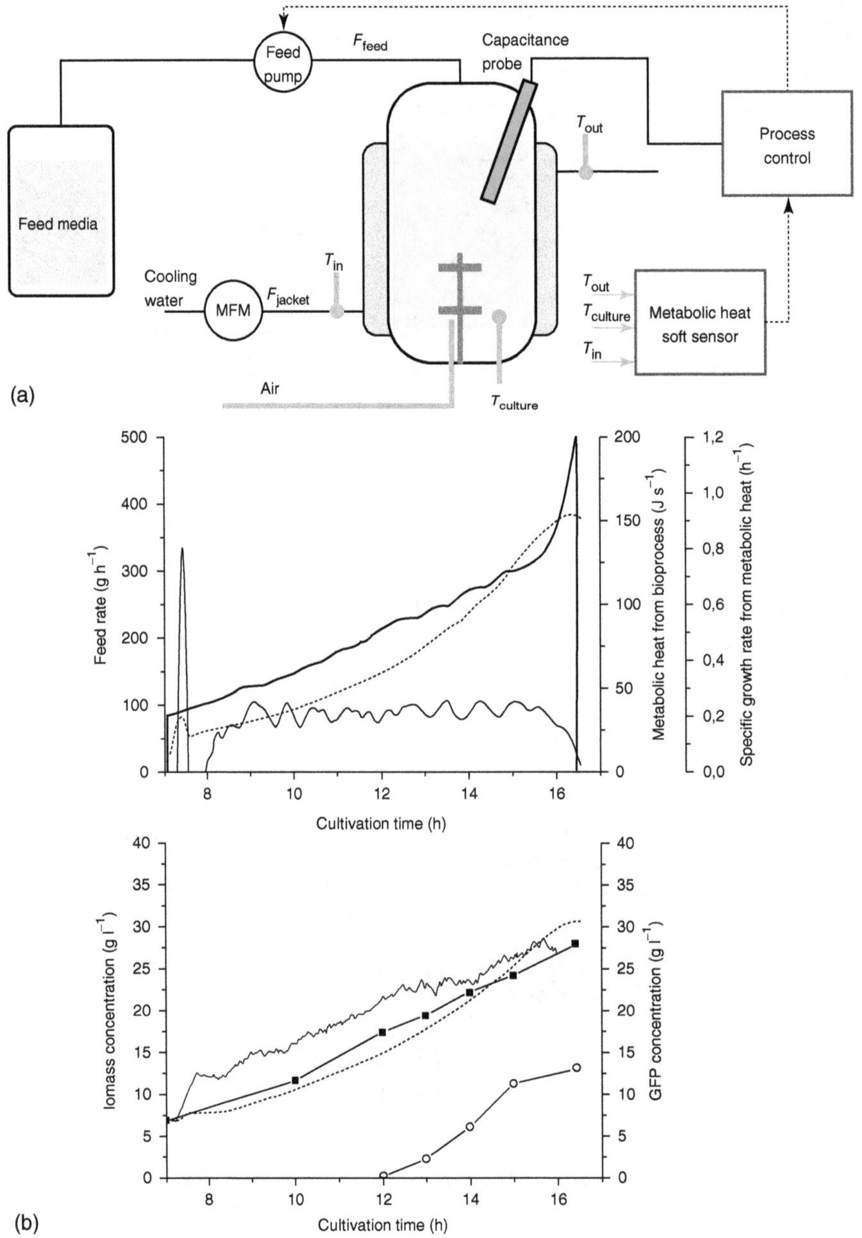
Feed pump
$F_{feed}$
Capacitance probe
$T_{out}$
Process control
Feed media
Cooling water
MFM
$F_{jacket}$
$T_{in}$
$T_{out}$
$T_{culture}$
$T_{in}$
Metabolic heat soft sensor
Air
$T_{culture}$
(a)
Feed rate (g h$^{-1}$)
Metabolic heat from bioprocess (J s$^{-1}$)
Specific growth rate from metabolic heat (h$^{-1}$)
Cultivation time (h)
Iomass concentration (g l$^{-1}$)
GFP concentration (g l$^{-1}$)
Cultivation time (h)
(b)

**Figure 14.6** Soft sensor configuration where the temperature signals from the cooling jacket of the reactor to estimate the heat production of the cell. This information is used for deriving the specific growth rate and then used that for controlling the dosage of the feed. (a) Experimental setup of the soft sensor. (b) Induced fed-batch cultivation producing recombinant GFP showing in upper panel controlled feed rate (___), metabolic heat production (----), and $\mu_{metabol}$ (—), and in lower panel biomass from heat (·····), optical density (■), and capacitance measurements (—). The production of GFP (○) is induced at 12 h (reproduced with permission from [110]).

energy balances to estimate online the specific growth rate during the whole fed-batch phase. The feeding of media is controlled from the estimated $\mu$ relative to a set $\mu$ value. The figure shows an inset of a control diagram where the feeding is adjusted by a PI controller to maintain the $\mu$ value. The $\mu$ estimate can then be easily converted to biomass concentration. The advantage with this setup is that the instrumentation with temperature sensors is there already; it is just to acquire the signals and implement the algorithm. The shortcoming of this approach is that not only the growth produces heat in a bioreactor culture, but also agitation, aeration, and titration. Consequently, this must be compensated for correctly. If metabolic heat is small, the accuracy suffers.

By applying sequential digital signal filtering, the soft sensor was made more robust for industrial practice with cultures generating low metabolic heat in environments with high noise level [110]. The estimated specific growth rate signal obtained from the three-stage sequential filter allowed controlled feeding of substrate during the fed-batch phase of the production process. The biomass and growth rate estimates from the soft sensor were also compared with an alternative sensor probe, a capacitance online sensor, for the same variables. The comparison showed similar or better sensitivity and lower variability for the metabolic heat soft sensor suggesting that using permanent temperature sensors of a bioreactor is a realistic and inexpensive alternative for monitoring and control. However, both alternatives are easy to implement in a soft sensor alone or in parallel.

Thus, the study demonstrates how the estimation model can be implemented as a soft sensor in a standard computer control software in parallel with other soft sensor functions, and by that showing a high degree of operability in typical laboratory or small-scale production environments.

The results also show that the soft sensor model with the filtering function is comparable, or even better, than alternative online hardware sensors for biomass concentration and specific growth rate [110].

### 14.7.3 Base Titration for Estimating the Growth Rate in a Batch Bioreactor

In Figure 14.7, a soft sensor utilizes the addition of base to control the pH of the culture [75, 116]. It is assumed that there is a direct correlation between the cells' production of $H^+$ and growth. Thus, the pump run time and the pH electrode signal for the culture are acquired. A simple algorithm converts the signals to biomass. Provided the assumption is correct and no other $H^+$-generating

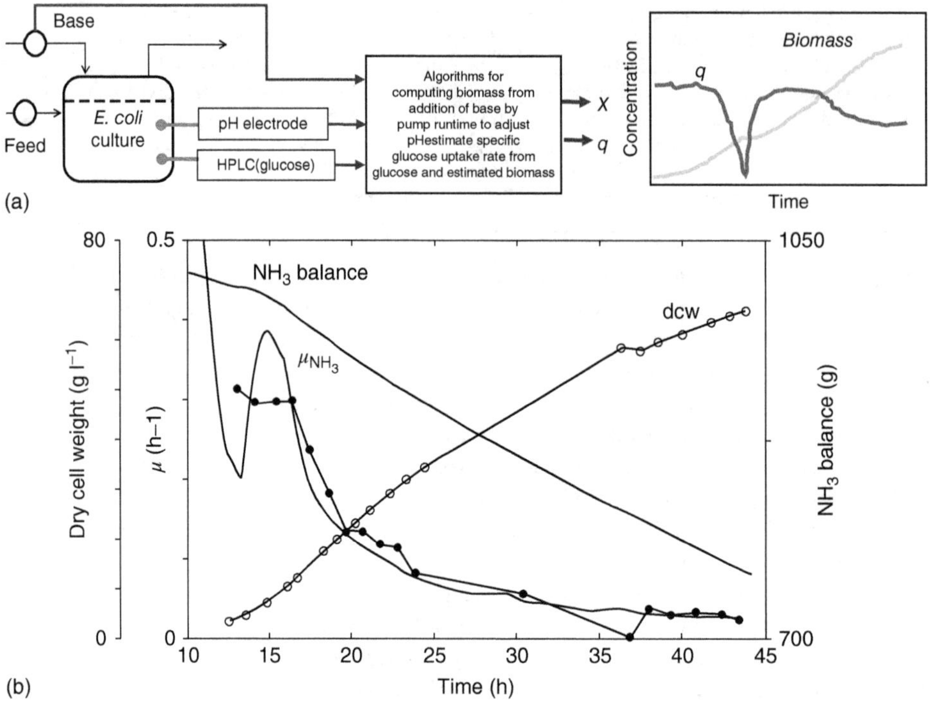


**Figure 14.7** (a) Soft sensor configuration that estimates the biomass concentration from calculating the addition of titrand. (b) The $\mu_{NH3}$ sensor signal (thick continuous line) from an *E. coli* fed-batch cultivation on minimal medium compared with $\mu$ calculated on dcw data (filled circle). Also shown are the measured data for $\mu_{NH3}$ calculation, that is, the $NH_3$ vessel balance data (____) and the cell dry weights (○). The exponential glucose feed was started at 12.6 h and was constant from 16.5 h (reproduced with permission from [116]).

processes occur in the culture the relation holds. However, if the medium has an unknown composition of amino acids the correlation is more complex. The example in the figure is taken from a culture where the medium is defined. The correlation between biomass reference sample and the estimates is high in the investigated region. Again, the advantage with the soft sensor is that any auxiliary instrumentation is not required; the soft sensor is directly implemented using existing signals in an algorithm added in the control software of the bioreactor system.

The study was extended with additional soft sensors based on standard online data from fermentation processes and models to other state variables in *E. coli* fed-batch processes: the biomass concentration, the specific growth rate, the oxygen transfer capacity of the bioreactor, and the new $R_{O/S}$ sensor, which is the ratio between oxygen and energy substrate consumption. The $R_{O/S}$ variable grows continuously in a fed-batch culture with constant glucose feed, which reflects the increasing maintenance demand at declining specific growth rate. The $R_{O/S}$

sensor also responded to rapid pH shift-downs reflecting the increasing demand for maintenance energy. It is suggested that this sensor may be used to monitor the extent of physiological stress that demands energy for survival.

This study shows that simple models can be used to gain additional information from standard logged fermentation process variables. The applications may be either for offline process characterization, as in this study, or for online feedback control. The latter raises increased demand on the primary measurement data or a demand for online data filtering. The sensors may under certain conditions be quite sensitive to errors in the input data, but this generates a systematic error in the output values and should not prevent their use for monitoring changes in the process performance.

### 14.7.4 Online HPLC for the Estimation of Mixed-Acid Fermentation By-Products

In Figure 14.8, online NIR and HPLC are used with a soft sensor based on first principles for the estimation of specific conversion of mixed acids, which is taken as set-point for feed control [74]. The soft sensor uses an *in situ* NIR probe with a fixed wavelength window to acquire an absorbance signal that is related to total biomass concentration. This signal is combined with data from an online HPLC where chromatogram peaks are analyzed for selected metabolites in the medium important for surveying the metabolic state of the protein expressing *E. coli* culture. After evaluation of peak areas, the HPLC data are converted to concentrations of acetate, glucose, lactate, and formate. By transforming the HPLC and biomass signals to specific concentrations (g metabolite/g cell h), the physiological state is monitored and used for determining the feeding rate in the fed-batch. The advantage in this case is that the instrumentation in the soft sensor is well established, and robust methods and the HPLC analyzes several analytes per run. The drawback is the relatively long delay of the HPLC (up to 25 min) compared with the NIR probe that responds in seconds.

In particular, this soft sensor approach provides the opportunity of controlling metabolic overflow from mixed-acid fermentation and glucose overflow metabolism in a fed-batch cultivation for the production of recombinant green fluorescence protein (GFP) in *E. coli* (Figure 14.8b). The computational part of the soft sensor used basic kinetic equations and summations for estimation of specific rates and total metabolite concentrations. Two control strategies for media feeding of the fed-batch cultivation were evaluated: (i) controlling the specific rates of overflow metabolism and mixed-acid fermentation metabolites at a fixed pre-set target values, and (ii) controlling the concentration of the sum of these metabolites at a set level.

The results indicate that the latter strategy was more efficient for maintaining a high titer and low variability of the produced recombinant GFP protein.

The presented study shows how the variability of a recombinant protein product can be reduced by soft sensor-based control. The soft sensor reduces the need for instrumentation and offline analysis. The main PAT objective, to increase the

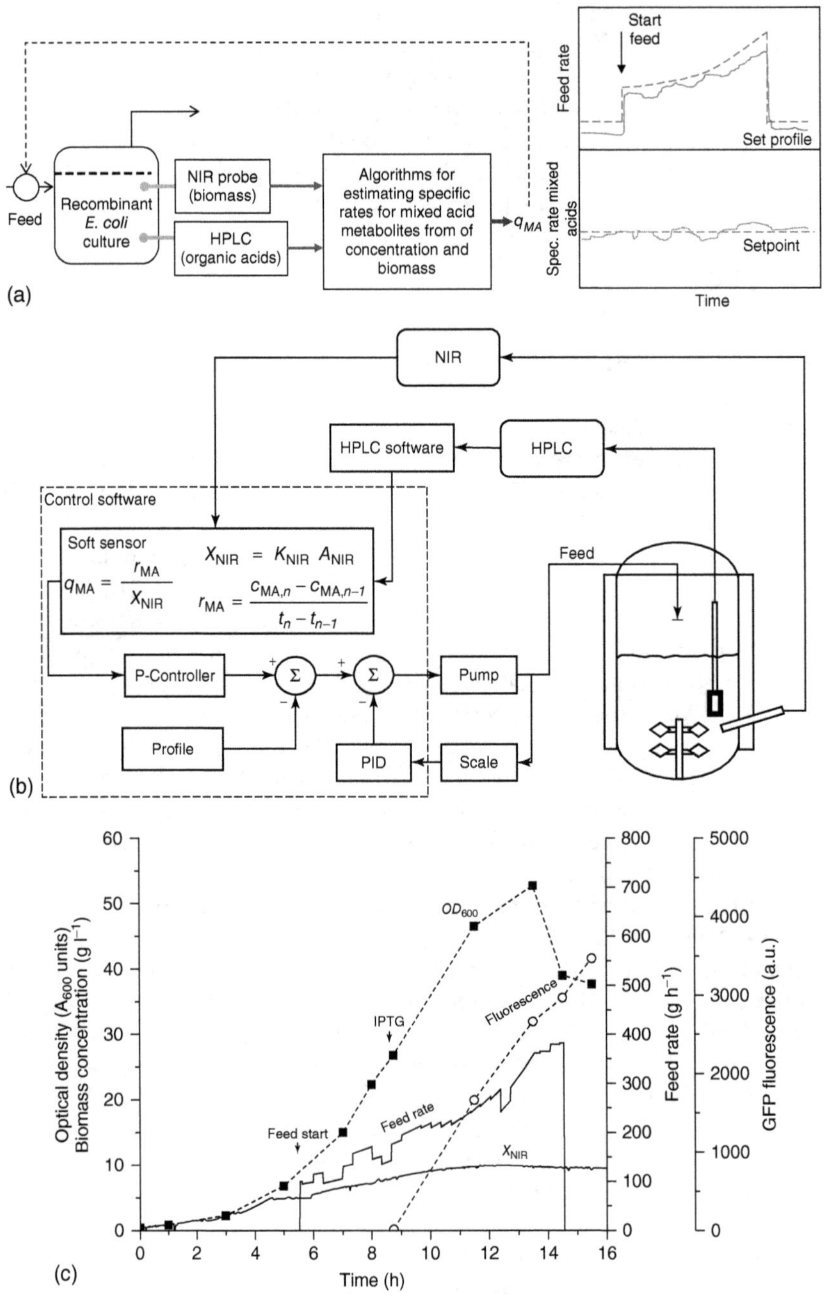
Start feed
Feed rate
Set profile
Spec. rate mixed acids
Setpoint
Time
Feed
Recombinant E. coli culture
NIR probe (biomass)
HPLC (organic acids)
Algorithms for estimating specific rates for mixed acid metabolites from of concentration and biomass
qMA
(a)
NIR
HPLC software
HPLC
Control software
Soft sensor
qMA = rMA / XNIR
XNIR = KNIR ANIR
rMA = (cMA,n − cMA,n−1) / (tn − tn−1)
Feed
P-Controller
Pump
Profile
PID
Scale
(b)
Optical density (A600 units) Biomass concentration (g l−1)
Feed rate (g h−1)
GFP fluorescence (a.u.)
OD600
IPTG
Fluorescence
Feed start
Feed rate
XNIR
Time (h)
(c)

**Figure 14.8** (a) Configuration approach where feeding is controlled around the specific production rate of mixed acid side-production. (b) Soft sensor control strategies of the feeding of the fed-batch cultivation were applied using the sum of specific production rates of the metabolites acetate, lactate, ethanol, and formate. (c) Control of the sum of MAF metabolite concentrations in the fed-batch GFP cultivation. Feeding starts at 5 h and induction is carried out at 9–10 h using 0.03 g $l^{-1}$ IPTG. Cultivations with identical control settings (data for biomass concentration from NIR measurements in (b) is not available) (reproduced with permission from [110]).

process stability and reduce the variability of key process parameters, in particular, to product titer, can be accomplished with the soft sensor approach.

Furthermore, it can be noted that the controller must not be very precise (i.e., show short declination, minor stationary fault) to actually accomplish low variability.

Moreover, the result of the study indicates that the physiological control approach is not necessarily the most successful control strategy.

However, further simplification of the soft sensor configuration would be desirable. If a lesser number of hardware sensors can accomplish the control strategy, the industrial applicability would be enhanced and maintenance and supply are further reduced.

### 14.7.5 Electronic Nose and NIR Spectroscopy for Controlling Cholera Toxin Production

In Figure 14.9, the soft sensor configuration is based on *in situ* near-infrared (NIR) spectroscopy and inline electronic nose (EN) sensor [102]. The NIR and EN were used to monitor and control a cholera toxin B-producing *Vibrio cholerae* fed-batch cultivation carried out at laboratory and production scales. Figure 14.9b shows the detailed setup of the experimental system.

A set of prediction models for biomass, glucose, and acetate using NIR spectroscopy were developed based on spectral identification and PLS regression resulting in high correlation to reference data (standard errors of prediction for biomass, glucose, and acetate were 0.20 g $l^{-1}$, 0.26 g $l^{-1}$, and 0.28 g $l^{-1}$). A compensation algorithm for aerated bioreactor disturbances was integrated in the model computation, which in particular improved the prediction by the biomass model. First, the NIR data were applied together with EN inline data selected by principal component analysis (PCA) for generating a trajectory representation of the fed-batch cultivation (Figure 14.9c). A correlation between the culture progression and EN signals was demonstrated, which proved to be beneficial in monitoring the culture quality. It was shown that a deviation from a normal cultivation behavior could easily be recognized and that the trajectory was able to alarm a bacterial contamination. Second, the NIR data indicated the potential of predicting the concentration of formed cholera toxin with a model prediction error of 0.020 g $l^{-1}$. Third, the online biomass prediction based on the NIR model was used to control the overflow metabolism acetate formation of the *V. cholerae* culture.

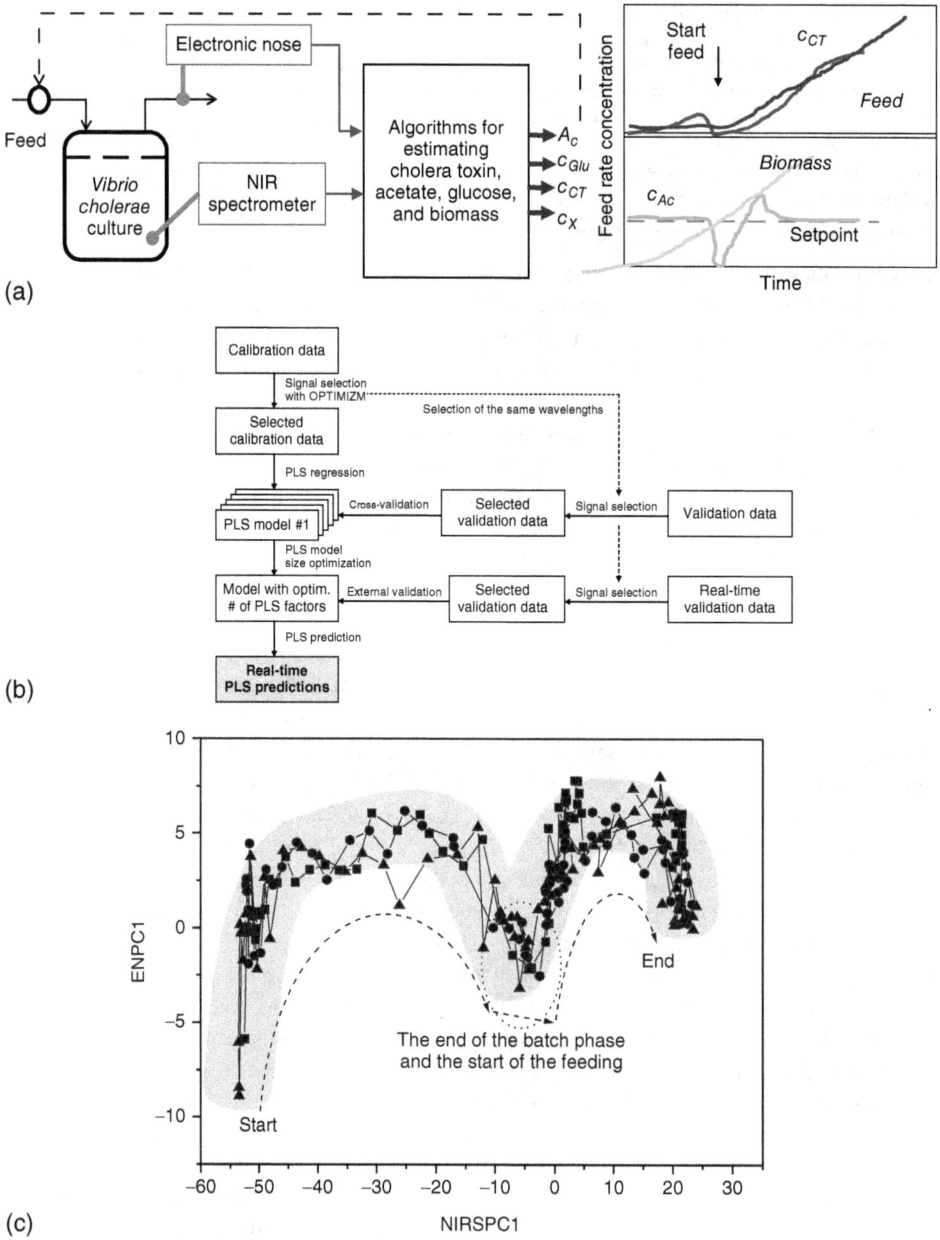


**Figure 14.9** (a) Configuration of the experimental setup of a soft sensor with an electronic nose and NIR spectroscopy. (b) A detailed flow chart of the system. (c) Trajectory score plots generated from the first principal component of preselected NIR and EN signals for the scaled-down *V. cholerae* cultivations (■ and •, calibration data sets; ▲, validation data set). The shaded area represents the range of values, in which the culture is in its "normal" state (reproduced with permission from [102]).

The controller compared actual specific growth rate as estimated from the prediction with the critical acetate formation growth rate, and from that difference adjusted the glucose feed rate.

These five examples are all PAT related as they exploit online operation and use scientific knowledge about the culture that is applied in simple computations. But the process parameters derived are not the most urgently needed, at least not when comparing needs listed in Table 14.2. This indicates that there is an urgent need to increase the matching of scientific research efforts and industrial PAT-related needs. The methodological prerequisites, both in terms or instrumental as well as access to modeling capacity, seem to exist. However, successful validations of viable and cost effect solutions are still lacking.

## 14.8 Concluding Remarks and Outlook

Soft sensors may provide the necessary analytical tools that are necessary for accomplishing important bioprocess objectives where the bioreactor is the key unit operation. In particular, the soft sensors can contribute to

- the exploitation of inline and online analytical instruments such as biosensors, spectrometers, and other optical methods in a more efficient or elaborate way in order to measure process parameters in the bioreactor for deriving critical process and quality attributes
- the utilization of the information obtained through multivariate data acquisition and data analysis
- the utilization of first principle models in order to take advantage of known mechanisms of the biotechnological system in use in the bioreactor.

One of the messages of this chapter is to show the relationship of these tools with soft sensor concept.

Moreover, it is advocated that soft sensor should preferably utilize already existing robust sensor devices for this purpose (Table 14.6). The soft sensors should by that reduce the analytical instrumentation and avoid extending the process control of a plant with additional equipment items.

However, the ambition of using simple robust hardware sensors in the soft sensor setup and from these derive critical process and quality parameters related to the PAT goals is rarely met for important biotherapeutic products. This could actually be done with existing techniques.

Online monitoring of small drug molecules and their side-products requires relatively advanced analytical equipment. Even when the target product is present at high concentration, which should be the case in most biotherapeutic production processes, hardware alternatives such as HPLC or capillary electrophoresis need to be reconfigured in an online injection setup close to the process. Thereby, the sensor system would risk deviating from simplicity and easy-use requirements. Larger product molecules, such as therapeutic proteins, would suffer from the same shortcoming. However, immune-specific methods in sensitive flow injection setups, such as surface plasmon resonance, capacitive immunosensors, and calorimetric chips, could provide robust soft sensors online.

In bioprocessing, monitoring of the bioreactor process is much more highlighted than downstream processing. Without doubt it is advantageous to monitoring and control early in the process train. However, precision and measurability may be easier in the purer media appearing downstream. From a quality control perspective, monitoring of critical quality attributes in the downstream stages are as important. Soft sensors could therefore play a profound role for improving process performance [122].

Consequently, further development of bioanalytical methods for online use would enhance the prospect for soft sensors in biotherapeutic processes. A very interesting opportunities prospect for this is the miniaturization and microfluidic approaches that have great potential for this.

For cell therapy manufacture, robust soft sensor methods would have significant impact if they could provide online identification of cell types for selection and further cell expansion. The microscopic techniques for *in situ* cell culture monitoring mentioned earlier may contribute to this [93, 94]. However, a more tailored technique combining biomarkers for the cell phenotypes in combination with cell selection would be even more beneficial. With such instrumental setups, an algorithm able to interpret multiple optical read-outs and combine these with other sensor signals related to the culture vessel or bioreactor environment would perfectly comply with the soft sensor concept.

## References


1. Etien, E. (2013) Modeling and simulation of soft sensor design for real-time speed estimation, measurement and control of induction motor. *ISA Trans.*, **52**, 358–364.
2. Godoy, J.L., Minari, R.J., Vega, J.R., and Marchetti, J.L. (2011) Multivariate statistical monitoring of an industrial SBR process. Soft-sensor for production and rubber quality. *Chemom. Intell. Lab. Syst.*, **107**, 258–268.
3. Fortuna, L., Graziani, S., Rizzo, A., and Xibilia, M.G. (2007) *Soft Sensors for Monitoring and Control of Industrial Processes*, Springer-Verlag, Berlin.
4. Lin, B., Recke, B., Knudsen, J.K.H., and Jørgensen, S.B. (2007) A systematic approach for soft sensor development. *Comput. Chem. Eng.*, **31**, 419–425.
5. Kadlec, P., Gabrys, B., and Strandt, S. (2009) Data-driven soft sensors in the process industry. *Comput. Chem. Eng.*, **33**, 795–814.
6. Chéruy, A. (1997) Software sensors in bioprocess engineering. *J. Biotechnol.*, **52**, 193–199.
7. Bogaerts, P.H. and Vande Wouwer, A. (2003) Software sensors for bioprocesses. *ISA Trans.*, **42**, 547–558.
8. Luttmann, R., Bracewell, D.G., Cornelissen, G., Gernaey, K.V., Glassey, J., Hass, V.C., Kaiser, C., Preusse, C., Striedner, G., and Mandenius, C.F. (2012) Soft sensors in bioprocesses: status report and recommendation. *Biotechnol. J.*, **7**, 1040–1047.
9. Bosca, S., Barresi, A.A., and Fissore, D. (2013) Use of a soft sensor for the fast estimation of dried cake resistance during a freeze-drying cycle. *Int. J. Pharm.*, **451**, 23–33.
10. Chen, J., Yu, J., and Zhang, Y. (2014) Multivariate video analysis and Gaussian process regression model based soft sensor for online estimation and

prediction of nickel pellet size distributions. *Comput. Chem. Eng.*, **64**, 13–23.
11. Ding, J., Qu, L., Hu, X., and Liu, X. (2011) Application of temperature inference method based on soft sensor technique to plate production process. *J. Iron Steel Res. Int.*, **18**, 24–27.
12. Kortela, J. and Jämsä-Jounela, S.L. (2013) Fuel moisture soft-sensor and its validation for the industrial BioPower 5 CHP plant. *Appl. Energy*, **105**, 66–74.
13. Jia, R., Mao, Z., Chang, Y., and Zhao, L. (2011) Soft-sensor for copper extraction process in cobalt hydrometallurgy based on adaptive hybrid model. *Chem. Eng. Res. Des.*, **89**, 722–728.
14. Li, Y.G., Gui, W.H., Yang, C.H., and Xie, Y.F. (2013) Soft sensor and expert control for blending and digestion process in alumina metallurgical industry. *J. Process Control*, **23**, 1012–1021.
15. Zhang, S., Wang, F., He, D., and Chu, F. (2013) Soft sensor for cobalt oxalate synthesis process in cobalt hydrometallurgy based on hybrid model. *Neural Comput. Appl.*, **23**, 1465–1472.
16. Wang, J. and Guo, Q. (2013) D-FNN based soft-sensor modeling and migration reconfiguration of polymerizing process. *Appl. Soft Comput.*, **13**, 1892–1901.
17. Novak, M., Mohler, I., Golob, M., Andrijić, Z.U., and Bolf, N. (2013) Continuous estimation of kerosene cold filter plugging point using soft sensors. *Fuel Process. Technol.*, **113**, 8–19.
18. Pani, A.K., Vadlamudi, V.K., and Mohanta, H.K. (2013) Development and comparison of neural network based soft sensors for online estimation of cement clinker quality. *ISA Trans.*, **52**, 19–29.
19. Rogina, A., Sisko, I., Mohler, I., Ujevic, Z., and Bolf, N. (2011) Soft sensor for continuous product quality estimation in crude distillation unit. *Chem. Eng. Res. Des.*, **89**, 2070–2077.
20. Khalfe, N.M., Kumar, S., Sunil, L., and Sawke, K. (2011) Soft sensor for better control of carbon dioxide removal process in ethylene glycol plant. *Chem. Ind. Chem. Eng. Q.*, **17**, 17–24.
21. Teixeira, B.O.S., Castro, W.S., Teixeira, A.F., and Aguirre, L.A. (2014) Data-driven soft sensor of downhole pressure for a gas-lift oil well. *Control Eng. Pract.*, **22**, 34–43.
22. Ward, A.J., Hobbs, P.J., Holliman, P.J., and Jones, D.L. (2011) Evaluation of near-infrared spectroscopy and software sensor methods for determination of total alkalinity in anaerobic digesters. *Bioresour. Technol.*, **102**, 4083–4090.
23. Rani, A., Singh, V., and Gupta, J.R.P. (2013) Development of soft sensor for neural network based control of distillation column. *ISA Trans.*, **52**, 438–449.
24. Zamprogna, E., Barolo, M., and Seborg, D.E. (2004) Development of a soft sensor for a batch distillation column using linear and nonlinear PLS regression techniques. *Control Eng. Pract.*, **12**, 917–929.
25. Choi, D.J. and Park, H. (2001) A hybrid artificial neural network as a software sensor for optimal control of a wastewater treatment process. *Water Res.*, **35**, 3959–3967.
26. Fortuna, L., Graziani, S., and Xibilia, M.G. (2005) Soft sensors for product quality monitoring in debutanizer distillation columns. *Control Eng. Pract.*, **13**, 499–508.
27. Qin, S.J., Yue, H., and Dunia, R. (1997) Self-validating inferential sensors with application to air emission monitoring. *Ind. Eng. Chem. Res.*, **36**, 1675–1685.
28. Jose, L., Sanchez, G., Robinson, W.C., and Budman, H. (1999) Developmental studies of an adaptive on-line softsensor for biological wastewater treatments. *Can. J. Chem. Eng.*, **77**, 707–717.
29. U.S. FDA (2004) Guidance for Industry PAT - A Framework for Innovative Pharmaceutical Development, Manufacturing, and Quality Assurance.
30. European Medicines Agency (2011) EMA-FDA pilot program for parallel assessment of Quality by Design applications. Document EMA/172347/2011.
31. International Conference on harmonization - ICH Quality Guidelines (2005–2009) Pharmaceutical Development, Pharmaceutical Quality Systems

and Quality Risk Management. ICH Geneva, Switzerland

32. Rathore, A.S., Bhambure, R., and Ghare, V. (2010) Process analytical technology (PAT) for biopharmaceutical products. *Anal. Bioanal. Chem.*, **398**, 137–154.
33. Glassey, J., Gernaey, K.V., Oliveira, R., Striedner, G., Clemens, C., Schultz, T.V., and Mandenius, C.F. (2011) PAT for biopharmaceuticals. *Biotechnol. J.*, **6**, 369–377.
34. Carrondo, M.J.T., Alves, P.M., Carinhas, N., Glassey, J., Hesse, F., Merten, O.W., Micheletti, M., Noll, T., Oliveira, R., Reichl, U., Staby, A., Teixeira, A.P., Weichert, H., and Mandenius, C.F. (2012) How can measurement, monitoring, modeling and control advance cell culture in industrial biotechnology? *Biotechnol. J.*, **7**, 1522–1529.
35. Mandenius, C.F. and Björkman, M. (2009) Process analytical technology (PAT) and Quality-by-Design (QbD) aspects on stem cell manufacture. *Eur. Pharm. Rev.*, **14**, 32–37.
36. Ulrich, K.T. and Eppinger, S.D. (2007) *Product Design and Development*, 3rd edn, McGraw-Hill, New York.
37. Mandenius, C.F. (2012) Design of monitoring and sensor systems for bioprocesses using biomechatronic principles. *Chem. Eng. Technol.*, **35**, 1412–1420.
38. Mandenius, C.F. (2012) Biomechatronics for designing bioprocess monitoring and control systems: application to stem cell production. *J. Biotechnol.*, **162**, 430–440.
39. Abbasalizadeh, S. and Baharvand, H. (2013) Technological progress and challenges towards cGMP manufacturing of human pluripotent stem cells based therapeutic products for allogeneic and autologous cell therapies. *Biotechnol. Adv.*, **31**, 1600–1623.
40. Sonnleitner, B. (1999) Instrumentation of biotechnological processes. *Adv. Biochem. Eng. Biotechnol.*, **66**, 1–64.
41. Ge, X., Kostov, Y., and Rao, G. (2005) Low-cost noninvasive optical $CO_2$ sensing system for fermentation and cell culture. *Biotechnol. Bioeng.*, **89**, 329–334.
42. Aehle, M., Kuprijanov, A., Schaepe, S., Simutis, R., and Lübbert, A. (2011) Simplified off-gas analyses in animal cell cultures for process monitoring and control purposes. *Biotechnol. Lett.*, **33**, 2103–2110.
43. Martens, S., Borchert, S.O., Faber, B.W., Cornelissen, G., and Luttmann, R. (2011) Fully automated production of potential Malaria vaccines with *Pichia pastoris* in integrated processing. *Eng. Life Sci.*, **11**, 429–435.
44. Fricke, J., Pohlmann, K., Tatge, F., Lang, F., Faber, B., and Luttmann, R. (2011) A multi-bioreactor system for optimal production of malaria vaccines with *Pichia pastoris*. *Biotechnol. J.*, **6**, 437–451.
45. Kiviharju, K., Salonen, K., Moilanen, U., and Eerikäinen, T. (2008) Biomass measurement online: the performance of in situ measurements and software sensors. *J. Ind. Microbiol. Biotechnol.*, **35**, 657–665.
46. Sonnleitner, B. (2013) Automated measurement and monitoring of bioprocesses: key elements of the $M^3C$ strategy. *Adv. Biochem. Eng. Biotechnol.*, **132**, 1–33.
47. Liu, J., Mazumdar, D., and Lu, Y. (2006) A simple and sensitive "dipstick" test in serum based on lateral flow separation of aptamer-linked nanostructures. *Angew. Chem. Int. Ed.*, **45**, 7955–7959.
48. Hu, J., Wang, S.Q., Wang, L., Li, F., Pingguan-Murphy, B., Lu, T.J., and Xu, F. (2014) Advances in paper-based point-of-care diagnostics. *Biosens. Bioelectron.*, **54**, 585–597.
49. Sarrafzadeh, M.H., Belloy, L., Esteban, G., Navarro, J.M., and Ghommidh, C. (2005) Dielectric monitoring of growth and sporulation of *Bacillus thuringiensis*. *Biotechnol. Lett.*, **27**, 511–517.
50. Ehgartner D., Sagmeister P., Herwig C., and Wechselberger P. (2015) A novel real-time method to estimate volumetric mass biodensity based on the combination of dielectric spectroscopy and soft-sensors. *J. Chem. Technol. Biotechnol.*, **90**, 262–272.
51. Teeparuksapun, K., Hedstrom, M., Kanatharana, P., Thavarungkul, P.,

and Mattiasson, B. (2012) Capacitive immunosensor for the detection of host cell proteins. *J. Biotechnol.*, **157**, 207–213.

52. Labib, M., Hedstroem, M., Amin, M., and Mattiasson, B. (2009) A multipurpose capacitive biosensor for assay and quality control of human Immunoglobulin G. *Biotechnol. Bioeng.*, **104**, 312–320.
53. Heinzle, E., Furukawa, K., and Dunn, I.J. (1983) Quadrupole mass spectrometry for automatic liquid and gas phase analysis in continuous fermentation. *Int. J. Mass Spectrom. Ion Processes*, **48**, 273–276.
54. Axelsson, J.P., Mandenius, C.F., Holst, O., Hagander, P., and Mattiasson, B. (1988) Experiences in using an ethanol sensor to control molasses feed rates in baker's yeast production. *Bioprocess Eng.*, **3**, 1–9.
55. Bracewell, D.G., Brown, R.A., and Hoare, M. (2004) Addressing a whole bioprocess in real-time using an optical biosensor-formation, recovery and purification of antibody fragments from a recombinant *E. coli* host. *Bioprocess Biosyst. Eng.*, **26**, 271–282.
56. Bracewell, D.G., Gill, A., and Hoare, M. (2004) An in-line flow injection optical biosensor for real-time bioprocess monitoring. *Trans. IChemE*, **80**, 71–77.
57. Olsson, L., Volc, J., and Mandenius, C.F. (1990) Determination of monosaccharides in cellulosic hydrolysates using immobilized pyranose oxidase in a continuous amperometric analyser. *Anal. Chem.*, **62**, 2688–2691.
58. Kumar, M.A., Mazlomi, M.A., Hedstrom, M., and Mattiasson, B. (2012) Versatile automated continuous flow system (VersAFlo) for bioanalysis and bioprocess control. *Sens. Actuators, B*, **161**, 855–861.
59. Schubert, T., Breuer, U., Harms, H., and Maskow, T. (2007) Calorimetric bioprocess monitoring by small modifications to a standard bench-scale bioreactor. *J. Biotechnol.*, **130**, 24–31.
60. Biener, R., Steinkamper, A., and Hofmann, J. (2010) Calorimetric control for high cell density cultivation of a recombinant *Escherichia coli* strain. *J. Biotechnol.*, **146**, 45–53.
61. Biener, R., Steinkamper, A., and Horn, T. (2012) Calorimetric control of the specific growth rate during fed-batch cultures of *Saccharomyces cerevisiae*. *J. Biotechnol.*, **160**, 195–201.
62. Schuler, M.M., Sivaprakasam, S., Freeland, B., Hama, A., Hughes, K.M., and Marison, I.W. (2012) Investigation of the potential of biocalorimetry as a process analytical technology (PAT) tool for monitoring and control of Crabtree-negative yeast cultures. *Appl. Microbiol. Biotechnol.*, **93**, 575–584.
63. Ramanathan, K., Rank, M., Svitel, J., Dzgoev, J.A., and Danielsson, B. (1999) The development and applications of thermal biosensors for bioprocess monitoring. *Trends Biotechnol.*, **17**, 499–505.
64. Rank, M. and Danielsson, B. (1992) Implementation of a thermal biosensor in a process environment: on-line monitoring of penicillin V in production-scale fermentation. *Biosens. Bioelectron.*, **7**, 631–635.
65. Clemente, J.J., Monteiro, S.M.S., Carrondo, M.J.T., and Cunha, A.E. (2008) Predicting sporulation events in a bioreactor using an electronic nose. *Biotechnol. Bioeng.*, **101**, 545–552.
66. Mandenius, C.F., Eklöf, T., and Lundström, I. (1997) Sensor fusion with on-line gas emission multisensor arrays and standard process measuring devices in baker's yeast manufacturing process. *Biotechnol. Bioeng.*, **55**, 427–438.
67. Bachinger, T., Mårtensson, P., and Mandenius, C.F. (1998) Estimation of biomass and specific growth rate in a recombinant *Escherichia coli* batch cultivation process using a chemical multisensor array. *J. Biotechnol.*, **60**, 55–66.
68. Liden, H., Mandenius, C.F., Gorton, L., Lundström, I., and Winquist, F. (2000) On-line monitoring of a cultivation using an electronic nose. *Anal. Chim. Acta*, **361**, 223–231.
69. Buczkowska, A., Witkowska, E., Gorski, L., Zamojska, A., Szewczyk , Krzysztof, W., Wroblewski, W., and Ciosek, P.

(2010) The monitoring of methane fermentation in sequencing batch bioreactor with flow-through array of miniaturized solid state electrodes. *Talanta*, **81**, 1387–1392.

70. Favier, J.P., Bicanic, D., Helander, P., and van Iersel, M. (1997) The optothermal approach to a real time monitoring of glucose content during fermentation by brewers' yeast. *J. Biochem. Biophys. Methods*, **34**, 205–211.

71. Hansen, S.K., Jamali, B., and Hubbuch, J. (2013) Selective high throughput protein quantification based on UV absorption spectra. *Biotechnol. Bioeng.*, **110**, 448–460.

72. Petersen, C.D., Beck, H.C., and Lauritsen, F.R. (2004) On-line monitoring of important organoleptic methyl-branched aldehydes during batch fermentation of starter culture *Staphylococcus xylosus* reveal new insight into their production in a model fermentation. *Biotechnol. Bioeng.*, **85**, 298–305.

73. Plum, A. and Rehorek, A. (2005) Strategies for continuous on-line high performance liquid chromatography coupled with diode array detection and electrospray tandem mass spectrometry for process monitoring of sulphonated azo dye and their intermediates in anaerobic-aerobic bioreactors. *J. Chromatogr. A*, **1084**, 119–133.

74. Gustavsson, R. and Mandenius, C.F. (2013) Soft sensor control of metabolic fluxes in a recombinant *Escherichia coli* fed-batch cultivation producing green fluorescence protein. *Bioprocess Biosyst. Eng.*, **36**, 1375–1384.

75. Warth, B., Rajkai, G., and Mandenius, C.-F. (2010) Evaluation of software sensors for on-line estimation of culture conditions in an *Escherichia coli* cultivation expressing a recombinant protein. *J. Biotechnol.*, **147**, 37–45.

76. Tohmola, N., Ahtinen, J., Pitkaenen, J.P., Parviainen, V., Joenvaeaerae, S., Hautamaeki, M., Lindroos, P., Maekinen, J., and Renkonen, R. (2011) On-line high performance liquid chromatography measurements of extracellular metabolites in an aerobic batch yeast (*Saccharomyces cerevisiae*) culture. *Biotechnol. Bioprocess Eng.*, **16**, 264–272.

77. Luchner, M., Gutmann, R., Bayer, K., Dunkl, J., Hansel, A., Herbig, J., Singer, W., Strobl, F., Winkler, K., and Striedner, G. (2012) Implementation of proton transfer reaction-mass spectrometry (PTR-MS) for advanced bioprocess monitoring. *Biotechnol. Bioeng.*, **109**, 3059–3069.

78. Dorfner, R., Ferge, T., Yeretzian, C., Kettrup, A., and Zimmermann, R. (2004) Laser mass spectrometry as on-line sensor for industrial process analysis: process control of coffee roasting. *Anal. Chem.*, **76**, 1386–1402.

79. Skibsted, E., Lindemann, C., Roca, C., and Olsson, L. (2001) On-line bioprocess monitoring with a multiwavelength fluorescence sensor using multivariate calibration. *J. Biotechnol.*, **88**, 47–57.

80. Hagedorn, A., Legge, R.L., and Budman, H. (2003) Evaluation of spectrofluorometry as a tool for estimation in fed-batch fermentations. *Biotechnol. Bioeng.*, **83**, 104–111.

81. Ödman, P., Johansen, C.L., Olsson, L., Gernaey, K.V., and Lantz, A.E. (2009) On-line estimation of biomass, glucose and ethanol in *Saccharomyces cerevisiae* cultivations using in-situ multi-wavelength fluorescence and software sensors. *J. Biotechnol.*, **144**, 102–112.

82. Elshereef, R., Budman, H., Moresoli, C., and Legge, R.L. (2013) Fluorescence-based soft-sensor for monitoring $\beta$-lactoglobulin and $\alpha$-lactalbumin solubility during thermal aggregation. *Biotechnol. Bioeng.*, **110**, 448–460.

83. Scarf, M., Arnold, A., Harvey, L., and McNeil, B. (2006) Near-infrared spectroscopy for bioprocess monitoring and control, current status and future trends. *Crit. Rev. Biotechnol.*, **26**, 17–39.

84. Cervera, A.E., Petersen, N., Eliasson Lantz, A., Larsen, A., and Gernaey, K.V. (2009) Application of near-infrared spectroscopy for monitoring and control of cell culture and fermentation. *Biotechnol. Progr.*, **25**, 1561–1581.

85. Rhiel, M., Ducommun, P., Bolzonella, I., Marison, I., and von Stockar, U. (2002) Real-time in situ monitoring of freely suspended and immobilized cell cultures based on mid-infrared spectroscopic measurement. *Biotechnol. Bioeng.*, **77**, 174–185.
86. Roychoudhury, P., Harvey, L.M., and McNeil, B. (2006) The potential of mid infrared spectroscopy (MIRS) for real time bioprocess monitoring. *Anal. Chim. Acta*, **571**, 159–166.
87. Landgrebe, D., Haake, C., Hoepfner, T., Beutel, S., Hitzmann, B., Scheper, T., Rhiel, M., and Reardon, K.F. (2010) On-line infrared spectroscopy for bioprocess monitoring. *Appl. Microbiol. Biotechnol.*, **88**, 11–22.
88. Oh, S.K., Yoo, S.J., Jeong, D.H., and Lee, J.M. (2013) Real-time estimation of glucose concentration in algae cultivation system using Raman spectroscopy. *Bioresour. Technol.*, **142**, 131–137.
89. Ohlson, S., Jungar, C., Strandh, M., and Mandenius, C.F. (2000) Continuous weak-affinity immunosensing. *Trends Biotechnol.*, **18**, 49–52.
90. Jungar, C., Strandh, M., Ohlson, S., and Mandenius, C.F. (2000) Analysis of carbohydrates using liquid chromatography-surface plasmon resonance immunosensing systems. *Anal. Biochem.*, **281**, 151–158.
91. Mandenius, C.F., Wang, R., Aldén, A., Thébault, S., Lutsch, C., and Ohlson, S. (2008) Monitoring of influenza virus hemagglutinin in process samples using weak affinity ligands and surface plasmon resonance. *Anal. Chim. Acta*, **623**, 66–75.
92. Tahkoniemi, H., Helmja, K., Menert, A., and Kaljurand, M. (2006) Fermentation reactor coupled with capillary electrophoresis for on-line bioprocess monitoring. *J. Pharm. Biomed. Anal.*, **41**, 1585–1591.
93. Höpfner, T., Bluma, A., Rudolph, G., Lindner, P., and Scheper, T. (2010) A review of non-invasive optical-based image analysis systems for continuous bioprocess monitoring. *Bioprocess Biosyst. Eng.*, **33**, 247–256.
94. Bluma, A., Höpfner, T., Lindner, P., Rehbock, C., Beutel, S., Riechers, D., Hitzmann, B., and Scheper, T. (2010) In-situ imaging sensors for bioprocess monitoring: state of the art. *Anal. Bioanal. Chem.*, **398**, 2429–2438.
95. Zhao, R., Natarajan, A., and Srienc, F. (1999) A flow injection flow cytometry system for on-line monitoring of bioreactors. *Biotechnol. Bioeng.*, **62**, 609–617.
96. Abu-Absi, N.R., Zamamiri, A., Kacmar, J., Balogh, S.J., and Srienc, F. (2003) Automated flow cytometry for acquisition of time-dependent population data. *Cytometry*, **51**, 87–96.
97. Broger, T., Odermatt, R.P., Huber, P., and Sonnleitner, B. (2011) Real-time on-line flow cytometry for bioprocess monitoring. *J. Biotechnol.*, **154**, 240–247.
98. Brognaux, A., Han, S., Sørensen, S.J., Lebeau, F., Thonart, P., and Delvigne, F. (2013) A low-cost, multiplexable, automated flow cytometry procedure for the characterization of microbial stress dynamics in bioreactors. *Microb. Cell Fact.*, **12**, 100.
99. Krishnamoorthy, G., Carlen, E.T., Bomer, J.G., Wijnperle, D., de Boer, H.L., van den Berg, A., and Schasfoort, R.B.M. (2010) Electrokinetic label-free screening chip: a marriage of multiplexing and high throughput analysis using surface plasmon resonance imaging. *Lab Chip*, **10**, 986–990.
100. Goetz, J., Gross, D., and Koehler, P. (2003) On-line observation of dough fermentation by magnetic resonance imaging and volumetric measurements. *Eur. Food Res. Technol.*, **217**, 504–511.
101. Clementschitsch, F. and Bayer, K. (2006) Improvements of bioprocess monitoring: development of novel concepts. *Microb. Cell Fact.*, **5**, 19.
102. Navrátil, M., Norberg, A., Lembrén, L., and Mandenius, C.F. (2005) Online multi-analyzer monitoring of biomass, glucose and acetate for growth rate control of a *Vibrio cholerae* fed-batch cultivation. *J. Biotechnol.*, **115**, 67–79.
103. Cimander, C. and Mandenius, C.F. (2002) Online monitoring of a bioprocess based on a multi-analyser system and multivariate statistical

process modelling. *J. Chem. Technol. Biotechnol.*, **77**, 1157–1168.

104. Dietzsch, C., Spadiut, O., and Herwig, C. (2013) On-line multiple component analysis for efficient quantitative bioprocess development. *J. Biotechnol.*, **163**, 362–370.

105. Sagmeister, P., Kment, M., Wechselberger, P., Meitz, A., Langemann, T., and Herwig, C. (2013) Soft-sensor assisted dynamic investigation of mixed feed bioprocesses. *Process Biochem.*, **48**, 1839–1847.

106. Wechselberger, P., Sagmeister, P., and Herwig, C. (2013) Real-time estimation of biomass and specific growth rate in physiologically variable recombinant fed-batch processes. *Bioprocess Biosyst. Eng.*, **36**, 1205–1218.

107. Bachinger, T., Riese, U., Eriksson, R., and Mandenius, C.F. (2000) Monitoring cellular state transitions in a production-scale CHO-cell process using an electronic nose. *J. Biotechnol.*, **76**, 61–71.

108. Dai, X., Wang, W., Ding, Y., and Sun, Z. (2006) "Assumed inherent sensor" inversion based ANN dynamic soft-sensing method and its application in erythromycin fermentation process. *Comput. Chem. Eng.*, **30**, 1203–1225.

109. Veloso, A.C.A., Rocha, I., and Ferreira, E.C. (2009) Monitoring of fed-batch *E. coli* fermentations with software sensors. *Bioprocess Biosyst. Eng.*, **32**, 381–388.

110. Paulsson, D., Gustavsson, R., and Mandenius, C.F. (2014) A soft sensor for bioprocess control based on sequential filtering of metabolic heat signals. *Sensors (Basel)*, **14**, 17864–17882.

111. Teixeira, A.P., Oliveira, R., Alves, P.M., and Carrondo, M.J.T. (2009) Advances in on-line monitoring and control of mammalian cell cultures. Supporting the PAT concept. *Biotechnol. Adv.*, **27**, 726–732.

112. Teixeira, A.P., Alves, C., Alves, P.M., Carrondo, M.J.T., and Oliveira, R. (2007) Hybrid elementary flux analysis nonparametric modeling: application for bioprocess control. *BMC Bioinf.*, **8**, 30.

113. Sharma, S. and Tambe, S.S. (2014) Soft-sensor development for biochemical systems using genetic programming. *Biochem. Eng. J.*, **85**, 89–100.

114. Stanke, M. and Hitzmann, B. (2013) Automatic control of bioprocesses. *Adv. Biochem. Eng. Biotechnol.*, **132**, 35–64.

115. Arauzo-Bravo, M.J., Cano-Izquierdo, J.M., Gomez-Sanchez, E., Lopez-Nieto, M.J., Dimitriadis, Y.A., and Lopez-Coronado, J. (2004) Automatization of a penicillin production process with soft sensors and an adaptive controller based on neuro fuzzy systems. *Control Eng. Pract.*, **12**, 1073–1090.

116. Sundström, H. and Enfors, S.O. (2008) Software sensors for fermentation processes. *Bioprocess Biosyst. Eng.*, **31**, 145–152.

117. Ündey, C., Ertunc, S., Mistretta, T., and Looze, B. (2010) Applied advanced process analytics in biopharmaceutical manufacturing: challenges and prospects in real-time monitoring and control. *J. Process Control*, **20**, 1009–1018.

118. Dochain, D. (2003) State and parameter estimation in chemical and biochemical processes: a tutorial. *J. Process Control*, **13**, 801–818.

119. Salgado, A.M., Folly, R.O.M., Valdman, B., and Valero, F. (2004) Model based soft-sensor for on-line determination of substrate. *Appl. Biochem. Biotechnol.*, 113–116, 137–144.

120. Bosca, S. and Fissore, D. (2011) Design and validation of an innovative soft-sensor for pharmaceuticals freeze-drying monitoring. *Chem. Eng. Sci.*, **66**, 5127–5136.

121. Le, H., Kabburb, S., Pollastrini, L., Sun, Z., Mills, K., Johnson, K., Karypis, G., and Hu, W.S. (2012) Multivariate analysis of cell culture bioprocess data: lactate consumption as process indicator. *J. Biotechnol.*, **162**, 210–223.

122. Chhatre, S., Bou-Habib, G., Smith, M.P., Hoare, M., Bracewell, D.G., and Titchener-Hooker, N.J. (2009) Use of PAT principles for the open-loop control of laboratory and pilot-scale chromatography columns. *J. Chem. Technol. Biotechnol.*, **84**, 1314–1322.

# 15 Design-of-Experiments for Development and Optimization of Bioreactor Media

*Carl-Fredrik Mandenius*

## 15.1 Introduction

Design-of-experiments (DoE) has become a widely accepted methodology for investigating the statistical relationships between input and output variables of technical systems [1–5]. Process and production systems where physical, chemical, or biological transformations take place can benefit substantially by using DoE as a design and optimization tool [6, 7]. In particular, bioprocesses with the demanding task of design of complex bioreactor culture media have been able to take great advantage of this [8–10]. In the planning of bioreactor experiments for evaluating and optimizing the composition of a production media the DoE methodology is particularly useful for avoiding experimental biases and significantly reducing the required number of experiments.

The procedures for applying the DoE methodology to media design are straightforward (Figure 15.1). They investigate defined input *factors*, such as media component mixtures and concentrations, added to a transforming culture from which output factors or *responses* are obtained, such as the concentration and formation rates of the target product molecule(s) or cells or other entities of importance. However, the strength of DoE is that it also reveals how interactions between the input factors (i.e., the media components or possibly also other process parameters) influence the output responses. These interactions are often difficult to discover and interpret with other experimental methods.

However, the conversion in a bioreactor and of its culture medium is more complex than a balance between input and output factors. As Figure 15.1 illustrates, the majority of the inputs exert their effects as an initial load of multiplicity effects onto the transforming bioprocess. However, everyone with experience in bioprocessing is well aware of the fact that actions and effects are displaced over time in a cultivation procedure. Some factors are settled from the beginning of the process, while others appear later, for example, by feeding media components at certain time instances or periods (Figure 15.1). Onset of shifts in metabolism caused by activating or deactivating components, such as growth factors, vitamins, and

*Bioreactors: Design, Operation and Novel Applications*, First Edition. Edited by Carl-Fredrik Mandenius.

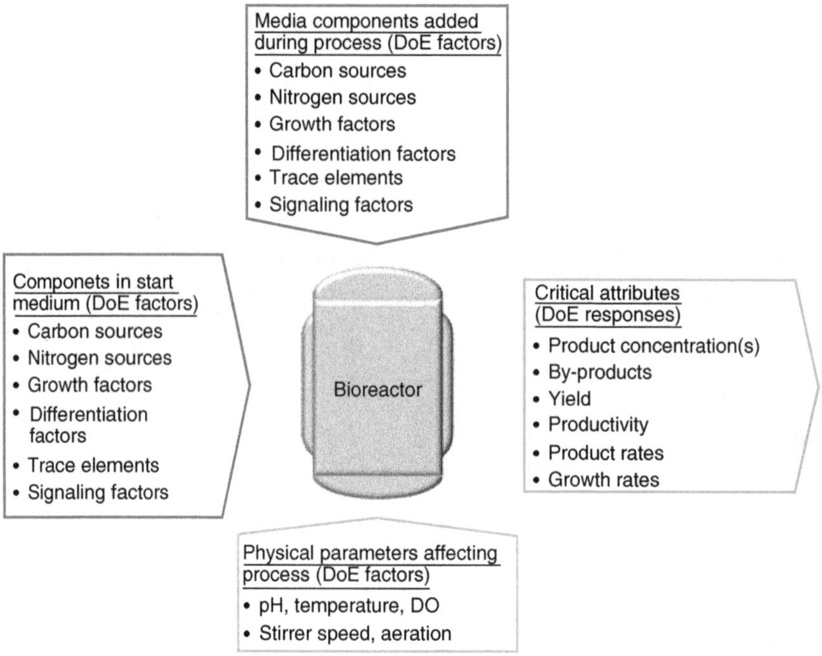


**Figure 15.1** Factors useful in DoE for optimizing media versus critical output parameters (termed responses in the DoE methodology). Note that not all factors are added in the initial culture media, some are added during processing. Also, note that other critical factors for the outcome of the process and that cross-interact with the media components can favorably be included in the DoE procedure.

microbial stressors, further complicate the course of events in the process. In recombinant bioprocesses, induction of expression of a recombinant protein is another time-dependent parameter of importance. In addition, the media factors are combined with operational process parameters such as temperature, agitation, and aeration. Thus, the interactions between the media parameters are therefore very relevant to investigate; the DoE methodology permits us to do that.

In this chapter, the state-of-the-art of media design and optimization by DoE are reviewed, and a few illuminating cases are discussed in more detail in order to show the possibilities for continuing use and further refinements of the methodology.

## 15.2 Fundamentals of Design-of-Experiments Methodology

The strength of the DoE methodology compared with optimization methods, which vary one variable at a time, is that a reliable result can be achieved with only a few experiments. In DoE, the most favorable direction to move in the

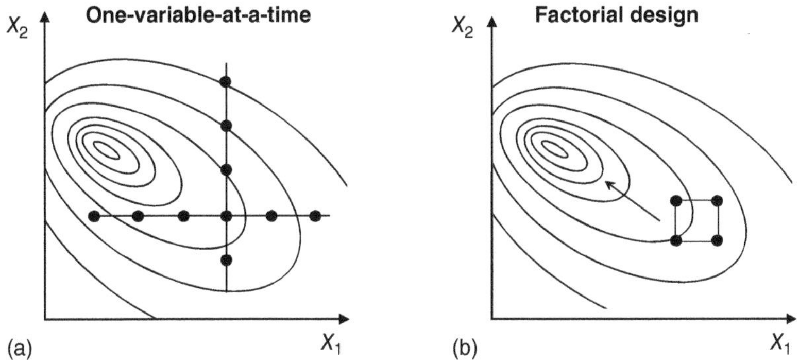


**Figure 15.2** (a) The figure shows how a quasi-optimum is achieved by varying one variable at a time. When keeping variable $X_1$ constant, five experiments varying variable $X_2$ are performed. Then, starting from the optimum (the center point), variable $X_1$ is varied in another five experiments. A correct optimum is never reached as there is a dependency between variable $X_1$ and $X_2$. (b) By simultaneous variations of variable $X_1$ and $X_2$, and analyzing the result in experimental design software, the direction of the true optimum can be found.

factorial design space to find a true optimum can be evaluated. The graphs in Figure 15.2 illustrate how easy it is to be trapped in a quasi-optimum by varying only one variable at a time even in the case of two variables when keeping variable $X_1$ constant; five experiments are performed while varying variable $X_2$. From the obtained optimum, variable $X_1$ is then varied in another four experiments. A correct optimum is never reached because there is a dependency between the two variables. Instead, by simultaneously varying both $X_1$ and $X_2$ and analyzing the result with DoE procedures, the direction of the true optimum can be revealed. When optimizing culture media or performing other bioprocess optimization studies, this approach is clearly preferable since with a single-variable approach it is likely that the experimenter ends up at quasi-optimal media concentrations.

### 15.2.1 Screening of Factors

The most efficient approach to start the DoE of a new bioreactor medium is with *screening* experiments of those media components considered important for the capacity of the medium. These experiments should proceed from selected corners in the experimental space. The procedure is commonly referred to a *fractional factorial design* (FFD) because only a fraction of the possible values of the corners in the design space are investigated. An example of a screening with three variables is shown in Figure 15.3a. The three selected factor variables are screened at two levels of values and each variable is tested twice at low and high levels. In media design optimization, the three variables could be three key nutrient components at two concentration levels each.

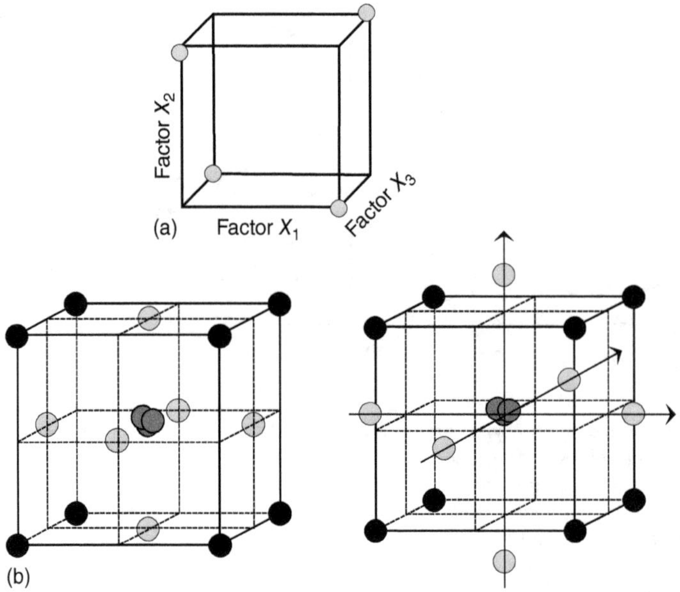


**Figure 15.3** (a) Experimental design performed as screening of important variables applied in a three-variable case. In addition, replicate experiments at the center point are recommended (not shown in the figure). (b) Two types of central composite designs: the central composite face-centered design (CCF, left) and the central composite circumscribed design (CCC, right) in a three-variable case including triplicate experiments at the center point.

This first step of the DoE plan requires careful consideration based on previous experience. It might be that cultivation protocols have already established well-defined factor variables, such as concentration of components in the media and settings of state variables in the fermenter. But the selection of variables and their levels is entirely up to the experimenter's judgment and preknowledge about the studied process. Inappropriate choices will limit the usefulness of the results and making it necessary to carry out new experiments with other variables and new levels.

The number of experiments in a two-level FFD can be described mathematically as $2^{n--k}$, where $n$ is the number of factors to be investigated at the low and high levels and $k$ is the number of steps to reduce the experimental design. If, for example, five variables are involved in the experiment, it will end up with 16 ($2^{5-1}$) or 8 ($2^{5-2}$) experiments, respectively, depending on by how many steps the design is reduced. The first of these designs is preferred since linear terms are not confounded with two-factor interactions. However, in current practice such a procedure is seldom done systematically – instead a level of appropriate reduction is chosen. The screening is improved by replicates in the *center point* of the experimental domain, with the purpose of collecting additional response values while also determining the experimental error for the response. The reduction of factor experiments decreases the statistical quality of the screening. This will be apparent in the statistical performance parameters *goodness of prediction* and *goodness of fit*.

The complete set of experiments should be performed in a random order, to avoid systematic error. A typical cause of systematic errors is when three experiments at the center point of a design are performed one after another using the same experimental procedure. Therefore, randomization should be applied in order to blind the experimenter to expected good or bad results.

The typical result of a screening experiment for culture media is that a few important media components are identified. These components should then be further investigated in a new experimental design with the purpose of determining the optimal factor values. This requires a more elaborate experimental plan, such as a *central composite face* (CCF)-centered design or *central composite circumscribed* (CCC) design. Figure 15.3b illustrates this for a three-component case of the CCF or CCC designs. In the CCF case, additional values (e.g., concentrations) of the variables are included in the surface center points between the corners of the experimental space. In the CCC, these are displaced outside the space at the same distance from the center point as the distance from the center point to the corners. Theoretically, the CCC design is somewhat better than the CCF design because the CCC design covers a larger volume in the design space.

The results of the CCF or CCC experiments are depicted in a *contour plot* or a *response surface* where the optimum is clearly visualized (Figure 15.4). This final step has coined the name also commonly for the procedure, the *response surface methodology* (RSM). Figure 15.5 summarizes the main steps of the whole procedure.

It should be emphasized that both quantitative and qualitative variables can be included among the factors, as can quantitative and qualitative multilevel factors. For example, in a case with two qualitative factors, $X_1$ (with four discrete levels) and $X_3$ (with three discrete levels), and one quantitative factor, $X_2$, a design can be generated as shown in Figure 15.6. However, in this case, it is not possible to use center points between the discrete levels of factor $X_1$ and $X_3$. Instead, a center point can be positioned at one of the discrete levels of the qualitative variables or on an intermediate level of the quantitative variable. A typical example from media design is when comparing two different media component alternatives.

Several commercial software are available for the planning and evaluation of the experimental designs. Frequently used software packages are Modde™ (Umetrics AB, Umeå, Sweden; www.umetrics.com), MiniTab™ (Minitab Inc., State College, PA, USA), and Design-Expert™ (www.statease.com). Most of the examples mentioned below use one of these packages, although other software programs can also be used, or special algorithms can be built in, for example, MATLAB or Excel.

## 15.2.2
## Evaluation of the Experimental Design

Once the experiments have been carried out according to the experimental design plan and the results of response variables have been compiled, *multiple linear*

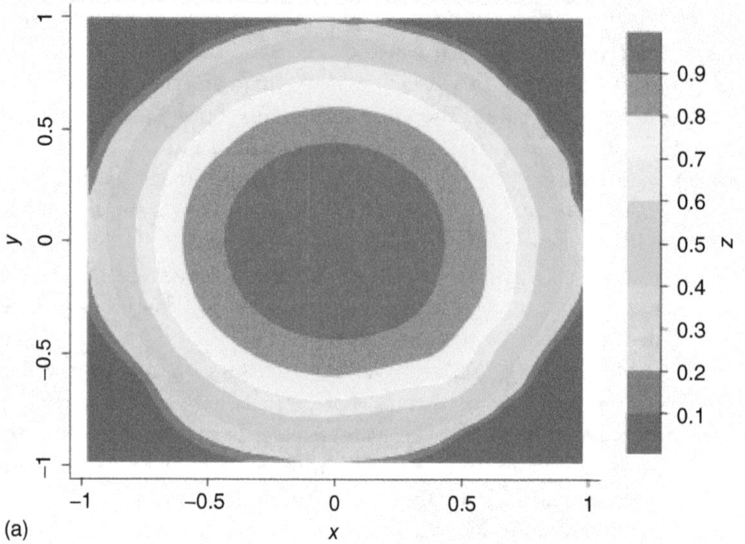

1
0.5
0
−0.5
−1
y
−1
−0.5
0
0.5
1
x
0.9
0.8
0.7
0.6
0.5
0.4
0.3
0.2
0.1
z

(a)

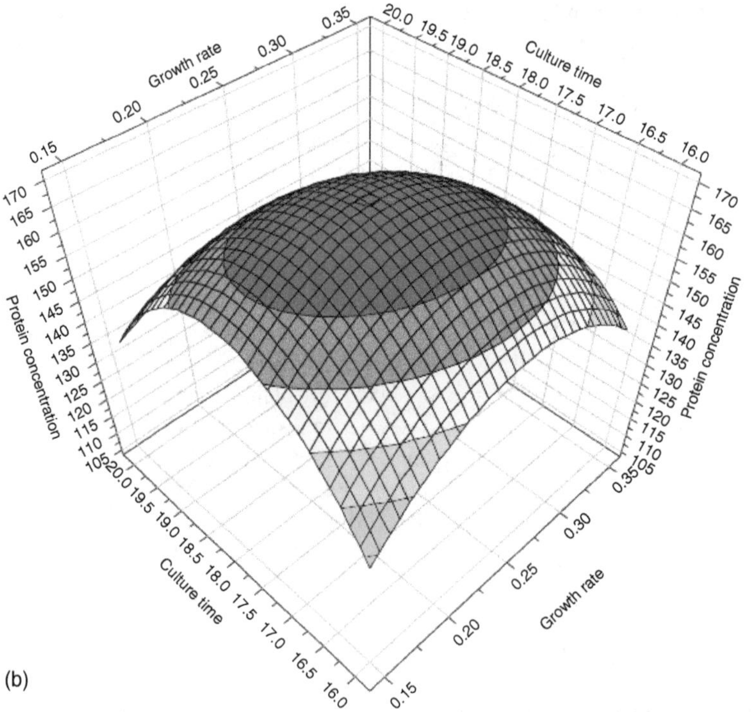

Growth rate
0.15
0.20
0.25
0.30
0.35
Culture time
20.0
19.5
19.0
18.5
18.0
17.5
17.0
16.5
16.0
Protein concentration
170
165
160
155
150
145
140
135
130
125
120
115
110
105

(b)

**Figure 15.4** The sequential working procedure of DoE and RSM when optimizing growth and production media in bioreactors: In the figure, also media components and factors added during processing are included. DoE procedures for addition of components are, however, seldom reported in literature while combinations of media factors with other physical factors controlled in the bioreactor are common.

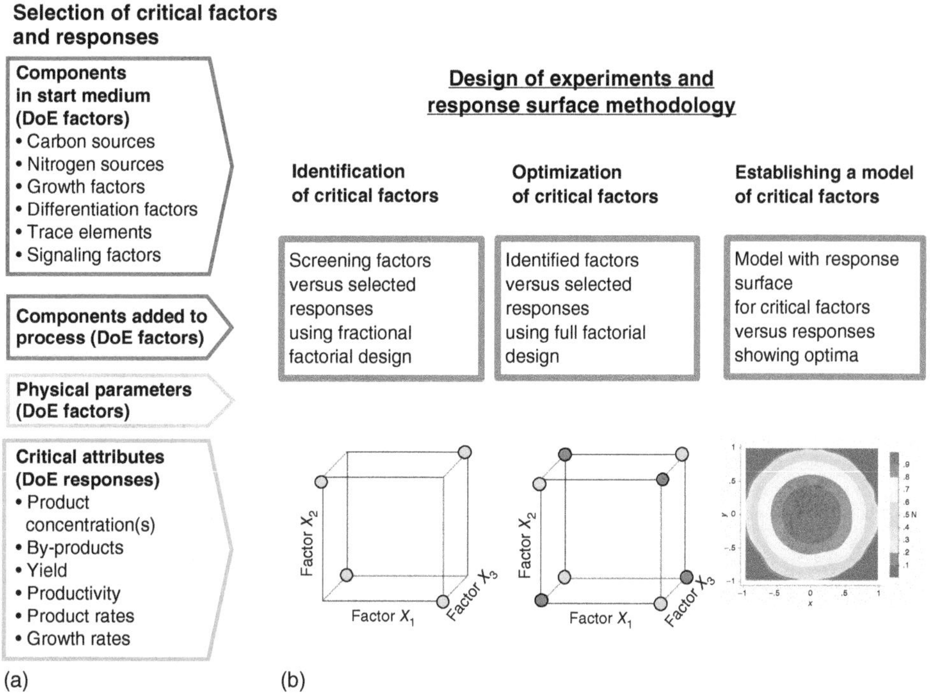


**Figure 15.5** (a) Typical representations of optimization in a DoE study. A contour plot showing selected factors on *x* and *y* axes and one response represented as topographic bars. (b) A response surface plot showing two factors and a response in a 3D graph representation.

*regression* (MLR) is used. The purpose of the MLR is to evaluate whether the experimental space has been orthogonally designed and whether it remains the same after the experiments have been performed.

In MLR, the mathematical model describes a relationship between one or more independent variables and a response variable as $Y_i = \beta_0 + \beta_1 X_{i1} + \beta_2 X_{i2} + \cdots + \beta_p X_{ip} + \varepsilon_i$ where $i = 1,2,\ldots,n$. More complex interaction due to covariation are described by linear or quadratic terms. However, the interaction terms or other complex terms must be assessed to see whether they give acceptable contributions to the model in comparison to the uncertainty of the contribution (signal-to-noise ratio).

In the evaluation of the contribution of each coefficient and the subsequent optimization of the model, the aim is to reduce the model deviation factor $\varepsilon_i$. When

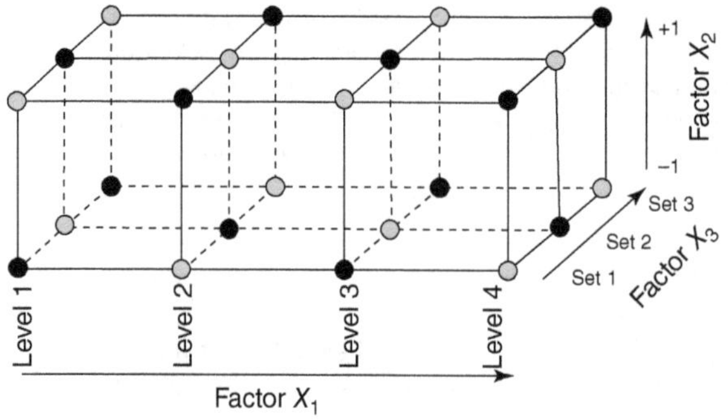


**Figure 15.6** A three-variable design including two qualitative variables at three and four discrete levels, respectively, as well as a third quantitative variable.

characterizing a response surface, the covariation terms describe its *skewness*, while quadratic terms describe its *curvature*.

In a complete design, due to the number of degrees of freedom, it is likely that the terms describing interaction and curvature can be included in the MLR model. When evaluating an FFD by MLR it is common practice to support the linear terms with a limited number of complex terms. In an FFD, such terms are often confounded with each other.

If the design is not orthogonal, *partial least squares regression* (PLS) is a better alternative. This evaluation method can also be used when there are several correlated responses in the data set. PLS may then provide a more robust evaluation method that is useful even if there are a limited number of missing data in the response matrix. The model evaluations in the media design applications discussed or cited in the following are mostly the examples of these analytical procedures.

For a case with five factors, we need either 16 ($2^{5-1}$) or 8 ($2^{5-2}$) factors in the experiments. The latter design is mainly used in robustness testing, where the sensitivity of the variation of the response variable is tested in a reduced set of experiments around a response variable optimum. The variation of the variables is usually made in very small steps around a specific field of interest, such as a preferred composition of a culture medium, in order to evaluate the process stability.

Commonly, the model is validated by two diagnostic residuals. The first of these, the $R^2$ value, is the fraction of the variation of the responses that can be explained by the model. This describes the *goodness of fit*, or how well current runs can be reproduced in a mathematical model. The other residual, the $Q^2$ value, is the fraction of the variation of the response predicted by the model according to cross-validation. This describes the *goodness of prediction*, or how well new experiments can be predicted using the mathematical model. The $R^2$ and $Q^2$ testing are sometimes neglected, which unfortunately reduces the possibility to fully appreciate the results. Typical values indicating good quality of robustness are $R^2 > 0.75$ and $Q^2 > 0.75$. Values less than 0.25 would normally be considered useless.

### 15.2.3
### Specific Design-of-Experiments Methods

In the following and in the cited studies, a number of common DoE design methods are referred to, such as the Plackett–Burman, Taguchi, Box–Behnken, and Box–Wilson methods (Table 15.1). These four designs are explained briefly in the following; however, other or similar design approaches are also applied in common practice (e.g., [4]).

Plackett–Burman method: This method refers to the original work from 1946 by Plackett and Burman, where they presented experimental designs for investigating

**Table 15.1** DoE methods overview.

| DoE method | Description | Utility | References |
|---|---|---|---|
| Plackett–Burman design | Two-level fractional designs for studying $k = N - 1$ variables in $N$ runs, where $N$ is a multiple of 4. Higher numbers of $N$ are nongeometric design. The levels are often presented as (+) and (−), and values systematically varied in an experimental plan. P–B designs are typically carried out as Resolution III or IV designs ($2_{III}{}^{k-p}$, $2_{IV}{}^{k-p}$). | The resolution III and IV P-B designs are commonly applied when screening a high number of media components (see details in Tables 15.2–15.5). In particular, when the media contains >20 components the lowering of the resolution saves much work. | [11] |
| Taguchi design | An alternative similar to PB design based on tables with level (1) and level (2). Simple Taguchi designs are very similar to PB designs but could also be extended to more levels. | The Taguchi designs, being commonly applied in industrial quality control, are preferred in media design. Due to its adaption to many quality parameters it is also very useful in culture media design with many components. | [12] |
| Box–Behnken design | Three-level designs for use with response surfaces. The BB designs are formed by combining $2^k$ factorials with block designs. | The BB designs allow reduction of number of experiments. As media design could benefit for testing at more than two concentration levels this is very useful. | [13] |
| Box–Wilson design | A design equivalent with a common CCD as described earlier with center points that are augmented with a group of "star points" that allows the estimation of the curvature. | | [14] |

the dependence of measured quantity on a number of independent variables or factors [11]. In this work, they describe the two-level factorial designs where the number of experiments is a multiple of 4. They discuss in particular the basic principles of DoE where $L$ levels are tested in a limited number of experiments in order to minimize the variance of the estimates of these dependencies. However, interactions between the factors were considered negligible. The solution to this problem is to find an experimental design where *each combination* of levels for any pair of factors appears the *same number of times*, throughout all the experimental runs. A complete factorial design would satisfy this criterion, but the idea is to find smaller designs.

Taguchi method: The method is similar to the Plackett–Burman method in its basic version and proposes several approaches to experimental designs that utilize two-, three-, and mixed-level FFDs [12]. The Taguchi method is in the literature occasionally described as a quality control method because it was the original purpose of the method. In some cases, the method can be used online while the process is running. Taguchi has published several guidelines for applying the design methods [15].

Box–Wilson method: This is in principle equivalent with a common CCD as described earlier. The Box–Wilson method investigates the impact of the experimental variables on the response output that use CCD to use a response surface design, which are commonly chosen for the purpose of response optimization [14]. Box and Wilson suggested the use of a second-degree polynomial model [14].

Box–Behnken method: This refers to experimental designs for RSM described by Box and Behnken [13]. In the method, each factor is placed at one of three equally spaced values, usually coded as −1, 0, +1. The design should also be sufficient to fit a quadratic model. The ratio of the number of experimental points to the number of coefficients in the quadratic model should be reasonable. The estimation variance should more or less depend only on the distance from the center, and should not vary too much inside the smallest cube containing the experimental points. The Box–Behnken designs are considered efficient in terms of number of required experiments.

The carrying out of the DoE is facilitated by convenient experimental equipment. For example, the multi-bioreactor setups (Figure 15.7) as well as microbioreactors, as further described in Chapter 2 by Lattermann and Buchs (this book).

The fundamentals and guidelines described in this section are all applicable to the design of culture media. The media components shown in Figure 15.1 are in the DoE; the input factors in the procedures described earlier and the critical process targets, or quality attributes, are the DoE responses. The DoE principles could then be applied to any media design task, broad or narrow, and it can easily be expanded to other key process parameters in the bioreactor. The examples accounted for in the following illuminate the wide flexibility of the DoE methodology.

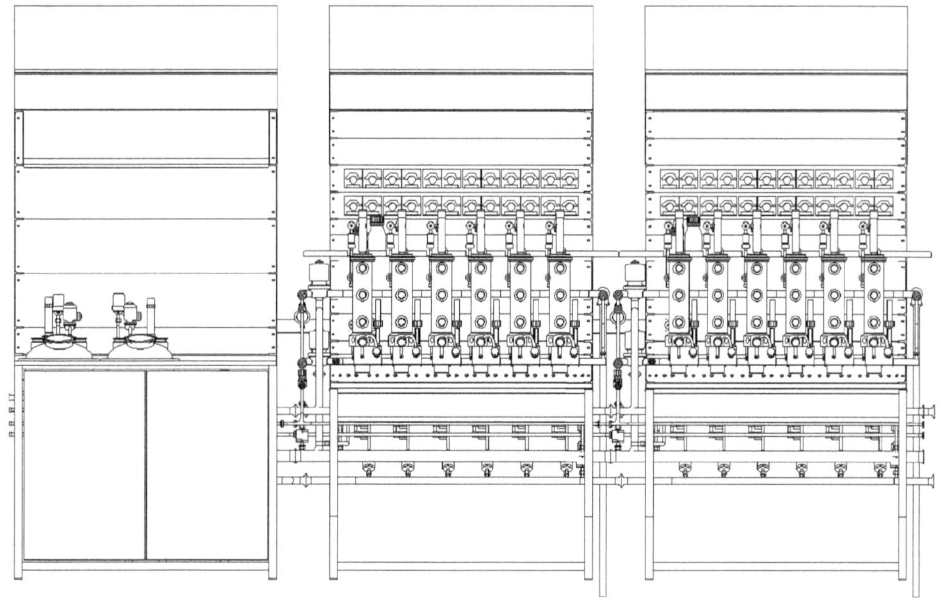

**Figure 15.7** A multi-bioreactor setup (Greta System, Belach Bioteknik AB, Stockholm, Sweden) suited for factor studies.

## 15.3 Optimization of Culture Media by Design-of-Experiments

The following section provides an overview of DoE applications for the design and optimization of media for microbial and cellular growth and production. The section is subdivided into four parts devoted to media for (i) primary and secondary metabolite production, (ii) production of monoclonal antibody and other proteins in mammalian cultures, (iii) production of cells where the cell *per se* is the product, and (iv) applications to other conditions or systems where the media composition is critical for the process. Quite often the DoE applications extend beyond just the design and optimization of the media, for example, when other factors such as physical process variables (e.g., agitation and temperature) are included.

The typical objective in the majority of the examples is to identify a better selection and quantitative composition of media components in order to maximize production and process economy. This is mostly done either by using the product titer, final concentration, yield, or productivity as the DoE response in the design. Still, an even better response metric could be the production media cost per target product volume, or other combinations of cost parameters in the production process, such as effect on separation and purification cost, labor cost versus amount of target product, and so on. However, so far, reports appearing in the current literature have in the majority of cases their focus on target product yield and productivity.

### 15.3.1
### Media for Production of Metabolites and Proteins in Microbial Cultures

The production of metabolites and proteins in a microbial bioreactor culture are the most typical applications of DoE. The objectives for these processes coincide with the overall objective described earlier.

Due to wide and long-term use, as well as due to tight production cost margins, many of the DoE examples reported in the literature involve the fermentation part in the manufacture of antibiotics.

A typical example is the production of the antibiotic clavulanic acid from *Streptomyces clavuligerus*. In a DoE study from 2005 by Wang *et al.*, the composition medium was optimized by first screening a variety of media ingredients in a Plackett–Burman two-level FFD ($2^n$), which was subsequently followed by determining the optimal levels of the screened out media components by RSM [16]. In the initial screening by FFD experiments, soy meal powder, $FeSO_4{\cdot}7H_2O$, and ornithine were identified as the most influential factors, while potassium phosphate and magnesium phosphate were not, in terms of the yield of clavulanic acid (Table 15.2). In the subsequently used RSM model based on a CCC design, the optimal concentrations of these three factors were determined for highest yield of clavulanic acid (the optimal concentration of soy meal powder was determined as 38.10 g l$^{-1}$, $FeSO_4{\cdot}7H_2O$ as 0.395 g l$^{-1}$, and ornithine as 1.18 g l$^{-1}$; see also Table 15.3). The correlation factor was 0.98, and the coefficient of variation

**Table 15.2** Fractional factorial design matrix for screening of medium components for clavulanic acid production by *S. clavuligerus*.

| Run | Glycerol | Soy meal | Potassium phosphate | Magnesium sulfate | Iron sulfate | Ornithine | Clavulanic acid |
|---|---|---|---|---|---|---|---|
| 1 | −1 | −1 | −1 | −1 | −1 | −1 | 328 |
| 2 | 1 | −1 | −1 | −1 | 1 | −1 | 217 |
| 3 | −1 | 1 | −1 | −1 | 1 | 1 | 316 |
| 4 | 1 | 1 | −1 | −1 | −1 | 1 | 379 |
| 5 | −1 | −1 | 1 | −1 | 1 | 1 | 247 |
| 6 | 1 | −1 | 1 | −1 | −1 | 1 | 256 |
| 7 | −1 | 1 | 1 | −1 | −1 | −1 | 391 |
| 8 | 1 | 1 | 1 | −1 | 1 | −1 | 455 |
| 9 | −1 | −1 | −1 | 1 | −1 | 1 | 370 |
| 10 | 1 | −1 | −1 | 1 | 1 | 1 | 250 |
| 11 | −1 | 1 | −1 | 1 | 1 | −1 | 380 |
| 12 | 1 | 1 | −1 | 1 | −1 | −1 | 439 |
| 13 | −1 | −1 | 1 | 1 | 1 | −1 | 330 |
| 14 | 1 | −1 | 1 | 1 | −1 | −1 | 526 |
| 15 | −1 | 1 | 1 | 1 | −1 | 1 | 380 |
| 16 | 1 | 1 | 1 | 1 | 1 | 1 | 369 |

−1 is low concentration and +1 is high concentration of the factors.
Reproduced from [16] with permission.

**Table 15.3** Experimental design and responses for clavulanic acid production by *S. clavuligerus*.

| Run | Coded values and actual values | | | Actual yield | Predicted yield |
|---|---|---|---|---|---|
| | Soy meal | Iron sulfate | Ornithine | | |
| 1 | −1 (42) | −1 (0.2) | −1 (1) | 500 | 485 |
| 2 | −1 (42) | −1 (0.2) | 1 (1.6) | 539 | 523 |
| 3 | −1(42) | 1(0.4) | −1(1) | 630 | 631 |
| 4 | −1(42) | 1(0.4) | 1(1.6) | 590 | 566 |
| 5 | 1(62) | −1(0.2) | −1(1) | 123 | 150 |
| 6 | 1(62) | −1(0.2) | 1(1.6) | 335 | 337 |
| 7 | 1(62) | 1(0.4) | −1(1) | 98 | 117 |
| 8 | 1(62) | 1(0.4) | 1(1.6) | 183 | 201 |
| 9 | −1.682(35.18) | 0(0.3) | 0(1.3) | 580 | 613 |
| 0 | 1.682(68.818) | 0(0.3) | 0(1.3) | 61 | 24 |
| 11 | 0(52) | −1.682(0.131) | 0(1.3) | 436 | 439 |
| 12 | 0(52) | 1.682(0.468) | 0(1.3) | 454 | 447 |
| 13 | 0(52) | 0(0.3) | −1.682(0.131) | 365 | 347 |
| 14 | 0(52) | 0(0.3) | 1.682(1.805) | 435 | 449 |
| 15 | 0(52) | 0(0.3) | 0(1.3) | 545 | 555 |
| 16 | 0(52) | 0(0.3) | 0(1.3) | 526 | 555 |
| 17 | 0(52) | 0(0.3) | 0(1.3) | 558 | 555 |
| 18 | 0(52) | 0(0.3) | 0(1.3) | 576 | 555 |
| 19 | 0(52) | 0(0.3) | 0(1.3) | 547 | 555 |
| 20 | 0(52) | 0(0.3) | 0(1.3) | 580 | 555 |

Reproduced from Wang et al. [16] with permission

was 6.6%. When running the antibiotic fermentation with these settings in a 72 h batch the product yield increased by 50%. Thus, by simply using the standard DoE protocol as described earlier, a profound improvement of the productivity of the production was accomplished. The R&D cost for reaching this result can be estimated as very low in comparison to its effect. Of course, a new production media composition will require further verification, scale-up, and additional validation followed by regulatory efforts due to the requirements for pharmaceutical drug products.

Another example involving the production of antibiotic molecules is the fermentation process for the polyaromatic peptide bacteriocide nisin *A* produced by a native *Lactococcus lactis* strain. By following a typical DoE approach with RSM, González-Toledo *et al.* [17] optimized the production of nisin in a bioreactor culture using supplemented sweet whey. In their first screen, an FFD of the order $2^{5-1}$ with three center points was used. The effect on nisin production by air flow, sweet whey, soybean peptone, $MgSO_4/MnSO_4$ mixture, and Tween 80 was investigated in this screening. From these FFD data, it was possible to sort out soybean peptone and sweet whey as the most significant factors. In the second round, a three-factor central composite design (CCD) with two central points was applied with these two (sweet whey at $7-10\,g\,l^{-1}$ and soybean peptone at $7-10\,g\,l^{-1}$) as well as with small quantities of added nisin as self-inducer (34.4–74.4 nisin activity unit $l^{-1}$). In the subsequent RSM with a second-order model, the optimal produced nisin activity was determined to be 180 nisin unit $ml^{-1}$ at 74.4 unit $ml^{-1}$ of added nisin inducer, $13.8\,g\,l^{-1}$ of soybean peptone, and 14.9 or $5.11\,g\,l^{-1}$ of sweet whey. The optimized conditions were tested in a 12 h batch fermentation and at pH 6.5 a nisin activity of 575 unit $ml^{-1}$ was reached. After the fermentation nisin was processed downstream reaching a final activity of more than 100 000 nisin units $g^{-1}$.

Other methodologically similar examples of applying DoE to optimization of media composition for antibiotics are neomycin production by solid-state fermentation with *Streptomyces marinensis* [18] and meilingmycin production by *Streptomyces nanchangensis* [19] and another example with nisin production by *L. lactis* [20] where fractional design methodology in combination with response surface was applied in a similar manner as described earlier (for more details of the design conditions used, see Table 15.4).

Another category of products from microbial cultures are primary metabolites. For instance, DoE was used to facilitate the media design in the production of mixed alcohols by a *Clostridium* strain immobilized on silica support [21]. A medium based on glycerol was used as carbon source and supplemented by additional media components, which were known to have great impact on product yield. After initial screening using one-factor-at-a-time, six media components ($KH_2PO_4-K_2HPO_4$, yeast extract, $MgSO_4{\cdot}7H_2O$, $CaCl_2{\cdot}2H_2O$, $FeSO_4{\cdot}7H_2O$, $(NH_4)_2SO_4$) were selected for optimization experiment at two different concentration levels using the orthogonal Taguchi design. The optimal composition of the medium was determined to be $2\,g\,l^{-1}$ of $KH_2PO_4-K_2HPO_4$, $5\,g\,l^{-1}$ of yeast extract, $0.1\,g\,l^{-1}$ of $MgSO_4{\cdot}7H_2O$, and $0.1\,g\,l^{-1}$ of $(NH_4)_2SO_4$. Subsequent fermentation with the optimized medium was carried out in a stirred-tank bioreactor. The accumulated alcohol production in the bioreactor after 10 days was $17.7\,g\,l^{-1}$, which was even higher than the value predicted by the model. Hence, this modified Taguchi plan helped in determining the appropriate medium composition for obtaining an optimal mixed alcohol yield.

Another example of a successful optimization of a medium is the production of the biosurfactant rhamnolipid by *Pseudomonas aeruginosa* [22]. The DoE was applied on an air isolate of the strain by screening a variety of responses including hemolytic activity, emulsification activity, drop collapsing test, and oil displacement test, as well as lipase activity. It was found that the isolate was able to reduce the surface tension of the medium and could form stable emulsions with tested vegetable oils. When the authors used an unoptimized medium containing sucrose as the carbon source, the culture produced $0.3\,mg\,ml^{-1}$ of rhamnolipid at 37 °C and pH 7 after three days in a shake-flask batch. When the media composition was optimized the yield of rhamnolipid increased 15-fold (Figure 15.8). Another report on rhamnolipid production showed similar results [23].

Experiences of media optimization reported other primary and secondary microbial metabolite processes including mannitol fermentation by *Lactobacillus*

**Table 15.4** Media for production of metabolites and biopolymers in microbial cultures.

| Application/purpose | Culture system | DoE method | Outcome | References |
|---|---|---|---|---|
| Metabolites | | | | |
| Optimization of production medium for antibiotics | *S. clavuligerus* producing clavulanic acid (shake flasks) | Fractional factorial design: two-level ($2^{6-2}$) and three-level ($3^3$) | Factors: six media components<br>Response: clavulanate<br>Result: yield 50% increase | [16] |
| Optimizing antibiotic production versus process conditions | *L. lactis* producing nisin A (lab bioreactor) | Fractional factorial design: two-level ($2^{5-1}$) | Factors: soya bean, Tween, peptone, minerals, air flow<br>Response: nisin<br>Result: optimized | [17] |
| Optimization of a production medium for antibiotics | *S. marinensis* producing neomycin (solid-state shake flasks) | Factorial design: two-level ($2^3$) | Factors: three media components<br>Response: neomycin yield<br>Result: optimum determined | [18] |
| Optimization of a production medium for antibiotics | *L. lactis* producing nisin (shake flasks) | Fractional factorial design: two-level ($2^{4-1}$) | Factors: four media components<br>Response: nisin yield, specific productivity<br>Result: increased yield | [20] |
| Optimization of production medium for antibiotics | *S. nanchangensis* producing meilingmycin (shake flasks) | Fractional factorial design: ($2^{9-5}$) + full design ($2^2$) | Factors: nine factors and two factors<br>Response: meilingmycin yield<br>Result: 4.5-fold increase | [19] |
| Optimizing media for extracellular metabolite production | *Clostridium* culture producing mixed alcohols (bioreactor) | Taguchi orthogonal design | Medium composition optimized for mixed alcohol production with six components | [21] |
| Optimizing culture media for producing biosurfactant | *Pseudomonas* isolate producing rhamnolipid (plates) | Fractional factorial design: two-level ($2^{11-2}$) + RSM | Factors: 11 factor components<br>Response: yield<br>Result: 15-fold increase | [22] |
| Optimizing culture media for producing biosurfactant | *Pseudomonas aeruginosa* producing rhamnolipid (lab bioreactor) | Full factorial design: ($2^4$) + RSM | Factors: four media and culture parameters<br>Response: titer<br>Result: fivefold increase | [23] |
| Microbial proteins | | | | |
| Optimization of media for production of exo-polymer | *Grifola frondosa* producing exo-polymer (lab bioreactor) | Fractional factorial design: three-level ($3^3$) + RSM | Factors: three components<br>Responses: exo-polymer, biomass<br>Result: increase in yield and biomass | [24] |

**Table 15.4** *Continued.*

| Application/purpose | Culture system | DoE method | Outcome | References |
|---|---|---|---|---|
| Optimization of media and feeding profile | *E. coli* producing rec. alkaline phosphate (lab bioreactor) | CCD | Induction time of media addition | [25] |
| Optimization of culture parameters and procedure | *P. pastoris* producing rec. laccase (shake flask) | Plackett–Burman design | Factors: 11 culture parameters<br>Response: laccase activity<br>Result: optimized | [26] |
| Optimizing media in microtiter plates | *Bacillus* culture producing MAb (microtiter plates/100 l scale) | Screening, optimization and genetic algorithm | Factors: media factors and induction time<br>Response: titer<br>Result: improvements | [27] |
| Optimization of production media | *Coprinus* sp. producing peroxidase | Fractional factorial design: two-level ($2^7$) | Factors: six medium factors, pH<br>Response: enzyme activity yield<br>Result: twofold increase | [28] |
| Optimization of production media bacterial culture | *Bacillus* sp. producing of polyglutamic acid (shake flasks) | Fractional factorial design: two-level ($2^{15-k}$) | Factors: 13 medium factors, pH; impeller speed<br>Response: PGA yield<br>Result: threefold increase | [29] |
| Optimization of production media | *Serratia marcescens* producing chitinolytic enzymes (lab bioreactor) | Full factorial design: two-level ($2^3$ ) | Factors: chitin; temp; pH<br>Response: enzyme activity<br>Result: ranking of chitins | [30] |
| Optimization of production media for bacterial culture | *S. lavendulae* producing penicillin acylase (shake flasks) | Full factorial three-level design ($3^3$) | Factors: yeast extract, olive oil, and so on<br>Response: enzyme yield<br>Result: optimization | [31] |
| Optimization of production media for fungal culture | *Aspergillus niger* producing rec. lysozyme (shake flasks) | Fractional factorial design: two-level ($2^{5-1}$) | Factors: starch; peptone; medium comp<br>Response: lysozyme yield<br>Result: optima determined | [32] |
| Optimization of production media yeast culture | *P. pastoris* producing rec. human cystatin C (lab bioreactor) | Factorial design: two-level ($2^3$) | Factors: three medium factors<br>Response: glycosylation<br>Result: 14% increase in glycosylation | [33] |

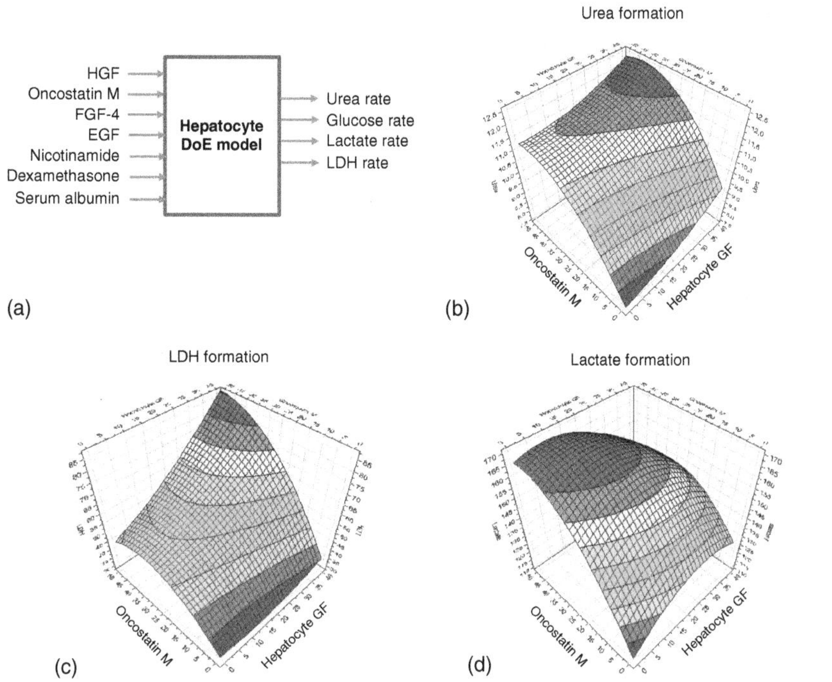


**Figure 15.8** Optimization for the production of the hepatoma cell line C3A. (a) Media factors (HGH, hepatocyte growth factor, oncostatin M; FGF-4, fibroblast growth factor; EGF, epidermal growth factor, nicotinamide, dexamethasone, human serum albumin) and responses (urea rate, glucose rate, lactate rate, lactate dehydrogenase rate) in the hepatocyte DoE model. Response surface graphs for optimization of formation of (b) urea, (c) lactate dehydrogenase, and (d) lactate and at varying oncostatin M and hepatocyte growth factor a constant FGF4 concentration of 20 μg $ml^{-1}$ (with permission from [34]).

*intermedius* [35], carotenoid production by *Pediococcus pentosaceus* and *Lactobacillus acidophilus* in semisolid maize-based cultures [36], and $\gamma$-polyglutamic acid fermentation by *Bacillus* sp. [29, 37] (Table 15.4).

Production of proteins in microbial cultures quite often concerns recombinant proteins. An example is the study of the effect of the media composition on the expression of recombinant lysozyme from hen's egg white by *Aspergillus niger* [32]. The influence by the ingredient factors, such as starch, peptone, ammonium sulfate, yeast extract, and $CaCl_2\ 2H_2O$, were screened in a $2^{5-1}$ FFD with five center points. The lysozyme response revealed that peptone, starch, and ammonium sulfate were the most influential factors, while the other factors had little effect at the levels tested. In this case, optimization was accomplished by a steepest ascent procedure followed by response surface modeling using a CCD. The medium composition of 34 $g\,l^{-1}$ for starch and peptone, 11.9 $g\,l^{-1}$ for ammonium sulfate, 0.5 $g\,l^{-1}$ for yeast extract, and 0.5 $g\,l^{-1}$ for $CaCl_2{\cdot}2H_2O$ was determined to be optimal for the lysozyme production. Using this medium in a

7-day fermentation resulted in a lysozyme yield of 209 $g\,l^{-1}$, which is close to the theoretically estimated yield (212 $mg\,l^{-1}$).

Li *et al.* [26] reported on the optimization of recombinant laccase produced in a *Pichia pastoris* culture. The enzyme was used for decolorization experiments with dyes (Reactive Blue KN-R and Acid Red 35), showing excellent decolorization capacity. Initially, the microorganism expressed an activity of 2357 units $l^{-1}$ of the enzyme using the substrate standard of laccase. To achieve a higher laccase activity, multiple key factors were systematically screened using a Plackett–Burman design for expected critical fermentation conditions including both media components as well as other fermentation parameters. The large number of factors used in the initial screen, which also included physical parameters such as pH and air flow, can be noted here. Subsequently, RSM based on Box–Behnken design was used for further optimization. An optimal enzyme activity of 5235 unit $l^{-1}$ was reached when 0.60% methanol was added into the culture every 24 h, at an initial pH of 7.1. Thus, the RSM resulted in a further doubling of the activity and yield of laccase.

Several other examples of optimization of production media for proteins in microbial cultures along the lines outlined earlier can be mentioned, such as penicillin acylase production by *Streptomyces lavendulae* [31], fungal peroxidase production by *Coprinus* species [28], production of chitinolytic enzymes by *Serratia marcescens* [30], exopolymer production by *Grifola frondosa* [24], and production of cellulose-degrading enzymes from *Aspergillus oryzae* [38] (Table 15.4).

So far, few attempts have been made to optimize parameters related to the efficiency of more specific functions of the cellular machinery. Such an application is the attempt to influence and optimize the glycosylation pattern at the production of a glycoprotein [33]. The degree of glycosylation for three glycoforms of cystatin C glycoprotein were studied in a full factorial design experiment with three different nitrogen media in recombinant *P. pastoris* cultures. A maximization of glycosylation was reached, but at the expense of productivity of the glycoprotein.

### 15.3.2
### Media for the Production of Monoclonal Antibodies and Other Proteins in Mammalian Cell Cultures

Media design for mammalian cell cultures by DoE has become increasingly popular due to the widespread use of these cultures in the biotechnology industry. The typical products are antibodies, enzymes, and other proteins. The culture media are quite often complex with a larger number of ingredients and growth factors than those used in media for microbial cultures. The interactions between the different components of the media then become more difficult to understand and untangle. This makes DoE a particularly useful tool for approaching this media design task.

As for most media optimization applications, it is also here typically the aim of the design experiments to optimizing final product yield or final concentration with the intention to establish the most favorable mix of nutrient factors to maximize the cellular productivity by supplying a well-balanced composition of nutrients that enhances the maximum yield of the product molecule. Thus, the principle outlined in Figure 15.1 is the basis for the DoE.

Production of monoclonal antibodies in hybridoma cell cultures has been the predominant application. A representative example is the study by Sen and Roychoudhury [39] who applied DoE to optimize the medium recipe of a hybridoma cell culture producing a monoclonal antibody (mouse $IgG_1$) specific to a breast cancer oncoprotein. The optimization was carried out by screening a relatively large number of nutrient components to allow selection of a limited number of these for subsequent experiments followed by determining the optimal concentrations for the selected media components. As many as 29 media components were included in the study (essential and nonessential amino acids, glucose, serum, and six salts, namely NaCl, KCl, $CaCl_2$, $NaH_2PO_4$, $MgSO_4$, and Na pyruvate were chosen in screening phase). A Plackett–Burman design was used to screen for those components that influenced the monoclonal antibody production. Seven of these (glucose, serum, asparagine, threonine, serine, NaCl, and $NaH_2PO_4$) enhanced the monoclonal antibody production with high confidence level. In the second step, RSM was used to determine the optimal level of the seven components. The monoclonal antibody production titer and specific productivity were both more than doubled compared with the control medium.

Similar illuminating examples of media studies with monoclonal antibody production are reported by Johnson *et al.* [40] and Rouiller *et al.* [41] (Table 15.5).

The other very common application of mammalian cell cultures is the production of biotherapeutic proteins. Predominately, these proteins are produced in recombinant production systems [44, 45, 48]. Again, the high number of components in the mammalian cell culture media is a major consumer R&D work in cell culture development, and also with DoE. The demanding workload was approached by Rouiller *et al.* [46, 47] by applying a high-throughput method based on media blending. They improved the performance of a Chinese hamster ovary fed-batch medium using 96-deepwell plates. Starting from a proprietary chemically defined medium, 16 formulations containing 43 of 47 components at three different levels were designed and tested. Media blending was performed according to a custom-made mixture of binary blends, which resulted in 376 different blends. These were tested during both the cell expansion and fed-batch production phases in a single experiment. Three approaches were chosen to provide the best output of the large amount of data obtained. A simple ranking of conditions was first used as a quick approach to select new formulations with promising features. Then, prediction of the best mixes was done to maximize both growth and antibody titer using a commercial software package. Finally, a multivariate analysis enabled the identification of individual potential critical components for further optimization. This high-throughput approach can be applied on a fed-batch process as well, thereby allowing the identification of an optimized process in a short time frame.

**Table 15.5** Media for production of monoclonal antibodies and other biopolymers in mammalian cell cultures.

| Application/purpose | Culture system | DoE method | Outcome | References |
|---|---|---|---|---|
| *Monoclonal antibodies* | | | | |
| Optimization of cell culture media composition | CHO cells producing mAb (96-plate and fed-batch bioreactor) | Factorial design: three-levels (quadratic model) | Factors: 43 media components<br>Response: viability and mAb titer<br>Result: optimal medium | [41] |
| Optimization of a serum-free media for cell culture | Hybridoma culture for production of mAb | Fractional factorial design: two-levels (PB) + RSM | Factors: 27 components<br>Response: mAb titer<br>Result: optimal medium composition | [39] |
| Optimization of a serum-free media for cell culture | Hybridoma culture for production of recombinant mAb | Fractional factorial design: five-levels (CCD) + RSM | Factors: five factor groups of 43 components<br>Response: mAb titer<br>Result: optimal composition | [42] |
| Optimization of a media for a fed-batch cell culture | Murine hybridoma for production of mAb | PCA and PLS regression model | Factors: all amino acids<br>Response: mAb titer<br>Result: improvement of medium | [43] |
| *Other proteins* | | | | |
| Optimization of a serum-free medium for HEK cell culture | HEK cells with baculovirus for production of alkaline phosphatase | Full factorial design | Factors: initial cell density, FBS and TSA concentration, MOI<br>Response: recombinant protein titer<br>Result: optimized condition | [44] |
| Optimization of a serum-free medium for CHO cell culture | CHO cells for production of recombinant macrophage colony-stimulating factor | Fractional factorial design: two-level ($2^{7-3}$) + RSM | Factors: insulin, peptone, yeast extract, SerEx, BSA, linoleic acid, dextran<br>Response. M-CSF titer<br>Result: faster development | [45] |
| Optimization of media for CHO cell culture | CHO cells producing MAb (96-plates) | Full factorial design: three-levels + RSM | Factors: 27 conditions<br>Response: six culture parameters<br>Result: optima identified | [46, 47] |
| Optimization of a serum-free medium for CHO cell culture | Recombinant CHO cell for production of target protein | Fractional factorial design: two-level | Factors: 5–7 media components<br>Response: protein titer<br>Result: increased titer | [48] |

The overview in Table 15.5 compares the methodology applied in these recent applications of media optimization. As the table illustrates, most researchers have applied quite similar DoE approaches.

In the very recent years, the number of DoE applications for biotherapeutic culture media optimization has increased. Especially, applications to cell culture media including additional growth and/or differentiation factors are reported frequently and it is not possible to completely describe in detail all applications. Applying DoE in the development of a culture could significantly facilitate and speed up the search for an efficient media protocol. This can of course be done in connection to other culture conditions as discussed earlier.

The typical purpose of these cell culture procedures is to investigate the levels of media components on productivity and product yield. Consequently, the experimental planning is straightforward: set levels of high and low concentrations of media components, decide responses to be measured and making an experimental screening plan as was described earlier.

### 15.3.3
### Media for Differentiation and Production of Cells

The object for the media optimization can also be the production of the cells *per se*. Production of cell materials from cell line collections for cell therapy purposes are examples of products. This includes the growing interest for stem cells and their derived cells. These applications can take considerable advantage of the DoE methodology (Table 15.6).

Thus, optimization of cell culture media using experimental design methodology is an attractive approach for improving efficiency in cultivation. Dong *et al.* [34] applied the methodology to refine the composition of an established culture medium for growth of the human hepatoma C3A cell line. Selection of nutrient components and growth factors was systematically screened according to standard DoE procedures. The results of the screening indicated that hepatocyte growth factor, oncostatin M, and fibroblast growth factor-4 significantly influenced the metabolic activities of the C3A cell line. RSM revealed that the optimal levels for these factors were $30\,\mathrm{ng\,ml^{-1}}$ of hepatocyte growth factor and $35\,\mathrm{ng\,ml^{-1}}$ of oncostatin M. Additional experiments on primary human hepatocyte cultures showed high variance in metabolic activities between cells from different individuals, making determination of optimal levels of factors more difficult. Still, it was possible to conclude that hepatocyte growth factor, epidermal growth factor, and oncostatin M had decisive effects on the metabolic functions of primary human hepatocytes.

Much attention has been devoted to media for differentiation and proliferation of embryonic stem cells. The fact that culture media for stem cell expansion contain undefined substances makes the task demanding. Considering the potential for future clinical work with such cells, use of more well-defined media is highly desirable. Knöespel *et al.* [49], therefore, investigated the detailed composition of a serum-free chemically defined culture medium for efficient expansion

**Table 15.6** Media for differentiation and production of cells.

| Application/purpose | Culture system | DoE method | Outcome | References |
|---|---|---|---|---|
| *Cell production* | | | | |
| Optimizing cell culture serum-free media growth factors | Hepatoma cell line C3A and primary human hepatocytes grown in (plate culture) | Two-level fractional factorial design | Factors: four medium factors<br>Response: C3A growth<br>Result: Two media factors identified as decisive | [34] |
| Optimizing serum-free medium for stem cell proliferation | Mouse embryonic stem cells (plate culture) | Factional factorial design screening (resolution IV) + full factorial design + RSM | Factors: 11 media factors<br>Response: cell growth<br>Results: insulin and leukemia inhibitor factor positive influence. Zn and cysteine low | [49] |
| Optimization of a serum-free medium for viral vaccine | Vero cell line producing vaccine (culture flasks) | Factorial factor design | Factors: SMF factors screened<br>Response: cell growth<br>Result: SFM factors identified | [50] |
| *Stem cell differentiation* | | | | |
| Optimization of hematopoietic cell expansion (mega-karyocytes) | Hematopoietic progenitor cells (lab culture) | Factorial factor design ($2^4$) screening | Factors: four differentiation factors<br>Response: Cell expansion yield<br>Results: identification of factors | [51] |
| Optimization of erythropoiesis from hematopoietic stem cells | Human cord blood $CD34^+$ cells to red blood cells (plate culture) | Fractional factorial factor design (resolution IV) + full design (CCD) + RSM | Factors: seven growth factors<br>Response: red blood cells<br>Results: 26-fold expansion | [52] |
| Optimization of the erythroid phases from hematopoietic stem cells | Human cord blood $CD34^+$ cells to erythroid cells (plate culture) | Taguchi orthogonal design (L8) + RSM (CCF) | Factors: three growth factors<br>Response: yield erythroid phases<br>Results: improved expansion | [53] |
| Optimization of extracellular matrices for osteogenesis | Human mesenchymal stem cells undergoing osteogenesis (plate culture) | Full factorial design ($2^4$) + RSM | Factors: four differentiation factors<br>Response: three biomarkers<br>Results: twofold improvement | [54] |

of embryonic stem cells from mouse (mESC). They started their study with a serum-free standard medium with 11 additional media factors (carnosine, cysteine, C1 components, transferrin, transferrin supplement, leukemia inhibition factor, insulin, BMP4, $CaCl_2$, $ZnSO_4$, and lipids). The growth of the stem cells was strongly influenced by the balance of media components. Screening using a Plackett–Burman design showed that insulin and leukemia inhibitory factor had a significant positive influence on the proliferation activity of the cells, while zinc and l-cysteine reduce cell growth. Further optimization using a "minimum run resolution IV" design showed that the leukemia inhibitory factor was the main factor for the survival and proliferation of cells. Thus, DoE screening assays are applicable to develop and to refine culture media for stem cells and could also be employed to optimize culture media for stem cells (Figures 15.9).

Several DoE studies have been carried out on the differentiation of hematopoietic cells. One example is the optimization of erythropoiesis from human cord blood cells ($CD34^+$) to red blood cells [52]. Seven cytokines (interleukin-3, interleukin-6, stem cell factor, erythropoietin, colony stimulating factor, thrombopoietin, and Ftl3 ligand) known to affect red cell maturation in humans were selected for screening with a DoE protocol. Stem cell factor (SCF) and erythropoietin (EPO) were found to be the most significant factors in red cell production. The methodology helped in defining the differentiation characteristics and interactions of the SCF and EPO factors on total cell expansion and maturation of cord blood cells toward the erythroid lineage and delineated optimal *in vitro* concentrations of each cytokine (75 ng $ml^{-1}$ and 4.5 unit $ml^{-1}$, respectively). The optimized cytokine cocktail increased 26 460-fold the total cell expansion and accelerated erythroid maturation, yielding considerable purity of the differentiated cells (over 90% glycophorin-A-expressing cells). The use of DOE was considered effective and efficient for the analysis, characterization, and optimization of *in vitro* erythropoiesis and provided a systematic platform for the use of other growth factors for *in vitro* expansion of cell products.

Also, Panuganti *et al.* [51] applied an FFD with hematopoietic stem and progenitor cells cultured with a variety of cytokines (e.g., interleukin-3, interleukin-6, interleukin-9, high- or low-dose stem cell factor, in conjunction with thrombopoietin, interleukin-11) to promote differentiation into megakaryocytic cells, the precursors to platelets. The study demonstrated that a three-phase culture process with increasing pH and $pO_2$ and different cytokine cocktails greatly increases megakaryocyte production. Thus, this study clearly shows the benefits of untangling the complexity of differentiation process of stem cells by DoE.

### 15.3.4
### Other Applications to Media Design

A few other applications of DoE related to design of media that do not fall into the above-mentioned categories can be noted. Several refer to the degradation of waste, especially wastewater from factories, where toxic or otherwise harmful

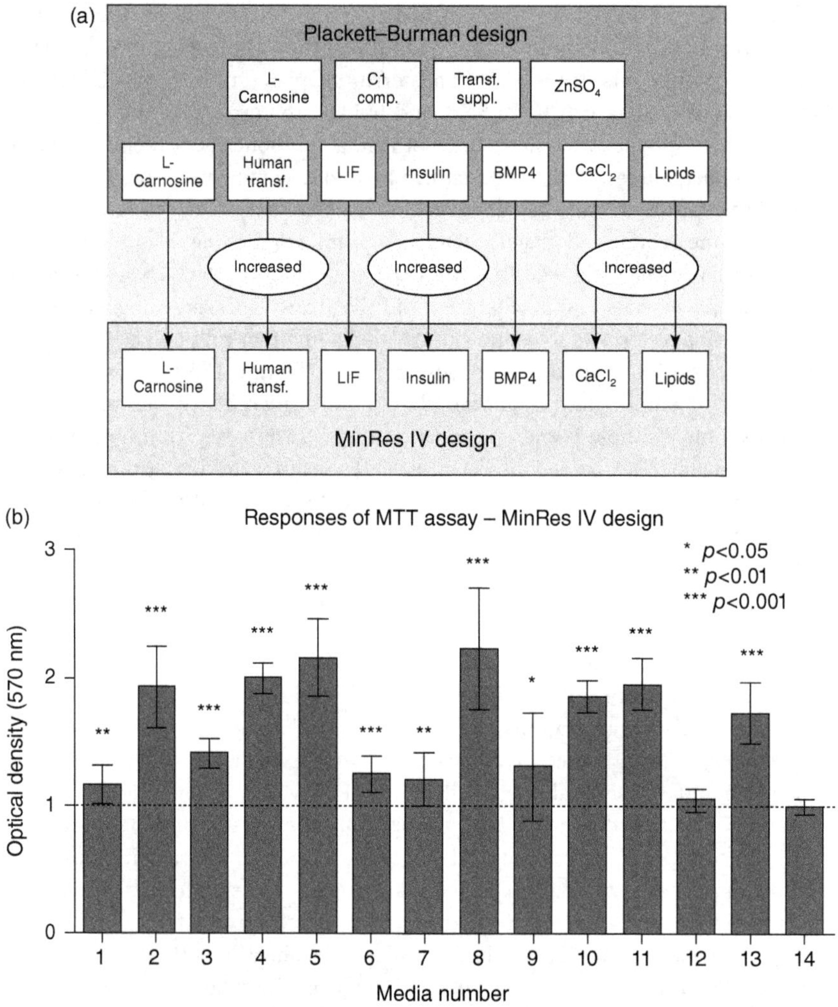


**Figure 15.9** (a) Application of the two-step strategy using the Plackett–Burman design in a first screening followed by factor selection and further adjustment using the MinRes IV design. (b) MTT responses (optical density) of mESC cultured without or with feeder cells (MEF) in 14 different media composed using a MinRes IV design. Values (means ± SD) were normalized to the reference medium no. 14. The unpaired *t*-test was performed to compare each group with the reference medium no. 14 (with permission from [49]).

composition of compounds should be removed or reduced from an effluent (Table 15.7).

For example, Yuan *et al.* [55] optimized the degradation of nicotine in waste streams in an *Ochrobactrum intermedium* culture where a medium based on yeast extract, glucose, and Tween 80 was used. They apply RSM to study the effects of the degradation rate by the media components in shake flask experiments

**Table 15.7** Other media design applications.

| Application/purpose | Culture system | DoE method | Results | References |
|---|---|---|---|---|
| *Degradation processes* | | | | |
| Optimizing degradation of nicotine in a microbial culture | *O. intermedium* assimilating nicotine (culture flasks) | Full factorial design (CCD) | Factors: three media components<br>Response: uptake nicotine<br>Result: media factors ranked | [55] |
| Optimizing treatment waste-water with organic compounds | Active sludge containing cresol and phenol | Full factorial design: two-level (CCF) + RSM | Factors: phenol, cresol, time<br>Response: removal<br>Results: Complete removal | [56] |
| Optimizing degradation of utter diacetyl flavor compounds | *Pediococcus,* and *Lactobacillus* in semisolid maize-based culture | Full factorial design ($2^4$) | Factors: four pretreatment factors<br>Result: diacetyl yield decreased, optima determined | [36] |
| Optimizing decolorization of azo dyes in industrial waste | *Phanerochaete* culture assimilating azo dyes (Direct Red-80, Mordant Blue-9) | Full factorial design: two-level (CCD) + RSM | Factors: conc. of two dyes, time<br>Response: degree of decolorization<br>Result: full decoloured | [57] |
| *Food processing* | | | | |
| Optimizing yoghurt production based of media composition | Yoghurt culture of *B. longum* | Fractional factorial design: two-level ($2^{8-4}$, CCF) | Factors: seven media factors, pH<br>Response: growth rate, glucose rate<br>Result: 50% reduction, 160% increase | [58] |
| *Other conversions* | | | | |
| Optimizing yield of mannitol in an anaerobic fermentation | *L. intermedius* producing mannitol | Fractional factorial design: two-level ($2^{4-1}$ ) | Factors: four media components<br>Response: mannitol yield<br>Result: increased yield | [35] |
| Optimizing uptake of $CO_2$ by a culture | Chemoautotrophic $CO_2$ fixating microorganism | Fractional factorial design three-level ($3^{4-1}$) | Factors. four media components<br>Response: $CO_2$ fixation<br>Result: $CO_2$ fixation improved | [59] |
| Optimizing at conversion of CO to hydrogen | *Petrobacter* consortium assimilating CO | Screening + full factorial design (2CCD) + RSM | Factors: six mixed factors<br>Response: hydrogen production<br>Result: 8% improvement | [60] |

using a full factorial CCD. The results showed that the most significant variable influencing nicotine degradation was yeast extract, followed by glucose, and then Tween 80. Moreover, these three factors interacted and by that enhanced the nicotine degradation. The experimental data also allowed the development of an empirical model describing the relationship between independent and dependent variables. By solving the regression equation, the optimal values of the variables were determined. When validating the optimized medium $1.22\,g\,l^{-1}$ of nicotine was degraded from a nicotine extract at a rate of $0.12\,g\,l^{-1}\,h^{-1}$ using a 30 l bioreactor. Here again, RSM proved to be reliable in developing the model, optimizing factors and analyzing interaction effects, by that providing better understanding on the interactions between yeast extract, glucose, and Tween 80 for nicotine biodegradation.

Processes related to environmental protection with the purpose of removing a harmful substance and transforming it to a product of commercial valuable by-product are other examples where DoE is useful for optimizing the culture's conditions including its media composition. For example, Pakshirajan and Mal [60] studied the conversion of carbon monoxide (CO) to hydrogen by an anaerobic microbial consortium consisting of *Petrobacter* species isolated from sewage treatment. The consortium was initially grown on an acetate-containing medium and later adapted to utilize CO as the sole carbon source for hydrogen production. Statistically designed experiments were then applied to optimize the CO conversion to hydrogen by the consortium. To determine the factors that significantly influenced the hydrogen production, a Plackett–Burman screening was used, which showed that temperature and $Fe^{2+}$ ions predominantly influenced hydrogen production; pH and $Ni^{2+}$ ions also influenced, but to a lesser degree. Concentration of $Fe^{2+}$ and $Ni^{2+}$ ions in the medium was then subsequently optimized by using a CCD followed by RSM, which yielded the optimum value of $213\,mg\,l^{-1}$ for $Fe^{2+}$ and $2.2\,mg\,l^{-1}$ for $Ni^{2+}$. At these optimum conditions, 60.8 mole hydrogen production was achieved, which was 8% higher than that observed from the screening experiment.

One more example of optimization-related environmental protection is the $CO_2$ fixation by a chemoautotrophic microorganism by a factorial design setup [59]. Statistical analysis was performed for predicting the levels of four factors, $H_2$, $O_2$, $CO_2$ concentration, and pH required for obtaining optimum culture conditions. For the FFD carried out on three levels ($3^{4-1}$), 27 experiments were run by combination in a 5 l jar fermenter, and the three-dimensional response surfaces were plotted to find out the optimum level of each factor for the maximum of $CO_2$ fixation. The obtained dry cell weight in optimized culture condition was predicted to be $11.4\,g\,l^{-1}$ after 48 h at $37.5\times10^{-3}\,mol\,min^{-1}$ of $H_2$, $9.4\times10^{-3}\,mol\,min^{-1}$ of $O_2$, and $4.9\times10^{-3}\,mol\,min^{-1}$ of $CO_2$ at pH 6.7. As compared with $CO_2$ fixation rate before optimizing culture conditions, the improvement of $CO_2$ fixation rate under optimized culture conditions was experimentally confirmed.

There are many examples where fermented food processing can also benefit from DoE. In dairy manufacture, the production goal is the quality of the task, which is the result of the fermentation conditions. Modern manufacturing methods rely on production cultures and procedures where optimization is a requirement. For example, bifidobacteria are used as probiotics mainly as cell suspensions or as freeze-dried additives. In order to optimize the culture, a $28^{--4}$ FFD was used in determining the critical parameters influencing bioreactor cultivations of the strain *Bifidobacterium longum* [58]. Glucose, yeast extract, and cysteine concentrations were found to be critical for the cultivation of this strain. Glucose and yeast extract concentrations were further optimized together with temperature in a three-factor CCD (40 °C, 35 g $l^{-1}$ yeast extract and 20 g $l^{-1}$ glucose). Freeze-drying of frozen cell suspensions of *B. longum* was studied as well with DoE resulting in an optimized temperature gradient from −10 to 0 °C, a 10 h temperature gradient from 0 to +10 °C and a 12 h temperature hold at +10 °C. Temperature programming reduced drying times by more than 50% and improved the product activity by more than 160%.

Another food-related example is the production of mannitol from a *Lactobacillus* fermentation process [35]. The effects of four salt nutrients (ammonium citrate, sodium phosphate, magnesium sulfate, and manganese sulfate) on the production of mannitol by *L. intermedius* in a simplified medium containing 300 g fructose, 5 g soy peptone, and 50 g corn steep liquor per liter in pH-controlled fermentation (5.0) at 37 °C were evaluated using an FFD. Only manganese sulfate was found to be essential for mannitol production. Added manganese sulfate concentration of 0.033 g $l^{-1}$ was found to support maximum production. The bacterium produced 201 g mannitol, 62 g lactic acid, and 40 g acetic acid from 300 g fructose per liter in 67 h.

An inventive step toward accelerating media optimization by DoE was taken by Deshpande *et al.* [61]. They used a microtiter plate with online measurement of dissolved oxygen for the optimization of a cultivation of Chinese hamster ovary cells in a culture medium with selected factors. By a dynamic liquid phase mass balance, the oxygen uptake rates were calculated from the dissolved oxygen level and used to indicate cell viability. Using a full factorial design with CCF, the optimum medium composition could be identified and determined for glucose, glutamine, and inorganic salts in one single microtiter plate experiment. The concentration of inorganic salts was found to have the most significant influence on the cultivation. The method seems to have good potential for medium optimization of cell culture media.

Recently, elaborate DoE protocols for 96-multiwell plates have been developed. By applying these rational protocols for experimental design, the optimization can be made more efficient and less time-consuming. The technique can also be applied to larger formats such as 384-well and 1536-well plates [62].

## 15.4 Conclusions and Outlook

The review has provided a focused overview of the recent ascents in applying DoE for bioreactor culture media design. It has mainly discussed such applications where only the media composition is optimized. However, it should be remembered that DoE methodology allows combinations of other factors, where other aspects of the design of the bioreactor can be encompassed, such as operational procedures and bioreactor geometries.

Certainly, many good examples of DoE applications are housed inside the industry and are available only for internal use. Luckily, some companies have shared their experiences of DoE for media development, as well as other bioprocess applications, some accounted for in this review.

General scientific articles are by now many and could be expected to mirror also the state-of-the-art techniques in the industry.

Moreover, supply companies most likely access wide application databases from customers.

It is realistic to assume that information from these sources will be gradually available for the public domain. By that, a more accurate view over the maturity of the DoE methodology will be possible to assess for the benefit of an increasing number of appliers.

Progress in development micro-bioreactors, multi-bioreactor system, bioreactors-on-chip, and sensor technology (cf. other chapters in this book on these topics) may contribute to advance and facilitate the use of DoE in bioreactor media design. An example where this has already been exploited is the robotic AMBR(R) (R) = sign for registered trademark system developed by TAP Biosystems (www.tapbiosystems.com) Such technical designs seem to have particular utility for further progress in this field for speeding up the use and improve precision and necessary accuracy.

## References


1. Box, G.E.P., Hunter, W.G., and Hunter, J.S. (1978) *Statistics for Experimenters, An Introduction to Design, Data Analysis, and Model Building*, 1st edn, Wiley-Interscience, New York.
2. Box, G.E.P., Hunter, W.G., and Hunter, J.S. (2005) *Statistics for Experimenters, Design, Innovation, and Discovery*, 2nd edn, Wiley-Interscience, New York.
3. Brereton, R.G. (2003) *Chemometrics, Data Analysis for the Laboratory and Chemical Plant*, John Wiley & Sons, Ltd., Chichester.
4. Montgomery, D.C. (2005) *Design and Analysis of Experiments*, 6th edn, John Wiley & Sons, Inc., New York.
5. Yates, F. (1937) Design and Analysis of Factorial Experiments. Technical Communication No. 35, Imperial Bureau of Soil Science, London.
6. Carlson, R. and Carlson, J. (2005) *Design and Optimization in Organic Synthesis*, 2nd edn, Elsevier, London.
7. Pilat, P., Votruba, J., Dobersky, P., and Prokop, A. (1976) Application of mathematical optimization methods in microbiology. *Folia Microbiol.*, **21**, 391–405.
8. Connors, N.C. (2003) in *Handbook of Industrial Cell Cultures* (eds V.A. Vinci and S.R. Parekh), Humana Press, Totowa, NJ, pp. 171–193.
9. Mandenius, C.F. and Brundin, A. (2008) Bioprocess optimization using design-of-experiments methodology. *Biotechnol. Progr.*, **24**, 1191–1203.
10. Weuster-Botz, D. (2000) Experimental design of fermentation media development, statistical design or global random search. *J. Biosci. Bioeng.*, **90**, 473–483.

11. Plackett, R.L. and Burman, J.P. (1946) The design of optimum multifactorial experiments. *Biometrika*, **33**, 305–325.
12. Taguchi, G. (1987) *System of Experimental Design: Engineering Methods to Optimize Quality and Minimize Cost*, UNIPUB, White Plains, NY.
13. Box, G.E.P. and Behnken, D.W. (1960) Some new three-level designs for the study of quantitative variables. *J. Qual. Control*, **31**, 16–29.
14. Box, G.E.P. and Wilson, K.G. (1951) On the experimental attainment of optimal conditions. *J. R. Stat. Soc.*, **B13**, 1–45.
15. Taguchi, G. and Konishi, S. (1987) *Orthogonal Arrays and Linear Graphs*, ASI Press, Dearborn, MI.
16. Wang, Y.H., Yang, B., Ren, J., Dong, M.L., Liang, D., and Xu, A.L. (2005) Optimization of medium composition for the production of clavulanic acid by *Streptomyces clavuligerus*. *Process Biochem.*, **40**, 1161–1166.
17. Gonzalez-Toledo, S.Y., Dominguez-Dominguez, J., Garcia-Almendarez, B.E. *et al.* (2010) Optimization of nisin production by *Lactococcus lactis* UQ2 using supplemented whey as alternative culture medium. *J. Food Sci.*, **75**, M347–M353.
18. Adinarayana, K., Ellaiah, P., Srinivasulu, B., Devi, R.B., and Adinarayana, G. (2003) Response surface methodological approach to optimize the nutritional parameters for neomycin production by *Streptomyces marinensis* under solid-state fermentation. *Process Biochem.*, **38**, 1565–1572.
19. Zhuang, Y.P., Chen, B., Chu, J., and Zhang, S. (2006) Medium optimization for meilingmycin production by *Streptomyces nanchangensis* using response surface methodology. *Process Biochem.*, **41**, 405–409.
20. Penna, T.C.V. and Moraes, D.A. (2002) Optimization of nisin production by *Lactococcus lactis*. *Appl. Biochem. Biotechnol.*, **98–100**, 775–789.
21. Khanna, S., Ranjan, A., Goyal, A., and Moholkar, V.S. (2013) Medium optimization for mixed alcohols production by glycerol utilizing immobilized *Clostridium pasteurianum* MTCC 116. *Chem. Biochem. Eng. Q.*, **27**, 319–325.
22. Vanavil, B., Perumalsamy, M., and Rao, A.S. (2013) Biosurfactant production from novel air isolate NITT6L: screening, characterization and optimization of media. *J. Microbiol. Biotechnol.*, **23**, 1229–1243.
23. Al-Araji, L.I.Y., Rahman, R.N.Z.R.A., Basri, M. *et al.* (2007) Optimisation of rhamnolipids produced by *Pseudomonas aeruginosa* 181 using response surface modelling. *Ann. Microbiol.*, **57**, 571–575.
24. Cui, F.J., Li, Y., Xu, Z.H., Xu, H.Y., Sun, K., and Tao, W.Y. (2006) Optimization of the medium composition for production of mycelial biomass and exo-polymer by *Grifola frondosa* GF9801 using response surface methodology. *Bioresour. Technol.*, **97**, 1209–1216.
25. Wechselberger, P., Sagmeister, P., Engelking, H., Schmidt, T., Wenger, J., and Herwig, C. (2012) Efficient feeding profile optimization for recombinant protein production using physiological information. *Bioprocess Biosyst. Eng.*, **35**, 1637–1649.
26. Li, Q., Xie, J., Zhao, L. *et al.* (2013) Optimization of fermentation conditions for laccase production by recombinant *Pichia pastoris* GS115-LCCA using response surface methodology and its application to dye decolorization. *Bioresources*, **8**, 4072–4087.
27. David, F., Steinwand, M., Hust, M. *et al.* (2011) Antibody production in *Bacillus megaterium*: strategies and physiological implications of scaling from microtiter plates to industrial bioreactors. *Biotechnol. J.*, **6**, 1516–1531.
28. Ikehata, K., Pickard, M.A., Buchanan, I.D., and Smith, D.W. (2004) Optimization of extracellular fungal peroxidase production by 2 *Coprinus* species. *Can. J. Microbiol.*, **50**, 1033–1040.
29. Soliman, N.A., Berekaa, M.M., and Abdel-Fattah, Y.R. (2005) Polyglutamic acid (PGA) production by *Bacillus* sp. SAB-26: application of Plackett-Burman experimental design to evaluate culture requirements. *Appl. Microbiol. Biotechnol.*, **69**, 259–267.
30. Green, A.T., Healy, M.G., and Healy, A. (2005) Production of chitinolytic enzymes by *Serratia marcescens*

QMB1466 using various chitinous substrates. *J. Chem. Technol. Biotechnol.*, **80**, 28–34.

31. Torres-Bacete, J., Arroyo, M., Torres-Guzman, R., De La Mata, I., Aceba, lC., and Castillon, M.P. (2005) Optimization of culture medium and conditions for penicillin acylase production by *Streptomyces lavendulae* ATCC 13664. *Appl. Biochem. Biotechnol.*, **126**, 119–131.
32. Gheshlaghi, R., Scharer, J.M., Moo-Young, M., and Douglas, P.L. (2005) Medium optimization for hen egg white lysozyme production by recombinant *Aspergillus niger* using statistical methods. *Biotechnol. Bioeng.*, **90**, 754–760.
33. Pritchett, J. and Baldwin, S.A. (2004) The effect of nitrogen source on yield and glycosylation of a human cystatin C mutant expressed in *Pichia pastoris*. *J. Ind. Microbiol. Biotechnol.*, **31**, 553–558.
34. Dong, J., Mandenius, C.F., Luebberstedt, M., Urbaniak, T., Nüssler, A.K.N., Knobeloch, D., Gerlach, J.C., and Zeilinger, K. (2008) Evaluation and optimization of hepatocyte culture media factors by design of experiments (DoE) methodology. *Cytotechnology*, **57**, 251–261.
35. Saha, B.C. (2006) Effect of salt nutrients on mannitol production by *Lactobacillus intermedius* NRRL B-3693. *J. Ind. Microbiol. Biotechnol.*, **33**, 887–890.
36. Escamilla-Hurtado, M.L., Valdes-Martinez, S.E., Soriano-Santos, J., Gomez-Pliego, R., Verde-Calvo, J.R., Reyes-Dorantes, A., and Tomasini-Campocosio, A. (2005) Effect of culture conditions on production of butter flavor compounds by *Pediococcus pentosaceus* and *Lactobacillus acidophilus* in semisolid maize-based cultures. *Int. J. Food Microbiol.*, **105**, 305–316.
37. Shi, F., Xu, Z., and Cen, P. (2006) Optimization of $\gamma$-polyglutamic acid production by *Bacillus subtilis* ZJU-7 using a surface-response methodology. *Biotechnol. Bioprocess Eng.*, **11**, 251–257.
38. Szendefy, J., Szakacs, G., and Christopher, L. (2006) Potential of solid-state fermentation enzymes of *Aspergillus oryzae* in biobleaching of paper pulp. *Enzyme Microb. Technol.*, **39**, 1354–1360.
39. Sen, S. and Roychoudhury, P.K. (2013) Development of optimal medium for production of commercially important monoclonal antibody 520C9 by hybridoma cell. *Cytotechnology*, **65**, 233–252.
40. Johnson, J.M., Hartshorn, J., McNorton, S., Padilla, J.A., Mohabbat, T., Luo, S., and Etchberger, K. (2005) Process optimization for an NS0-derived hybridoma cell line in a chemically defined protein-free medium, in *Proceedings of the18th Meeting of the European Society of Animal Cell Technology (ESACT)* (eds F. Godia and M. Fussenegger) Springer, The Netherlands, pp 633–636.
41. Rouiller, Y., Perilleux, A., Collet, N., Jordan, M., Stettler, M., and Broly, H. (2013) A high-throughput media design approach for high performance mammalian fed-batch cultures. *MAbs*, **5**, 501–511.
42. Parampalli, A., Eskridge, K., Smith, L., Meagher, M.M., Mowry, M.C., and Subramanian, A. (2007) Development of serum-free media in CHO-DG44 cells using central composite design. *Cytotechnology*, **54**, 57–68.
43. Selvarasu, S., Kim, D.Y., Karimi, I.A., and Lee, D.Y. (2010) Combined data preprocessing and multivariate statistical analysis characterizes fed-batch culture of mouse hybridoma cells for rational medium design. *J. Biotechnol.*, **150**, 94–100.
44. Jardin, B.A., Zhao, Y., Selvaraj, M., Montes, J., Tran, R., Prakash, S., and Elias, C.B. (2008) Expression of SEAP (secreted alkaline phosphatase) by baculovirus mediated transduction of HEK 293 cells in a hollow fiber bioreactor system. *J. Biotechnol.*, **135**, 272–280.
45. Liu, C.H. and Chang, T.Y. (2006) Rational development of serum-free medium for Chinese hamster ovary cells. *Process Biochem.*, **41**, 2314–2319.
46. Rouiller, Y., Solacroup, T., Deparis, V., Barbafieri, M., Gleixner, R., Broly, H., and Eon-Duval, A. (2012) Application of quality by design to the characterization

of the cell culture process of an Fc-Fusion protein. *Eur. J. Pharm. Biopharm.*, **81**, 426–437.

47. Rouiller, Y., Perilleux, A., Marsaut, M., Stettler, M., Vesin, M.N., and Broly, H. (2012) Effect of hydrocortisone on the production and glycosylation of an Fc-fusion protein in CHO cell cultures. *Biotechnol. Progr.*, **28**, 803–813.
48. Lao, M.S. and Schalla, C. (1996) Development of a serum-free medium using computer-assisted factorial design and analysis. *Cytotechnology*, **22**, 25–31.
49. Knoespel, F., Schindler, R.K., Luebberstedt, M., Petzolt, S., Gerlach, J.C., Zeilinger, K. *et al.* (2010) Optimization of a serum-free culture medium for mouse embryonic stem cells using design of experiments (DoE) methodology. *Cytotechnology*, **62**, 557–571.
50. Rourou, S., van der Ark, A., van der Velden, T., and Kallel, H. (2009) Development of an animal-component free medium for Vero cells culture. *Biotechnol. Progr.*, **25**, 1752–1761.
51. Panuganti, S., Schlinker, A.C., Lindholm, P.F., Papoutsakis, E.T., and Miller, M.W. (2013) Three-stage *ex vivo* expansion of high-ploidy megakaryocytic cells: toward large-scale platelet production. *Tissue Eng. Part A*, **19**, 998–1014.
52. Lim, M., Panoskaltsis, N., Ye, H. *et al.* (2011) Optimization of *in vitro* erythropoiesis from CD34(+) cord blood cells using design of experiments (DOE). *Biochem. Eng. J.*, **55**, 154–161.
53. Lim, M., Shunjie, C., Panoskaltsis, N. *et al.* (2012) Systematic experimental design for bioprocess characterization: elucidating transient effects of multi-cytokine contributions on erythroid differentiation. *Biotechnol. Bioprocess Eng.*, **17**, 218–226.
54. Decaris, M.L. and Leach, J.K. (2011) Design of experiments approach to engineer cell-secreted matrices for directing osteogenic differentiation. *Ann. Biomed. Eng.*, **39**, 1174–1185.
55. Yuan, Y.J., Lu, Z.X., Huang, L.J., Bie, X.M., Lü, F.X., and Li, Y. (2006) Optimization of a medium for enhancing nicotine biodegradation by *Ochrobactrum intermedium* DN2. *J. Appl. Microbiol.*, **101**, 691–697.
56. Dey, S. and Mukherjee, S. (2013) Performance study and kinetic modelling of hybrid bioreactor for treatment of bi-substrate mixture of phenol-m-cresol in wastewater: process optimization with response surface methodology. *J. Environ. Sci.*, **25**, 698–709.
57. Pakshirajan, K., Sivasankar, A., and Sahoo, N.K. (2011) Decolourization of synthetic wastewater containing azo dyes by immobilized *Phanerochaete chrysosporium* in a continuously operated RBC reactor. *Appl. Microbiol. Biotechnol.*, **89**, 1223–1232.
58. Kiviharju, K., Leisola, M., and Eerikäinen, T. (2005) Optimisation of a *Bifidobacterium longum* production process. *J. Biotechnol.*, **117**, 299–308.
59. Kwak, K.O., Jung, S.J., Chung, S.Y., Kang, C.M., Huh, Y., and Bae, S.O. (2006) Optimization of culture conditions for $CO_2$ fixation by a chemoautotrophic microorganism, strain YN-1 using factorial design. *Biochem. Eng. J.*, **31**, 1–7.
60. Pakshirajan, K. and Mal, J. (2013) Biohydrogen production using native carbon monoxide converting anaerobic microbial consortium predominantly *Petrobacter* sp. *Int. J. Hydrogen Energy*, **38**, 16020–16028.
61. Deshpande, R.R., Wittmann, C., and Heinzle, E. (2004) Microplates with integrated oxygen sensing for medium optimization in animal cell culture. *Cytotechnology*, **46**, 1–8.
62. Olsson, I.M., Johansson, E., Berntsson, M., Eriksson, L., Gottfries, J., and Wold, S. (2006) Rational DOE protocols for 96-well plates. *Chemom. Intell. Lab. Syst.*, **83**, 66–74.

# 16
# Operator Training Simulators for Bioreactors

*Volker C. Hass*

## 16.1
## Introduction

Starting in the early 1990s, the application of training simulators in the process industry rose due to different reasons, which are summarized by Reinig *et al.* [1]. It is important to note that, due to the integration of processes, plant-(facility-) operation has become more and more complex. The higher complexity entailed a higher degree of automation and more complex operational procedures that created new requirements to the operators' skills (Figure 16.1). At the beginning of the twentieth century up to the early 1940s, the plant operators had to perform all necessary functions of the process on their own and directly on-site. With the beginning of the middle of the century, the processes were controlled with control panels in analog technique. Still some functions had to be carried out by the operators on-site, but many of the functions were realized by the automation system. Modern control and automation today is done using digital control systems. Most decisions concerning process operation are deduced from computer monitor information. Control actions by the operators are realized by inputs to a computer via graphical user interfaces. The operation of industrial processes has become a rather abstract task. In this context, one has to consider that the process efficiency and product yields are – to a great extent – dependent on the decisions the operators make [3].

Also, modern biotechnological production processes using bioreactors require development and realization of complex but robust operational strategies in order to achieve maximum yields and process efficiencies while minimizing the energy and resource needs. Even apparently well-known processes such as the production of baker's yeast do require sophisticated concepts for substrate dosage and other process control strategies, due to the high degree of interaction between the fundamental processes in a bioreactor, such as mixing, heat and mass transfer, growth of organisms, and biochemical reaction kinetics.

It can clearly be seen that bioreactor operators need to be able to safely perform a number of operational procedures such as standard operations for

*Bioreactors: Design, Operation and Novel Applications*, First Edition. Edited by Carl-Fredrik Mandenius.

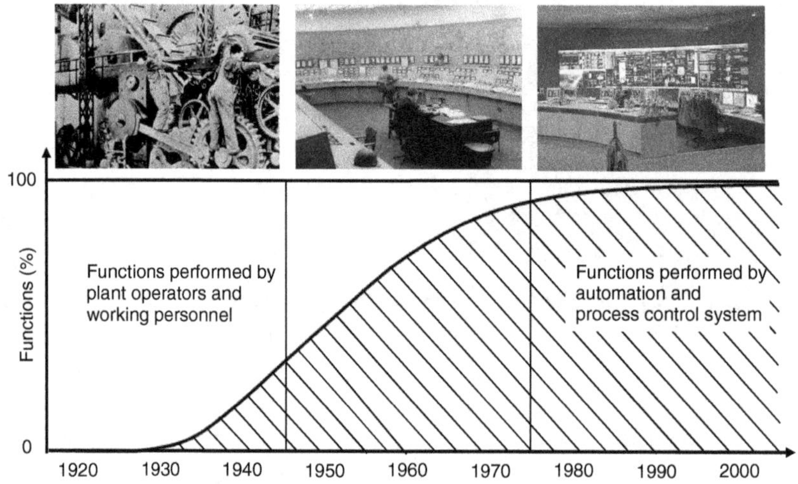


**Figure 16.1** Changing operator tasks during the last century (with kind permission from Kuntzsch [2], adapted from Bergmann [3]).

optimal production, start-up and shut-down sequences, troubleshooting in case of unexpected growth and production behavior, actions related to equipment malfunction and incidents, safety-related actions and adaptation of operational procedures to different organisms, and substrates or products including possibly controller parameter adjustment.

The growing complexity of operational and control strategies leads to increasing demands for the training and education of bioprocess operators, biochemical engineers, and members of other engineering disciplines. A steady increase in the abstraction capability of process operators is expected. These growing demands in the training are faced with limited financial and human resources and the desire to shorten the training time. At the same time, it is generally accepted that activating teaching methods contribute significantly to a good learning success. Against this background, training simulators are an increasingly important tool in the industrial training of process operators and engineers, in academic education and research as well as in the continuing professional education. They are currently applied in aviation [4], navigation [5], conventional, and nuclear power plants [6–8], in the petrochemical industry [9], in the fine chemical production [10], and for batch process operation in the chemical industry [11]. Some applications have been reported for waste water treatment plants [12] and for biotechnology [13].

Some reviews concerning training simulators have been published in recent years [1, 14, 15], indicating their increasing importance.

In this chapter, an overview on training simulators in the process industries is given. Their different structures and fields of applications are explained. The general structure of training simulators and briefly the types of mathematical models implemented in the simulators are illustrated. Subsequently, development strategies and tools for the development of training simulators are described. In an application section, some bioreactor training simulators and their integration in full plants are illustrated. The chapter concludes with some remarks concerning future development and use of such simulators in the bioprocess industry.

## 16.2 Simulators in the Process Industry

In the process industry and related research, different types of simulators are used. Due to the tremendous increase in the performance of personal computers and the constantly improving computational capabilities, the conditions for the use of simulation software have improved significantly. Simulation software packages are used with different purposes and in different areas of application. Without claiming to be exhaustive, major areas are computer-aided engineering (CAE) and computer-aided process engineering (CAPE), production planning, design of process control and automation, research and development, and training and education.

CAE and CAPE mainly comprise planning and design of stationary, time invariant processes close to optimal operating points. Further developed CAE and CAPE tools allow for dynamic, time variant process simulation. These packages may be used for the design or optimization of operational strategies as well as the prognosis, diagnosis, and interpretation of the dynamic process behavior (i.e., for batch processes). Often, CAE and CAPE tools comprise model libraries with standard models and provide tools for individual model development.

With the aid of simulators for production planning, complete production and manufacturing processes may be developed and optimized. These simulators frequently do not include mechanistic models of unit operations but focus on the procedural sequences and durations of individual production steps.

Another group of simulators allow for the design of controllers and automation systems, the optimization of operational procedures and the planning of start-up and shut-down procedures. These simulators support the modeling of sequential functions in a process and the dynamic step response functions of subsystems to be controlled and automated, as well as the controllers applied in the processes.

Research and development require simulator software with a very high flexibility for the formulation of new models and algorithms for parameter estimation or process optimization.

A fundamental purpose of simulation and modeling in universities, research institutes, and industrial research departments is to support the acquisition of new knowledge and scientific insight. Often the simulators are used for the design of new experiments.

Eventually, simulators in education and training are used to convey basic process knowledge or advanced process insight for process optimization. Another field is the actual training of operational competences and competence to act.

To process these tasks, numerous commercial or university simulation systems are available. Very powerful simulation systems include ASPEN and

Hysis (Aspentech Inc.), ChemCad (Chemstations Inc.), SIMBA (Ifak System GmbH), GPS-X (Hydromantis Environmental Software Solutions Inc.), Matlab/SIMULINK (The Mathworks Inc.), WinErs (Ingenieurbüro Schoop GmbH), Berkeley Madonna (University of California at Berkeley), DIVA (Stuttgart University), and DIANA (MPI Magdeburg) as well as the package C-eStIM, developed at the University of Applied Sciences Bremen [16]. These software systems are used in the chemical industry, environmental technology, and biotechnology as well as universities.

## 16.3 Training Simulators

Operator training simulators (OTS) and interactive simulators for education are a special class of simulators. Although both OTS and interactive simulators for education are used for training and educational purposes and can share much in common from a technical perspective, it is useful to clarify the differences between these two simulator types. OTS are basically used for the training of plant operators with respect to process operation and handling using the process control system. The training frequently comprises start-up and shut-down procedures as well as the training for normal operation and troubleshooting strategies for critical process situations. Within the course of the training, it is especially crucial how a process is operated (know how) and to a lesser extent why a particular strategy is chosen (know why) [14]. A special feature of OTS is that the timing of the simulation corresponds to the timing of the real process (real-time operation). In addition, the trainees can operate the simulator via a process control system user interface in analogy to the real process.

The essential task of interactive simulators for education is to impart and deepen the process understanding by students and trainees (know why). Based on the simulation results of an interactive simulator for education, certain effects or phenomena of a process can easily be demonstrated and illustrated as compared with a rather abstract mathematical model. Examples of simulators for education can be found occasionally in plants and more frequently in higher education and universities [17–22].

A special class of simulators for education are those that are used for the training of instrumentation and control engineers in the process industry. With these simulators, industrial production processes are particularly displayed in their dynamic behavior. Provided the simulator is capable of real-time calculations and possesses appropriate interfaces, standard process controllers can be linked to the simulator. In this configuration, instrumentation and control trainees can safely perform an analysis of the plant behavior and practise the setting of control parameters [17, 20]. More elaborate combinations of OTS and interactive simulators for education may be used for the planning and commissioning of complete automation and process control systems.

In the subsequent text, the expressions "operator training simulator" and "interactive simulator for education" will not strictly be distinguished. Instead, the term "training simulator" is used as it seems likely that both simulator types continue to grow together in the future.

## 16.3.1 Training Simulator Types

Different training simulator types exist, which might be grouped according to their fields of application or the training purpose.

### 16.3.1.1 Simulators for "Standard" Processes

Because many processes appear to have similar structures, complex training simulators may be developed for these "standard" processes in a cost-effective manner. A wide variety of learning tasks – including the training of the typical response to process disturbances or malfunctions – may be trained when using standard processes. Examples of the use of standard processes (e.g. methanol synthesis and polymerization) in training simulators may be found both in internal company training centers [23, 24] as well as in training centers for several companies [25, 26]. The PowerTec Training Centre Essen, Germany, runs different power plant training simulators such as a fossil-fired power plant simulator and a combined cycle gas turbine (CCGT) plant simulator with or without heat extraction. Since many power plants are very similar, different energy companies can train their personnel in the PowerTec Training Centre. Many details have been implemented in the simulators. Up to 300 malfunctions may be simulated and more than 2000 process values may be changed.

A special position has the training simulator at the Europäisches Bildungswerk für Beruf und Gesellschaft, Germany, because the training program with this simulator particularly focuses on small- and medium-sized companies in the chemical industry [27, 28].

### 16.3.1.2 Company-Specific Simulators (Taylor-Made Simulators)

Even similar processes in different companies exhibit differences in plant configuration, the apparatus used, the control system technology, and/or the specific substances processed. Simulators that have been specially adapted to very specific processes can offer advantages over the standard process simulators, if the trainees are to be trained on the specific process characteristics and if the teaching of specific knowledge and skills is required. Examples of training simulators, in which entire company-specific systems or subsystems were modeled have been published by Leins *et al.* [29] as well as Schaich and Friedrich [15]. With this simulator, processes can be simulated, in which 2 reactors, 9 separation columns, and 18 substances are handled. Training objectives of these simulators are exercising cold and warm start-up or shut-down, the processing of formulations, and the incident/malfunction training.

An important element of the company-specific simulators is the possibility to operate the simulator using an already existing process control system. The main advantage of company-specific simulators is the highly practical and targeted training. A key disadvantage of these simulators is the high cost of model creation. However, several authors assume that this effort can be reduced in the future by an improved corporate model management or suitable model management systems [14, 15, 30].

#### 16.3.1.3
#### Process Automation and Control

Since training simulators are capable of reproducing the dynamic behavior of complex industrial processes quite precisely, they can also be used for the training of employees in the field of measurement and control technology. Using typical example processes, for example, a simulated stirred-tank reactor, control system analysis or control parameter setting can be practiced [17]. In addition, training simulators may be used as a tool for the engineering of specific process automation concepts and their test before installation, including tests of sequential control actions and tests of newly developed control structures [31].

#### 16.3.1.4
#### Training Simulators in Academic Education

Besides the use of training simulators in the process industry the potential of education and training simulators is increasingly recognized in the context of academic education. However, the training objectives in the context of higher education differ from those of in-company training. For this reason, simulators for academic education at colleges and universities have a special position within the group of training simulators.

While in industry a rather special training for handling certain processes (know-how) is often the main purpose, the universities focus on teaching fundamental insights into basic process engineering operations (know-why). Therefore, in universities lower requirements exist in terms of detail (accuracy of process dynamics, the process image (operating options, etc.)) as compared with industrial training simulators. However, in universities a strong requirement is that the underlying fundamental processes are described correctly with respect to the defined teaching aim. The complexity of the simulators used in universities varies considerably, ranging from LabView-based simulators [19] to the most complex web-based training simulators [21].

In higher education there seems to be a focus on training simulators for process control and automation. Since control is a rather abstract discipline to many engineering students training simulators for the training of control basics are also increasingly developed [18, 20, 22, 32], supporting action-oriented learning.

### 16.3.2 Training Simulator Purposes

From the above-mentioned compilation two groups of competence and tasks can be derived, where training simulators can be applied as an efficient training tool in industrial and academic training and education. Group I comprises applications where the improvement of process handling or operation is the predominant target. Group II deals with applications, where the engineering and development of processes is supported by training simulators.

#### 16.3.2.1 Training of Process Handling

**General understanding**

Provided appropriate models have been included, training simulators can be run harmlessly, even into unusual or extreme operating conditions. They offer deep insight into participating subprocesses of the overall process for operating or planning personnel so that their awareness on the whole process and terms such as energy saving or sustainability can be raised. Training simulators have the potential to deepen the general understanding of the regarded process.

**Basic training of unit operation handling**

In the process industry, a rather small number of key unit operations is frequently applied. One of the most energy consumptive unit operations in the chemical and petrochemical industries is the distillation process. Training simulators for this basic operation were developed, for example, by Gilles *et al.* [33], Wozny and Jeromin [34] and Bao [35]. Due to the coupling of heat and mass transfer processes with the fluid dynamics within a separation column, the distillation process dynamics is complex and difficult to understand for plant operators. Poor process control may directly lead to increased production costs. The aim of the training simulators for such basic operations is to especially sensitize plant operators for the complexity of the process and practice key strategies of process operation.

**Standard process operation and training of full plant operation**

Besides mastering important basic unit operations, plant operators must be able to operate complete systems and plants. For safe standard process operation, the operators are trained to observe the process signals, take care of alarms, correct minor changes by turning on appropriate actuators such as pumps or valves, invoke recipes, and so on. This training particularly aims at the prevention of negative or even harmful operation. This implies misuse of pumps, valves, motors, and other aggregates, which could lead to bursting pipes or the breakdown of complete technical subsystems.

#### Start-up and shut-down procedures

Safe, and optimized start-up and shut down procedures may result in shorter starting times of operation, for example, after plant revision, which has positive economic effects. Start-up and shut-down of a plant always is a special situation for plant operators. In order to exactly know the sequence of these procedures, start-up and shut-down can be trained. However, in order to train these phases the simulator models must cover a wide range of operating points of the unit operations involved.

#### Exceptional situations and malfunctions or incidents

Exceptional situations, disturbances, malfunctions, or even dangerous incidents may occur in most chemical or biochemical processes. They may lead to damages at technical devices, injure member of staff, people living close to industrial sites, and harm the environment. While temporary failures primarily cause economical losses, disturbances of the machinery or the peripheral system in combination with faulty operation often cause serious accidents – and subsequently the corresponding economic losses. With training simulators, plant personnel can be trained to cope with such disturbances and to prevent accidents.

#### Handling of the process control system

Process control systems are usually complex software systems, connected to the industrial plant via input/output devices. They comprise functions such as data monitoring, graphical and numerical visualization of process variables, operation of actuators such as valves, pumps, compressors, and stirrers, and invoking recipes and recipe administration. In the engineering station of the process control system, controller parameters may be set, new recipes and automatic sequence control schemes may be designed, the graphical representation of process variable may be adjusted to actual requirements, and so on. Even this short summary of functions illustrates the complexity of the control system and the need of training for the process control system handling. The training may be done using very simple simulators, just changing process variables randomly. An alternative is the use of training simulators in which the real process behaviour is modeled to a certain degree. The advantage of the latter is the improved realism of the system and the increased relevance of the training.

#### Training before plant/process start-up

One particular field of application for OTS is the training of the process handling (standard operation, start-up and shut-down procedures, situations with high risks, handling of the process control system, etc.) even before starting up a new plant. The major goal in this training is to increase the operators' knowledge about the specific process features, the process control system used and train complex standard situations, which are commonly used during start-up of a plant. In this way, training simulators are used to reduce the overall time required to bring a new process or plant to safe and productive operation. A major challenge for the training simulator development is that there might be a lack of reliable data concerning specific process or material properties. Thus, the training simulator behavior has to be estimated from first principles plus practical experiences of the training simulator developers.

#### 16.3.2.2 Training Simulators Supporting Engineering Tasks

**Controller design and adjustment**

Training simulators may be used to assist the development of control loops and the controller setup as well as transferring the knowledge about the developed control strategies to the plant operators. For controller design, only a small part of the full plant needs to be represented in the simulator. However, the dynamic model needs to describe the process very precisely. Embedding the model and the respective controllers into a training simulator has a big advantage: the performance of the newly designed controllers as well as their handling may easily be demonstrated to plant operators, using a well-known and understood graphical user interface that is ideally very similar to the process control system already in use.

**Test of automation systems**

The models within OTS describe a very broad range of operating conditions as well as the process dynamics. This knowledge may be used to develop and test automation systems. The newly developed control system may be interconnected to the simulator instead of the real plant, using appropriate I/O-interfaces. In this way, the fundamental functions as well as the applicability of the automation and process control system may be tested and improved, using the I/O-signals sent to and calculated by the training simulator model. In this form of application, the training simulator is used as a digital control system tester (DCS-tester, [36]).

**Development of overall control strategies and process optimization**

An even more advanced application for training simulators is the development of new overall control strategies as suggested by Kuntzsch [2] using a bioethanol plant simulator. For this type of application it is advantageous if the models used in the training simulator are mechanistic models, covering a wide range of operating conditions. The simulator is interconnected with programmable logic controllers, which follow a predefined list of instructions or the operational sequence defined by a sequential function chart (SFC). In this way, improved operational procedures may be developed, programmed, and tested using the simulator before they are applied to the real process.

## 16.4 Requirements on Training Simulators

As described before a training simulator provides experienced operators, technicians and apprentices with the opportunity to develop and maintain several operational skills. To ensure efficiency in training the simulator should fulfill several requirements, such as providing (i) precise simulation of the chemical, physical, and biological events of the process, (ii) realistic simulation of automation and control actions, (iii) real-time and accelerated simulation, (iv) user interfaces that realistically recreate important parts of the actual plant system, (v) multipurpose usage, (vi) maintainability for user-friendly model updates, and (vii) adaptability to modified or different processes.

### 16.4.1
### Precise Simulation of the Chemical, Biological and Physical Events

In the context of training simulators, precise simulation means that the simulator must be able to describe the trajectories of key state variables of the process as a function of the key actuating variables. The stationary values of the state variables must be reflected by the simulator as well as their change in time. Furthermore, the simulator needs to be able to simulate events such as disturbances, malfunction of pumps, measurement devices, or even breakage of pipes, and the subsequent changes in the state variables in a realistic manner.

It is important to note that the models underlying the simulator should be able to reflect a big range of operating conditions with reasonable preciseness, in order to simulate start-up und shut-down sequences or also suboptimal process performance, maybe as a result of suboptimal process handling. Also in these cases, the time course of the state variables in transitionary states must be simulated with high accuracy.

### 16.4.2
### Realistic Simulation of Automation and Control Actions

Most industrial processes are operated using process control systems. The operators adjust actuators such as pumps and stirrers directly via the process control system or by changing set-points of a number of controllers in the process. In addition, manually operated pumps, valves, and so on may have to be handled by the plant operators. In some cases, the plant engineers or operators need to adjust controller parameters according to the actual operating conditions. From this, it can easily be deduced that a training simulator must be able to simulate the automation and control system in a realistic manner. This means that the change of state variables in time as a result of changes in the controller set-points or the actuator settings has to be simulated and is of particular importance, as the overall process dynamics may be determined by process control loops or actuator dynamics. In addition, it might be required to simulate malfunctions of the process due to suboptimal control parameter settings in a realistic way.

### 16.4.3
### Real-Time and Accelerated Simulation

One of the key advantages of training simulators is their detailed knowledge representation concerning a particular process or plant. In order to convey this knowledge to the plant operators, engineers, or students, a fairly high number of experiments at different process conditions may be required. Subsequently, it is very desirable that a training simulator can perform accelerated simulations. However, some processes are very fast (i.e., the dissolved oxygen change in bioreactors) and thus require real-time simulation for a sensible training.

### 16.4.4
### Realistic User Interfaces

Safe and optimal plant operation requires a good knowledge about the functions offered by the process control system. These functions are accessible via a graphical user interface. In order to train the handling of the process control system, it is highly desirable that the user interface of the training simulator is very similar to one of the real process control systems. It should represent the structure of the process using interactive process flow charts, must show the numerical values of key measurements, and must offer buttons to switch actuators on or off as well as windows to input process values and set-points. In addition, it is desirable to graphically represent online measurements and show graphs of historical data. In cases where the handling of the process control system is a key training issue, the user interface of the simulator should be more or less identical to the real process control system user interface.

In cases where the general understanding of the process behavior is the key training issue, a graphical user interface should be used supporting the intuitive operation of the process also for trainees with different educational background.

### 16.4.5
### Multipurpose Usage

A key cost factor in training simulator development is mathematical modeling and model parametrization. The reasons for this are that the models represent a big part of the process knowledge that must be collected, the model formulation itself is rather difficult and the model implementation into a simulation software may be prone to error and requires intensive testing.

This necessitates a long life cycle for the models and a wide range of training simulator applications (with only minor changes in the models), which should be taken into consideration in the specification list. As mentioned earlier, possible applications for training simulators are the test of automation systems

(DCS-tester), training before first plant start-up, training of new operators, training for incidents, training for general process understanding, controller adjustment training, development of optimized process control strategies, and so on.

### 16.4.6 Maintainability for User-Friendly Model Updates

In many processes educts, products or catalysts change during the life cycle of a production plant.

With respect to bioreactors, improved organisms may be used within an existing plant. These changes need to be reflected by adapted mathematical models. Furthermore, during plant operation new data may be gained for new operating conditions, allowing for model improvement and adaptation to real case data.

An important requirement on training simulators is that the mathematical models may easily be updated and implemented. If the models cannot be updated, the simulator may show nonrealistic plant behavior, which may seriously jeopardize the trainees' acceptance of the tool.

### 16.4.7 Adaptability to Modified or Different Processes

Industrial processes undergo permanent improvements and modifications. Certain process equipment may be exchanged, modified, or supplemented. For example, more efficient use of energy often is achieved by a new arrangement of the heat exchanger system within a process.

In order to keep a training simulator updated, it should be as easy as possible to exchange, withdraw, or add additional process models to it. Thus, it is desirable to build up a model library containing standard process units models, which can be parameterized in order to reflect the behavior of the unit operation under consideration. In this way, a high adaptability of the simulator to process changes is ensured. Furthermore, training simulators for new processes may be developed, using the model library.

## 16.5 Architecture of Training Simulators

Training simulators in the process industry frequently exhibit very similar structures. In principle, a training simulator combines dynamic process models and computer graphics. The computer graphics emulate a laboratory, plant, or machine that allows the user to operate and control the process by using virtual actuators and controllers and other automation equipment. From a technical perspective, training simulators frequently consist of (one) a process simulator, (two) process-related control components (PRC), (three) (at least one) monitoring

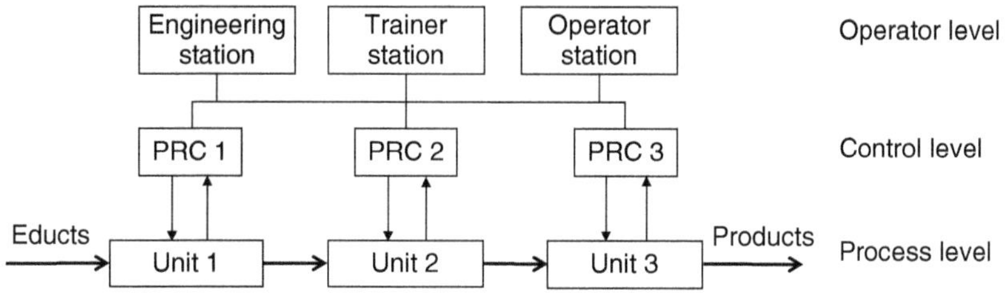


**Figure 16.2** Structure of training simulators.

and control station (MCS). In addition, (four) a trainer console and/or (five) an engineering-station may be connected to the training simulator (Figure 16.2).

The process simulator (one) contains the process models. Thus, it unites the expert knowledge concerning the process, for example, in the form of mathematical equations or algorithms. The simulator also comprises (sub)programs for solving the equations numerically in order to calculate process conditions as a result of operating actions.

The process-related components (two) include D/A and A/D converters as well as a field bus, if a real digital control system is used as the automation system in the training simulator. In this text, the PRC also comprise instruments, operating elements, controllers, and actuators (valves, pumps, etc.), which are not directly operated from the operator console but manually in the real process.

The monitoring and control station (three) usually consists of one or more screens and keyboards. Here, operators observe process variables and initiate automation system actions. As a result of the introduction of local area networks (LAN), the Internet and programming languages such as Java the MCS can also be implemented at farther computers. However, the special features of Internet-enabled training and process simulators shall not be covered here (see Ref. [21]).

In particular, for larger training simulators, it may be useful to operate a trainer console (four) with special functions in addition to the MCS on which the trainees will be trained. On this trainer console the coach may invoke special conditions or scenarios, initiate malfunctions or incidents, and so on. In addition, functions to monitor the trainees operating actions may be integrated into the trainer console. At this level, the expertise of mediating learning content can be implemented particularly supportive. From a scientific point of view, the development of the functions of trainer consoles cannot be considered to be complete.

Finally, training simulators may have an engineering station (five), upon which further functions such as the configuration of the OTS or the model refinement can be realized. If necessary, the engineering station may also have access to model libraries or material databases.

With respect to their application and the effort for development, it is sensible to distinguish between the three different structures of the training simulators [37]:

(I) The simulator stimulates the process-related components as well as the operating and monitoring station. This means, that in the simulator just the (bio)chemical process and the unit operations are simulated. The operation and control of the simulated process is carried out using the existing process control system including its process-related components.

This type of training simulators is particularly useful for training based on the actually applied process control system as well for thorough testing the process control system functions.

(II) The simulator directly stimulates the monitoring and control stations of an existing process control system, the process-related components are emulated as part of the simulator. The operation of the simulator training is performed via the monitoring and control stations of the process control system. This type of simulator usually requires reduced investments, as the process-related components are not real but emulated by the simulator.

(III) The training simulator contains all components including an emulation of the monitoring and control station. Such training simulators are referred to as full-scope simulators.

It is recommended to decide for one of the variants in the early stages on the definition of a training simulator project. A promising way to find the most suitable training simulator variant is to carefully define the training targets and application aims.

In summary, it can be stated that the basic structure of a training simulator is very similar to that of a process control system. The main differences are essentially the process simulator, the trainer console, and the extent to which the training simulator does emulate control system functions.

## 16.6 Tools and Development Strategies

As the development process for training simulators is of very high complexity, it should be performed in a structured way. A training simulator development project may be divided into five main phases (Table 16.1). Each of them comprises different tasks and ends with a well-defined result.

In the preliminary project phase (phase 0), feasibility studies are performed, taking into account the general ideas of applications and the training aims, the location(s) where the simulator shall be used and its general structure. The tools that might be used for the development, the time scale, and potential costs and benefits (return on investments) also need to be checked. The result of the preliminary phase is usually a feasibility study, which forms the basis for the project phase I.

In project phase I, the requirements on the simulator are defined and the basic planning is performed. Here, decisions are made concerning the definite training targets as well as the process that shall be simulated. It is recommended to define the structure of the training simulator according to the previous section during this basic planning phase. Main result of phase I should be the specification booklet including a process description, process flow charts, and a good description of training scenarios as well as the definition about the training simulator basic structure. It is also helpful to define a full list of state variables that shall be simulated as well as a full list of actuating variables used to manipulate the process. Finally, it has to be specified which (at least how many) graphs should be shown, whether historical data shall be accessible, and how the graphical user interface on the monitoring and control stations should look like in principle.

During project phase II, the detailed planning is performed. As a result, all necessary information required to program and build up the simulator should be available. Thus, the process structure needs to be defined in detail including all unit operations, actuators, measurement devices, state variables, control loops, and so on. Furthermore, the structure and functionality of the automation system within the simulator needs to be specified. It is helpful to use sequential flow charts (SFC), describing the main operational procedures, which shall be implemented according to the training scenarios and the standard operations, required. In phase II, the training scenarios may be revised or receive a more detailed specification. A detailed description of the cause-and-effect relationships within the process as well as the underlying principles in combination with real process data (at least from pilot plant experiments) forms the basis for the development of the process model. Eventually, the mathematical process model needs to be developed from the information given in project phase II. Project phase II is also concerned with the specification of the structure of the training simulator itself and thus the required hardware, interfaces, and software systems. Here, also the required software tools need to be identified. Suggested deliverables of project phase II are a verbal process description, process flow charts, P&ID – flow charts, sequential flow charts for standard and training procedures, revised training scenarios and mathematical process models as well as a list of computer hardware and software tools, required for the realization of the training simulator. Finally, a realization schedule for project phase III should be delivered.

In project phase III (realization and start up), the training simulator is built, tested, started up, and handed over for customer use. Important elements of this phase are the programming of the mathematical models, the realization of the automation and graphical user interfaces on the monitoring and control stations, the hardware implementation, and the testing of the simulator. The training scenarios are now tested and finalized, a documentation of the system is written. Eventually, the start-up of the simulator is performed and the customer is trained to use the simulator.

The importance of a well-organized postproject phase (phase IV - system operation and maintenance) in the life cycle of simulators may be underestimated. Here, system maintenance, revisions, improvements as well as adaptations to modified plant structures are performed; new staff members are trained in the usage and maintenance of the simulator. In addition, training scenarios may be adapted to actual needs within the company, which in turn might require an additional adaptation of the training simulator (Table 16.1).

**Table 16.1** Project phases for training simulator build up.

| Project phase | Deliverables |
|---|---|
| 0 – Preliminary project phase | Feasibility study |
| I – Basic planning | Specification booklet: process description, process flow charts, training scenarios, OTS basic structure; list of state and actuating variables; principle GUI and graphs |
| II – Detail planning | Specification booklet II: detailed verbal process description (cause and effects, principles), P&ID – flow charts, sequential flow charts (SFC) for standard and training procedures, revised training scenarios, mathematical process models, computer hardware and software tools |
| III – Realization | Schedule for project phase III: completed training simulator, test protocols start-up, customer training, finalized training scenarios, documentation |
| IV – Operation and maintenance (postproject) | Utilization documentation, updated training scenarios, revision, and update documentation |

## 16.7
## Process Models and Simulation Technology

Modeling and simulation technology have a special impact on training simulators, since modeling itself may be accounting for up to 50% of the total development costs [14]. Hence, the goal should be to reduce these costs significantly. This aim may be approached by improvements on the areas of (i) models and modeling strategies, (ii) software systems for model development, (iii) multiple use strategies for models, enhanced model update and exchange and model libraries. On top of that, a training simulator should be build up, using efficient tools for model calibration/parameterization and the development of individual, highly specialized models.

### 16.7.1
### Process Models

During the development of training simulators, a broad spectrum of different models may be used. In this text, we use a rather broad definition of the term "model":

> *A model is an abstract or concrete representation of the reality, illustrating the relevant relationships, properties, or sequences of events of a (technical) system (process, plant, facility) with respect to the defined model utilization.*

Without claiming for completeness, the following abstract model types are used in the different phases of a training simulator development project: (i) verbal process descriptions, (ii) graphics to represent process structures (e.g., PI&D) and process sequences (e.g., sequential flow charts), (iii) tables (e.g., decision tables in control), (iv) several types of mathematical models, (v) algorithms (vi) expert systems, and (vii) computer programs.

In the context of training simulators, it is useful to define the terms descriptive and prescriptive models, empirical and mechanistic models as well as stationary and dynamic models, because all these models may be used in a beneficial way within training simulators.

Descriptive models describe an existing original by reduction onto the properties, relevant for the specific purpose. Examples for descriptive models are concrete or computer 3D-plant models (often reduced in size), which are used to explain the process structure to visitors or new employees. However, also some mathematical models are descriptive models, such as interpolation splines fitted through certain sets of data.

Prescriptive models describe a planned original in the form of a supporting project or process planning and realization. Examples are technical drawings for constructing a machine or P&ID for a planned plant.

With respect to bioreactors and chemical engineering processes, it is fairly easy to achieve good agreement between measured data and descriptive models but extrapolation to new process conditions is hardly possible. In contrast, prescriptive models are used for engineering. They are often based on first principles. With prescriptive models, new process conditions may be predicted. However, a good agreement between the model and reality also requires a significant modeling effort. In various areas of training simulators, prescriptive as well as descriptive models are used.

Mathematical models are often grouped into empirical and mechanistic models, stationary, and dynamic models. Empirical mathematical models simply describe experimental data by suitable mathematical structures – possibly without any physical, biological, or chemical meaning. Examples for these models are polynomials, splines, or neural networks. Their strength is the high adaptability to many data sets, trajectories, and curves by suitable sets of model parameters. Their disadvantage is the very poor predictive capability and their limited use for understanding cause–effect relationships and mechanisms. However, in training simulators, empirical models can be used with high benefit, if existing training data cover the full range of operating conditions.

Mechanistic models are based on first principles and describe physical, biological, or chemical mechanisms such as heat or mass transfer, biochemical kinetics, and so on. Their disadvantage is that effective modeling requires a high understanding of the mechanisms underlying a certain process or unit operation. Furthermore, model parameter estimation can turn out to be very difficult, often due to a lack of appropriate data. The advantages of mechanistic models are their good predictive capabilities, also into new ranges of operating conditions. They also comprise the mechanistic knowledge of a process, and thus are kind of a knowledge base within a training simulator and may, therefore, be used for teaching and training.

Stationary models are time invariant models. From a training simulator point of view, they directly describe input–output relationships, mostly in the form of algebraic equations. Stationary models are beneficially used to describe process characteristics.

Dynamic models describe the variation of variables with time, often in the form of systems of differential equations. Dynamic models in training simulators are frequently derived from fundamental principles such as dynamic mass, energy, or pulse balances.

Of course, there are also mixed forms of models. Occasionally, also mechanistic models based on first principles are used as empirical models, such as the mathematical structure of the Michaelis–Menten enzyme kinetics within the Monod growth model. In the context of training simulators, it may be appropriate to use first-principle models and combine them with empirical functions. For instance, empirical step response functions can be used to describe the dynamic transition of a state variable from one point in a characteristic to another due to the change of the actuating variable.

Frequent aims for modeling are to gain understanding of cause–effect relationships, time courses, or sequences of events; the optimization of process structures, equipment, and process operations as well as supporting communication during planning, project management, and project realization; and finally process operation.

Within training simulators, mathematical process models such as systems of (non)linear algebraic equations and (non)linear differential equations form the basis for process simulation. They are combined with other model types such as decision tables or sequential flow charts (e.g., for the realization of recipes).

Different submodels for unit operations in (bio)chemical engineering exist, describing (i) microkinetics (biochemical reaction kinetics), (ii) macrokinetics (heat and mass transfer processes) and (iii) phase equilibria, (iv) mass and energy balances, (v) reactors and apparatus (e.g., CSTR or plug flow reactors). In addition, substance data, data for mixtures, and construction material data form a fundamental basis for training simulator models. Training simulators often represent a full plant, rather than a single-unit operation. Thus, two other modeling levels shall be pointed out: (vi) the plant structure model in which the individual unit operations are linked to each other and (vii) the process automation and control system model.

It has been proven useful to implement training simulator models according to the structure shown in Figure 16.3 Blesgen [38], Blesgen and Hass [39], and Kuntzsch [2]:

- biological–chemical submodel (i)
- physicochemical submodel (ii) and (iii)
- reactor or apparatus model (iv) and (v)
- (sub-)plant model (vi)
- process automation model (vii).

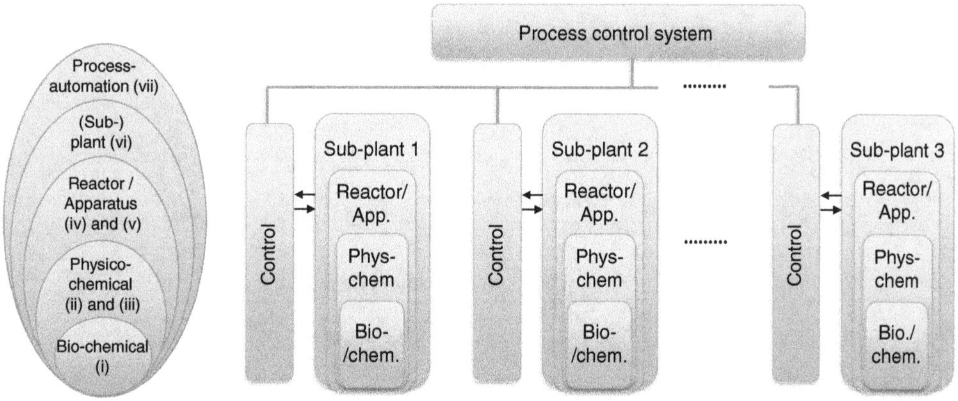


**Figure 16.3** Modeling levels for training simulators – numbers refer to the model types pointed out in the text (adapted from [2, 38, 39]).

## 16.7.2 Modeling Strategy

Also for training simulators a general strategy for model development may be applied (Figure 16.4). Model development should start by defining the system, which shall be integrated into the simulator. Equally important is the definition of the modeling targets. In the case of training simulators, it is recommended to define the expected training outcomes and desirable training scenarios in this phase.

In the second phase, the process under consideration should be described, using adequate forms of verbal and graphical process descriptions. Here, the process structure, first-principles as well as cause-and-effect relationships, the operational ranges and operational sequences, possible malfunctions, and so on are described according to the defined targets.

The third phase comprises the mathematical modeling. Often, mathematical models for the system under consideration already exist. Thus, modeling for training simulators very often is a process of model search, selection, adaptation, modification, and extension. If no adequate mathematical model exist, it may be necessary to formulate an entirely new model.

After a model has been selected or formulated it needs to be implemented into a simulation system. Subsequently, the model needs to be parameterized according to the process dimensions and characteristics. It shall be pointed out that operator training simulator models usually contain a rather high number of model parameters. In general, it is impossible to identify these parameters solely from original process data or specifically designed experiments or process runs. However, model parameters may be derived from the unit operation geometry found in the literature or database information on material and substance properties, published kinetic parameters and to a lesser extent from parameter estimation, using specific process experimental data.

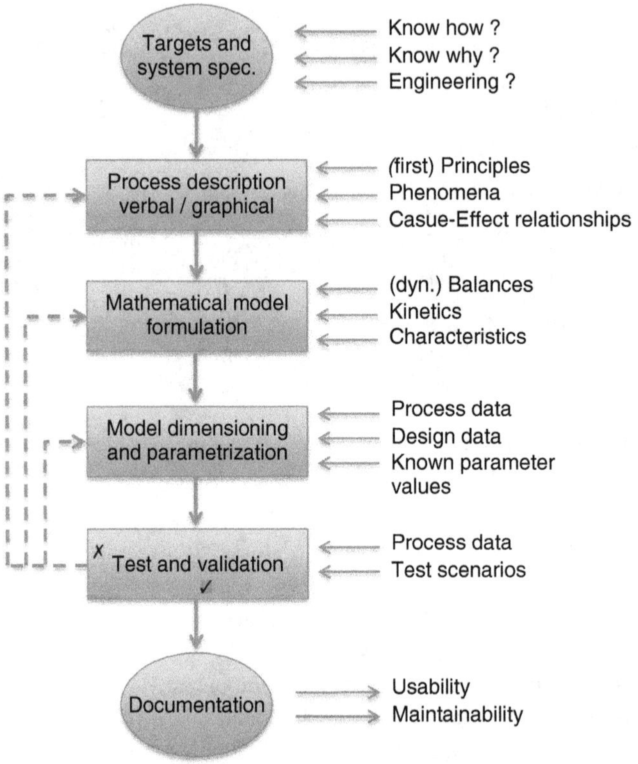


**Figure 16.4** Modeling cycle for training simulator modeling.

Due to the lack of data and the training simulator model complexity, model validation and test often is a combination of plausibility checks and comparison of the simulation results against actual process data. The tests of the simulator model should cover the full range of operational conditions, even if no experimental data is available for the full range. Even if a precise quantitative process representation may not be absolutely necessary, the model must at least represent the cause-and-effect relationships in a process in a "qualitatively correct" manner and thus be in accordance with the given process description.

The final and very important step is the documentation of the model and its implementation. This documentation is the key to the applicability of the model as well as its maintenance, further development, and improvement as well as further internal or external use. The documentation is a prerequisite to an extended life cycle of the models developed.

In contrast to most models published in the scientific literature, a full training simulator for a biotechnical plant is build up from a fairly large number of unit operation submodels with different complexity and substructures. The described modeling strategy is applicable for each submodel itself but also for the combination of submodels to form a model on the next level.

### 16.7.3 Software Systems for Model Development

As already mentioned, model development is one of the main cost factors in training simulator development. Thus, powerful software tools are required to support modeling, which must offer the subsequent functionality:

- numerical solution of systems of nonlinear differential equations with (embedded) systems of nonlinear algebraic equations
- parameter estimation algorithms
- graphical representation of simulation results
- easy input of initial conditions and parameter values
- interface to other software systems, such as process control systems
- interface to model libraries
- documentation and maintenance tools

It would be desirable if the system supports the buildup of model libraries.

Existing tools for training simulator development may be grouped into (i) software for the numerical solution of systems of mathematical equations such as Matlab/Simulink [40] or WinErs [41], (ii) process engineering simulators such as Aspen [42] or ChemCad [43] and (iii) compiler-based software packages using mathematical libraries such as IMSL Numerical Libraries (Rogue Wave Software, Inc.), NAG Numerical Library [44], CERN Program Library [45], or Numerical Recipes [46]. Besides these, some in-house software systems have been developed for training simulator development (e.g., Ref. [47]).

All these systems allow to develop individual models that may be linked to process control systems by appropriate interfaces. In addition, process engineering simulators offer libraries of ready-to-use stationary and dynamic models for several unit operations. Compiler-based systems offer the largest flexibility and transparency to the model developers, because not only the models are known but also the algorithms for their solution and parameter estimation. Software systems for the numerical solution of systems of mathematical equations are user friendly and require only little specific training for usage or particular programming knowledge.

Especially in large companies, training simulators are usually tailored to the processes to be trained and are often offered by process control manufacturers or simulator manufacturers such as Honeywell [48], Foxboro [24], ABB [14], and Siemens [23] or simulator developers (AspenTech, Star simulation, Hyprotech, Simtronics, Trident computer Resources [23], and WinMod [31]).

### 16.7.4 Multiple Use of Models

Because of the high modeling effort, it is sensible to number up the model applications. This requires a good separation of process specific and general properties in the model. Generic models for standard subprocesses or unit operations can easily be summarized in model libraries. They are also commercially available in the libraries of the process engineering simulators.

Frequently, processes contain specific process units. The models for these processes represent particular knowledge of the process owner. To build up a knowledge database containing knowledge in the form of models, it is recommended to build up an internal model library. The models in this library may also be used internally for different purposes such as those described in the previous sections of this chapter.

Furthermore, the set of model parameter values is specific process knowledge. They could also be a part of an internal model library and should be described and documented in an appropriate form.

Major challenges for multiple use of process models are their documentation and the interfaces between different models, model libraries, and software systems. This is why standards such as CAPE-open-LaN [49] have been developed and should be applied, if possible.

## 16.8 Training Simulator Examples

In this section, three examples of training simulators with bioreactors are presented: A bioreactor training simulator for suspended cultivations of *Saccharomyces cerevisiae* (baker's yeast) and *Escherichia coli*, a training simulator for the anaerobic digestion process and a bioethanol pilot plant training simulator.

### 16.8.1 Bioreactor Training Simulator

A training simulator for bioreactors has first been published by Hass [13], using the cultivation of baker's yeast and the ethanol fermentation as examples. This simulator was then further developed by Gerlach *et al.* [50, 51] to describe recombinant protein production with *E. coli*. A derivate of the simulator has been developed by Hass and Pörtner [52] and was included into a textbook for bioprocess engineers and biotechnologists.

Target groups for the bioreactor training simulator are technicians and engineers as well as students in biotechnology, biochemical engineering, and related fields. The training aims with this simulator are supporting the education of biochemical engineering fundamentals and mainly the industrial and academic training for the operation of stirred-tank bioreactors using modern process control systems, the training for the operation of real cultivation experiments, and the planning of operational strategies for bioreactors (Figure 16.5). The training simulator was designed to train the start-up of bioreactors, batch, fed-batch, and continuous operation. Finally, it can be used in controller adjustment training for bioreactor control loops.

The yeast *S. cerevisiae* exhibits metabolic features, which are particularly suitable to demonstrate the interdependence between microbial metabolism,

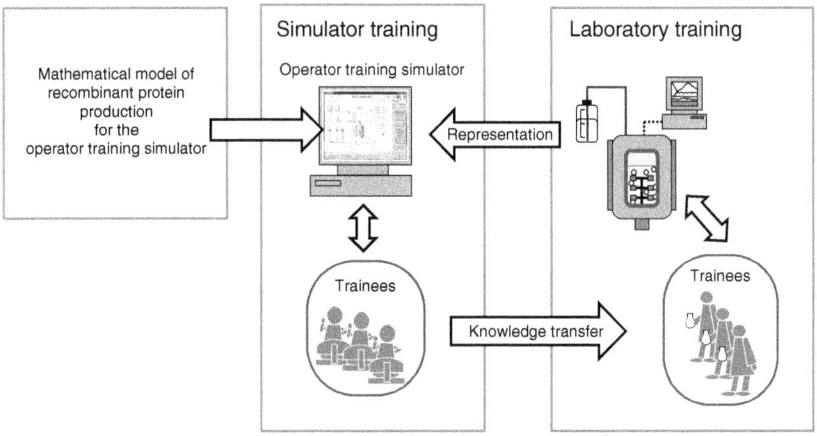


**Figure 16.5** Virtual pretraining with an operator training simulator for recombinant protein production with *E. coli* in a fed-batch bioreactor can support the training effectiveness in a subsequent real cultivation experiment (from [51] with kind permission).

bioreactor operational strategies, and overall process performance. The cultivation process is of industrial relevance and can also be performed in university laboratories and practicals. Yeast gains its energy for growth by metabolizing glucose under aerobic (respiration) or anaerobic (fermentation) conditions. *S. cerevisiae* produces carbon dioxide and ethanol as products of the glycolysis during anaerobic fermentation. Ethanol may be used as substrate under aerobic conditions, if no glucose is available (diauxic growth). Of particular interest for process operation and control is the Crabtree effect. *S. cerevisiae* produces ethanol even under aerobic conditions, if more glucose is available than can be metabolized via the respiratory chain. Furthermore, the product ethanol shows inhibitory effects.

From the biological properties of the organism, operational strategies can be deduced. In order to maximize biomass yield from the carbon source substrate and produce a maximum biomass concentration, a fed-batch strategy for substrate feeding should be applied, keeping the glucose concentration low. At the same time, the gassing and stirring strategy must ensure sufficient oxygen levels in the cultivation broth; pH and temperature need to be adjusted and controlled at optimal growth conditions. The overall process dynamics requires well-adjusted process controllers in a wide range of operating conditions.

Genetically engineered strains of *E. coli* are used to produce recombinant proteins, such as green fluorescent protein (GFP). Gerlach *et al.* [50] extended the bioreactor training simulator by a model for GFP production using *E. coli*. GFP production is induced by the addition of isopropyl_β-D-1-thiogalactopyranoside (IPTG) at specified biomass concentrations. During aerobic cultivation, *E. coli* produces several organic acids, which may show growth inhibitory effects at certain increased concentrations. Also, this process is of industrial relevance and can be performed in university laboratories. Here, feeding and induction strategies can be investigated with respect to the GFP yield, monitoring, and control of a rather complex process using online and offline data may be trained.

Both organisms are cultivated in a 20 l stirred tank bioreactor (stainless steel). The reactor is equipped with feed, acid, base, antifoam, and product tanks and appropriate pumps. Oxygen supply is realized via a gas mixing station (air, oxygen, nitrogen) where gassing rates may be adjusted. Temperature control is realized via a heating jacket. The vessel is equipped with probes for temperature, pH, dissolved oxygen, and foam level. Balances indicate mass changes within tanks and bioreactor. Samples may be taken via a sample valve for offline analysis of substrate and product concentrations. PID controllers are available for temperature, pH, and level. Dissolved oxygen PID control was realized in the form of a cascade (stirrer speed, air flow rate, oxygen flow rate). The operation of the bioreactor is performed via a process control system graphical user interface. The bioreactor can be operated in batch-, fed-batch, and continuous mode.

The models implemented in the bioreactor training simulator are the biological submodels, a physicochemical submodel, a reactor model, the plant model, and the model for automation and control. The biological yeast model was first an unstructured growth model from [53], which was later replaced by a structured model with four compartments [50]. It describes aerobic and anaerobic growth on glucose, nitrogen source consumption, diauxic growth, maintenance metabolism, oxygen uptake, and carbon dioxide evolution. It is based on kinetic expressions, taking into account the influence of substrate and product concentrations, temperature, dissolved oxygen, and pH. The *E. coli* model is also a structured model with five compartments, additionally accounting for recombinant protein production, induction mechanisms, and mixed acid production.

The models are complemented by a physicochemical model describing oxygen transfer and pH as well as foaming in the process. The viscosity of the broth is dependent on temperature. The stirred-tank reactor model consists of dynamic balances. It also describes the influence of stirrer speed and gas flow rates on the oxygen transfer coefficient. The stirrer may show flooding effects at gas flow rates above critical gas flow rates. The heating jacket and heat transport kinetics are also part of the reactor model. The heat transport depends on the stirrer speed as well as the flow rate of the thermal fluid through the heating jacket.

The plant model comprises functions for the tanks, pumps, and valves as well as for the dynamics of measurement devices, exhibiting measurement noise.

The control and automation model comprises controllers for the above-mentioned control loops. In addition, it monitors simulated measurements and actuating variable values and offers different graphical displays for online data observation and analysis of historical data. The training simulator can be run in real time or in up to 20-fold acceleration modes.

Both versions of the training simulator have been validated, using real data obtained from cultivations in bioreactors of similar size and configuration [13, 50].

Different studies have been performed with this bioreactor training simulator with bachelor students in the second and third year as well as with master students.

Major training scenarios were the start up of bioreactors, inoculation, and operating fed-batch processes. Some training scenarios were used to directly prepare the trainees for real laboratory cultivations with similar equipment and organisms. Other training scenarios focused on the adjustment of PID controllers. In particular sessions, the trainees were asked to develop their own process operational strategies in order to maximize biomass or product.

All studies showed that the training simulator was accepted by the trainees as a valuable and interesting training tool. The tool supports imparting skills, such as the analysis of online gas signals during the process, which are difficult to teach otherwise. Trainees could also learn to develop and test operational strategies, which is almost impossible in conventional academic or industrial laboratory training. From the studies, it may be deduced that trainees were well prepared for real cultivations, due to a previous training with the simulators.

### 16.8.2
### Anaerobic Digestion Training Simulator

A training simulator for the anaerobic digestion process has been developed by Blesgen [38] and Blesgen and Hass [39]. Industrial anaerobic digestion is taking place in rather large bioreactors with volumes up to 20 000 $m^3$. In contrast to other bioprocesses, in anaerobic digestion, a population of different organisms need to interact to hydrolyze the basically undefined substrates to form biogas – a gas mixture mainly containing methane and carbon dioxide. Complex biological, chemical, and physical processes take place in the reactor that is influenced by suitable process control strategies. A challenge for training simulator development is the combination of the rather slow biological with the rapid technical processes.

The biogas training simulator is a good example for multipurpose training simulators. It was developed in the form of a digital control system tester (DCS-tester) to support the build up and testing of control loops and a process control system for a laboratory biogas plant. The defined training aims of this simulator were rather similar to those of the bioreactor training simulator described in the previous section. On top of that, the simulator was used to develop new control strategies and to investigate the interdependence of biological/physicochemical with technical effects within a research project.

From this specification, the demands on the mathematical process model were deduced. The model had to be able to simulate anaerobic digestion processes of various substrates in batch, fed batch, and continuous mode with adequate accuracy while maintaining the mass balance. Modeled state variables had to include methane and carbon dioxide concentrations, biogas production rate, pH, temperature in the reactor/medium, and pressure in the headspace of the reactor. Apart from the relevant biological, biochemical, and physicochemical processes, dynamics of sensors and actuators had to be modeled, to enable controller development. As the biogas process is very slow, a 200-fold accelerated simulation mode had to be implemented. Computation of the model had to be possible on conventional PC systems maintaining numerical stability and high computing speed.

Also, this training simulator is based on four interacting submodels describing the biological, physicochemical, reactor, and plant subsystems of the overall process. The simulator was adapted to simulate a 10 l lab-scale reactor processing three different well-defined substrates as well as varying mixtures of these compounds. The biological and physicochemical submodels were implemented in the programming language FORTRAN; the reactor and plant submodels were directly implemented in an industrial process control and simulation program. The combined submodels were connected to various control loops, automations, data acquisition systems, and graphical user interfaces.

The biological submodel is based on a model described by Bernard *et al.* [54] and was adapted to process complex substrate mixtures of carbohydrates, proteins, and lipids, which are degraded by acidogenic bacteria to produce volatile fatty acids (VFA) and yield byproducts such as carbon dioxide and new biomass. The VFA are further degraded by methanogenic bacteria to form methane ($CH_4$), carbon dioxide ($CO_2$), and new biomass. Monod-type kinetics were used to model substrate uptake rates as functions of substrate or intermediate concentrations. Furthermore, functions describing the inhibition of the biological process through suboptimal temperature, pH, and VFA concentration were implemented.

The physicochemical submodel describes the pH in the reaction medium, the fractionation of inorganic carbon (TIC) into hydrogen carbonate ($HCO_3^-$), carbonic acid ($CO_3^{2-}$), and carbon dioxide ($CO_2$), partial pressures of carbon dioxide and methane in the liquid phase and mass flows of these gases into the headspace of the reactor. The pH is influenced by some of the initial substrate components (i.e., proteins), VFAs, and influent acid and alkali flows to allow for pH control.

The ideal stirred-tank reactor model basically comprises a system of 13 nonlinear differential equations modeling biomass, substrate, and product concentrations within the culture broth. Its structure is similar to that of the bioreactor training simulator.

The plant submodel comprises all plant components. The plant model can be adapted to simulate different plant configurations. Modeled aggregates include valves, pumps, tanks, conduits, and sensors. Actuator and instrument dynamics were accurately modeled to enable design of control loops and automations. Measuring noise was added to the calculated state variables.

The described models were parameterized and validated as described by Blesgen [38].

The training simulator was applied for the development and test of basic controllers and control system functions, the development of new fuzzy-logic control schemes for biogas reactors, the investigation of the influence of technical aspects on measurement signals used for process analysis and control and finally for university training and education. It was reported that important control loops, for example, for level, temperature, and pH could properly be designed and tested with the simulator and were then applied on a real laboratory plant. Furthermore, a complex substrate feeding strategy has been successfully developed with the simulator, avoiding inhibitory effects, typical for biogas processes. The 200-fold accelerated simulation not only proved to be very useful in training and education of students but also for the development of a fuzzy controller, as the controller parameterization required a rather high number of simulation experiments.

An additional benefit of the simulator arose from the analysis of process measurement dynamics with respect to biological and technical influences. This analysis illustrated the sensitivity of the accuracy and dynamics of exhaust gas measurements to gas flow rates and dilution effects and thus contributed to improve the pilot plant setup.

### 16.8.3
### Bioethanol Plant Simulator

Usually, bioreactors are integrated in complex plants, where upstream as well as downstream processes are used. These processes show complex and dynamic interactions between the process units. These complex process dynamics may have a significant influence on the resource and energy efficiency of a full biotechnological production plant.

The bioethanol plant training simulator has been developed to investigate the influence of process control and automation strategies on the resource and energy efficiency of a bioethanol plant [2, 55].

The simulator consists of two bioreactors: a cross-flow filtration unit and a distillation column (Figure 16.6). *S. cerevisiae* is used as the ethanol producing organism. In bioreactor 1 (B-8) the starter culture is produced aerobically and then transferred to an anaerobically operated bioreactor 2 (B-17) for ethanol production. Both bioreactors have a maximum working volume of 15 l. The ethanol–yeast–water suspension is fed to a cross-flow-filtration unit (F-1), which is used to recycle the biomass to the anaerobically operated bioreactor. The filtered ethanol–water solution is fed to a normal pressure distillation unit (C-1). Two streams are obtained from this unit. The top stream mainly contains ethanol and the bottom stream mainly contains water plus residual substrates and side products with low vapor pressure. The bioethanol plant simulator may be operated in batch, fed-batch, and continuous mode.

The bioreactor simulator model is similar to the one described in Section 16.8.1. More detailed descriptions of the distillation simulator and the cross-flow filtration simulator may be found in [56] or [2]. All models for the individual unit operation simulators have been validated on the basis of experimental data [2, 13, 50, 56]. Graphical user interfaces for the monitoring and control stations have been developed as shown in Figure 16.7.

As the simulator should be used to investigate the impact of process control strategies in the bioethanol plant on its resource efficiency, the simulator had to be capable of describing the dynamic behavior of the full plant, including the (bio)reaction kinetics, the kinetics of heat and mass transfer, the dynamics of the bioreactors, cross-flow filtration unit, or the rectification column as well as the dynamic behavior of the valves, pumps, compressors, heat exchangers, measurement devices, and other plant elements.

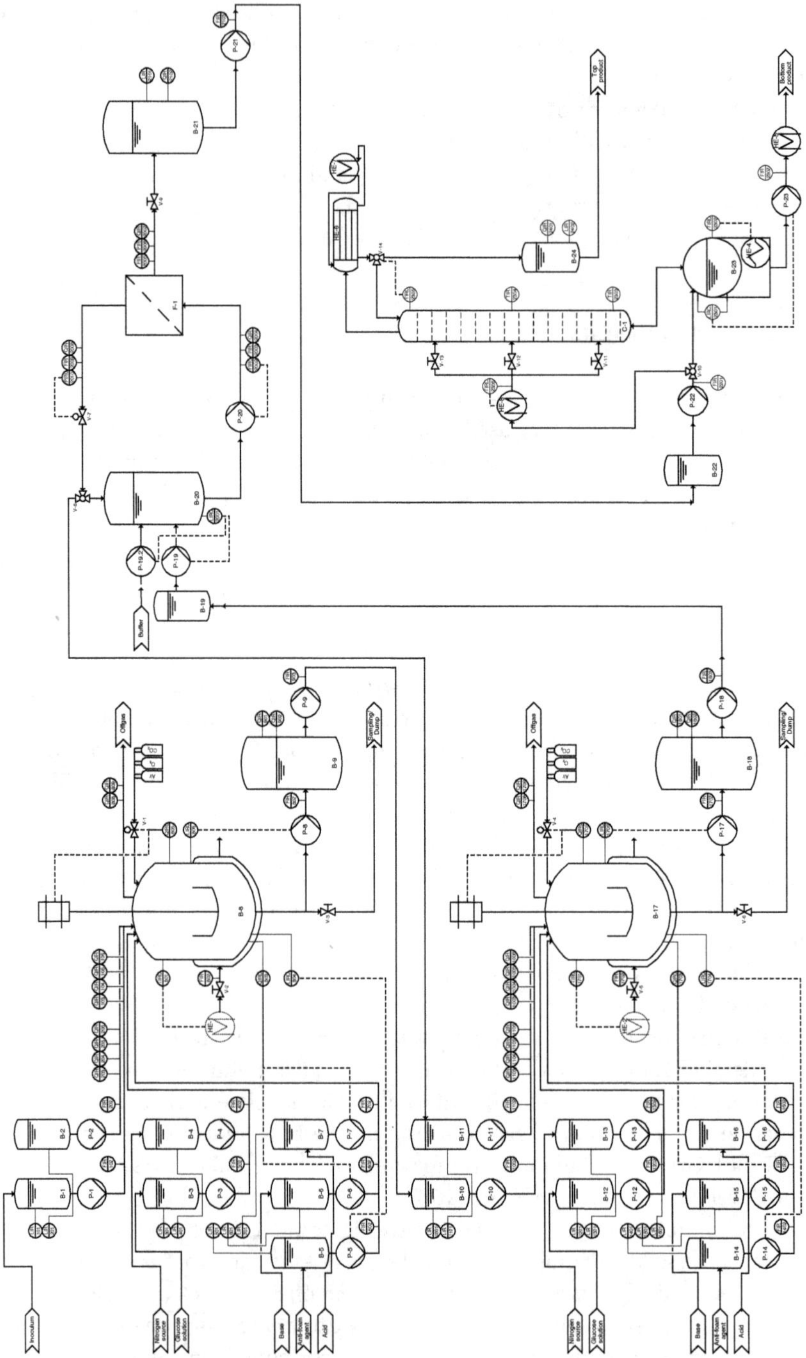


**Figure 16.6** P&I diagram of a bioethanol plant training simulator with two bioreactors (B-8; B-17), a microfiltration unit (F-1) and a rectification column (C-1) (from [2] with kind permission).

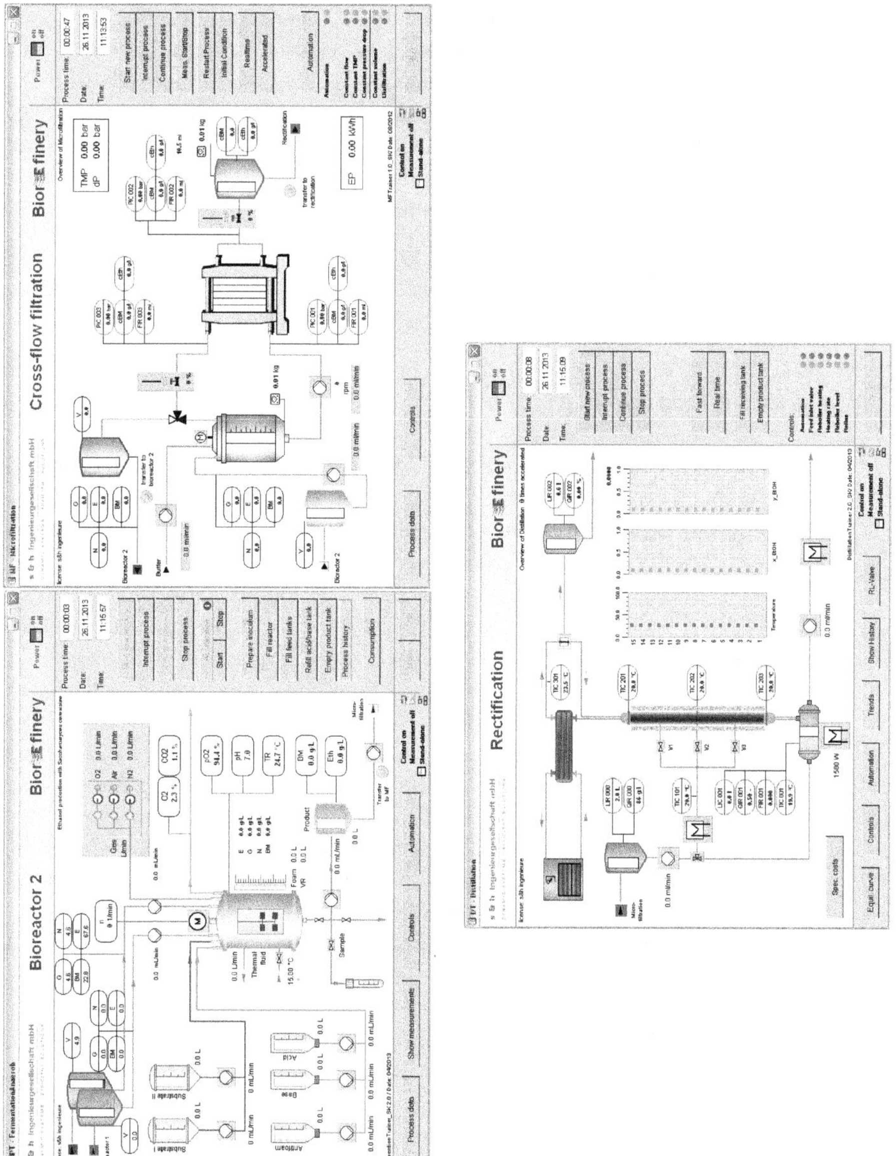


**Figure 16.7** Graphical user interfaces on the monitoring and control station for the bioethanol plant simulator (bioreactor, cross-flow-filtration, distillation column) (from [2] with kind permission).

In addition, a model of the full automation system of the process was required, which was realized on two levels. First, the required PID-control loops were implemented similarly to the real plant. Second, the operational sequence for plant operation was modeled using GRAFCET (Graphe Functionnel de Commande Etapes/Transitions, IEC 60848, IEC 1131-3), which is a SFC [57] and can be used to describe any operational sequence for a process graphically (Figure 16.8).

The briefly described bioethanol plant simulator was applied to perform studies on the impact of different operational strategies on the resource efficiency of the plant. GRAFCET plans with different sets of parameter values have been used to realize different operational procedures. As a result, the authors conclude that operational strategies and conditions for the overall process may lead to a variation of the energy demands for ethanol production of 48% [55]. This example illustrates the importance of a well-conducted training of plant operators, as their decisions concerning operational procedures can have a major impact on product yields as well as the resource and energy efficient operation of complex biotechnological plants and bioreactors.

## 16.9 Concluding Remarks

Industrial processes with bioreactors exhibit complex dynamic behavior. This leads to high demands on the operational skills of plant operators. Also in research, cultivation experiments in pilot-scale bioreactors are of increasing dynamic complexity, thus requiring very good operational skills of research staff and students. The high interdependency of complex biological, chemical, physical, and technical subprocesses makes it difficult to educate and train operators and technicians as well as biochemical engineering students for the efficient operation and analysis of bioprocesses. However, several studies show that training simulators offer an interesting tool to effectively improve operational skills of plant operators, technicians, engineers, and students of biochemical engineering and technology, which in turn can lead to a significant reduction of production costs.

In addition, training simulators can be used as digital control test systems and support the development of control systems as well as the development of new automated operation and advanced control strategies.

Despite their complexity, successful modeling of bioprocesses has been demonstrated by scientists for many years and the models now form a very powerful kind of knowledge base for these processes. Powerful personal computers, simulation tools, and modeling strategies in combination with efficient project management make it possible to use these models as a basis for training simulators. As modeling is a major cost factor in training simulator development, however, it is recommended to build up model libraries and facilitate multiple use of mathematical process models.

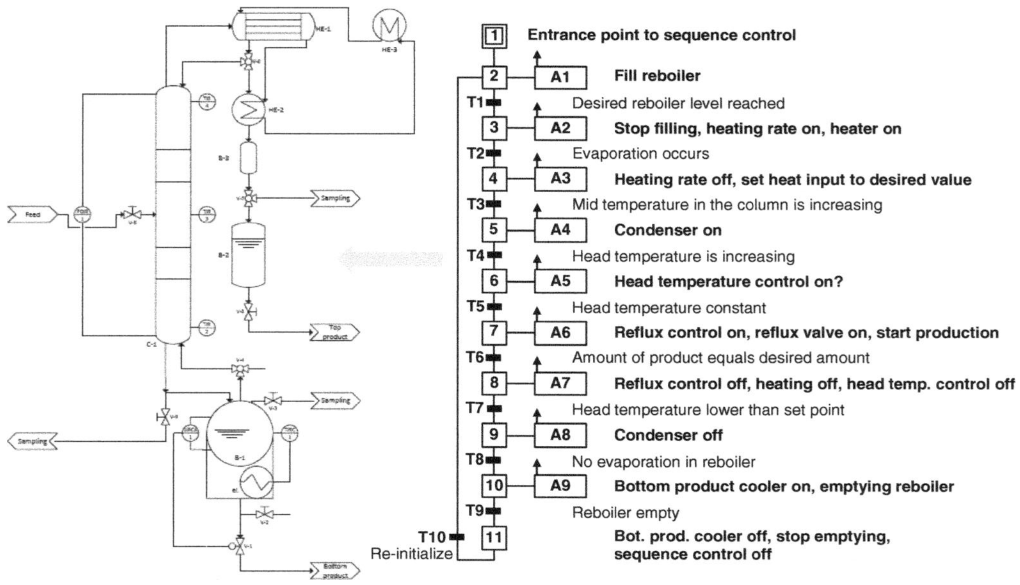


**Figure 16.8** The bioethanol plant training simulator is automated using GRAFCETS as sequential function charts (SFC), realizing well defined operational sequences. Left: P&ID distillation column; right: GRAFCET for distillation operation ( ).from Kuntzsch [2] with kind permission

Future developments will most certainly focus on the easy exchange of mathematical models for specific processes in training simulators. Furthermore, the simulators fields of applications will even be increased. Training simulators might be used in combination with scale-up strategies and in process transfer. Training simulators with scalable unit operations could facilitate an efficient scaling of process control strategies.

## References


1. Reinig, G., Winter, P., Linge, V., and Nägler, K. (2004) Trainingssimulatoren: engineering und Einsatz. *Chem. Ing. Tech.*, **69** (12), 1759–1764.
2. Kuntzsch, S. (2014) Energy efficiency investigations with a new operator training simulator for biorefineries. PhD thesis. Jacobs University Bremen.
3. Bergmann, J. (1999) *Lehr- und Übungsbuch Automatisierungs- und Prozessleittechnik – Eine Einführung für Ingenieure und Wirtschaftsingenieure*, Fachbuchverlag Leipzig im Hanser-Verlag, München, Wien.
4. Lee, Y.-H. and Liu, B.-S. (2005) Inflight workload assessment: comparison of subjective and physiological measurements. *Aviat. Space Environ. Med.*, **74**, 1078–1084.
5. Murai, K., Okazaki, T., and Hayashi, Y. (2011) A few comments in visual systems of a ship handling simulator for sea pilot training: training for entering a port. *Electron. Commun. Jpn.*, **94**, 10–17.
6. Díaz, M., Garrido, D., Romero, S., Rubio, B., Soler, E., and Troya, J.M. (2006) Experiences with component-oriented technologies in nuclear power plant simulators. *Softw. Pract. Exp.*, **36**, 1489–1512.
7. Jayanthi, T., Seetha, H., Narayanan, K., Jasmine, N., Nawlakha, R., Sankar, B., Chakraborty, J., SatyaMurty, S., and Swaminathan, P. (2011) Simulation and integrated testing of process models of {PFBR} operator training simulator. *Energy Procedia*, **7**, 653–659.
8. Park, J., Jung, W., Ha, J., and Shin, Y., (2004) Analysis of operators' performance under emergencies using a training simulator of the nuclear power plant. Reliability Engineering and System Safety, **83** (2) 179–186.
9. Madhavan, S. (1984) Ammonia process simulator. *Plant/Oper. Prog.*, **3**, 14–18.
10. Zhiyun, Z., Baoyu, C., Xinjun, G., and Chenguang, C. (2003) in *Process Systems Engineering 2003, 8th International Symposium on Process Systems Engineering*, Computer Aided Chemical Engineering, vol. **15** (eds B. Chen and A.W. Westerberg), Elsevier, pp. 1447–1452.
11. Balaton, M., Nagy, L., and Szeifert, F. (2013) Operator training simulator process model implementation of a batch processing unit in a packaged simulation software. *Comput. Chem. Eng.*, **48**, 335–344.
12. Dharmappa, H.B., Corderoy, R.M., and Hagare, P. (2000) Developing an interactive multimedia software package to enhance understanding of and learning outcomes in water treatment process. *J. Cleaner Prod.*, **8** (5), 407–411.
13. Hass, V.C. (2005) Verbesserung der verfahrenstechnischen Ausbildung durch einen "Virtuellen Bioreaktor". *Chem. Ing. Tech.*, **77** (1-2), 161–167.
14. Kroll, A. (2003) Trainingssimulation für die Prozessindustrie: Teil 1+2. *atp*, **45** (2), 50–57, **45**(3), 55–60.
15. Schaich, D. and Friedrich, M. (2003) Operator-Training Simulation (OTS) in der chemischen Industrie – Erfahrungen und Perspektiven. *atp*, **45** (2), 38–48.
16. Kuhnen, F. (2008) *Modellierungssystem C-eStIM. Kurzeinführung*. Hochschule Bremen, internal communication.
17. Birk, J. and Holl, P. (1995) Schulungssimulator zur Reglereinstellung. *Chem. Ing. Tech.*, **67** (2), 190–192.
18. Feyo de Azevedo, S., Oliveira, F., and Capelo Cardoso, A. (1994) Teacon – a simulator for computer-aided teaching of process control. *Comput. Appl. Eng. Educ.*, **1** (4), 307–319.

19. Hoffmann, U. and Kotter, M. (2002) Simulation eines geregelten Rührkessels mit LabVIEW. *Praxiswissen Elektronik Industrie*, Bd. **3**, Hüthig-Verlag, Heidelberg, pg 65–71.
20. Liefeldt, A., Löhl, T., Pegel, S., Fritsch, C., Engell, S., and Schmid, C. (2002) DYNAMIT & Learn2Control – Neue Wege in der regelungstechnischen Ausbildung. *Chem. Ing. Tech.*, **74** (5), 648.
21. Urbas, L. (1999) Entwicklung und Realisierung einer Trainings- und Ausbildungsumgebung zur Schulung der Prozessdynamik und des Anlagenbetriebes im Internet. VDI Fortschrittberichte Reihe 10; Nr. 614, VDI-Verlag, Düsseldorf.
22. Young, B.R., Mahoney, D.P., and Svrcek, W.Y. (2001) Real-time computer simulation workshops for the process control educations of undergraduate chemical engineers. *Comput. Appl. Eng. Educ.*, **9** (1), 57–62.
23. Chin, K. (2000) Learning in a virtual world: training simulators sharpen plant operators' critical skills to ensure smooth operation. *Chem. Eng.*, **12**, 107–110.
24. Schulze, K. and Arnold, M. (1997) *Trainingssimulatoren für die chemische Industrie. Tagungsband, Fortschritte in der Simulationstechnik*, Vieweg Verlag, Wiesbaden, pp. 486–490.
25. Küppers, L., Schlegel, G., and Stürenburg, H.-G. (2000) *Simulation eines Gas- und Dampfturbinen-Kraftwerkes zur Ausbildung des Betriebspersonals*. VDI-Berichte, Bd. **1534**, VDI-Verlag, Düsseldorf, S., pp. 67–80.
26. PowerTec Training Centre Essen (2014) www.kraftwerksschule.de (accessed 30 November 2014).
27. Schmidt, F. (1998) Verbesserung der Bediensicherheit durch qualifikationsgerechtes Trainieren von Anlagenfahrern in der Chemischen- /Prozessindustrie. *Konferenz Einzelbericht: 4. Fachtagung Anlagen-, Arbeits- und Umweltsicherheit*, Preprints, Köthen, S. V20.1–V20.13.
28. EBG (2014) www.ebg.de/de/standorte/sachsen-anhalt/halle-biotechnologie-chemie-und-umwelt.html (accessed 27 November 2014).
29. Leins, R., Eul, J., and Lojek, R. (1989) Trainingssimulator für eine Ethylen-Anlage. *atp*, **31** (6), 259–263.
30. Eggersmann, M., von Wedel, L., and Marquardt, W. (2002) Verwaltung und Wiederverwendung von Modellen im industriellen Entwicklungsprozess. *Chem. Ing. Tech.*, **74** (8), 1068–1078.
31. Mewes, J. (2001) Systematische Sicherheit: echtzeitsimulatoren für das professionelle Qualitätsmanagement der Automatisierungssoftware. *Verfahrenstechnik*, **35** (10), 54–56.
32. Bartz, R., Engell, S., Schmid, C., Roth, H., Becker, N., and Schaedel, H.M. (2002) Project-oriented internet-based learning in the field of control engineering. *Proceedings of the 2002 ASEE/SEFI/TUB Colloquium*.
33. Gilles, E.D., Holl, P., Marquardt, W., Mahler, R., Schneider, H., Brinkmann, K., and Will, K.-H. (1990) Ein Trainingssimulator für die Ausbildung von Betriebspersonal in der chemischen Industrie. *atp*, **32** (7), 343–350.
34. Wozny, G. and Jeromin, L. (1991) Dynamische Prozesssimulation in der industriellen Praxis. *Chem. Ing. Tech.*, **63** (4), 313–326.
35. Bao, G. (1997) SIS series simulator and its application in petro-chemical plant operations training. *IEEE International Conference on Systems, Man and Cybernetics, Computational Cybernetics and Simulation*, Orlando, FL, October 12–15, 1997, pp. 3361–3365.
36. Klatt, K.-U. (2009) Trainingssimulation - Erfahrungen und Perspektiven aus Sicht der chemisch-pharmazeutischen Industrie. *atp*, **1–2**, 66–71.
37. Holl, P. and Göttmann, O. (1995) Management von Trainingssimulatorprojekten. NAMUR-Arbeitsblatt NA 60.
38. Blesgen, A. (2009) Entwicklung und Einsatz eines interaktiven Biogas-Echtzeit-Simulators. PhD thesis. Universität Bremen.
39. Blesgen, A. and Hass, V.C. (2010) Efficient biogas production through process simulation. *Energy Fuels*, **24** (9), 4721–4727.

40. The Mathworks, Inc. (2014) uk .mathworks.com (accessed 27 November 2014).
41. Ingenieurbüro Schoop GmbH (2014) www.schoop.de (accessed 27 November 2014).
42. Aspen Technology, Inc. (2014) www .aspentech.com (accessed 27 November 2014).
43. Chemstations Europe GmbH (2014) www.chemstations.eu (accessed 27 November 2014).
44. The Numerical Algorithms Group Ltd. (2014) www.nag.co.uk (accessed 27 November 2014).
45. CERN (2014) cernlib.web.cern.ch/ cernlib/libraries.html (accessed 27 November 2014).
46. Press, W.H., Flannery, B.P., Teukolsky, S.A., and Vetterling, W.T. (1988) *Numerical Recipes*, Cambridge University Press, Cambridge.
47. Rheinmetall Defence Electronics GmbH (2014) www.rheinmetall-defence.com/ en/rheinmetall&uscore;defence/systems&uscore;and&uscore;products/simulation&uscore;and&uscore;training/power&uscore;plant&uscore;simulation/index .php (accessed 27 November 2014).
48. Reinig, G., Winter, P., Linge, V., and Nägler, K. (1998) Training simulators: engineering and use. *Chem. Eng. Technol.*, **21**, 711–716.
49. The CAPE-OPEN Laboratories Network (2003) http://www.global-cape-open.org, Co-LAN – The CAPEOPEN Laboratories Network (accessed 12 2003).
50. Gerlach, I., Hass, V.C., Brüning, S., and Mandenius, C.F. (2013) Virtual bioreactor cultivation for operator training and simulation: application to ethanol and protein production. *J. Chem. Technol. Biotechnol.*, **88**, 2159–2168.
51. Gerlach, I., Brüning, S., Gustavsson, R., Mandenius, C.F., and Hass, V.C. (2014) Operator training in recombinant protein production using a structured simulator model. *J. Biotechnol.*, **177**, 53–59.
52. Hass, V.C. and Pörtner, R. (2011) *Praxis der Bioprozesstechnik mit virtuellem Praktikum*, Spektrum Verlag.
53. Bergter, F. and Knorre, W.A. (1972) Computersimulation von Wachstum und Produktbildung bei Saccharomyces cerevisiae. *Z. Allg. Mikrobiol.*, **12**, 613–629.
54. Bernard, O., Zakaria, H.-S., Dochain, D., Genovesi, A., and Styer, J.-P. (2001) Dynamic model development and parameter identification for an anaerobic wastewater treatment process. *Biotechnol. Bioeng.*, **75** (4), 424–438.
55. Hass, V.C., Kuntzsch, S., Schoop, K.-M., and Winterhalter, M. (2014) in *Chemical Engineering Transactions*, vol. **39**, Part 1 (eds S. Pierucci, J.J. Klemes, P.S. Varbanov, P.Y. Liew, and J.Y. Yong), pp. 541–546, AIDIC Servizi S.r.l., Italy, ISBN: 978-88-95608-30-3; ISSN: 2283–9216, doi: 10.3303/CET1439091.
56. Kuntzsch, S. and Hass, V.C. (2014) Einsatz eines neuen Trainingssimulators zur Untersuchung des Energiebedarfs der batch-Rektifikation bei unterschiedlichen Fahrweisen. *Chem. Ing. Tech.*, **86** (5), 714–724.
57. Gernaey, K.V., Glassey, J., Skogestad, S., Krämer, S., Weiß, A., Engell, S., Pistikopoulos, E.N., and Cameron, D.B. (2000) *Process Systems Engineering, 5. Process Dynamics, Control, Monitoring, and Identification*, Chapter 5, Wiley-VCH Verlag GmbH & Co. KgaA, pp. 1–60.

# Index



*Bioreactors: Design, Operation and Novel Applications*, First Edition. Edited by Carl-Fredrik Mandenius.

CPSIA information can be obtained
at www.ICGtesting.com
Printed in the USA
JSHW051051300422
25110JS00002B/42

9 783527 337682